Life on Display

Life on Display

*Revolutionizing U.S. Museums of Science
and Natural History in the
Twentieth Century*

KAREN A. RADER
VICTORIA E. M. CAIN

The University of Chicago Press   Chicago and London

The University of Chicago Press, Chicago 60637

The University of Chicago Press, Ltd., London

© 2014 by The University of Chicago

Published 2014

Paperback edition 2018

Printed in the United States of America

27 26 25 24 23 22 21 20 19 18   1 2 3 4 5

ISBN-13: 978-0-226-07966-0 (cloth)

ISBN-13: 978-0-226-59873-4 (paper)

ISBN-13: 978-0-226-07983-7 (e-book)

DOI: https://doi.org/10.7208/chicago/9780226079837.001.0001

Library of Congress Cataloging-in-Publication Data

Rader, Karen A. (Karen Ann), 1967– author.

  Life on display : revolutionizing U.S. museums of science and natural history in the twentieth century / Karen A. Rader, Victoria E. M. Cain.

      Pages ; cm.

  Includes bibliographical references and index.

  ISBN 978-0-226-07966-0 (cloth : alk. paper) — ISBN 978-0-226-07983-7 (e-book) 1. Natural History museums—United States—Exhibitions— History. 2. Science museums—United States—Exhibitions—History. 3. Museums exhibits—United States—History. 4. Museum techniques— United States—History. 5. Displays in Education—History. I. Cain, Victoria E. M. 1974– author. II. Title.

  QH70.U6R33 2014

  508.074'73—dc23   2014008021

For Samuel and Grace Rader Powers

and for Vivian and Caroline Cain-Weingram

Contents

Acknowledgments

Writing this book took nearly a decade of collaborative work, and we want to thank the many colleagues, support personnel, and friends and family members who encouraged us through the process. We have known one another since 2004, and first thanks go to Gregg Mitman for urging us to make contact when we were both based in the New York area. Rader, then at Sarah Lawrence College, was working on a new project titled "Biology on Display: Museums and the New Life Sciences in America," supported by the National Science Foundation under grants 0134486 and 0734859. (Any opinions, findings, and conclusions or recommendations expressed herein are ours alone and do not necessarily reflect the views of the National Science Foundation.) Cain, then in graduate school at Columbia University, was working on her dissertation, "Nature under Glass: Popular Science, Professional Illusion and the Transformation of American Natural History Museums," supported by the American Association of University Women and the Institute for Social and Economic Research and Policy at Columbia. We made our first public presentation of what became this book's core argument in a joint paper for the Manchester Museum's "Nature behind Glass" conference in September 2007; we thank organizer Samuel J. M. M. Alberti for his feedback, advice, and encouragement. That conference paper represented the kernel of chapter 3, parts of which were initially published in our coauthored *Museums and Society* article ("From Natural History to Science: Display and the Transformation of American Museums of Science and Nature," vol. 6, no. 2 [June 2008]: 152–71).

ACKNOWLEDGMENTS

We researched and wrote (and rewrote and rewrote) this book together, and we both claim full responsibility for the content and prose of each chapter. Cowriting was a rewarding, exhausting experience, and we believe our radically different academic training and career experiences resulted in a richer work than either of us could have produced alone. Rader developed her general thinking about interactive museums and public versus popular science through a Woods Hole History of Biology Seminar titled "What Is the Value of the History of Science?"—which informs important parts of chapters 6 and 7. Some of these ideas were first developed for and became part of another collaborative publication with fellow seminar attendee Katherine Pandora ("Science in the Everyday World: Why Perspectives from the History of Science Matter," *Isis*, vol. 99, no. 2 [June 2008]: 350–64). Rader's multiyear international collaboration with the "Animals as Things, Animals as Signs" project, headed by Liv Emma Thorsen and based at the University of Oslo, shaped elements of our discussion of various animal displays; these also appear in the introduction and chapter 8 of the Oslo project's anthology, *Animals on Display*, coedited by Thorsen, Rader, and Adam Dodd and published by Penn State University Press (2013). Cain examined the history of tensions between scientists and exhibit makers while a postdoctoral fellow at the American Academy of Arts and Sciences, and refined her conclusions during a symposium titled "Lay Participation in the History of Scientific Observation" at the Max Planck Institute for the History of Science and in a subsequent article for a special issue of *Science in Context* called "Lay Observation in the History of Scientific Observation," ed. Jeremy Vetter (vol. 24, no. 2 [June 2011]). Some ideas from this article appear in chapters 2 and 3. Participation in the 2009 works-in-progress symposium, "Capturing the Witness: Eyewitnessing and Visual Evidence," led by Vanessa Schwartz and Lynn Hunt and hosted by the University of California, Los Angeles, and the University of Southern California, allowed Cain to reconsider how museum staffs both borrowed from and contributed to ideas about visual pedagogy in early twentieth-century United States. Cain's resulting article for the *Journal of Visual Culture* ("'The Direct Medium of the Vision': Visual Education, Virtual Witnessing and the American Museum of Natural History, 1890–1923," in the special issue titled "Capturing the Moment: Visual Evidence and Eyewitnessing," ed. Lynn Hunt and Vanessa Schwartz, vol. 10, no. 3 [December 2010]) influenced chapters 1 and 3. A Mellon Postdoctoral Fellowship in visual history at the University of Southern California and a Spencer Postdoctoral Fellowship afforded Cain the time and space to consider how and why museum staffs attempted to use contemporary commercial aesthetics in the service of education and institution building. Those conclusions, published in part as "'Attraction, Attention, and Desire': Consumer Culture as Pedagogical Paradigm, 1900-1930" in *Paedagogica Historica* (vol. 68, no. 5 [October 2012]), shaped our discussion of this topic throughout the book, especially chapter 7.

Collectively, we are deeply indebted to the communities of scholars from various disciplines whose insightful feedback on various portions of this manuscript, at various preliminary stages, improved our thinking and our writing immeasurably. From the history of science, these scholars include Mary Anne Andrei (who also provided some valuable Smithsonian Archive hints), Jim Endersby, Pamela Henson, Sally Kohlstedt, Susan Lindee, Lynn Nyhart, Brian Ogilvie, Paolo Palladino, and Mark Rothenberg; from the history of education, John Rudolph and Noah Sobe; from the history of visual and material culture, Jason Hill, Henry McGhie, Sarah Miller, Vanessa Schwartz, and Katja Zelljadt. We are especially grateful to Jeremy Vetter, Jonathan Zimmerman, and Sally Kohlstedt for reading early drafts of the manuscript in its entirety. Their suggestions were invaluable. We would also like to thank our two anonymous University of Chicago Press reviewers, whose careful attention to the book's broad arguments and detailed evidence challenged us to clarify our thinking and writing in important ways.

Any project focusing on multiple institutions and diverse actors over a century-long period is going to require resourceful library and research support—and ours is no exception. Barbara Mathé was a wonderful guide through the thickets of material at the American Museum of Natural History, as was Smithsonian historian Pamela Henson regarding the National Museum of Natural History. Rader thanks Sylvie (Gassaway) von Duuglas-Ittu for the diligent research and oral history–taking assistance she provided for the Exploratorium and Boston Museum of Science trips made in 2006. Archivists, curators and administrators at the Bancroft Library at the University of California, assistant curator Violetta Wolf and library dame extraordinaire Carolyn Kirdahy of the Lyman Library of the Boston Museum of Science, and John Beckman and Rob Kent of the Chicago Museum of Science and Industry collectively did everything from helping Rader find and sort through materials to connecting her with people to interview. Exploratorium staffers, especially Megan Bury, Charles Carlson, Rose Falanga, and Ron Hipschman, provided important interviews and discussions about that museum's history, as well as facilitated access to their colleagues and to illustrations. The staff of the Alfred M. Bailey Library and Archives of the Denver Museum of Nature and Science, particularly archivists Kris Haglund, Rene Payne and Samuel

Schiller, and librarian Kathie Gully, offered Cain invaluable assistance, institutional knowledge, and good cheer. Claudia Jacobson, Susan Otto and Judy Turner at the Milwaukee Public Museum, Eunice Schlictling at the Putnam Museum, Cathy McNassor at the Natural History Museum of Los Angeles County, Armand Esai at the Field Museum, and William Peniston and Jeffrey Moy at the Newark Museum, helped Cain locate important evidence and useful images. Cain also thanks Felicia Maffia, the manager of exhibit development at the Pacific Science Center for the interviews and information about that institution. Hilary Corbett, John Glover, Jean Scott, and John Ulmschneider of Virginia Commonwealth University (VCU), along with other librarians at Columbia University, New York University, Sarah Lawrence College, and the University of Southern California performed heroic feats of citation sleuthing and obscure material sourcing through interlibrary loan. Wendy Wasman of the Cleveland Museum of Natural History tracked down elusive pictures and dates for us, and Evora Swopes of *Science News* and Bahadir Kavlakli and Tom Blake of the Boston Public Library Digital Services respectively helped us locate and scan some of the book's many illustrations.

We count ourselves very fortunate for having had strong institutional homes and wonderful local administrators, colleagues, and students over the course of this project. At Sarah Lawrence College, President Michele Tolela Myers, Dean Barbara Kaplan, fundraising diva Sophia Primps, and many faculty members, especially Mary Dillard, Marsha Hurst, Leah Olson, Marilyn Power, Shahnaz Rouse, Lyde Sizer, and Charles Zerner, endured endless informal discussions in the early stages of the project and made helpful suggestions that resulted in Rader gaining more formal support and a broader public audience for her research. The Sarah Lawrence Writing Group gave her valuable feedback on mere skeletons of some chapter drafts. Students in Rader's year-long service learning seminar—"Nature on Display"—provided an engaged forum for important discussion of museum history and historiography: their enthusiasm about working in museums and talking about displays was boundless and sustaining. Former College of Humanities and Sciences Dean Robert Holsworth, as well as Associate Deans Catherine Ingrassia and Fred Hawkridge at Virginia Commonwealth University, encouraged and supported Rader's expanded forays into Science and Technology Studies teaching, programing, and research and eased her transition to working at a large public academic institution. Current Dean James Coleman and Associate Dean Alison Baski, History Department chairs Bernard Moitt (2006–11) and John Kneebone (2012 onward), and VCU faculty colleagues J. Brian Cassell, Deirdre Condit, J. Clifford Fox, Njeri Jacskon, Brianna Mezuk, William Newmann,

Jill Rowe, and Judyth Twigg served similarly supportive roles as this project neared an end. Wanda Clary provided critical administrative support. At Columbia University, professors William Leach, Betsy Blackmar, and Eric Foner provided Cain with sharp insights and thoughtful guidance throughout rounds of dissertation drafts, grant proposals, and job applications. Cain's fellow graduate students, including Jeremy Derfner, Monica Gisolfi, Julia Heck, Betsy Herbin, Emily Lieb, Pardis Mahdavi, and Karin Zitzewitz, offered keen editorial eyes. Members of the University of Southern California's Visual Studies Graduate Certificate Program, especially its director, Vanessa Schwartz, and Douglas Greenberg, then the Executive Director of the USC Shoah Foundation Institute for Visual History and Education, expanded and invigorated Cain's understanding of visual culture. Other visual culture scholars working in Los Angeles at the time, including Sarah Goodrum, Katja Zelljadt, Sarah Miller, strengthened Cain's commitment to the field. Cain is indebted to the faculty, students, and staff of New York University's Museum Studies program, especially Bruce Altshuler, Tatiana Kamorina, Hima Mallampati, and Miri Young, for their gracious and unstinting support. Their knowledge of and enthusiasm for the work and world of museums was crucial to framing this project; their kindness and collegiality was essential to its completion. Other members of that university's thriving intellectual community should also be recognized. Jonathan Zimmerman and the members of the History of Education Writing Group eased Cain into the scholarship on the history of education, and insights resulting from their lively discussions provided essential background for the manuscript. We are grateful to New York University's Humanities Initiative's Grants-in-Aid program for funding the illustrations in this book, and we thank Aysa Berger for administering this grant.

The University of Chicago Press's senior editor Karen Darling patiently shepherded the manuscript through the review and production process and, at many points along the way, offered insightful comments and sage advice. We also appreciate the important work of the production team, including Sophie Wereley, Abby Collier, Yvonne Zipter, and Micah Fehrenbacher.

While we lived with this book project, our friends and families lived with us, helping us stay sane and grounded. Rader would like to thank friends Soni and Tom Beecher, Claire and Dave Gregory, and Tobe and Ralph Sevush for everything from their well-timed expressions of faith in her to their mundane acts of kindness, especially for emergency childcare help. Cain would like to thank friends Jon Axelrod, Madeleine Elfenbein, Sarah Thomas, and Joel Derfner for their sympathy, considerable editorial talents, and hilariously cynical advice. Our families got smaller and sadder as we wrote this book—we lost too young a parent, Holly Rader, and a sister, Alexandra Cain—but those members that remained, especially Douglas and Constance Cain, Charles and Kathy Powers, and Richard Andrew Rader continued to sustain and encourage us. Our spouses, John Powers and Seth Weingram, have steadfastly supported our lives and work, and we are grateful for the surprise, delight, and balance they bring to our lives. We dedicate the book to our children, Samuel and Grace Rader Powers, who were around when we conceived this project, and Vivian and Caroline Cain-Weingram, who were born during its development. You inspired and distracted us in unpredictable and essential ways.

The Mission of Display

After touring the United States in 1931, British critic G. K Chesterton reported that he had glimpsed there a stark symbol of modernity: "We have passed," he wrote, "from the age of monuments to the age of museums." Natural history museums in the United States had become an "entirely modern thing," he proclaimed. The description was not a compliment. Unlike monuments, museums inspired no awe or dumbstruck wonder. Rather, he explained, they "faintly fatigued" with their "cold and compulsory culture." Today's natural history museum, he concluded, was "meant for the mere slave of a routine of education to stuff himself with every sort of incongruous intellectual food in one indigestible meal." "Dragged on from one thing to another; from one thing that interests him mildly to another that bores him stiff," the museumgoer was no longer a scholarly pilgrim, but a sightseer.[1]

As Chesterton acknowledged in his essay, American museums of science and nature looked and functioned very differently in the twentieth century than they had in earlier eras. Even the definition of the word "museum" had changed dramatically in the ages Chesterton compared.[2] From the Renaissance through the long eighteenth century, the term conjured up quiet study, spiritual contemplation, and, most important of all, collections of rare or wondrous objects.[3] When the philosophical and political revolutions of the eighteenth century forced open the doors of royal museums, palaces, and gardens, the word took on an additional meaning. A museum became a physical experiment in republican ideals, a space that ordered natural objects and fostered tentative mingling between the social classes who came to view them.[4] Museum founders in the young United States, eager for audiences, further broadened the connotations of the word by making museums synonymous with entertainments.[5] Impresarios ranging from Charles Willson Peale to P. T. Barnum easily combined educational, spectacular, and sickening specimens for purposes of public education, national or local pride, and—especially in Barnum's case—profit.[6] And as loose local coalitions of wealthy amateur naturalists, boosters, professional scientists, and politicians joined forces to establish new museums of natural history in the nation's burgeoning cities in the 1870s and 1880s, the word "museum" came to imply a diligent middle-class respectability and civic ambition.[7]

Despite the remarkable diversity of their audiences and aims, American museums of nature and science in the nineteenth century shared a common characteristic: they all possessed collections of specimens. However expansive the definition of "museum" might have seemed, a collection remained its binding thread, its defining feature. Collections of objects distinguished museums from other kinds of institutions, and museums' varying missions—scientific, educational, historic, and recreational—all revolved around their physical holdings. Even in the 1930s, when Chesterton was writing, it was difficult for Americans to imagine a museum without a collection. It wouldn't have seemed like a museum at all.

Yet by the time of Chesterton's tour, American museums of nature and science were already well into the process of piecing together a new institutional paradigm, one in which displays were just as important as collections, and sometimes even more so. In the early twentieth century, influential museum leaders came to believe that improving displays, not collections, was the best way to transform their institutions into important places for noncompulsory public education. Museum workers' kaleidoscopic efforts to enhance exhibits ultimately led to colorful, fragmented halls, and Chesterton cheerfully deplored the chaotic results. The galleries of modern museums, he wrote, featured "totally different things, with totally different types of interest, including a good many of no apparent interest at all."[8] Museum displays, he concluded, had become "not a product of popular imagination, but of what is called popular education."[9]

This book chronicles the prehistory and the legacy of the revolution Chesterton witnessed: how ongoing efforts to create popular educational displays ultimately compelled public natural history and science museums in the United States to develop new institutional roles and identities, in both twentieth-century science education and American culture.

In the antebellum United States, cultural battles had raged openly over the content and shape of the young nation's museums. These clashes, occasioned by a dramatic proliferation of diverse public and private institutions calling themselves museums, represented tensions between competing ideas of the roles that museums could and should play in American life. Should museums be closed ceremonial shrines to scientific knowledge and authority or should they be open places for public confrontation, experiment, and debate? Should they be, as Brooklyn museum director Duncan Cameron famously put it, temples or forums?[10] In the 1890s, leaders of the nation's public natural history museums advocated a kind of détente between these extremes, a set of ideals and practices that they dubbed "the New Museum Idea."[11] The New Museum Idea, as they interpreted it, called for museums to serve the public in two ways—through the simultaneous production and dissemination of scientific knowledge.

We begin our story here, arguing that reformers' embrace of the New Museum Idea inspired a century-long renegotiation of the relationship between display, research, and education in American museums of nature and science. Over the next hundred years, conflicts over what museums should do and be continued, even escalated. Museum workers' struggle to balance multiple missions profoundly altered institutions' appearances and functions. But it was their continual debate over display that drove decisions to restructure museums and their work. Indeed, it was exhibits' transformation that produced the most dramatic organizational innovation in older institutions and, also, led to the establishment of new types of museums, many of which eschewed collecting in the name of science education. Thanks to this century-long investment in exhibition, we argue, the museum landscape looked very different in 1990 than it did in 1890.

More than other kinds of exhibits, the history of biological display illuminates the nature and range of the intellectual and institutional transformations that took place in twentieth-century museums. Biological displays crisscrossed and connected different kinds of scientific and educational institutions. Following the creation and re-creation of these exhibits across different contexts forces us to scrutinize the knowledge-making and organizational practices that legitimated various biological sciences, museums, and displays. In tracing the multiplicity and contingency of these concurrent developments, our history further enriches the practice-oriented literature on twentieth-century shifts in biological ideas, materials, and research.[12] But we also demonstrate the potential of applying what sociologists, economists, and political historians have called a "new institutionalist" sensibility to existing scholarship: by moving beyond specific contexts and individual displays and examining the broader history of these institutions, we explore how changing configurations of museums and exhibits had important collective effects on one another.[13] We have explained, to borrow a phrase from sociologist Everett Hughes, how these "institutions chose their environments."[14]

By investigating the history of biological displays, we have been able to document the origins and development of museum work's most contentious aspects. Americans pinned great social hopes on biology throughout the twentieth century, investing heavily in its study, applications, and popularization.[15] As museums assumed the fraught task of displaying such an important subject, a diverse crowd of actors, scientists among them, debated what aspects of the biological sciences exhibits should present to the public and how they should go about doing it. The relatively high cost of biological displays, which often involved preserving dead specimens or sustaining live ones, generated intense discussions about the proper balance of research, display, and other forms of public outreach in museums.

Such debates, which we have mined from museum workers' personal correspondence, annual reports, professional journals, and the popular press, reveal when, why, and how museums struggled to define and redefine their public missions and their role in science education. These disputes give us new insights into how and why museums' institutional identities have changed over time and the consequences of those changes for contemporary practitioners. By foregrounding the history of these fraught discussions, we build on existing museum studies scholarship that recognizes museums' development as contingent and uncertain rather than predetermined and hegemonic. We, too, see natural history and science museums as dynamic "contact zones" but of a different sort than the ones James Clifford and Mary Louise Pratt famously described.[16] Museums of science and nature gradually became communities with unique conventions and values, even as they remained spaces where varying producers and consumers of scientific knowledge negotiated power relations.[17]

The collaborations that biological displays demanded from museum workers throughout the twentieth century highlight significant intersections between professional and popular science. Throughout the twentieth century, both scientists and exhibit makers struggled to render the science of life in publicly compelling fashion, and the two groups worked closely with one another to represent its wonders convincingly. Eavesdropping on the resulting conversations—and quarrels—affords a more nuanced analysis of relationships between generators and popular-izers of scientific knowledge in an age of mass communication. While university-based scientists may have transitioned more or less seamlessly to disciplinary professionalism, we show that museum workers—cura-tors, exhibit designers, and educators—struggled to define and wield their expertise as the civic sphere of their institutional work gradually transformed.[18]

Few scholars have attempted to give an overview of this important century in the institutional and social development of museums of sci-ence and nature. Admittedly, the diversity of these institutions is daunt-ing, and the idea of chronicling their respective histories in a single volume even more so. We do our best not to flatten out the ideologies and idiosyncrasies of individual institutions; at the same time, instead of assuming reified boundaries between them, we focus on the shifting meanings and values that were assigned to display and its public func-tions, and how these circulated while institutional differences were still very much in the making. In this way, our broad-brush approach makes visible some common social dynamics that arose both within and be-tween different types of institutions—natural history museums, science museums, and science centers—as a consequence of their decision to "go public." By examining these overlaps, we do not wish to deny what ultimately became important institutional differences.[19] Rather, we seek to elucidate how museums both drew from and participated in what was, in hindsight, a broader conversation about the scientific and social im-portance of exhibition practices.

Working on such a large chronological and institutional canvas has enabled us to revise some traditional historical narratives about muse-ums. Historians have often dismissed natural history museums in the period between 1930 and 1960, describing these decades as static.[20] While we found that display did languish in some institutions, in others, cu-rators and exhibit makers experimented wildly in response to the dy-namic displays of life science and medicine emerging from newer science museums. The century-long span of our narrative has also meant we can track the continuous development of particular activities or preoccupa-tions. Whereas some scholars and practitioners have argued that mu-seums only began to pay real attention to education during the Cold War or the 1980s, when museum educators fully embraced constructiv-ism, we chronicle the widely varying ways museum staffs defined and pursued educational goals over the course of a longer period of time. We describe ideas and actions that look nothing like education from the perspective of a contemporary practitioner but were, at the time, considered cutting-edge and proved crucial to the historical development of museum education. Likewise, although some dimensions of the tension between professionalism and popular education had diminished in natural history museums by the 1890s, others, like conflicts between research curators and professional exhibit makers, had yet to bloom fully.[21] When they did, they produced changes in the organization of museum work that had inter- and intra-institutional ramifications throughout the next century.

However ambitious our institutional and chronological scope, this work is in no way comprehensive. It does not address how the history of public museums converges or diverges from those of university museums or private science or technological museums. Nor does it address how museums' missions, educational programs, and displays developed in the twenty-first century, when dramatic cultural and technological shifts once again transformed these institutions, moving them into an era that practitioners have dubbed "Museum 2.0."[22] And while we have conducted research in and on what feels like a vast number of institutions, using available archival and primary source evidence, we focus in detail on just a handful—often, exceptionally influential museums that set the terms of debate and practice for museums around the nation.

Historian of science Sally Gregory Kohlstedt has noted that scholarly approaches to understanding natural history and science museums in the twentieth century have varied widely over the last three decades, resulting in interdisciplinary battles and tensions. Historians of science initially viewed museums as noncontroversial institutional backdrops for biographical or intellectual histories of scientific ideas; later, a few of these same historians, now under the influence of cultural and postcolonial studies, began to highlight the "constraining authority of museums, foregrounding the effects they have on individuals and cultures." Both approaches, Kohlstedt argues, tender important insights, but both have drawbacks. Cultural histories emphasize museums' educational efforts but tend to characterize them as mere "mechanisms of authority and exploitation," while scientific histories articulate the research ambitions of museum scientists but ignore the importance of the educational and collecting work in which curators also participate.[23]

We have turned to social history, which allows us to explore how and why scientific and social trends intersected in the museum, and we examine the interpersonal and political dynamics that resulted. By making visible the experiences and practices of those struggling to transform museum display, we have been able to complicate a number of long-held assumptions. For example, historians and practitioners have often assumed research and display were at arms' length from the early twentieth century onward. But our close study of staff correspondence and institutional budgets reveal that, in fact, scientists were extraordinarily invested in the success of museums' displays and saw display as an integral element of their own public outreach work. Many capitalized on exhibition in order to expand funding for and strengthen interest in their own research agendas. Many saw public education as a natural extension of their own efforts to improve American society through biology and hoped to further this work in any way they could. Likewise, we were surprised to find that exhibit designers were periodic participants in the research process, supplementing and sometimes prompting research projects through display work. In turn, we have been able to detail how science, education, and exhibit work professionalized in new and different ways in the museum context, expanding scholars' understanding of the history of professionalization in these broader fields.

As a dynamic account of how museums negotiated their multiple roles in science and society, our story speaks to varied schools of science and technology studies—to cultural historians and sociologists of science, in particular. We hope that scholars of museum studies, American studies, and visual culture will also find it useful, for it offers new evidence and interpretations of major turning points in the historical practices and policies of these popular institutions. We believe that the passage from the age of monuments to the age of museums was indeed, as Chesterton wrote, a crucial historical moment and that understanding this history will provide museum practitioners with a deeper understanding of how they are still living with its legacy today.

ONE

"A Vision of the Future": The New Museum Idea and Display Reform, 1890–1915

After young Russian immigrant Mary Antin visited the museum of the Boston Society of Natural History in the 1890s, she could not bring herself to return. The specimens in the museum's dim cases "failed to stir my imagination," she recalled, "and the slimy things that floated in glass vessels were too horrid for a second glance." "The exhibits were so dingy, so overcrowded and so revolting [that] recalcitrant children . . . were dragged in and walked through the Natural History rooms to strike terror in their hearts," remembered zoologist Thomas Barbour.[1] The halls of late nineteenth-century American natural history museums routinely elicited similar reactions, for they were "neither a delight to the eye of the unprofessional visitor nor a satisfaction to the specialist," wrote National Museum curator Robert Ridgway. At the museum of the Davenport Academy of Sciences, for instance, Iowans found the thousands of insects, pressed plants, arrowheads and copper axes, and animal skulls so "fearfully dull" that its galleries were almost entirely deserted, volunteer curator William Pratt confided to a friend in 1890. New York City's public schoolteachers considered the exhibits at the American Museum of Natural History so incomprehensible in the 1890s that they refused to take their students to visit, believing the trip wasn't worth the effort.[2]

Between the 1890s and the 1910s, a young generation of

reformers inspired by a transnational movement known as the New Museum Idea came to believe that natural history museums could no longer ignore these audiences. Museums should, they argued, reach and teach all citizens, and display seemed the most expedient way to do so. Displays could bring a museum visitor "into direct contact with Nature, show him her infinite resources and establish between him and the outdoor world an intimacy through which he will derive not only pleasure, but also physical, mental, and spiritual strength," explained curator and museum reformer Frank Chapman in 1899.[3] Convinced that displays could and should become powerful pedagogical tools, reformers persuaded their colleagues to reconsider the contents of museums' shelves and the configuration of their halls.

An infectiously enthusiastic faction of reformers, known in the museum world as "museum men," worked to realize exhibits' educational potential in these decades, encouraging fellow staff members to join them as they experimented with new approaches. At museums new and old, staff members rooted through collections, steadily sorting, labeling, and tossing aside specimens to compile collections that could spark interest in scientists and schoolchildren. They worked to build displays that would, in the words of the National Museum's George Brown Goode, transform museums from "a cemetery of bric-a-brac into a nursery of living thoughts." In this way, museum men succeeded in imposing new order on the nation's miscellany of natural history museums, establishing an institutional template that reformers eventually embraced, reproduced, and improved on while still hewing to a unified vision of the museum as a major force in popular education. By 1907, Brooklyn Museum curator Frederic A. Lucas could confidently declare that "the character of museums has changed so greatly . . . those who know these institutions only as they exist today may not realize how widely they differ from what they were even twenty-five years ago."[4]

This transformation was not realized without struggle. Ideals, practices, and people clashed as reformers reorganized displays and the institutions that housed them. Attempts to make room in museum halls for museumgoers' tastes, needs, and interests collided with efforts to discipline unruly collections and displays. Staff members bickered over the shape and extent of display reform, some bucking against the notion that public education merited such attention, others insisting that curators should prioritize display making over their other duties. By the mid-1910s, reformers had persuaded most staff members to rally behind their vision. The museum was not merely a place dedicated to the preservation of specimens and the production of scientific knowledge, wrote ichthyologist and director of the California Academy of Sciences Barton W. Evermann in the *Popular Science Monthly* in 1918. "The museum has come to regard itself, and to be regarded by the public, as an educational institution," he pronounced.[5]

Progressive Era museum reformers' interpretation of the New Museum Idea as a mandate for public education through display would undergird professional and public ideas about the institutional role of natural history and science museums for the next hundred years. The vision they articulated would spark decades of animated arguments about display practices, pedagogical principles, and the proper place of exhibits in museums. It would not only unsettle museums' institutional cultures but would also ensure that museum displays became battlegrounds for contentious cultural debates over American science and science education in the twentieth century.

The New Museum Idea and the Museum Men

To most Americans, the Smithsonian's U.S. National Museum stood first among the dozens of natural history museums scattered across the United States in the 1870s and 1880s. It was still a young institution and it was small compared to the museums that sprawled across Europe's capitals. But its size and scope awed Washington tourists and its scientific ambition staggered the nation's naturalists. Smithsonian curators hoped to assemble specimen collections that would present American scientists with a complete picture of the historical development of the globe's variegated life forms and rich natural resources.[6] So they purchased vast compendia of natural objects from well-known European collectors and persuaded American naturalists—professional scientists, avid amateurs, and young hobbyists—to flesh out the museum's holdings with their finds. They also worked closely with federal surveyors, who sent back railroad cars of specimens as they combed the country west of the Mississippi, and dispatched professional collectors to gather the nature of other nations. Curators systematically separated and analyzed the resulting mountains of specimens, categorizing animals, plants, and minerals into groups distinguished by similar physical features or functions. Over the course of the late nineteenth century, the scientific staff of the National Museum helped map the shape and capacities of the natural world, providing ever more detailed information to governments, investors, scientists, laborers, and the reading public as they did so.[7]

Though the National Museum officially welcomed all visitors in the

Figure 1   View of the natural history exhibits in the U.S. National Museum, as published in the
Smithsonian Institution's 1896 *Annual Report*. Visitors, surrounded by thousands of
specimens, sat at a small table and perused gallery guides. As was typical in natural
history halls devoted to paleontology, a skeleton of a *Zeuglodon*, an extinct whale
now known as a *Basilosaurus*, hung from the ceiling, as one of a handful of eye-
catching large natural specimens that framed the room. (Smithsonian Institution
Archives, Image NHB-9469.)

early 1880s, its halls were no place for the public. Its displays, with thousands of specimens massed in glass cases (see, e.g., fig. 1), were "a perfect chaos," admitted staff ornithologist Robert Ridgway. The galleries
appalled Frederic Lucas, an osteologist and taxidermist who had joined
the museum's staff in 1882. Lucas had trained at Ward's Natural Science
Establishment, a well-respected purveyor of natural history collections.
Accustomed to Ward's exacting aesthetic standards, committed to expanding the public's understanding of the natural world, Lucas found
the disorganized state of the Smithsonian's displays downright offen-
sive.[8] The young man was relieved to find that the museum's recently
appointed assistant director, the courtly thirty-year-old George Brown
Goode, felt the same way.

Goode's faith that a proper understanding of the sciences, and especially the biological sciences, could improve the lives and lots of all

11

citizens was no particular surprise to his colleagues—natural science had been one of the great love affairs of his life, after all, and his own lot had improved immeasurably as a result of its influence.[9] As a boy he had spent long hours roaming farmland and woodlot in Amenia, New York, paging through the Smithsonian Reports that lined his father's personal library. After a postgraduate year at Harvard's Museum of Comparative Zoology, he returned to his own alma mater, Wesleyan, to create a museum out of its scattered natural history collections. But the encounter that changed his life was meeting Smithsonian ichthyologist Spencer Baird while volunteering for the U.S. Fish Commission. Baird persuaded the twenty-two-year-old to join him at the commission and the museum, and over the next decade and a half, Goode's world widened. He collected his way through Bermuda, Florida, and Connecticut, publishing dozens of scientific treatises on the geographical distribution of fish and authoring popular volumes on game and food fishes and the state of the American fishing industry. He took charge of the Smithsonian's displays at the Philadelphia Centennial Exhibition in 1876, becoming a curator and, eventually, assistant director of the U.S. National Museum. Thrilled by the beauty of tropical fish and concerned with the health of the nation's fisheries, Goode appreciated how scientific knowledge could illuminate individual lives and drive whole economies.

As a result, the young administrator firmly believed the National Museum had a moral duty to educate the broadest public possible about scientific concepts and applications. The needs of "the mechanic, the factory operator, the day-laborer, the salesman, and the clerk" were as important as "those of the professional man and the man of leisure," Goode reminded his curators, and the museum should be arranged accordingly. But for the museum to serve all Americans present and future, Goode suggested, it needed to advance three missions simultaneously: specimen preservation, scientific research and publication, and public education.[10] Curators should adhere to contemporary standards of professional scientists, but they should also collect, arrange, and display specimens in ways that broader audiences would find useful and visually appealing. "The people's museum should be much more than a house full of specimens in glass cases," he wrote. Rather, "it should be a house full of ideas, arranged with the strictest attention to system . . . a series of objects selected with reference to their value to investigation, or their possibilities for public enlightenment."[11]

Goode was hardly the first to promote this vision, which contemporaries called the New Museum Idea, but he was its most vocal advocate in the United States, and he spent more than fifteen years attempting to implement its precepts at the National Museum.[12] Goode was a genial polymath, admired by colleagues and adored by subordinates, and the National Museum's staff readily embraced his vision. They worked hard for him, following the institutional blueprint he set out early in his tenure even after his untimely death in 1896. Goode asked Smithsonian curators to look beyond their areas of specialty and broaden their collecting agendas, and under his supervision, the museum's holdings grew from 200,000 to more than three million.[13] At Goode's direction, curators divided these collections in two, a practice known as "dual arrangement," placing choice specimens on display and reserving the rest for professional study.[14] By 1894, the museum displayed only about one-fiftieth of its ornithological holdings in the public galleries—the rest lay in cabinets accessible only to researchers. They affixed lengthy labels to those objects on display and encouraged the museum's newer staff members to "try experiments in installation and exhibition that would not have been feasible in an older institution," Lucas affectionately recalled. "Not unnaturally some defects appeared when [Goode's] theories were put into practice." Lucas later wrote, "But the underlying principles were sound."[15] Lucas's endorsement was an understatement: the principles Goode advocated proved sound enough to inspire a generation of museum workers to launch a reform movement that would transform the halls of American museums.

In the late 1880s and early 1890s, Lucas was not Goode's only disciple: a far-flung network of museum workers came to concur with Goode that, as Milwaukee Public Museum director Henry L. Ward put it, museums "owe a debt to all classes of citizens and must afford pleasure and instruction to the masses as well as means of research for the scientist."[16] Some had been persuaded by Goode's speeches, some by his personal influence, for the Smithsonian's staff had ballooned from thirteen to two hundred during his tenure at the National Museum, and a generation of museum staff had trained at his knee. Inspired by his exhortations to make the natural history museum into "one of the principal agencies for enlightenment of the people," as he put it in an 1889 speech, they began to consider how they might adapt the reforms Goode had put in place at the National Museum to other institutions.[17] Swept up in the democratic impulses of the Progressive Era, this determined young group of reformers joined the host of men and women in libraries, public schools, agricultural and technical colleges, and other institutions striving to make scientific information accessible, open, and free to all Americans.

Reformers championed public education in museums for reasons that ranged as widely as their job descriptions. To a certain extent, museum workers' support for new ideals and practices was a matter of professional self-interest. Curators, whose job was to tend, arrange, and study specimens, were pleased at the prospect of revamping collections and displays along lines that would command greater interest from other biological scientists.[18] Directors, responsible for institutions' scientific and educational agendas, personnel, and finances, recognized that reforms might compel nature-loving patrons and municipal officials into funding their work.[19] Preparators, who readied specimens for study and display, appreciated the untapped pedagogical potential of their work.[20]

But support for museum reform did not always flow according to predictable channels. The New Museum Idea and its associated reforms accommodated all manner of interests and ideologies, so reformers' motivations for supporting museum-based public education varied considerably even within fairly narrowly defined groups. At the American Museum in the early 1900s, for instance, curators' support for museum reform spanned the political spectrum. Henry Fairfield Osborn, the museum's dominating curator of paleontology and eventual chairman, viewed the museum as an instrument of racial and social control, wielded in order to persuade potentially unruly immigrants to uphold the social status quo.[21] Joel Allen, its venerable mammalogist, believed that conservation was the single most important reason to promote museum reform.[22] Herpetologist Mary Cynthia Dickerson and Frank Chapman, the museum's curator of ornithology, had not only conservation but also personal liberation in mind, and they hoped to share their own deep pleasure in the natural world and broad knowledge of the biological sciences to enrich Americans' daily lives.[23]

Though museum reformers might disagree about the whys of educating this public, most were in perfect concert about the how of it, concurring that museums could best reach and teach broad audiences through careful displays of specimen collections.[24] This idea, so familiar nowadays that it seems inextricable from the very idea of a natural history museum, was still something of a revelation in the late nineteenth and early twentieth centuries. At most museums in the 1870s and 1880s, curators arranged collections to accommodate taxonomists and systematists, whose work required them to compare hundreds of similar specimens in order to identify species and trace their evolution over space and time. The resulting masses of objects made lay visitors' eyes glaze over, but the curators responsible rarely worried about losing museumgoers' attention. Museums should not cater to those who hoped "amusement may be lavished upon him" but to those "at least willing to put forth an effort to obtain information," Minnesota's state geologist N. H. Winchell announced in 1891, and most curators agreed. As late as 1898, Carnegie Museum director William J. Holland noted his profound irritation with people who "seemed to make no serious study" of his institution's holdings. As a result of their inattentive behavior, Holland felt quite justified in devoting more of his resources to building up the scientific collections than he did to reaching flighty "younger visitors, who came frequently but seemed rather to take a general view than to make a careful inspection."[25] Curators held that intense, even painful, concentration was prerequisite to learning from specimen displays and expected visitors to rise to the occasion.

But by the early twentieth century, reformers refused to dismiss the gawking, the ignorant, or the bored. This didn't mean they didn't get frustrated with the public, of course. "Distract them for a moment by a mechanical carriage, a street procession or an April shower, and I doubt if the distinct recollection of one shell, one bird, one gem or one mineral could be satisfactorily traced upon their minds," American Museum curator Louis Pope Gratacap had written in 1898. But nowadays, he explained in 1900, "we become thoughtful when we witness the visitor's vacuity of expression as he passes before cases devoted to the phylogeny of the arachnids."[26] Whereas earlier displays "expressed what the official occasionally put into words: 'the public be d——,'" the Milwaukee Public Museum's wry director, Henry L. Ward, explained in 1907, "within a very few years it seems to have come to most museums that they were on the wrong track . . . that they utterly failed to reach the public."[27]

By the 1900s, Gratacap, Ward, Lucas, and other reformers were urging their colleagues to revamp collections and displays with public education in mind. Though most reformers now believed the average museum visitor wanted to learn ("provided it can be done without too much exertion to eye or brain," wrote Ward), they also intended to "teach the public[,] whether they are looking for instruction or not."[28] The easiest way to reach indifferent visitors, they agreed, was through exhibition. The modern natural history museum "depends upon its exhibition series for its influence" on the broader public, Lucas explained, for displays were the only reliable method by which they could "reach a large constituency that has little or no book learning."[29]

Contemporary ideas about vision and attention seemed to confirm such assertions. Progressive Era scholars and scientists, natural history museum curators among them, considered vision the lowest and the most essential of the senses, and so touted the visual as the best means for attracting and holding the eyes—and minds—of the wider public.[30] "It is generally conceded that seventy-five percent of all the intelligence of the world comes through the eyes, and twenty-five per cent through the ears," Detroit Institute of Arts curator A. H. Griffiths wrote in 1908.[31] Consumer culture's powerful influence offered persuasive evidence for this theory. Museum staff knew well that thoughtfully designed, visually compelling displays could attract and hold the attention of all Americans—they witnessed the power of advertisements, shop windows, and other colorful commercial theatrics in the street every day.[32] The brightly colored illustrations cavorting across the pages of magazines, books, and newspaper inserts further testified to the power of the visual and pictorial, "barbarous as it is in many cases, in imparting information quickly and clearly," admitted biologist and museum director Edward S. Morse.[33] If they were serious about educating the broadest audience possible, Morse and other reformers believed, museums could ill afford to ignore the power of the visual, so they urged their peers to pay proper attention to the educative potential of attractive exhibitions.

Naturalists all, reformers found further justification for their commitment to display in one of the pedagogical traditions that had helped them most: specifically, observational training. Not only could specimen displays convey information efficiently, argued reformers, but they also afforded visitors precious opportunity to hone their visual skills. In the decades after the Civil War, educators, philosophers and scientists considered the structured exercise of eye and memory essential to the healthy development of human faculties.[34] Looking hard, long, and well did not come naturally, they argued, but required practice, so sense training became an important component of science teaching on all levels. Most staff members had been formally trained in the rigors of disciplined looking—some through elementary school object lessons, others by professors' insistence that they glean scientific insights by staring at coral or hen's eggs for days on end, still others during apprenticeships with collecting expeditions or taxidermy shops—and they hoped to offer visitors a similar kind of education.[35] Visitors might not have had similar chances to practice concentrating, observing, and deducing, curators pointed out, and it was "in museum study that one of the best remedies for this is to be found," Field Museum curator Oliver Farrington wrote in 1902.[36]

Reports from educators convinced museum reformers of the worth of this approach, as teachers noted that displays held the attention of visitors from all backgrounds. Contemplating museum displays stimulated the weak eyes and bored minds of wealthy city children, one private school teacher declared in 1904, interesting those "from homes where city conditions are always present and where, I fear, the materialism of life is at times pronounced."[37] Museum display also riveted laborers, immigrants, and their children in a way books and words could not, wrote William Maxwell, New York City's Superintendent of Public Schools, awakening "interest that will carry them into a new language, into the knowledge of the grade and at the same time into a more wholesome, more sanitary life."[38]

Reformers urging American natural history museums to adopt the ideals of New Museum Idea and adapt displays accordingly made their platform official in 1905, when a small group of museum staff met at the Carnegie Museum in Pittsburgh to organize the American Association of Museums (AAM). (Though the AAM was open to staff members from all kinds of museums, natural history museum staff dominated the organization. Five natural history museum directors and curators, the director of a botanical garden, the director of a museum of commerce and technology, and the director of a general museum with an emphasis on anthropology and natural history composed its first board, and they issued invitations to join the AAM through the columns of *Science* magazine.) Founding members of the AAM venerated Goode as a patron saint and regularly invoked his ideas about museums' obligations to the public, as well as his conviction that museums' new missions were best carried out by a well-trained, professionalized work force. Through annual meetings and correspondence between the organization's members, these reformers established Goode's vision as the prevailing professional ideal in early twentieth-century natural history museums.[39]

Reformers also relied on more informal professional networks to promote their gospel. Clustered in the country's largest natural history museums—the American Museum of Natural History, the Field Museum of Natural History, and the Smithsonian's National Museum—the reformers instilled new ideas about scientific standards and educational possibilities in younger colleagues, who eventually fanned out across the nation to take positions at other museums.[40] Energetic young staff members forcefully transplanted these ideals to the hinterlands or, at the very least, tailored them to fit the needs of newer and smaller institutions. The head of the Milwaukee Public Museum used the Field Museum as his model, while directors of the Colorado Museum of Natural History, the California Academy of Sciences, and the Los Angeles County Museum of History, Science, and Art did their best to make their museums' halls resemble those of the American.[41] Though sorry to see their compatriots leave, reformers working in large institutions admitted they were delighted by the prospect of "forward[ing] the work of museums wherever they may be located," wrote American Museum curator Harlan Smith in 1910.[42]

Within this emerging professional community, the most ardent advocates for improving display were dubbed "museum men," a band of resourceful, versatile museum workers who argued, as Goode had, that "no pains must be spared in the presentation of the material in the exhibition halls."[43] The term "museum man" was imprecise at best, sexist at worst, for women actively worked alongside men to educate the public through innovative displays throughout these decades. (But for their gender, American Museum curator Mary Cynthia Dickerson and Fairbanks Museum director Delia Griffin would surely have been counted among the nation's leading museum men.)[44]

Still, the moniker stuck, and contemporaries widely used the term as shorthand for a person who was, as Lucas wrote, "gifted with a natural aptitude for museum work, one who can plan, install and label exhibits, care for the collections in his charge, and do something with the means and materials at his disposal, no matter how limited these may be."[45] Many museum men were talented educators, plainspoken popularizers with a gift for explaining scientific concepts in straightforward prose or pictures. Many possessed real artistic ability or, at least, competence: they could deftly render nature through sketching, painting, or the less exalted arts of taxidermy and wax modeling.[46] But their most distinctive trait was the conspicuous satisfaction they took in the growth of their institutions. Though museum men were ardent naturalists, colleagues often described them as Samuel Pierpont Langley described George Brown Goode: "He loved the museum and its administration above every other pursuit, even, I think, above his own special branch of biological science."[47]

Bonded by common experiences, eager to learn from (or boast to) one another, museum men kept in close touch. By the 1900s, many had known each other for decades, working shoulder-to-shoulder on field expeditions or sweating through the sweltering July heat in taxidermy studios. Shared experience forged lasting friendships and a remarkable network of correspondence. Museum men wrote one another monthly, weekly, even daily, dispensing advice about new exhibition tricks and techniques, passing along bits of gossip or job opportunities. They ribbed one another about professional decisions and personal quirks. They pushed each other to experiment—often with great success—with different ways of presenting information visually, and they eagerly shared the results of their improvisations. In an era before annual meeting proceedings and newsletters, these letters provided a forum for debate over exhibitionary standards and practices and hardened their conviction in the importance of display.[48]

Beginning in the 1890s, this tight-knit group worked to implement the blueprint for reform that Goode had laid out in the 1880s, and as they gained confidence, the museum men improvised and improved on his original vision. Though they counted preparators, curators and directors among their ranks, their able enthusiasm often hoisted them into leadership positions, and by the 1910s, they were among the most powerful members of the nation's professional museum community. As a result, they were able to put display at the front and center of museums' agendas. They banished cluttered shelves of artwork, freaks, and miscellany from peripatetic general museums, replacing them with intelligible exhibits of animals, minerals, and vegetables. They propped up the failing academies of science freckling the nation, working to transform dusty specimen stashes into displays that would appeal to local students and their parents. And they helped found entirely new natural history museums, delighted with the chance to translate Goode's ideals into fresh physical form.

The history of the Colorado Museum of Natural History exemplifies how museum men imposed their vision on the nation's museums and their displays in the early twentieth century. When it opened its doors in 1908, the museum was a moth-eaten jumble of specimens, run by a well-meaning but untrained staff. Frustrated board members quickly appointed Jesse Dade Figgins, then head of the American Museum's department of preparation and display, to clean house. Born in the Piedmont two years after the Civil War ended, Figgins had intended to become a Methodist minister but found the tug of sky and swamp too insistent, so abandoned theology for natural history.[49] He collected his way down the coast to the Carolinas in the 1880s, selling his wares to whatever scientific institutions or collectors cared to purchase them, studying ornithology, mammalogy, and herpetology at night.[50] These efforts earned him a job collecting and preparing specimens at the U.S. National Museum and, eventually, a place at the American Museum, where he spent the better part of thirteen years. But he never lost his loyalty to the New York museum. "Once one has spent some time in the American Museum, I doubt if they are ever quite contented elsewhere, or lose the feeling that they belong there," he later wrote.[51]

Figgins was a good man to have on either end of an expedition, even though he took up a lot of space.[52] He stood six feet and carried his weight out in front of him. He adored the out of doors and good food in equal measure—dewberry pie and gin swizzles figured prominently in his letters home from the field.[53] Outgoing and confiding, he moaned cheerfully to friends about his ever-expanding weight and teased them in turn about their poor taste in women and their terrible aim in the field. But he could survey and track, hunt and trap, skin and salt, identify and catalog, stuff and mount, fundraise, publish, and make a mean stew.[54] Known by the end of his career as "the best all-round museum man in the entire country," he was an accomplished naturalist, and could build all manner of attractive displays from scratch, making him, wrote American Museum curator Joel Allen, "an exceptionally competent man for such a position."[55]

Figgins had absorbed not only wide-ranging expertise at the American Museum but also the professional ideals of its reform-minded staff members. So when he arrived in Denver, Figgins announced his intention to make its museum a "centralized place of learning in all branches of natural history," a thoroughly modern institution that took public education and scientific research as seriously as popular amusement. Working sixteen-hour days, he fired off long letters soliciting funds, directing fieldwork, and proposing new exhibits to friends, donors, information seekers, naturalists, trustees, and collectors. He initiated surveys of the state's flora, fauna, and natural resources. He relabeled and remounted the few salvageable specimens the museum possessed and began to collect new specimens for the museum's display and study collections.[56]

Throughout, he remained focused on what he saw as the single most important goal of twentieth-century museum work: using well-designed displays of specimens to educate the public about nature and science. To patrons and visitors, Figgins described the Colorado Museum's objectives in the vaguest of terms—"public education," "public instruction," or "serving the public" through what he called "ecologic" and "biologic" education—but his letters reveal a very specific vision. It was, he wrote a friend in 1916, "of vital importance" for natural history museums to teach visitors to understand the laws of the living world and to appreciate its miracles.[57] A generation of museum reformers shared his conviction. In the first two decades of the twentieth century, this faith in the personal and political importance of nature's study deepened reformers' commitment to the New Museum Idea and improved display.

"Of Vital Importance": Science Education in the Twentieth-Century Museum

Museum reformers' enthusiasm for science education was an understandable response to the profoundly disruptive changes American society and culture faced as the nineteenth century drew to a close. Homes and factories overtook fields and fens. Immigrants from Eastern and Southern Europe flooded rapidly growing cities, which were already afloat in country youths fleeing the drudgery of farm life. As they spent more and more time surrounded by windows and walls, Americans worried that they no longer appreciated the resources and rhythms of the natural world as they once had. In 1891, G. Stanley Hall shocked the nation when the published results of his tests on first graders in Boston revealed their profound ignorance of the natural world. Ninety percent of the children had no understanding of an elm tree or a field of wheat or the origin of leather, statistics that horrified Americans nostalgic for the farms on which they had been raised.[58] The children of farmers and laborers, the "people who live most in the environment of nature in its directest sense," now know nothing almost about their environs, declared social reformer and Episcopalian bishop Henry Codman Potter at a 1900 groundbreaking ceremony at the American Museum of Natural History. "How much have they been trained to observe of a stone, or of a flower, or of a sky?"[59]

American adults didn't limit such handwringing to the young; they worried about modernity's impact on their own souls and senses as well. "My grandfather, who traveled to new Orleans nearly one hundred years ago by boat, on horseback, and on foot, received a sense-training by the way which I miss when I take the journey today in a parlor car," lamented Boston educator Anna Slocum at a 1911 AAM meeting. "The advent of steam and electricity with the specialization which has come in their train, the movement of population from country to city, have driven out the old ways of travel, the everyday duties and home industries that gave . . . an education that before was provided in more natural ways."[60] To some Americans, this distance from the natural world seemed downright dangerous, threatening their health, culture, and society. To others, it was simply tragic. Americans had long found great joy in contemplating the natural world, and they were sorry others could no longer appreciate its fascinations.[61]

Reformers from a wide variety of educational and scientific institutions embraced nature and science education as a partial solution to this perceived national crisis. The study of natural history and its modern academic incarnation, biology, they proclaimed, was integral to Americans' health and happiness. Learning to appreciate nature would lift hearts and minds, they argued, and studying it scientifically would result in more efficient social organization and more rational use of resources. The study of nature would exalt Americans' daily lives, promised Cornell horticulturalist Liberty Hyde Bailey, increasing "the mystery of the unknown and enlarg[ing] the boundaries of the spiritual vision."[62] A thorough understanding of science and especially what they called "socio-biology," the social blueprint that biological lessons offered American culture, would lead to a healthier polity, announced biologist and science popularizer William Ritter. In the twentieth century, he added, "more intelligence must consist largely in more intelligence about science."[63]

Museum reformers, too, came to believe that educating the huddled masses about the natural world was "vital to the future of the country," as curator Frank Chapman put it in 1910, and this conviction added urgency to reformers' interest in public education.[64] "Science represents not only the highest point so far reached by human intelligence," wrote curator George Dorsey, but also "furnishes us with our one guide for the future." As a result, he added, "it is the privilege and duty of every scientific institution" to educate the public on the topic. Indeed, "the true value and importance of any institution to a community lies in the power which it may exert upon the people at large to call into consciousness the various activities of their life."[65] If reformers wished Americans to approach the world around them gratefully, as worshippers at the altar of nature would, and rationally, as participants in an orderly democracy should, argued museum reformers, then they needed to organize their institutions accordingly. "I sometimes think," Chapman wrote a friend in 1910, "that to accomplish this work is more important than doing scientific research."[66]

Early twentieth-century schools, libraries, and publishers pelted the public with scientific facts and concepts, but museum reformers were confident their institutions had something different to offer. Reformers pointed out that museums brought visitors into direct contact with fascinating natural objects, unlike schools, whose classes still largely relied on rote recitation to drill scientific knowledge into students.[67] Museum displays presented concrete examples of "what to the pupils are apt to be rather vague ideas," explained Lucas in 1907. "Mimicry, subspecies, influence of environment are all discussed in text-books, but their meaning is much better gathered from objects that illustrate these facts."[68] Scientists, social critics, and educators working outside museum walls tended to agree. School biology courses were often "so ponderous that they defeat their own ends," Slocum wrote, but museums had the potential to "inspire . . . rather than fatigue" and, in doing so, "lighten the burdens of the school."[69]

Reformers asserted that museum display could also serve as a necessary antidote to the corrupted versions of science proffered by popular culture. Seeing the study of nature as improving and wholesome recreation, early twentieth-century Americans consumed information about it at a brisk clip. In small-town classrooms and big-city auditoriums, Americans attended lectures on natural history, offered by explorers and popularizers eager to lure them "back to nature," and watched animals cavort across moving-picture screens.[70] They pored over popular field guides and paged through richly illustrated magazine articles about animal life. Nature essays and nature novels written by amateur naturalists were particularly popular in these decades: they topped best-seller lists, making their way into school curricula, newspapers, and the coat pockets of those hoping for refinement or refreshment through nature's study.[71] Scientists often derided this material, objecting to their sensationalism and sentimentalism.[72] Botanist Otis Caldwell was so worked up over the issue that in 1904 he demanded nature study educators abandon their "legendary, mythical, imaginary, and impersonating" approach and banish such books from their classrooms.[73] Museum reformers offered their help, proclaiming it was their duty to "enlighten the general public and especially those entrusted with the education of children" about such inaccuracies, as Chapman wrote in 1904. As a result, they promised to provide scientific information untainted by ignorance or sentimentality. At the same time, reformers agreed, scientific rigor should not mean dullness.[74] When educating the public, it was just as important to evoke a sense of wonder, of excitement about the living, breathing world. After all, explained Lucas, "if you cannot arouse the interest of visitors you cannot instruct them."[75]

Convinced that the nation's citizens would benefit from the study of science and nature, swayed by museum reformers' vision of their role in this work, civic pillars and wealthy philanthropists began to fund museums on an unprecedented scale. Many private donors were frank about their intentions: they hoped to purchase social order through science education. When financier Norman Harris donated $250,000—more than six million dollars in today's money—to the Field Museum in 1911, for instance, he "visioned larger things . . . than the teaching of natural history," the museum announced in a later press release. "He believed that every agency which increases the attractiveness of knowledge and the ease of acquiring it is an agency for better citizenship and for more stability in civic conditions."[76] But the study of nature had many social meanings, and some donors contributed out of "values which are more spiritual than material"—pure passion for nature and its "peaceful paths," as Connecticut industrialist J. Kennedy Tod explained in the note he enclosed alongside a thousand-dollar check to the trustees of the Colorado Museum.[77] City politicians were rarely as generous, but they, too, were convinced that their citizens would benefit from visiting, in Milwaukee museum director Henry Ward's words, "an educational museum that caters to the general public whom we desire to interest and instruct whether or not they visit us with this desire in their own minds."[78] By the 1900s, most municipal natural history museums received some help from local government, in the form of land, buildings, revenue earmarked for maintenance costs, or exemptions from various taxes, in exchange for admitting—and educating—all citizens for free.

Public financial support gave new heft to reformers' argument that museums owed, as Milwaukee's Ward put it, "a debt to all classes of citizens," so museum staff, believing that public education was the best way to discharge this debt, experimented aggressively with new educational efforts.[79] Museums reached out to classroom teachers in the 1890s, offering nature study training and illustrated talks for elementary school teachers, and by the 1910s, most sizable public natural history museums in the United States provided free weekend lecture series to local educators. Smaller institutions were on the job as well. The sole curator of the museum of the Davenport Academy of Sciences, for instance, developed science curricula for the city's schools and moonlighted as an elementary school nature study instructor when work at the museum was slow.[80] A handful of museums proffered more advanced instruction. The Chicago Academy of Sciences offered courses to students for university credit, and curators from the Brooklyn Institute of Arts and Sciences taught university professors, physicians, and high school teachers how to dissect fish and how to culture marine bacteria at its laboratory in Cold Spring Harbor.[81]

Yet museum reformers held fast to their belief that the display of specimens should be the spine of museum-based education and urged institutions to organize new educational programs around exhibits. After all, Ward explained, the chief work of the modern museum was "to take hold of a collection of inanimate objects and display them so as to catch and hold the attention of a previously indifferent public and make the collection tell a story so clearly and interestingly that the public is forced to give heed; and, leaving, do so with the pleasurable satisfaction of having been instructed while entertained."[82] So museum lecturers and educators, many of them the wives and daughters of curators, began to guide school-children past existing displays, to promote and interpret new exhibits in crowded auditoriums, or to create portable displays and accompanying teaching scripts for use in schools. In Buffalo and Denver, San Diego and Chicago, staff members built hundreds of small dioramas of chipmunks, sugar cane, and seashells, which trucks trundled from museum to school and back again. To cope with the crushing clerical burdens that accompanied the circulation of slides, scripts, and exhibits, museums established departments of public instruction.

But museum educators refused to limit themselves to scheduling field trips.[83] In hopes of cultivating broader audiences for their museums, they experimented with all manner of new programs. In-house educators organized illustrated lecture series for the public, persuading curators to describe exotic adventures in the field to buzzing crowds on Friday nights and Saturday mornings.[84] They founded collecting clubs and sponsored identification contests to help already enthusiastic hobbyists pursue birds, bugs, and botanical specimens more vigorously. They lectured to local gardening groups and delighted nature photographers by arranging for them to exhibit their collections in museums' echoing entry halls. A few worked with sympathetic curators to create rooms where children could bring in or touch specimens under the guidance of an adult. Reformers applauded these efforts. "The refining and humanizing influence of these gentler pursuits," Carnegie Museum director William J. Holland wrote in 1899, "cannot fail to leave a marked impression upon the life of our busy city."[85] Public response confirmed Holland's prediction. This work "is surely the live wire between the people and their Museum and I say bully for you for pushing it as you have," one delighted bird-watcher wrote the Milwaukee Public Museum in 1915.[86] Amid the fizz of pedagogical experimentation, however, museum reformers continued to urge education departments to focus on teaching the public through natural history museums' own collections.

Collecting with the Public in Mind

By the 1900s, most museum workers shared reformers' calm certainty that Americans of all backgrounds could learn about nature and science by contemplating well-arranged natural objects. They were less certain, however, about how to create such displays. To inform visitors about the mechanics and miracles of life often required an entirely different set of objects than museums possessed. What kinds of specimens could the Colorado Museum, whose half-empty shelves were filled with moth-eaten Rocky Mountain game and mineral samples, use to depict evolution? Did museums need to build entirely new kinds of collections for the public? And if museums did so, how, given institutions' limited funds, time, and space, could curators simultaneously maintain collections useful for scientific research?

Museum men, many of whom had become directors or department chiefs, saw opportunity in this challenge. If they could build collections that would at once satisfy the needs of professional scientists and lay visitors, they could solidify museums' stature as the preeminent public institutions for the study of science at all levels.[87] They experimented with a wide variety of collecting subjects and practices as a result, striving to balance scientific standards with public tastes and needs. Working with other reform-minded museum workers, they steered collections toward provinces of knowledge that interested scientists and retained considerable visual appeal and sociobiological importance. Not all collections needed to please all audiences—dual arrangement ensured this—but acquiring and maintaining specimens was costly, so museum men pushed to compile collections that could pull double duty—and charm donors to boot.

Hoping to build mutually rewarding relationships with budding amateur naturalists, museum men also urged visitors to bring in their own finds. Though museum men assumed these two approaches to education—top-down dissemination and bottom-up participation—would complement one another, they quickly revised this assumption. The thousands of donations that excited visitors brought in undermined their efforts to build comprehensive, carefully cataloged collections that would shore up museums' educational and scientific authority. Professional standards eventually trumped public participation, and museum men's early attempts to bolster public contributions to museum collections gave way to quiet policies that allowed curators to offer visitors a professionally sanctioned vision of science and nature without unwanted interference.

Early twentieth-century curators continued to compile collections with little or no display appeal, for even those museum men most committed to public education had no intention of sacrificing scientific research on the altar of display.[88] At the 1907 AAM meeting in Pittsburgh, for instance, attendees maintained that all natural history museums, even those devoted primarily to public education, should engage in a certain amount of fieldwork, investigation, and publication—this "more or less balances our accounts with the scientific world and places our small staff on the professional footing," explained Henry L. Ward. Museums' scientific importance had long depended on the accumulation of vast specimen collections, and scientists still relied on their ever-expanding holdings to identify, describe, and classify new species and subspecies. So curators continued to cultivate the kinds of collections that only a scientist could love. Philadelphians weren't exactly stampeding in to see Academy of Natural Sciences curator J. Percy Moore's collection of poly-chaetous annelids, or segmented worms, but zoologists found it awfully useful.[89] Acknowledging biology's turn toward experimentalism, curators at the nation's largest museums also began compiling records and collections intended to position museums as players in ecology, animal behavior, microbiology, and other emerging fields of biological inquiry.[90] The American Museum, for instance, developed a bacteriological collection in 1911, ensuring that the museum remained a crucial resource for university biologists, public health officials, and agricultural development experts, and curator Charles-Edward Amory Winslow collaborated happily with all manner of researchers in microbiology.[91]

At the same time, museum men urged curators to build collections that were of little value to scientists but illustrated those aspects of the biological sciences the public found most exciting—and that curators believed were most important for them to know. Sometimes called teaching collections or display collections, these specimens spoke directly to the real and perceived interests of schoolchildren and their teachers, lay naturalists, and those people who visited museums "for no other purpose than amusement," as Lucas put it in 1909.[92] Museum men worked with fellow curators and educators to compile display collections teachers found useful, creating exhibits illustrating evolution, variation, and adaptation to natural environments. According to New York City biology teacher George W. Hunter, the museum's staff did a masterful job constructing exhibits for "filling in and rounding out certain biological concepts" presented in high school science courses.[93] At the Milwaukee Public Museum, for instance, Ward and his curators collected dead birds from local pigeon fanciers for a display demonstrating how artificial selection had resulted in new breeds.[94] At the National Museum, head of the Department of Biology Leonhard Stejneger built display collections intended to illustrate the global geographical distribution of different faunal groups.[95]

Local nature was one perennially popular subject for display and teaching collections. Riding on widespread public enthusiasm for nature study, museum staff accumulated local specimens for amateur bird-watchers, bug hunters, and botanizers of all ages, and happily loaded tables with specimens of local flora and fauna for children to examine and touch. The living and breathing populated these collections, and museum men effused over visitors' enthusiasm for them, despite the considerable labor they required. Though its maintenance was partly responsible for "the added grey hairs and deeper creases in my face," Figgins admitted to his board, he was also pleased to report that the Colorado Museum's collection of live animals and plants "attracts much interest and attention."[96] Museum men frequently added aquaria of local fish and terraria crawling with sleepy-eyed lizards, snakes, and other reptiles to their halls, for the animals were relatively easy to care for and far more satisfying to view than fish and reptiles preserved in alcohol.[97] "These specimens are very attractive," Ward wrote a colleague in 1905, "so much so that last Sunday I was informed that it was almost impossible for visitors to get near the aquarium on account of the crowd of people about it."[98]

At museum men's behest, curators began to compile display collections of something they referred to as "economic nature," specimens used to illustrate the science surrounding the production of food, coal, clothing, and other essential goods.[99] For this, they plumbed older collections of natural resources, compiled with assistance from mid-nineteenth-century government surveyors, and harnessed local businesses to donate new specimens illustrating local bounty. The result was collections specific to surrounding regions. With the help of Mississippi Valley's thriving pearl button industry, the Davenport Academy of Sciences in Iowa accumulated a vast study collection of mussels and other pearl shells, displaying a choice selection of the gleaming bivalves near an exhibit illustrating the transformation of shell to shirt button.[100] The Colorado Museum prominently displayed examples of the state's sugar beet varieties alongside wax models of the insects that threatened them. Ever attentive to opportunities to improve their institutions' standing, museum men hoped aloud that visitors weren't the only ones who would profit from such displays. "You can readily see that collections of this nature would not only be of help to the citizens of Denver, but would be of vast assistance to the agricultural and horticultural interests throughout the state," Figgins wrote in 1911. "With work of this nature in progress there would seem a reasonable prospect of gaining the state's support for its continuance."[101]

This entrepreneurial didacticism prompted museum men to establish display collections pertaining to public health, a topic that held tremendous public, pedagogical, and philanthropic appeal. Americans regarded tuberculosis and other epidemic diseases with fascinated dread and eagerly consumed information about their spread and toll. Progressive reformers working in public agencies and private foundations capitalized on this interest. To teach Americans and particularly the immigrant poor about the relationship between personal hygiene, urban sanitation, and disease, they sponsored public health expositions throughout the early twentieth century.[102] Museum men promptly volunteered their institutions as participants and hosts for the exhibitions. Temporary public health exhibits were a phenomenal success: according to one *New York Times* estimate, the American Museum's 1910 *International Tuberculosis Exhibit* attracted ten thousand visitors on the first day it opened.[103] Eager to disseminate socially significant biological knowledge while attracting new support from donors, museum men commissioned permanent displays depicting bacterial life cycles and the transmission of germs. The botanical halls of Chicago's Field Museum boasted enormous illustrations of pneumonia, tuberculosis, syphilis, meningitis, and other "such microscopic plants," while the American Museum of Natural History went even further, founding a Department of Public Health to create collections for exhibits on the topic.[104]

Though they appreciated the diverging needs of scientists and lay visitors, budgets were limited and space was tight, so museum men often steered collecting agendas toward categories that appealed to the widest possible spectrum of museum users. Vertebrate paleontology was one subject that hit this sweet spot.[105] The massive size and the mysterious disappearance of prehistoric animals intrigued researchers and sightseers alike, and fossils' rare and fragmentary state ignited scientific curiosity and patrons' competitive instincts. So museum men indulged the ruthless generosity of robber barons, urging them to sponsor new field expeditions and exhibitions of the resulting specimens.[106] Delighted curators identified new species, corrected earlier anatomical misinterpretations and taxonomic confusion, and explored topics preoccupying contemporary biologists: biostratigraphy and biogeography, the process and pattern of evolution, and the relationship between inheritance and development.[107] Visitors were equally gratified. When curators mounted the massive beasts in poses based on recent research, hoping awed visitors would come away with a better understanding of evolutionary biology and historical geology, the public thronged the museum to gawk at the reconstructions.[108] "These creatures were then absolutely new to the vast majority of the American public," recalled artist Charles Knight, "and they went wild over the new mounts."[109] Paleontology enhanced museums' credibility as places of science and popular education, vividly illustrating curators' ability to make the earth yield scientific secrets, and museum men's capacity to produce exhibits that addressed public preoccupations with extinction and progress.[110]

Knowing animals likewise appealed to all audiences, museum men

29

encouraged the expansion of vertebrate zoology collections. Museum staff's intimate knowledge of the impending extinction of various species intensified these efforts.[111] Consequently, museums' commitment to collecting in vertebrate zoology intensified throughout the early twentieth century. The Field Museum sent out nearly seventy-two official vertebrate zoology expeditions between 1894 and 1928; the National Museum eighty-eight between 1886 and 1927, and the Academy of Natural Sciences in Philadelphia, somewhere around fifty-seven. The American Museum alone fielded 206 between 1890 and 1940.[112] Curators of zoology and other researchers were thrilled by the millions of specimens that resulted, for more exacting taxonomic practices and the hunt for evidence of natural variation within and between varying biomes and geographic locations required ever-larger collections. The new collections satisfied museum reformers, for visitors were more willing to learn about environmental adaptation, sexual reproduction, and interspecies relationships—the sociobiological concepts museum staff most hoped to disseminate—when contemplating animals with visual appeal.[113] As for visitors, they voted with their feet, crowding the halls to see recently acquired grizzly bears, blue herons, and peccaries.[114]

While they were glad to see visitors' enthusiasm, when it came to collection building, museum men were careful not to give public preferences too much weight. Intent on distinguishing their holdings from those of dime museums and freak shows, which still dominated many an American's understanding of a museum, they hoped to build collections so unimpeachable that no one could consider museum halls "places for the care of mere curiosities," as curator Harlan Smith put it in 1910—even if that meant excluding those objects the public was most eager to see.[115] This could be difficult, for museums' funding depended in part on attendance levels, and it was tempting to capitulate to museumgoers' expectations in order to attract visitors to museums. But in conferences and correspondence, museum men agreed they would not allow visitors' wayward preferences to distract them from the institutional task at hand. Ward was insistent on this point. "It has been my constant purpose to vigorously exclude from exhibition at the museum all objects that might merely incite wonder," Ward wrote in a 1905 letter to the *Evening Wisconsin*.[116] He happily banished all freaks from Milwaukee's halls, for, he declared, they "would be liable to distract from the prime aim of the collections, which is a systematic exposition of the fundamentals of natural history."[117]

Museum men's evolving approach to collecting practices also made evident the challenges of engaging and educating the public while shoring up professional standards. Museums had long relied on the donations of amateur naturalists to fill their shelves, and their emphases on survey collection and local nature ensured that all citizens, regardless of educational background, could make useful contributions. "The specialist's research is largely dependent, at least in America, upon the gifts of amateurs," wrote zoologist Thomas Montgomery, and museums relied heavily on the donations and observations of enthusiastic visitors.[118] In 1903, for instance, Brooklyn Institute of Arts and Sciences purchased no specimens, depending instead on donations from amateur naturalists and local collectors. Donations accounted for well over 90 percent of the objects in American natural history museums during this period. "Very seldom is the Academy called upon to purchase a specimen," noted the 1907 annual report of the Davenport Academy of Sciences. "Almost our entire museum has grown up through donations."[119] This situation reflected the ongoing importance of lay participants in the biological sciences. Avid amateurs continued to be critical participants in the museum world, and curators often worked closely with them to develop their own understandings of species' life histories or geographical distribution. Indeed, many of the nation's amateurs still knew, wrote, and researched as much as their officially professional peers.

But starting in the early 1900s, museums more consciously encouraged citizens to contribute to collections, stocking museum libraries with identification lists and hunting manuals, urging student groups to bring in their finds. In Milwaukee, Ward ordered dozens of field guides for visitors' use.[120] To museum staff, collecting natural specimens seemed an ideal educational activity, the single best way for people to learn about the natural world and a healthy remedy for the ills of industrial labor and cramped urban living. People of all ages and backgrounds responded, sending in specimens and hopeful queries. "Did you receive the reptile I mailed to the Museum? If there is any way of finding out, I would be glad to hear from you, what it is," one young doughboy wrote the Milwaukee Public Museum in 1917. "I found it under my cot, when I went to put on my shoes one morning."[121] Visitors daily brought dozens of dead birds to the Milwaukee Public Museum, asking when they would enter the museum's cases, while curators at the Carnegie Museum regularly received donations from Pennsylvanians living a hundred miles beyond Pittsburgh.[122] Donors often came straight from the field, live specimens in hand, and curators graciously accepted hundreds of struggling frogs, mice, and fledgling birds each year as a result.[123] (Curators periodically had to turn away such donations outright because they were too difficult to care for—Milwaukee's curators sadly declined the donation of a sleepy snowy owl in 1914, and the Colorado Museum adamantly refused five live wolf cubs a few years later.)[124] Firms and companies sent along manufactured goods to illustrate what plants and animals could ultimately become, and many economic collections consequently included random assortments of cotton socks and breakfast cereal, chewing tobacco and tar.[125]

Museum men initially welcomed these gifts, viewing such exchanges as opportunities for public outreach. Though they considered select donors to be scientific peers and others to be helpful contributors to broader scientific knowledge, most they seem to have deemed willing students, and they did their best to supply them with informative details in order to live up to their commitment to educate the larger public. "If I am sending you too many of them let me know," D. B. Barlow of Casa Grande, Arizona, wrote after sending yet another unsolicited package of horned toads, live rattlers, and gila monsters to the Los Angeles museum.[126] Instead, director Frank Daggett cultivated a close relationship with Barlow, providing him with information about the species as well as photographs of Barlow's credited contributions to the museum's exhibits and collections. When fourteen-year-old Jessie Kehl of Almond, Wisconsin, donated what she believed to be a petrified face to Milwaukee's museum in 1914, Ward gently explained that although he would conduct a "careful examination" of her donation, "no human face or any other fleshy part of the human body has ever been known to petrify."[127] Such exchanges not only promoted science education, Ward wrote a friend in 1915, but also brought "into close contact with the Museum of a large number of citizens." "This has made them feel that they had something to do with the building up of the Museum, and you would be very much surprised and interested, and probably even touched, to see the pride with which some of these children bring around their friends and show them the birds which they had given and which had been mounted and has their name as donor upon the label."[128]

Yet museum men soon wished citizens would be a little more restrained in their generosity. Annual reports listed frequent gifts of dog fleas, old shoes, and dried leaves. Sometimes people sent along their dead pets.[129] The secretary of Denver's Fire and Police Board offered the Colorado Museum an assortment of weapons previously used by criminals. Figgins politely declined the gift, citing the museum's lack of space. Not even the offer of nooses used to hang convicted criminals could induce him to accept objects he believed would detract from the museum's primary purpose of educating people about natural history.[130] Determined to build clean and comprehensive collections, museum men briskly turned away hundreds of deformed specimens. "I have your favor of yesterday regarding the bird 'half chicken and half duck' which you have," Ward wrote to one potential donor. "The museum does not care to exhibit anything of that nature."[131] Even those specimens curators wished to keep were often worthless for scientific purposes because donors had failed to note when and where they were collected. Such donations forced museum staff to implement new standards of stewardship to decide whether they were willing to accept items of uncertain provenance.[132]

Museum men eventually came to believe that this avalanche of donations furthered neither science nor education but simply drained museums' resources. Coloradans made so many donations to Denver's natural history museum that their acceptance threatened to "seriously interfere with routine exhibition work," Figgins wrote in 1914.[133] Unsolicited donations took not only staff time but also precious storage and display space. Where to put all the donations, particularly if one was trying to catalog and organize thousands of specimens into orderly collections? "Ingenuity is," Carnegie Museum director William J. Holland wrote, "at times sorely taxed in assigning a proper location to the very miscellaneous objects which come to us."[134]

To protect their time, their space, and the integrity of their scientific and educational missions, museum men quietly developed policies that allowed them to exercise greater control over their collections. For new specimens, they increasingly turned to their own staff members or hired experienced professional collectors, who sent along whatever objects and identifying records were requested—and, to curators' great relief, nothing else.[135] For teaching collections, too, reformers increasingly relied on professional assistance. Rather than depend on the drooping wildflowers and dried specimens visitors offered up, museums hired artists to render flowers in glass, wax, and watercolor and relegated the spoils of visitors' gardens to annual flower shows.[136] At museum men's urging, institutions adopted "no strings" donation policies whereby they could "receive specimens with a smile and thanks at the front door and send it out to the ash barrel at the rear, in case it happens to be something we don't wish," explained Ward in a 1902 letter to a former colleague. "In this way, we keep everybody happy."[137] The museum would no longer be "a dumping ground and a junk shop," he wrote, yet visitors could still feel as if their efforts were appreciated.[138]

Though museum men were entirely committed to educating the public, they increasingly concluded that they could more easily build institutions that did so if they kept the public at arm's length. As Ward wrote his board of trustees in 1902, "the museum can only grow as it should through the work of professionals rather than amateurs."[139] With such rules in place, curators and directors could expand their holdings in rational, measured ways, and plan future expenditures and expansions accordingly. They could develop collections that bridged popular and curatorial interests and establish professional standards across institutions. And, at least within museum walls, they could shape public perceptions of what could be found in the natural world. But such policies increasingly excluded the contributions of ordinary citizens and reduced burgeoning public participation in museums. Donors could no longer enjoy the certainty that their contributions, however humble, would be made magnificent by display in a museum hall or, at least, by inclusion within its scientific collections.

The evolution of collections and collecting practices in early twentieth-century natural history museums made plain the paradox at the heart of Progressive Era museum reform. Museum men at once opened natural history museums up, catering to visitors' tastes and encouraging amateur participation, and closed them down, overriding visitors' preferences and excluding amateur contributions in order to establish coherent educational messages and professional standards. These competing impulses quickly collided, and when it came to collecting, professionalism proved the stronger force.

Vitalizing Display: Popular Science and New Exhibition Trends

Thanks to Goode's influence, curators in established natural history museums had begun to separate study and display collections in the late 1880s and 1890s, carefully ordering and labeling those specimens that remained in the galleries. While their halls became more intelligible as a result, many visitors still found them deadly dull. Faced with poorly lit cases, awkwardly posed animals, and esoteric labels, Saturday afternoon museumgoers bordered "on collapse," observed the *Outlook* in 1909, "dragging their weary limbs through rooms full of things in which they had not the faintest spark of interest."[140] In recently founded museums, many of them understaffed and flailing for funding, curators could not even claim their exhibits were comprehensible, for their shelves were half empty.

Museum men found this situation intolerable, for to them, display was the keystone of museum reform. "It is not enough merely to show objects, not enough even if they are well labeled," Lucas fumed.[141] If they hoped to educate the public, they declared, the layout and contents of museums' halls should be more than merely coherent. Curators needed to make the museum, "the whole and all of its parts, inviting," argued Newark Public Museum director John Cotton Dana.[142] "We can no longer place our birds, insects, and shells in serried ranks, with the enlightening information that this or that specimen is *Planesticus migratorius*, that it lives in Chicago, that it was shot on July 13, and that it was given by Mr. Tom Jones," director of the Chicago Academy of Sciences Frank Baker wrote in 1907. Baker was no museum man, and he disdained many of the reforms they urged as unnecessary pandering, but he concurred in 1907 that museum staffs needed to "vitalize our exhibits."[143]

Yet the specifics—exactly how and why exhibits should be "vitalized"—remained unclear. Some museum workers maintained that cleanliness and clarity in displays were sufficient, citing Goode's earlier aphorism that a museum should consist of a collection of instructive labels illustrated by specimens. Most museum men disagreed, arguing that this placed too much responsibility on visitors, who were hardly equipped to know how or what to learn. "In the exhibition of specimens, the aim should be not merely to furnish information to the man who is looking for it, though this should assuredly be done, but to attract and interest the chance or indifferent visitor and to arouse in him a desire for further knowledge," wrote Lucas.[144] To elicit appreciation for nature and convey basic scientific facts, they argued, exhibits needed to offer a version of nature that a preoccupied public would find exciting or, at the very least, excitingly familiar. If museums wanted visitors to pay attention long enough to learn, they concluded, public tastes and visual appeal should be considered just as important as scientific standards.

Marrying scientific and public understandings of the natural world was a tricky venture, however, so early twentieth-century museum workers experimented a good deal before settling on a loose set of best practices. Nowhere was this more evident than in specimen arrangement. In the late 1880s and in the 1890s, curators intent on bringing museums up to scientific snuff had arranged specimens according to the dictates of Linnaean taxonomy. The resulting displays worked well for practiced naturalists and poorly for everyone else. They neatly situated specimens within a larger spatial map of species' relations, but visitors who wished to know what fish were best to eat, which birds lived in the area, or how the horse had come to look as it did found such displays entirely unhelpful. By the late 1890s, museum men agreed that museums should "avoid the incongruous effects of purely technical classification," as Figgins put it, and arrange specimens along lines more relevant to visitors' own lives.[145]

With these ideals in mind, curators fumbled for new organizational paradigms throughout the early twentieth century, merging the biological categories of Linnaeus with humbler but more familiar groupings. They ordered displays around categories of hybridization, wildness and domestication, animals featured in fairy tales, books, or contemporary news stories, industrial resources, pride of place, even humanized narratives. This mix of systems, they believed, enriched the meaning of objects and made museums more accessible to lay visitors. As with collections, the resulting mixtures varied widely from museum to museum, for curators arranged specimens in response to the needs and interests of local nature lovers. At the Davenport Academy of Sciences in 1901, for instance, curator and science educator Jurgen Paarmann reorganized the museum's holdings into organizational structures that might have seemed arbitrary to outsiders but were entirely logical to Iowans and in keeping with the particular missions of the museum. Paarmann's new bird exhibit, for instance, was partly scientific, governed by Linnaean categories of species and family, knowledge of geographic distribution, and the newly emerging science of animal ecology, which emphasized both behavior and the relations between species living in a particular geographic locale.[146] But more homegrown ideas about the natural world also influenced his museum displays. Though "game birds" was not a recognized scientific category, Iowans found the term useful, and he incorporated it into the exhibit. By rearranging the museum's holdings along vernacular lines, Parrmann believed he would reach large numbers of local nature lovers who lacked knowledge of—or interest in—perplexing scientific nomenclature. "Persons having collections of their own will come . . . to compare unknown specimens with those exhibited, & identify their own," he assured the board.[147] (He was right. Throughout the 1890s, annual attendance had hovered around four hundred visitors. In 1902, thanks in large part to the new displays, attendance shot up to 3,505, and the year after that to 9,598.)[148]

Museum men also tried to merge popular preferences and scientific standards in specimen labels. "Simple, homely facts," wrote Figgins, "[should crowd] out the scientific names that to most visitors are both meaningless and useless."[149] But curators shouldn't go too far in the other direction either, Lucas observed. "Time and space should not be wasted in stating that which is perfectly obvious; as for example, that the Scarlet Tanager is a beautiful red and black bird," he insisted in 1909. "Neither is it worth while to record unimportant measurements, for the visitor is not interested in knowing that a certain animal . . . is sixteen and one-half inches long."[150] Visitors, he wrote, wished to know about the bird's common name, relation to local life and culture, and to hear about its peculiar behaviors—particularly those behaviors that directly affected label readers.[151] As a result, museum men encouraged their colleagues to abandon scientific language and thoroughness for shorter, sweeter descriptions in vernacular English. "The extended text-book system will weary and discourage one not an enthusiast," Ward asserted at the 1907 meeting of the American Association of Museums. "The repellently pedantic impression that it makes upon the average citizen does, I believe, more than offset its ideal usefulness. For some reason which people have ceased to attend school few of them like to go back to it."[152]

Though most curators eventually agreed to rewrite labels along these lines, few were excited by the prospect, for such revisions took an astonishing amount of time.[153] As National Museum geology curator George Merrill observed, "There are few forms of literary work that require greater care than label writing."[154] Over the next two decades, in private correspondence and professional talks, curators groaned about the late nights they spent researching the revision of basic taxonomic and geographical labels into labels with more accessible, nonscientific details—for instance, the blue jay label shown in figure 2.[155]

Museum men found their colleagues more accommodating when it came to improving the appearance of specimens. Sensitive to visitors' expectations, most museum workers appreciated the disappointment visitors felt when contemplating animals that hardly resembled the beasts and birds they had hoped to see. So they quickly agreed that specimens "robbed of their natural appearance, in a greater or less degree, by insufficient methods of preservation and preparation" needed to be replaced with more realistic mounts, as Field Museum ichthyologist Seth Eugene Meek wrote a 1902 essay in the *American Naturalist*. This way, Meek explained, "museums are constantly asking less of the fancy or imagination."[156] Unsurprisingly, museum men, many of whom had long experience in specimen preparation, led the charge in replacing these specimens with more visually pleasing mounts. At their encouragement, preparators tried out newly developed taxidermy techniques, consulting recently published papers and handbooks on how to mount elk and elephant or traveling to Chicago, San Francisco, or New York to train with acknowledged experts in the field. Museums that lacked skilled preparators began to contract out, hiring specialized firms to provide mounts that made visitors gasp with delighted recognition.[157]

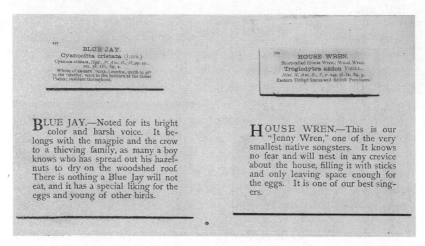

BLUE JAY.—Noted for its bright color and harsh voice. It belongs with the magpie and the crow to a thieving family, as many a boy knows who has spread out his hazelnuts to dry on the woodshed roof. There is nothing a Blue Jay will not eat, and it has a special liking for the eggs and young of other birds.

HOUSE WREN.—This is our "Jenny Wren," one of the very smallest native songsters. It knows no fear and will nest in any crevice about the house, filling it with sticks and only leaving space enough for the eggs. It is one of our best singers.

Figure 2   Throughout the 1910s and early 1920s, museum men provided their colleagues with examples of labels revised to appeal to broader audiences. In these before-and-after labels, Frederic Lucas suggested changes for two bird specimens. Friendly descriptions of the birds' most recognizable traits replace journal citations, habitat coordinates, and other details that were useful to scientists but confusing to laypeople. (The LuEsther T. Mertz Library of the New York Botanical Garden, Bronx, New York.)

Perhaps the most important contemporary development in this field was the "Akeley method," named after preparator and inventor Carl Akeley, which enabled taxidermists to mount large mammals more convincingly than in the past. Taxidermists, using bones, measurements, and plaster impressions from the field, created life-sized clay sculptures of animals. They cast these sculptures in plaster, then pasted and sewed skins over the plaster manikins. The results were startlingly realistic and allowed preparators to position animals in poses that were both characteristic and dramatic.[158]

But museum men believed specimens needed to be more than realistic. They needed to be magnificent. Before they could be expected to observe or interpret specimens, "the uninformed public must first be 'taken into camp' as it were, with a visual impression which gives pleasure," Brooklyn Institute director William Henry Fox wrote in 1915.[159] If visitors were going to be educated about an entire species through a single example, agreed museum men, they deserved to see the loveliest example science and art could produce. Awareness of popular culture's impact on American visions of the natural world sharpened this ambition. Lantern slides, illustrations, and moving pictures exposed early twentieth-century Americans to more images of the natural world, and the mere presence of an object in a museum no longer ensured an audience. Specimens had to be impressive enough to compete with the representations in popular circulation. Museum audiences would have been sorely disappointed if the elephants in the museum did not live up to advertisements and illustrations of Barnum's Jumbo, their memories of circus or zoo elephants, and the elephants they saw in Colonel Selig's sensational films.[160]

Consequently, in the late 1900s and into the 1910s, some museum men began to exaggerate in the direction of magnificence rather than stick to the bald facts of their specimens. Occasionally, this led museums to exhibit outright inaccuracy. In 1905, after he returned from Africa with several elephants for the Field Museum, the famous Carl Akeley posed the animals in a fighting stance, with the single tusker in the group appearing to attack the larger elephant. Though he later admitted that he had "never seen or heard African elephants fighting each other," except for mock tussling among adolescents who would later "stand there, wiping each others' heads with their trunks in an affectionate way," he decided fighting would excite the public more than playing. He posed the group accordingly and titled it, *The Fighting Bulls*.[161] This kind of blatant manipulation occurred infrequently, however, and museum men usually called their colleagues to account when it did.[162]

More often, museum men blurred science and art in order to meet— or exceed—popular expectations. They did not shy away from improving specimens, for instance, feeling no compunction about substituting choicer parts of one animal for those of another. In 1912, Henry L. Ward wrote N. Annandale, superintendent of the Natural History Section of the Indian Museum in Calcutta complaining that the Milwaukee Public Museum's menagerie-born rhino had but a nub of a horn. Annandale sent along a more impressive version, and Ward promptly sawed off the worn horn and glued the larger one in its place.[163] Museum men frequently engaged in this kind of Frankenstein-like tailoring to ensure their museums could display the most impressive animals imaginable. They were unabashed about it, believing beauty was crucial to their mission. In 1918, Figgins notified Alfred M. Bailey, curator of birds and mammals at the Louisiana State Museum, that the elk specimen he was sending had been slightly altered. "Instead of sending the skull and antlers that belong to the skin I am going to substitute a much better pair of antlers," he wrote. "For a scientific specimen, I would not do this, of course, but for exhibition, I view it as entirely 'legitimate,'" he wrote.[164] Specimens difficult to preserve required less substitution but more artifice. Preparators found frogs, for instance, nearly impossible to mount; their skin shredded and shriveled outside of a moist environment. Museums had traditionally stored frogs in glass jars, but curators doubted the educational power and public appeal of pale specimens bobbing in discolored liquid. As a result, preparators began to fashion realistic versions of frogs out of wax, celluloid, or plaster.[165] These models appeared ready to hop or croak, but no one could call them real.

Such techniques, argued Lucas, "led to much friendly discussion as to whether it is better to show a stuffed specimen that does not at all resemble the living animal or a cast that cannot be distinguished from it."[166] Curators and directors' opinions on the matter varied widely, depending primarily on whether they believed museum exhibits should be more accountable to scientists or to the lay public. Museum men, believing public education should take precedence in matters of display, argued that upholding scientific standards of truth too diligently could stifle visitors' interest in nature. If specimens were not made exciting or at the very least appealing, they argued, audiences would simply stop paying attention, and museums would fail to educate the public. The point of exhibition, Ward declared, was to move museum visitors, to give them "all sorts of impressions and thrills . . . be they right or wrong impressions and thrills."[167]

Some reformers objected that the resulting displays did museumgoers a disservice, arguing that visitors would be disappointed when confronted with the periodic disfigurations of the outdoors. "A tendency to prefer imitation to reality is not one which those interested in the progress of science at least should seek to promote," Field Museum curator of geology Oliver Farrington protested in his 1915 presidential address to the AAM.[168] Other reformers worried specimens' perfection would become distracting, that interest in scientific fact would pale in the face of fascinating artistic accomplishment. As German ornithologist and Dresden museum director A. B. Meyer wrote after his 1903 visit to the American Museum, exhibits' "leaves and flowers are carefully cast in wax and consequently transparent and very true to nature, but somewhat obtrusive. There is too much of this good work, for the attention is diverted from the object of the exhibition."[169]

Museum men ultimately persuaded the professional community that such attitudes were outdated. "The object of exhibits is to hold the mirror up to nature and let it reflect an image of nature as she looks when alive, not as she appears when dead and shriveled," wrote Lucas. Taxidermists had long incorporated artificial materials in their mounts, he reminded his colleagues, "and if a cloth leaf and a glass eye are allowable, why not a wax frog and a celluloid fish?"[170] By the mid-1910s, most AAM members had come around to museum men's point of view and accepted the partial or total reproduction of animals as an effective educational technique.[171] (In private correspondence to one another, museum men were less diplomatic. "How infinitely little educative sense these studious cusses have!" Ward exploded to a friend.)[172]

Intrigued by what art history professor and AAM speaker Homer Keyes described as the "potent educative success" of early twentieth-century consumer culture, museum men convinced their colleagues to experiment with commercial techniques of display.[173] So they consulted with the engineers of its bright aesthetic—department store window designers, theater directors, magazine illustrators—and urged preparators and curators to use a more explicitly commercial look for educational ends.[174] This was a very real departure from past practice. Protective of their scientific reputations, many curators at natural history museums in the 1880s and 1890s had taken umbrage whenever the press or public grouped their institutions with commercial entertainments, but museum men pushed them to abandon this defensive stance.[175] "Knowing the advantages of Natural History Museums from the educational standpoint and the vital importance of some branches of the subject as related to human interests, I have no hesitancy in employing any reasonable means of getting the interest of the public," Figgins wrote in 1916. "Some fossilized members of old eastern museums have condemned me for such methods. . . . I accept such criticism as a compliment."[176]

Impressed by department store windows' persuasive power, museum men pressed curators and preparators to incorporate more glass, light, and color into their exhibits. They placed objects against brightly hued backgrounds and consulting with colleagues at art museums to help them choose tints. Unlike their nineteenth-century predecessors, who relied predominantly on architectural placement of windows to light displays, they made more dramatic use of controlled interior lighting, borrowing lighting techniques and color palettes from theater and retail establishments.[177] Museums of all sizes gradually adopted the kind of plate glass display cases department stores had made famous, and museum men pleaded with colleagues to illuminate them from within, so as to avoid distracting reflections or glare.[178] Visitors, wrote New York State Museum director Arthur Parker, must be "conscious of the exhibit and see nothing else. Our imagination must not be disturbed."[179]

Museum men, acknowledging the pull of the pictorial, also began to hang vivid illustrations of specimens in context in their halls. In the American Museum's Jesup Hall of Woods, curators hung framed watercolors of leaves and foliage above each log (fig. 3). They also placed potted palms by the benches and, above each case, mounted paintings of the trees themselves in a natural environment, in essence creating a

Figure 3   Cases in the American Museum's Jesup Hall of Woods in 1903. The dictates of
scientific illustration, traditions of Romantic landscape painting, and attempts to
make visually dull displays more interesting often merged in natural history museum
exhibition. Here, curators prompted visitors' imaginations by hanging scientifically
accurate paintings of leaves, stems, or flowers above the specimens, and by
placing paintings featuring trees in their natural environments over the cases.
(Image #0042124, American Museum of Natural History Library.)

gallery of scientifically accurate landscape painting. Their colleagues in
the museum's paleontology department hired artists to create pictorial
"restorations" of prehistoric animals, displaying the lively watercolors
and oils near the bleached fossil mounts. In the 1890s, the few paintings
that accompanied these specimens measured less than a foot or two in
length; by the 1910s, paintings frequently accompanied group displays
of mounted specimens and often ran a third of the length of the gallery.
Artist Charles Knight's commissioned canvases, for instance, measured
seventeen inches by twenty inches in 1900. By 1925, they were often five
feet by seven feet.[180]

The culmination of these new techniques and the most successful
marriage yet between professional and popular science, according to
museum men, was what they called "life groups" or "group displays":
skillful arrangements of animal mounts characteristically posed among
moss, plants, and branches. Inspired by the work of amateurs like Mar-

tha Maxwell and commercial taxidermists like the Jonas Brothers and Ward's Natural Science Establishment, these new displays sought to replicate scenes of nature within museum walls.[181] Between 1887 and 1888, Mrs. E. S. Mogridge, Jenness Richardson, and John Rowley had created delicate displays of birds in nests or on tree boughs for the American Museum. William T. Hornaday's buffalo group for the National Museum, unveiled in 1888, and Carl Akeley's muskrat group for the Milwaukee Public Museum furthered group display techniques, and attracted considerable public attention.[182] "Once intrenched [sic] behind the bulwarks of high scientific authority," Lucas later wrote, group displays "began to find their way into all museums."[183] Dedicated museum men began to enlarge these early experiments in the early 1900s, positioning groups of male, female, and immature animals in full pelage and plumage amid real or wax flora, then placing the groups in front of painted landscape backdrops and framing them with proscenia.[184] The resulting dioramas aimed to convey typical movements and behaviors of the featured animals, as well as the tangled ecological relationships that made up species' habitats. Habitat dioramas seamlessly combined scientific observations, Romantic artistic conventions, and popular ideas about animals' appearance and behavior.[185]

Museum men—and museum reformers more generally—trusted that such displays would rouse a deep and sincere interest in the marvels of nature. To many museum reformers, the experience of emotional engagement between human and nature was a fact just as important as genus and species and could be communicated only by displaying a natural world that looked realistic. They believed, as Smithsonian taxidermist William Temple Hornaday put it in 1891, that the contemplation of birds and beasts should be "an affair of the heart," and so their exhibition should excite "feelings of admiration that often amount to genuine affection and . . . delight" in the natural world.[186] The naturalism of group displays made this possible, resonating with viewers' experiences, reproducing what they had seen in the wild, on the page, or on the screen.

Such aspirations were vividly embodied in the American Museum of Natural History's early bird groups and habitat dioramas, built by Frank Michener Chapman, the museum's young curator of ornithology. Renowned in museum circles for his vigor, competence, and sensitivity, Chapman was an immensely accomplished scientist and a committed museum man.[187] Professional biologists and amateur nature lovers alike admired his astute studies of avifauna, which unabashedly mixed rigorous scientific analysis of animal behavior with an empathetic Romanticism. The prematurely balding, impressively mustachioed curator believed natural history museums had not only a duty to stir public interest in the natural world but also a responsibility to encourage its protection. Chapman believed that life groups and habitat dioramas would accomplish just this, affording city dwellers regular glimpses of nature at its most gorgeous. Even more important, he believed dioramas would rouse urgent interest in conservation. Unlike most Americans, Chapman refused to accept the destruction of habitats and species as an inevitable by-product of economic development. Lawmakers and American consumers could save American wildlife, he argued, if only they set their hearts and minds to it.[188] But, he believed, Americans would do so only if they could visualize the inherent dignity and romance of wildlife. To check the slaughter of birds for millinery and sport, Chapman began to devote his considerable energies to transforming the American Museum's bird halls into places that would arouse both "a desire for information, and a love for the nature presented."[189]

To build displays that persuaded visitors of "the importance of the sentiment, as well as of the science" of ornithology, Chapman worked closely with the American Museum's preparators and painters to construct one of the nation's first large-scale habitat groups, depicting an endangered colony of large birds on Canada's Bird Rock Island, then considered one of the ornithological wonders of the Atlantic Coast.[190] Based on photographs taken by Chapman while he dangled from the end of a rope over the mouth of the Saint Lawrence River, the group featured seventy-three examples of seven species nestled into grasses on a rocky promontory nearly seven feet long and seven feet high.[191] But it was in 1902, when Chapman attempted to capture the bird life of Cobb's Island, Virginia—a shrill breeding ground for terns and summer gulls decimated by hunters eager to sell birds to milliners—that he constructed the American Museum's first habitat diorama. Chapman and preparator John Rowley mounted the birds as if they were in flight and hung them over a sandy beach, in front of a scenic painting of the island's coastline.

The resulting display, with its nod to the power of commercial panoramas and other theatrical entertainments, shocked curators more accustomed to neatly labeled rows of specimens. Though colleagues had praised his specimen groups, many regarded the addition of painted backgrounds "as too dramatic, a violation of the museum tradition which made an exhibit presented to the public as formal as the papers they read before their confreres," Chapman later reported.[192] Because the *Cobb's Island* exhibit departed so radically from the museum's existing exhibits, Chapman's supervisor, zoologist Joel Allen, insisted on getting approval from the president of the museum's board before opening it to the public. Nervous staff members steered the sickly but still formidable board president Morris K. Jesup through the museum halls in his wheelchair to see it. Jesup contemplated the diorama, paused, and announced "That is *very* beautiful." "No words could have pleased me more," Chapman later wrote.[193] Between 1898 and 1909, Chapman supervised the installation of thirty-four such exhibits in the museum's Hall of North American Birds.[194]

The public adored group displays. "It was a real delight to observe the thousands of strangers moving along the museum halls enjoying the exhibitions of specimens of which the mammal and bird groups seemed to attract the greater mass of people," Ward wrote in 1903. Milwaukee's groups proved so popular that during the Wisconsin State Fair, Ward opened the museum at 8 A.M., an hour earlier than usual, "so that everyone from outside the city could get a chance to see the exhibitions."[195] The American Museum's bird groups attracted so many inquiries from visitors that, in 1901, Jesup asked Chapman to publish accompanying illustrated pamphlets to "meet a popular want, for there is constant demand for them on the part of visitors."[196]

Life and habitat groups straddled an active fault line between art and science, between public education and scientific expectations, so museum men had to work to reassure the public—and the larger museum community—that there was no real conflict between the two. "Beauty is not incompatible with scientific value," explained curator Mary Cynthia Dickerson in a 1914 issue of the *American Museum Journal*. In fact, "it may strengthen the appeal of science."[197] Curators and directors who supported display reforms took care not to break ranks, reiterating similar lines to colleagues suspicious of these changes. Field Museum geologist Oliver Farrington's 1916 AAM address confirmed that museum men had succeeded in persuading their peers that art and science were, in fact, inseparable in natural history museums interested in public education. "If science alone were presented by the science museums, the exhibition of rows of specimens would fulfill their entire function. But the visitor wants life, vividness, and action, or the semblance of it," he told his audience. The kind of visual appeal and popular interest that artistic exhibitions created should now be considered integral to museum display. "If subordinate," he added, "their function is a desirable one, no one, I think, will deny."[198]

Inreasing numbers of administrators and curators came to the same conclusion, and they began to evaluate exhibits by aesthetic standards. By the 1910s, "ugliness was sufficient to condemn any exhibit, no matter how scientific," recalled one curator.[199] Discussion of new display techniques began to dominate the dialogue at the annual meetings of the American Association of Museums throughout the late 1900s and 1910s as a result. Though the AAM's members were a fractious lot, and disagreed vehemently over other aspects of museum work, they unanimously praised these strides in display, especially habitat dioramas. Members concurred that such exhibits had a tremendously positive impact on attendance and were key to promoting sentiment for the natural world in the public.[200] Consequently, in the first decade of the AAM's existence, nearly a quarter of the papers delivered at its annual meetings addressed how museums—small and large—could create group displays and incorporate them into their halls.[201] Other aspects of exhibition commanded similar attention.

Thanks to reformers' vision and museum men's influence, the exhibits of American natural history museums changed dramatically over the course of the early twentieth century, shifting from drab specimen collections to two- and three-dimensional spectacles of the natural world. Reformers' pedagogical goals shifted as well. Though museum reformers had earlier boasted that mute specimen displays could teach visitors to look closely and think for themselves, by the 1900s and 1910s, they concluded that the public needed more assistance, not less. Learning to look and mastering other scientific skills were all very well and good, but learning basic facts and attracting larger audiences, who would presumably be drawn in by the promise of more comprehensible exhibits, seemed an even more attractive option. As a result, most curators and preparators began to build displays with the goal of teaching visitors of all backgrounds, "whether they are looking for instruction or not," as Ward wrote a Cincinnati naturalist in 1906.[202] By design, these exhibits did not require significant intellectual or visual labor from visitors, and their clear narratives and compelling visual presence encouraged viewers to absorb whatever interpretation curators had decided to hand them.

Early twentieth-century discussion over what values should prevail in natural history museums' displays abated somewhat in the 1910s, as museum men's efforts to popularize science through the intentional adoption of vernacular approaches to the natural world and an overtly commercial visual aesthetic became the professional norm. Museum men's adoption of popular ideas co-opted a certain measure of public control. As displays became more accessible and attractive to the public, the power of curators to shape public ideas about the natural world increased significantly. At the same time, the introduction of popular categories of knowledge into labels and specimen arrangements subtly eroded the primacy of professional science within the museum. When

curators placed Linnaean nomenclature on equal visual footing with human interest or local use, it became just one of many systems of value
competing for gallery space. Indeed, from the 1890s on, the wood and
glass of display cases enclosed not only specimens but also a central paradox in Progressive reform: the simultaneous attempts to employ centralized, professional expertise and to democratize participation.

The Aim of the Museum? Fissures and Fractures in Early Museum Reform

By the 1910s, museum reformers, and especially museum men, congratu-
lated themselves for establishing an entirely new institutional ethos. As
Lucas later reflected:

From being largely storehouses of material for the use of a few men of science, they
have become great institutions for the education and "rational amusement" of the
public at large; the exhibition of specimens, instead of being, so to speak, a side branch
of museum work, has become one of the most important functions of a museum; in
some cases even, exhibition is the main function or object, research and publication
being secondary considerations, and in the larger museums, the greater part of the
space is devoted to the exhibition of specimens for the benefit of the public.[203]

Lucas's own institution epitomized this shift. In 1904, when he first
assumed the position of curator in chief of the Brooklyn Institute of Arts
and Sciences, the museum's galleries had been so embarrassingly chaotic
that he forbade his friends from visiting the place. When he left seven
years later, he had set its halls to rights, boasting that he left behind "a
small but evenly balanced exhibit of the classes of vertebrates, a fine
beginning of a similar display of invertebrates, some groups such as the
guacharo birds, orioles and other as yet unsurpassed, and unique exhibits
illustrating evolution and the preservation of animals."[204] On weekday
mornings, schoolchildren gathered in front of earnest docents gesturing
at tidily labeled cases of stuffed birds and beasts, while trucks laden with
loan exhibits jolted toward city schools. On weekends, thousands of visitors arrived to gawk at new and permanent exhibits, clotting auditorium
staircases as they waited for illustrated lectures to begin.

Yet reformers' ongoing efforts to remake museums in the image of the
New Museum Idea were more fraught than Lucas's description suggested,
and quiet objections to museums' new mission revealed tiny fissures in
the foundations of museum reform. Though most museum professionals

concurred by the 1910s that their institutions should educate the public through enhanced display and supported the reforms taking place, consensus about the relative priority of public education proved more elusive. A significant faction of curators and directors—some of them avowed museum reformers—maintained that it was misguided to prioritize public education and display work above scientific research, even in smaller institutions. "We must divorce ourselves from the idea that we must of necessity try to please the public," asserted malacologist and Chicago Academy of Sciences director Frank Baker at a 1907 meeting of the American Association of Museums. "The principal function of museums, large and small, is the acquisition and preservation for future study of such material as will throw upon the great problems of life. . . . The exhibition of material is secondary to this great work."[205] "First and foremost it should be the aim of the museum to promote the advancement of the subject," agreed Field Museum anthropology curator George Dorsey. "By advancement I mean not so much what is commonly understood as the diffusion of knowledge, as the advancement which results from new investigation."[206]

More typical were those who believed that education should be on equal footing with scientific research and, in smaller museums, take precedence. A museum "is scarcely fulfilling its functions by alone maintaining a series of collections," secretary of the Smithsonian Institution and paleontologist Charles D. Walcott wrote in 1912. To assume its rightful place among American educational institutions and "perform its full duty as an educational factor," Walcott explained, it needed to pursue the production and dissemination of scientific knowledge simultaneously. An approach that combined "research and full publication with the display of collections, is the only proper one," he wrote. Otherwise, he concluded, "a museum must necessarily fall behind in the work which, with the present rapid strides toward 'the increase and diffusion of knowledge among men,' everyone expects of it."[207]

Debate over museums' ultimate purpose sharpened into argument when administrators adjusted institutional budgets to accommodate museums' new educational ambitions. From the 1890s on, resources migrated from scientific departments to departments of public instruction and preparation—the influx of work to which required a massive expansion of administrative staff, even while individual education salaries remained low; this shift in emphasis also entailed the regular purchase of costly supplies.[208] Museums with active education departments devoted a generous fraction of their budget to the production of loan exhibits and lantern slides, and large urban museums hired small armies of schedulers, messengers, and drivers to ensure these visual aids reached their intended locations on time and intact.[209] To reformers, the investment was worth every penny. Though expensive, these programs promised to pay for themselves, for they guaranteed that museums could cite large attendance and outreach numbers to current and potential donors. Other museum workers weren't so sure.

New responsibilities also raised some ire among museum workers, as they groped to define what museums should be—and what their staff members should be expected to do as a result. Long-time volunteers and dedicated scientific researchers often lacked their younger colleagues' enthusiasm for long, hard hours implementing a new vision for their institutions. Curators who had joined museum staffs before reformers took charge frequently resisted attempts to impose these new values, resulting in considerable conflict and institutional turmoil. And museum men's total conviction in the value of exhibition work, combined with their institutional pragmatism and growing power, set many of their colleagues' teeth on edge. A few curators became so annoyed with reformers' exhortations that they left natural history museum work altogether. At the American Museum, curator of ichthyology Bashford Dean resigned, declaring that he was a scientist, not "a janitor," after director and museum man Hermon Bumpus insisted he rearrange the fish exhibits to make them less scientific and more attractive to the public.[210] A handful of scientists working outside the museum world expressed their sympathies. "When the preparator and mounted rhinoceros are considered above the curator, and the exhibition collection above published research, then a museum is becoming senile," wrote University of Pennsylvania zoologist Thomas Montgomery in 1911. "Young naturalists starting out should be helped with fellowships and advice, mutually encouraged, not treated as preparators."[211]

Despite these periodic spats, by the 1910s, most professional museum workers embraced reformers' partially realized vision of museums as palaces of popular education. "Though we are in the very infancy of our conception of what the museum is and is going to be," wrote Paul Rea in 1923, president of the American Association of Museums and director of the Cleveland Museum of Natural History, "it seems to me that we must all realize that we are the fortunate possessors of a vision of the future, to a greater or lesser extent, and we have only begun to transmit our vision." Natural history museums had a long way to go, Rea acknowledged, but as a result of the efforts of a generation of reform-minded museum professionals, "the conception of museums as store-houses and research organizations, of interest principally to specialists, has been broadened to give equal importance to the interpretation of their subject matter to the public," he added. "This is the most significant advance in museum purposes and methods during the whole hundred and fifty years of museum history. It is making museums popular universities in a field of apparently endless possibilities."[212]

TWO

The Drama of the Diorama, 1910–1935

In 1890, visitors to the American Museum of Natural History wishing to learn about flamingos peered into one of a dozen large glass cases in order to view a specimen of the snake-necked bird, flanked by a stork and a pelican (fig. 4). In 1910, they looked instead through a large plate glass window onto a Bahamian beach surrounded by cypress trees, where a lifelike colony of mounted flamingos straddled turretlike nests as they fed their chicks.[1] And by 1930, the museum's ornithologists and exhibit makers were hard at work on the enormous Hall of Pacific Bird Life, in which museumgoers, standing under the dome of a glowing Pacific sky, could glimpse a flock of greater flamingos flying over the Galapagos Islands (fig. 5). Visitors and critics marveled over these changes to natural history museums' halls, delighted by new attempts to convince them that they were "looking upon a transported bit of the natural world," as mammology curator Harold Anthony put it.[2] Though museums married verisimilitude and visual pleasure in all sorts of displays in these years, no exhibits were more important or more influential than habitat dioramas.

Habitat dioramas proliferated in American natural history museums in the 1910s, 1920s, and early 1930s, transforming the appearance and appeal of halls in the process. Crowds flocked to them, murmuring at their spectacular background landscapes and depictions of life in the wild. This approach to exhibiting life had revolutionized natural history museums, declared cultural critic Lewis Mumford

Figure 4   This 1900 image of American Museum of Natural History's Bird Hall exemplifies the
gradual changes in museum galleries in the early twentieth century: heavy vertically
oriented museum cases filled with hundreds of taxonomically arranged specimens
are mixed with lower cases featuring life groups of single species. So that visitors
might see specimens more easily, reform-minded curators illuminated display cases
with electric light and placed shades over the windows to reduce glare. (Image
#00003409, American Museum of Natural History Library.)

in a 1918 issue of the *Scientific Monthly*, taking them "from storehouse to
powerhouse . . . from a passive 'showing off' to an active education, from
the informed miserly tradition of an earlier day to the directed socialized
spirit of the opening age." Museums' newly "artistic presentation of nature" and their inclusion of more environmental context, he announced,
made scientific ideas accessible to the broadest of audiences and embodied "the democratic sympathies of the present order."[3] Museum reformers reading Mumford's essay must have been delighted, for it seemed to
confirm that museum halls had finally become compelling places for
public education.

The impact of dioramas on the back of the house was just as dramatic
as the changes they wrought in the exhibit galleries: as museums constructed more and more dioramas, the displays began to influence museums' cultures in ways that surprised their early supporters. Dioramas were
expensive, but they garnered enthusiasm from all corners of the museum

Figure 5   Museums began to invest more heavily in the construction of life groups and habitat
dioramas in the 1900s and 1910s, hiring artists and preparators to help them build
more elaborate displays. Richard Dahlgren's 1905 photograph of the American
Museum of Natural History's *Nesting Colony of Flamingos on Andros Island* reveals how
collaborations between curators and artists—in this case, ornithologist Frank Chapman and painter Louis Agassiz Fuertes—created newly theatrical scenes of carefully
rendered nature. (Image # 00003939, American Museum of Natural History Library.)

community in the 1900s and early 1910s.[4] Exhibit makers and exhibit
making grew in stature as a result of dioramas' success with visitors, winning respect and resources from museum scientists, administrators, and
board members. But curators, tantalized by diorama-driven opportunities
for field collecting and observation, soon began to fume that the displays
robbed them of the time, funding, and labor necessary for scientific research. Directors, responsible for deciding just how much space and cash
to invest in the elaborate displays, struggled to manage the diverging
aspirations of increasingly specialized staff members. Rather than allowing museum workers to realize these disparate aims without conflict, dioramas awakened new animosity between staff factions. As the exhibits
grew in number and status, these clashes intensified. The financial strains
museums suffered during the early years of the Depression only added
fuel to this fire.

Habitat dioramas, then, embodied both the triumphs and the tensions
of the reformed modern museum. By attracting larger audiences and incorporating more current biological perspectives, the displays helped to
make the halls of American natural history museums more popular and
more pedagogically relevant. Yet the colorful exhibits also made plain
how difficult it would be to provide science education, protect and collect
specimens, and produce scientific knowledge simultaneously. Dioramas
forced interwar staff members to prioritize or declare their support for

one aim over the others, rather than pursue them with equal intensity. United only a few years earlier by an ambitious vision of the New Museum Idea, museum professionals now bickered among themselves about what museums should be and do. By the late 1920s and early 1930s, as staff members debated their institutions' practices and priorities, it had become clear that natural history museums faced an impending identity crisis.

All Things to All People: Early Support for Habitat Dioramas

For a brief and shining moment in the late 1900s and early 1910s, habitat dioramas seemed to offer all things to all people in the museum world. The colorful displays synthesized aspects of modern biological science with the latest pedagogical trends. They enticed more people to visit museums and convinced wealthy patrons to dig into their wallets to fund the institutions. They facilitated scientific research and furthered museums' reputations as places serious about public education. Persuaded of their worth, museum directors slowly set aside their reservations about dioramas' costs, hired the specialized staff members, and raised the considerable funds necessary to build the displays.[5] Their decision to introduce more dioramas, though not initially controversial, would ultimately transform American natural history museums' halls as well as their institutional cultures.

Scientists were among dioramas' most vocal advocates in these years, for the displays captured more contemporary aspects of the study of living nature than the specimen cases that still dominated the halls of natural history museums. Habitat dioramas embodied the German "biological perspective," an outlook that had originated in the work of practical naturalists and provided a broad foundation for both early twentieth-century scientific work in ecology and social reforms in German science education.[6] At the heart of this perspective was Karl Mobius's "community concept," which, like habitat dioramas, emphasized the importance of considering the relationships between animals in a specific location.[7] Whether highlighting relationships among animal species or the influence of altitude, climate, and terrain on a particular region's flora and fauna, the biological perspective reflected a shared belief that new attention to the dynamics of animal and plant life in nature defined modern natural history.

Though they did not always credit their European counterparts as inspiration, American scientists working in and outside museums whole-heartedly agreed that their exhibits should reflect this outlook, and dioramas did so better than any other kind of display.[8] Those museums building dioramas, ecologist Charles Adams asserted in 1907, should be lauded for their "special effort" to display specimens in relation "to their environment, their associates and their life histories," rather than continuing "to arrange the geological specimens according to the system of some synoptic or text-book, without regard to the influence of local conditions or environment upon the material."[9] Dioramas successfully embodied scientists' "recognition of the fact that nothing of nature is of isolated origin; it is the product of the working of complex and varied forces," agreed Field Museum curator of geology Oliver Farrington in 1915. "The food of an organism, the various stages of its development, its habitat and its habits must be represented before the organism itself can be thoroughly understood."[10] Dioramas were, Harvard entomologist and former museum curator William Morton Wheeler reflected only a decade later, "prophetic of a change which through the influence of the ethologists, behaviorists, physiologists and psychologists, has now pervaded the whole field of the biological science." The exhibits' holistic descriptions of the natural world afforded the public a level of information that taxonomic displays simply could not, he added, noting that "an organism cannot be isolated, even conceptually, from the peculiar environment to which it has become adapted during eons of geologic time, without a serious misunderstanding of its true nature."[11]

Many museum scientists, following American Museum curator Frank Chapman's lead, also believed dioramas could rouse public interest in wildlife conservation, a cause near and dear.[12] Birds and animals in their habitats aroused far more sympathy from viewers than synoptic displays, they maintained, and that sympathy led museumgoers to reflect on what they could do to preserve such creatures.[13] "The Museum's influence for the preservation of animal life is continually increasing as advances are made in methods of exhibition," observed American Museum president and curator of paleontology Henry Fairfield Osborn.[14] "Museums from the beginning have preached conservation," agreed Illinois State Museum curator and geologist Alja Crook. "The long quiet work of museums in this line" had awakened public awareness of its importance, he asserted, and "special exhibits"—habitat dioramas chief among them—should be recognized as one of the conservation movement's most useful tools.[15]

These curators also praised how dioramas combined the object-based pedagogy long associated with taxonomic display with more modern approaches to learning, especially by merging traditional exercises in visual discipline with an educational approach that verged on childhood play. Dioramas continued to encourage deduction of scientific concepts through close observation and disciplined analysis, curators explained, but did so in newly engaging fashion. They piqued visitors' interest via games of "I-spy," noted Milwaukee Public Museum director Henry Ward.[16] Dioramas allowed museumgoers to uncover details and relationships within a seemingly unified natural setting in the gradual fashion that field naturalists did, wrote American Museum curator of herpetology Mary Cynthia Dickerson. In the dioramas constructed by her own department, "the abundance of vegetation and the great array of animal life . . . have been subordinated to the effect of the whole," she explained. "All of the animals are directly before the eye yet are so chosen and placed as to be inconspicuous except on a more careful search, thus imitating the condition in nature."[17] As museumgoers busied themselves with locating, identifying, and discussing various organisms, elements of identifiable natural settings, and their interrelationships, Dickerson and other curators suggested, they would glean scientific concepts more readily and happily than they had when perusing synoptic arrangements in long glass cases.[18]

These arguments resonated with those contemporary American biologists who expressed anxiety that the study of life had become unpleasantly sterile. "Many persons who already had developed an interest in live animals were repelled, and even driven from this field of activity," argued ecologist Charles C. Adams in 1917, by the fast-paced intensity of laboratory courses.[19] Nowadays, oceanographer William Ritter wrote in 1911, professional specialization and "the metaphysics of biology," embodied by bounded, modern laboratories and microscope-aided study of isolated structures that so many academic scientists now favored, stand "as almost impenetrable screens between the perennially-interesting, everywhere-present, easily-seen facts of the living world, and the natural responsiveness of young learners."[20] By presenting reality in its most glorious state, dioramas presented an effective antidote to this problem. Inside the museum and out, scientists agreed that the exhibits successfully captured what zoologist Edmund B. Wilson called "the 'sentimental side' of natural history," which, they argued, engendered "that keen primary interest in biological phenomena for their own sake, apart from their scientific analysis, that was characteristic of so many of the earlier naturalists." Dioramas' scenes of natural beauty especially pleased field scientists and biologists who still appreciated the romantic outlook of nineteenth-century naturalists, believing, like Wilson, that "there is no more potent spring of scientific research" than aesthetic satisfaction in the ever-changing face of the natural world.[21] "A tinge of feeling, of senti- ment, toward organisms promotes interest, interest promotes attention, and attention is an essential prerequisite to the acquisition of knowl- edge and to sound reasoning," Ritter concurred. "We learn most quickly, most spontaneously, most comprehensively, most securely, things that interest us, and things interest us most toward which the affections go out."[22]

More and more, curators agreed with contemporary educational psy- chologists' argument that "attention does its best work when the feeling of effort is wanting," as Edgar James Swift put it in a 1911 issue of Sci- ence.[23] They praised dioramas' beauty on pedagogical grounds, certain that their visual appeal would induce visitors to learn unconsciously. Approaching a diorama, "the aimless visitor involuntarily pauses," Chap- man observed. "His imagination is stirred, his interest aroused, and the way is opened for him to receive the facts the exhibit is intended to convey."[24] Beauty, declared the American Museum Journal in 1913, had become mandatory for displays, for "if any non-compulsory educational institution is to prove effective, it must give the knowledge it wishes to impart a guise of unusual fascination." This strategy, added the jour- nal's editors, would also result in broader museum audiences, attracting "the masses of people, who are without scientific training and probably have no especial scientific interests."[25] Dioramas' visual and emotional appeal made learning effortless, which was a necessary and desirable quality, wrote the American Museum's director Frederic Lucas, given that "the majority of visitors to a museum do not seek anything beyond entertainment."[26]

Though museum directors across the nation shared scientists' enthusi- asm for the displays, they commissioned and introduced more dioramas into their halls very slowly. Their caution was warranted. Although the exhibits resulted in higher attendance, they were a serious investment, requiring space, specialized labor and knowledge, and a lot of money. In the early decades of diorama building at the American Museum, for instance, attendance doubled, but exhibit building costs tripled.[27] Group displays, the forerunners of habitat dioramas, could be constructed rel- atively cheaply by posing mounted adult specimens of the two sexes and their young on a branch or near a bush. But habitat dioramas were more expensive productions, integrating large background paintings and elaborate foreground accessories to create the illusion of looking at particular locations. Rather than simply presenting generalized ex- amples of animal behavior, the displays attempted to depict with total fidelity a location's plant communities, geomorphology, faunal habits and relationships, climate, and light. To acquire sufficiently impressive specimens and accessory materials for the foreground as well as thousands of photographs and sketches necessary for the background required fieldwork. Most museums believed it essential to send at least one artist, one taxidermist, and a good scientific collector to the place depicted. Once back at the museum, building a small habitat diorama could easily require four months of labor and a few thousand dollars, while largest ones could run as much as $15,000–20,000, taking a considerable staff to produce and several years to complete.[28] Even good-sized institutions found such exhibits a financial stretch.

As a result, most of the nation's museums only gradually introduced dioramas. Curators and directors often began by positioning existing groups in front of large background paintings, a temporary yet affordable gesture toward the more elaborate displays, or by inserting a diorama or two amid rows of the serial and diagrammatic exhibits that dominated museum galleries in the 1910s.[29] Using materials and cases from the natural history exhibit at the 1893 World's Columbian Exposition, the Field Museum commissioned a few large dioramas and a handful of smaller groups before 1915.[30] A handful of institutions, like the young Colorado Museum of Natural History, which was rich in space but poor in holdings, and the California Academy of Sciences, beginning fresh after the devastation of the 1906 earthquake, invested heavily in the displays. Their staff members reserved empty halls for the exclusive use of "environmental groups," as exhibit designer and taxidermist John Rowley described them, knowing that these galleries might take a decade or more to fill. But these institutions were the exception and not the rule, for most museums remained hybrid spaces well into the 1920s, featuring both taxonomic displays and dioramas. When the American Museum's new Hall of Reptile and Amphibian Life opened in 1927, for instance, students could spend their time in the middle of the hall, learning from the synoptic series of models and the systematic and biological diagrams, or visit the habitat groups lined along the walls, where children and more casual visitors tended to gather.[31]

To reconstruct such moments from the wild required a level of artistic expertise that most staff members did not possess, so from Connecticut to California, natural history museum directors hired what would become a proud new stable of painters, sculptors, modelers, and preparators.[32] The nation's largest and most influential museums set up separate preparation departments, but most museums could not afford a full-time exhibition staff, so they developed relationships with commercial taxidermists and well-known background artists, frequently involving these contractors in every stage of design, planning, and preparation. Before the turn of the century, the men who prepared specimens and constructed exhibit cabinets were considered little better than janitors. But the revolution in display and the introduction of dioramas in particular dramatically altered exhibit makers' status in the museum world. By the late 1900s, well-trained taxidermists were in high demand. A generation of talented preparators, many of them hailing from Ward's Natural Science Establishment, filled the Carnegie Museum, the Smithsonian, the American Museum, the Field Museum, the Milwaukee Public Museum, and other institutions with richly detailed visions of nature.[33] Accomplished commercial illustrators like Bruce Horsfall, well-known wildlife illustrators like Louis Agassiz Fuertes, photographers like Charles Carpenter, painters like William R. Leigh and Charles Corwin, and theatrical set designers like Albert Operti joined or contracted with museums, taking brush to background and creating the illusions of deep space for which dioramas were so acclaimed. Widely praised for their exacting work, these exhibit makers became as indispensable to dioramas as curators were to collections.[34]

Museums building dioramas needed not only skilled artists but also ready cash, so curators and directors found themselves turning to both long-standing and new patrons for sponsorship. The hunt for funding was not particularly arduous. Early twentieth-century patrons of natural history museums, most of them wealthy nature lovers, hunters, and ardent conservationists, found dioramas visually and politically easy to like—and therefore, easy to fund. The artistry of these displays appealed to well-to-do sportsmen, many of whom offered, without prompting, to support their construction and the larger ideological missions they represented. Built in the name of public education, dioramas embodied these donors' commitment to providing visitors a "vision of the whole world of nature," as American Museum president Henry Fairfield Osborn put it in 1911, "a vision which cannot be given in books, in classrooms or in laboratories."[35] To be sure, patrons' investments in the diorama exhibits were not entirely altruistic. Bagging diorama specimens often required travel to gorgeous locations, a tempting proposition for sportsmen eager for any excuse to head to field, forest, or bush. For the displays that resulted, credit could easily (and permanently, through the prominent placement of a tasteful bronze plaque) be given to those sponsoring their construction. And donors agreed that taking their children to visit the animals they had shot on museum-sponsored collecting expeditions was deeply satisfying.[36] A habitat diorama, Colorado Museum of Natural

History trustee John A. McGuire assured a potential donor in 1920, "would be a great monument to you and something you would be proud of all your life, and of which your posterity would also be proud, as you would be given credit in a very nice plate on the side of the case . . . which all the visitors naturally see."[37]

Natural history museum directors, intent on acquiring more of the expensive displays, carefully massaged patrons' enthusiasm, requesting specific specimens for dioramas they hoped to build. When Wyoming honey farmer and hunter B. W. Caraway offered to provide the Colorado Museum of Natural History with a series of mounted heads, for instance, director Jesse Figgins politely declined, suggesting Caraway might consider providing the museum with three brown bears instead. Trustee McGuire followed up Figgins's request, urging Caraway to shoot "three nice brown bears" then "purchase another specimen or two." If Caraway did this, McGuire promised, he would get full credit for sponsoring the resulting diorama. "The museum is depending largely upon sportsmen for these groups," McGuire concluded, "and as many sportsmen have already contributed such things to the museum we are anxious to get others to do the same thing—to the end that we may have one of the greatest—if not the greatest—museum on the continent some day."[38] When it came to wealthier donors, directors did not limit their requests to specific specimens but, instead, asked them to sponsor entire expeditions. Part of the appeal of dioramas lay in the specificity of place and time; to replicate a particular place as precisely as possible required museum staff to head to the field in order to collect the specimens, accessories, sketches, measurements, and photographs. Donors were more than happy to underwrite such expeditions, especially if it gave them an opportunity to explore the wild with knowledgeable companions.[39]

The resulting expansion of expeditions gave curators another reason to embrace habitat dioramas: new and unparalleled opportunities for field research. Before the twentieth century, professional collectors and hunters gathered specimens in the field, while curators remained at home, patiently waiting for collections to come to them. Too often, ecologist Charles C. Adams lamented in 1909, "museums make specimens of the curators and keep them out of the field."[40] This gradually changed over the course of the 1900s, as curators worked to bring their study collections up to scientific snuff, going into the field to replace specimens without provenance with newer, better-documented samples. Dual arrangement and subsequent efforts to acquire larger, lovelier specimens for public galleries also encouraged curators to trade the office for the out-of-doors.[41] But it was with the advent of diorama-driven expeditions that museums' scientific staff began to travel frequently.

Though scientific research was rarely the primary goal of expeditions launched to collect for displays, diorama-driven fieldwork yielded plenty of data.[42] Those curators interested in ecology and biogeography conducted extensive field surveys, collecting thousands of specimens and records crucial for modern systematic and taxonomic study. Curators found that the chance to collect in the field was a scientific revelation. Local and commercial collectors rarely kept field notes, making the specimens, according to Frank Chapman, "simply hopeless" for twentieth-century life scientists, whose research questions were far more ambitious than those of their nineteenth-century forebears. Chapman, for instance, sought to understand the full range of intraspecies' variability within a particular geographical area, information that would help him determine species and subspecies with greater precision and afford him insights into larger questions of zoogeography. Though he had tried to glean this information from the American Museum's vast study collections, he had trouble perceiving any patterns. The information he could gain from fieldwork, however, sharpened his scientific vision; after one of his first expeditions to South America to study Andean birdlife, Chapman happily reported that seemingly random variations in the museum's collections actually reflected geographical configurations and provided clues to species' dispersion and evolution.[43]

Research findings like these at once justified and shaped the aspirations of museum scientists and display designers. Curators interested in animal behavior reveled in the opportunities for in situ field observation provided by such expeditions. Exhibit makers routinely arranged striking specimens in dioramas' foregrounds, requiring curators and preparators to capture and convey species' "spirit." Museum scientists quietly parlayed this display imperative into a rationale to conduct even more extensive life history studies on collecting expeditions, peering into the lives, behaviors, and biotic communities of animals then barely known to science.[44] When the American Museum's Department of Herpetology decided to build habitat dioramas of the rhinoceros iguana and the *Hyla vasta*, or Hispaniolan giant tree frog, respectively the largest lizard in the Americas and the largest tree frog in the world, almost nothing was known about either animal—only three specimens of the frog had ever been collected by scientists. Herpetologist Gladwyn Kingsley Noble and his wife Ruth Crosby Noble, an assistant curator in the museum's Department of Education, went into the field to collect and observe the animals and considered themselves "fortunate in finding both forms and in working out their life histories," as Noble wrote in the museum's 1923 annual report.[45] (The Nobles admitted they had also brought along several snake bags, in case they should want to expand the expedition's mission.) Along with the usual casts and accessories for the displays, the couple shipped more than forty live iguanas and two hundred live frogs back to the museum, so they and other museum scientists might continue to explore the animals' bodies and behaviors. The resulting exhibit, Noble explained, "attempts to portray some incident in the life history of the creatures and at the same time to demonstrate certain biological principles." This display strategy, he argued, will "increase considerably the teaching value of the habitat groups."[46]

By the mid-1910s, museums' early investments in dioramas seemed to be paying off handsomely, improving science education and increasing scientific opportunities, certainly, but also augmenting museum attendance. "There seems to be no question but that the methods pursued have resulted in an increased interest being taken by the visitors to the museum," Henry L. Ward happily informed his board of trustees in 1907.[47] Visitors adored the new exhibits, crowding museum halls to see newly unveiled displays. Even the small bird dioramas at the Chicago Academy of Sciences were "emphatically popular," reported director Frank C. Baker.[48] Though museums' attendance records in these years are often imprecise, directors' correspondence, annual reports, and newspaper articles described ever-larger museum audiences and routinely credited this growth to habitat dioramas.[49] After director Jesse Figgins completed several habitat dioramas during his first year at Colorado Museum, he reported that attendance climbed sharply; more than 105,821 people, almost half of Denver's population, had visited the museum that year. The more habitat groups the museum opened, the more people came through its doors, he explained to a trustee in 1916, for dioramas "hold the attention of visitors," drawing them again and again to halls where they had previously "found little of interest."[50]

Better yet, dioramas attracted new and different kinds of museumgoers, a phenomenon that tickled museum reformers aiming to make their institutions more accessible to the public. "Some are farmers, others are members of nature clubs from neighboring towns visiting the place in a body, and still others are mothers with their children," gleefully reported museum man Homer Dill at the 1915 American Association of Museums meeting. "Here are all kinds of people! Students, men of science, and people who would never read a book or go to a lecture are here learning something."[51] Attendance numbers seemed to justify his enthusiasm.

When his museum's cyclorama of the bird life on Laysan Island opened in 1914, for instance, more than two thousand people—nearly 10 percent of Johnson County's population—came to see it in the first two days.[52]

Critics and journalists maintained that dioramas both comforted and inspired these new guests: visitors not only spent more but also more meaningful time in front of habitat dioramas than any other museum exhibits, they declared, for dioramas evoked genuine emotion, empathy, and interest. Habitat dioramas and painted backgrounds, the editor of *Field and Stream*, Warren Miller, wrote in 1915, reawakened the "pleasures that one experienced . . . when going into the woods in the spring time." When looking at a diorama of the American bullfrog, Miller wrote, he hardly felt like his "own weatherbeaten and battle-scarred self" but, rather, like a boy of ten "eager with the devouring eagerness of childhood; keen in his observation of every least detail of the pool beside which he is standing."[53] Dioramas prevented a latent love of nature from being "deep-buried in the dust of city's turmoil."[54] "They are an extraordinary production," agreed naturalist R. W. Shufeldt, "and in many instances produce a picture upon the mind of the beholder which is not easily effaced in a lifetime."[55]

Dioramas soon attracted student groups, as nature study and science teachers began to rely on the displays as visual aids for classwork.[56] Curators intended dioramas to convey the very social and scientific lessons currently central to school curricula: sympathy for animals, the efficient conservation and exploitation of natural resources, successful adaptation to changing environments, and the physical processes by which life was organized and advanced.[57] Several of the bird dioramas at the Colorado Museum, for instance, depicted local birds during different seasons, so that visitors might learn better how to encourage or conserve economically useful species. The museum's display of Denver's winter birds, director Jesse Figgins explained to the U.S. Commissioner of Education, "illustrates the advantages of planting fruit bearing shrubs that will attract desirable birds during the winter months; the group representing 'Fall' is now under construction and will include the many local species of birds that feed largely upon the seeds of undesirable weeds and grasses." Figures in an accompanying label "demonstrate the advantages of these birds in keeping such plants in check."[58]

But dioramas conveyed such messages in less than explicit fashion, for curators trusted that visitors would perceive the scientific and social lessons embedded in the new exhibits. The American Museum's diorama of birdlife in a tropical forest overlooking Mount Orizaba, Mexico, depicted the three different life zones of North American fauna, an attempt by curator Frank Chapman to educate the public about an iteration of natural selection, namely, altitude's influence on the evolutionary development and distribution of birds.[59] But no full explanation of this principle accompanied the exhibit. Nor were there labels pointing out where in the diorama visitors could see expressions of natural selection.

Curators' expectation that dioramas would prompt students to observe and deduce scientific principles quietly worked nicely with prevailing classroom culture and helped popularize natural history museums as a field trip destination.[60] Museums made the most of this interest, promoting new displays to schools, providing free lectures, showing relevant films to both teachers and schoolchildren. Curators and museum educators even provided teachers with related lesson plans to use before and after their visits.[61] Schoolteachers who couldn't bring their classes and smaller museums that could not afford to build dioramas relied just as heavily on the exhibits, borrowing thousands of lantern slides of dioramas from large city museums' well-organized loan and delivery systems. The Davenport Public Museum in Iowa, for instance, purchased several sets of these slides for weekend lectures, knowing that local residents as well as students would jump at the opportunity to see rare beasts on the big screen.[62]

Habitat dioramas' popularity buoyed the mood of museum workers in the late 1900s and early 1910s, for the displays seemed to combine museums' missions in ways that appealed to scientists, visitors, trustees, educators—nearly everyone with a stake in the museum world. Curators gloried in the fieldwork and expansion of collections afforded by diorama-driven expeditions and shared habitat and behavior findings with their newly hired colleagues in departments of preparation. Visitors thronged around the intricate exhibits, craning their necks to see every gorgeous detail, confirming curators' convictions in dioramas' pedagogical power as they did so. Museum administrators planned diorama-driven safaris with wealthy sportsmen over claret and canvas-back, resulting in generous donations to their institutions. The growing scale and complexity of diorama building in the late 1910s and the 1920s, however, meant that the displays could not comfortably remain all things to all people for very long.

Doubling Down on Dioramas: Museum Culture Transformed

Mounting excitement over habitat dioramas led exhibit makers, directors, and department heads to double down on the displays. As a result, throughout the late 1910s and 1920s, museums planned more—and ever more spectacular—dioramas, often building or setting aside entire wings for the exhibits. "The popularity of these realistic presentations with all visitors testifies to the wisdom of the plan to increase their number as rapidly as possible," announced Philadelphia Academy of Natural Sciences' director Charles Cadwalader, "especially in view of the fact that their artistic and educational value is immeasurably in advance of old methods of exhibit. The Academy should be as pre-eminent in this form of service as it is in the field of scientific research."[63] Delighted by the opportunity to construct more elaborate visions of nature, exhibit makers thrived during these years, commanding more respect and resources from other staff members, patrons, and the public. Curators, however, found the expansion of dioramas more complicated. Their initial enthusiasm often gave way to frustration and resentment, as the displays began to consume more and more of museums' money and labor and to alter curators' professional obligations. By the mid-1920s, many curators had begun to question the wisdom of museums' escalating investment in habitat dioramas, as well as the value of the broader institutional commitment to public education and service that the displays represented.

Visitors would never have guessed that dioramas aroused reservations among the scientific staff, for museums offered the displays a much larger stage throughout the 1910s and 1920s. Instead of integrating dioramas into existing halls, directors and boards of trustees often granted the displays their own vast spaces. John Rowley, a former chief of taxidermy at the American Museum, persuaded trustees of the California Academy of Sciences to let him design their new building's hall to accommodate his vision for the displays. For the museum's halls of North American mammals and birds, Rowley produced open floor plans, tucking large diorama alcoves, each twenty-five feet long and twelve and a half feet high, into their walls, and setting aside spaces for smaller dioramas in between. To reduce shadows and other distractions from his illusions of nature, each alcove had a curved and plastered rear wall; to eliminate reflections and allow exhibit designers to create mood, he inserted overhead skylights and electric bulbs.

Inspired by Rowley's work in California, other museums saw spectacular potential in geographically focused exhibits, and they, too, began to set aside or rebuild existing halls for dioramas, organizing them to depict the variety of flora and fauna of entire continents. Dioramas had always been relatively large—they had to be in order to contain groups of elk, deer, or timber wolves—but, liberated from preexisting architectural constraints, they now began to stretch from floor to ceiling or take up full walls at the dead ends of galleries of North American or African megafauna. In 1917, for instance, noting that he hoped to "create something similar to the new bird habitat dioramas in the American Museum," Milwaukee's Henry Ward began construction on a diorama twelve feet long and nine feet deep, which extended from floor to ceiling. This size became quite typical for dioramas. More spectacular were the dioramas that opened in the late 1920s and early 1930s; the *Water Hole Group* in the California Academy of Sciences' Simpson African Hall loomed more than thirty feet tall.[64] Urged on by Jesse Figgins, who had trained under Rowley in New York, the Colorado Museum created a Hall of North American Mammals in the late 1910s, and in 1926, sent curators to South America to gather more than thirteen hundred specimens for a hall of habitat dioramas, two of them to be 110 feet in length, intended to grace the museum's new wing.[65] The Los Angeles County Museum opened its own halls of North American and African animals in 1925, again designed by Rowley, who punctuated the end of each hall with panoramas measuring forty by one hundred feet.[66] Smaller museums, unable to afford collecting expeditions in Alaska or the Belgian Congo, relied on more local settings for inspiration. The resulting halls were no less ambitious, however: in 1919, for instance, Chicago Academy of Sciences director Frank Woodruff and his curators transformed a seventy-six-foot-long hall into the Illinois of forty years before, posing 110 different groups of animals along their re-creation of the banks of the Calumet River.[67]

By the late 1910s, a natural history museum's institutional ambitions could be read from the number, size, and aesthetic quality of its diorama displays—or, at the very least, from its plans for the displays. Dioramas became not only more common but also more visually dramatic in these years, as enthusiastic directors and emboldened exhibit makers pushed for more artistic renderings of the natural world. Chapman's earliest dioramas had tiptoed quietly into the terrain of art, drawing on the pre-cinematic practices of landscape painting, cycloramas, and scientific photography while preserving the stilled, staid mood of specimen cases.[68] But Chapman had closely supervised the construction of these exhibits and had demanded that background artists and modelers replicate his instructions and photographs exactly. The resulting dioramas were scientifically accurate but often poorly composed and static. Starting in the mid-1910s, however, preparators began to insist on more artistic control.[69] Acknowledging their new colleagues' talents, curators and directors gradually allowed museums' exhibition staff to take the lead in designing dioramas or to alter existing compositions to make the displays more alluring. Field Museum director Frederick Skiff, for instance, began to listen more closely to his preparation department in the 1910s; in 1913, when exhibit designer Charles Corwin personally requested funds to paint a new landscape background for the museum's capybara diorama, Skiff promptly acquiesced. Their conversation broke the ice between the director and exhibit staff. From that point on, Corwin spoke directly to Skiff, rather than communicating with him through his supervising curator.[70]

The newly spectacular displays indicated the readiness of curators, preparators, and directors to shake off earlier inhibitions about exhibits that packed the visual and emotional " 'punch' of the movies," as American Association of Museums leader Lawrence Vail Coleman described it.[71] Indeed, throughout the late 1910s and 1920s, museums increasingly borrowed aesthetic strategies employed by more exciting commercial displays and theatrical producers—all in the name of science education.[72] Administrators hired department store window designers and interior decorators to consult on exhibits, and the artists and designers who staffed preparation departments drew on their own experience in set design and advertising, introducing spot, colored, and indirect illumination, painting deep backgrounds that gave the sensation of vast and open space. At the 1916 AAM meeting, preparator Dwight Franklin, who became one of Hollywood's leading prop artists in the 1920s, reminded his colleagues that museums had a lot to learn from the theater. "Many of the scenes which so delight us at the play, might, with a few changes, be translated into charming museum groups," he observed.[73] Diorama designers should pay special attention to theatrical lighting, he added, for "the whole feeling of a group may be changed by throwing different colored lights on it and by increasing or decreasing the volume of light." "A warm orange light thrown on a desert group, for example, gives a surprising feeling of heat, while a blue light on an arctic scene makes one quite chilly," he pointed out. "These qualities all tend to make a group more convincing."[74]

Preparators and curators also worked to eliminate visual distractions. They recessed exhibits into the walls so viewers could only approach displays from a single, carefully planned perspective. They tilted display glass, and curved the sides and ceilings of each diorama to create the illusion of ongoing landscape unbroken by reflections or right angles. They darkened museum halls and lit dioramas from above to make it seem as if natural light emanated from the dioramas' painted skies.

Exhibit makers and curators employed an equally theatrical approach to the animals that starred in their tableaux of nature. Relying on sophisticated sculptural and taxidermic techniques, taxidermists used eye contact and head position to make specimens seem self-aware and intentionally encouraged anthropomorphic associations.[75] Exhibit makers frequently incorporated family dynamics that visitors could easily recognize, so playfully scuffling bear cubs, mother does nuzzling fawns, and watchful male lions supervising prides filled dioramas in these decades. But they also tapped darker impulses. When Carl Akeley brought an African elephant back to the Milwaukee Public Museum, for instance, director and curator Henry Ward requested that Akeley make the elephant "look just as big and just as imposing as possible. . . . We want to have him look like a wild beast and not a tame circus elephant that one would feed peanuts to without danger."[76] The result was a belligerent-looking mount, whose impressively large ears and tusks were thrown forward to signal imminent aggression. Such poses were not unusual: the theme of struggle for survival in the wild possessed scientific, visual, and political appeal, so dioramas often featured animals poised to pounce on unsuspecting victims or claw eager challengers.[77] "The notion that intense action or even violence is inconsistent with scientific dignity is obsolescent," explained Lawrence Vail Coleman in 1921. "Even the fight has found its way into the American Museum through the agency of two Wild Boars."[78]

Exhibit makers capitalized on a variety of new techniques and technologies to render more dramatic visions of nature. Background painters daubed shiny patches with buttermilk, which dried to a flat finish, so no inadvertent glare would catch a visitor's eye. They relied on mathematical formulas and grids scaled up from stereoscopic photographs of the landscape to minimize the visual distortion that resulted from replicating a flat photograph on a curved background canvas. (These painters could be incredibly exacting—to ensure that visitors felt fully immersed in the Whitney Hall of Oceanic Birds' illusion of the Pacific isles, artist Francis Lee Jaques insisted that the horizon line be at the same level in every diorama.)[79] Foreground artists built landscapes out of wooden frames and wire mesh, covered in sisal or burlap dipped in plaster. To create convincing leaf litter, they relied on painted and molded cotton and beeswax, materials that captured real leaves' translucence. They then soaked the artificial leaves in wallpaper paste, shellacking them to make them look moist and applying glycerin to create dewdrops. Papier-mâché became bark, latex rubber held evergreen needles onto real tree-boughs, and mixtures of formalin, glycerin, and water preserved shrubs, small trees, mosses, and vines.[80]

Preparators also borrowed colorful materials and methods from theatrical productions, department store displays, and the more spectacular exhibits of world's fairs, acknowledging what one AAM speaker described as the "potent educative success" of early twentieth-century consumer culture.[81] Preparators at the Milwaukee Public Museum, for instance, routinely purchased materials from the Advance Theatrical Exchange ("Producers of Acts, Sketches, Monologues, etc. We Teach Elocution, Tango, All Up-To-Date Dances, Stage Dancing . . . Entertainers furnished for Charettes, Entertainments, Theatricals"); Wm. Potter and Sons, which were experts in "showcards, advertising novelties, plate glass show cases"; Chas. Polacheck & Bros., which provided artistic commercial lighting; and Robert Brand & Sons, which traded in fixtures for cigar and jewelry stores.[82] Increasingly, exhibit makers, curators, and directors now accepted fairs and department stores as respectable cultural and educational establishments and even began to compare their own work to retail, all the while insisting that the public amusement they provided was "of the highest type," as curator Arthur Crook put it.[83]

While dioramas incorporated twentieth-century technologies of appeal, they also drew on older Romantic and nationalist aesthetics of nineteenth-century landscape and animal painting.[84] Many exhibit makers had experience in panorama painting, which demanded academic precision in order to create illusions of nature's most sublime elements; others came from the world of commercial illustration, which rewarded delightfully specific depictions of dramatic background landscape.[85] Consequently, exhibit makers favored detailed and often shadowed foregrounds, highlighted middleground plains, and deep horizons with a single dramatic feature intended to draw viewers' eyes into the distance.

The vibrant results pleased culturally conservative trustees and wealthy sportsmen, who saw such depictions of wilderness as appropriately patriotic. They also caught the imaginations of conservation-minded curators eager for the public to appreciate natural beauty. As early as 1909, Frank Chapman noted that the illusions of landscape in the dioramas of the American Museum's Hall of North American Birds attracted just as much comment as the birdlife they illustrated. "The interest they arouse, particularly among foreigners, makes one wonder why we, in this country, have not already established a museum for the display of paintings of American scenery?" he wrote.[86] Exhibit makers' romanticized take on natural scenery did not stop at dioramas of North American nature, however. They readily applied this approach to African, Asian, and South American backgrounds as well.[87]

The Colorado Museum of Natural History's golden eagle diorama (fig. 6) embodied the influence of these new artistic approaches. Robert Niedrach and Fred Brandenburg built the exhibit, a depiction of an eyrie

Figure 6 The Colorado Museum of Natural History's *Golden Eagle* exhibit, constructed between 1937 and 1938, typifies the stylistic and visual conventions of habitat dioramas. Its deep background landscape and detailed foreground created a dramatic illusion of perspective, pleasing audiences hungry for thrilling scenic views. The exhibit also satisfied scientists' expectations, featuring adult specimens of both sexes and juveniles engaged in what field biologists considered typical natural behaviors. Here, a nesting female and her two chicks are paired with a male eagle, back from a hunt. (Photo by Rick Wicker, Image Archives, Denver Museum of Nature and Science.)

near Pikes Peak on a midspring afternoon; Charles Waldo Love painted its background.[88] The exhibit satisfied scientific expectations, featuring adult specimens of both sexes and juveniles engaged in what were considered typical behaviors: here, a nesting female and her two chicks are paired with a male eagle, back from a hunt. Its intentionally dramatic composition and contents were equally typical of dioramas' newly theatrical visual and narrative conventions. Poised on an outcropping, the male was positioned in profile. His yellow feet and terrifying talons straddled a freshly caught rabbit. The soft, almost Impressionist color palette of the lichen-covered rock, the middleground moraines, and the background peaks and skyscape set off the male's dark feathers and the shadowy protection of the ledge, while the chicks' white fluff stood out against their mother's breast. Love, Niedrach, and Brandenburg emphasized the ledge's dizzying height by pairing the foreground's looming rock with the middleground's tiny evergreens—acrophobic visitors encountering the exhibit for the first time must have fought an urge to step back. (The decision to capture this particular moment in the lives of these eagles made it easier for the preparators to meet the expectations of fellow museum scientists and artists, certainly, but it also naturalized a gender dynamic familiar to Americans in this era. As men did in so many human family portraits in this era, the male eagle stood well above his mate, tidily obscuring the fact that female golden eagles tend to be far larger than their male counterparts. He provided dinner, while she looked after the young.)[89]

Critics and visitors alike welcomed these new displays. Conservationist William Temple Hornaday, himself a former taxidermist at the National Museum, acknowledged the exhibits' long development time and considerable institutional cost but justified it by pointing to dioramas' educational, visual, and emotional power: "In comparison with their cumulative value their cost is utterly trifling," Hornaday concluded in a 1922 issue of *Scribner's*.[90] The unveiling of each new habitat diorama attracted hordes of eager children, observed the *New York Times*, though audiences ranged from "men of affairs to newsboys of the streets." The American Museum became so crowded in the winter that "solid waves of human beings may sometimes be seen moving towards the museum entrance," according to the *Times*, and "hardly a visitor comes to the museum without looking up the far-famed birds in their bushes and trees and swamps" exhibited in its Hall of North American Birds.[91] After it opened a handful of dioramas in 1930, the Academy of Natural Sciences of Philadelphia received more visitors per day than it previously had per week. When more dioramas went on view in 1931, the museum's attendance topped a hundred thousand for the first time, a 50 percent increase from the previous year.[92]

Museums' finances also reflected the success of these exhibits with a more rarified segment of the public. The Philadelphia museum's new exhibits not only drew foot traffic but also "brought a considerable increase in membership, especially among the classes of larger contributors," director Charles Cadwalader proudly noted.[93] Thanks to private patronage, the American Museum's general income steadily climbed from $446,000 in 1910, to $946,000 in 1920, to $1,827,000 in 1930. Indeed, the number of donations in the 1920s was more than ten times what it had been in the 1910s—an increase that museum administrators attributed almost exclusively to dioramas and expeditions.[94] Delighted administrators found themselves agreeing with Hornaday's assessment on pecuniary grounds. "No modern museum would dare" eliminate dioramas, observed Colorado Museum director Jesse Figgins, "for it would at once lose both attendance and financial support."[95]

Pleased with this response, curators and exhibit makers became more ambitious and began to experiment on an even grander scale, framing dioramas within thematic spaces that prompted visitors to experience, at least imaginatively, the outdoor exploration museum workers so enjoyed.[96] Natural history museums, taking advantage of the decade's economic boom, began to create entire halls in which visitors seemed to be transported outdoors. At the American Museum, preparators and artists worked to turn Chapman's original bird dioramas inside out in 1925, doming, painting, and lighting the bird hall's ceiling to resemble a softly luminous sunset. Canadian geese, brown pelicans, an Andean condor, an albatross, an eagle pursuing a fish hawk, a flight of frigate birds, and a flock of Scaup ducks all sailed motionlessly through forty-seven hundred square feet of this simulated sky. Preparators used every technique available to give a sense of open air and flight—"invisible" wires to suspend the birds, lighting that created "a suffused atmospheric effect forbidding the thought of a solid arch," even trompe l'oeil birds painted to seem as if they were flying high above the mounts. Taxidermists used aluminum rods to curve up the birds' feathers to make it seem as if their wings were beating against considerable air pressure and hung each bird by three wires to prevent swaying. The hall evoked an unparalleled level of public response. Sportsmen who visited were so taken with the illusion that they aimed imaginary guns at the ducks. Aviators compared the wingspreads of the albatross to gliders. According to *Museum News*, such enthusiasm made the hall's tremendous cost well worth the expense. "The experiment abundantly justifies itself," the paper's editors declared.[97]

More often, museums would design diorama halls so they seemed like carefully planned continental railway tours or visits to a grand hotel with a veranda that overlooked gorgeous scenery.[98] The American Museum's Vernay-Faunthorpe Hall of South Asiatic Mammals upped the ante in 1930, incorporating thematic decor to make museumgoers feel as if they were in South Asia.[99] Visitors immediately confronted an elaborate teakwood screen framing a map that depicted the location of each habitat diorama on the subcontinent then sat down on carved teakwood benches to contemplate the views into the jungle. Preparators recessed most of the hall's cases, trimming them with teak, topping them with maharajah spires, and filling the spaces in between dioramas with carved screens and Indian artwork. "No intrusive details are permitted to work against the creation of an illusion which will transport the visitor into the heart of India," reported curator of mammalogy Harold Anthony.[100]

Such halls were brick-and-mortar manifestations of museum personnel's unwavering belief that the dioramas represented the pinnacle of zoological and ecological exhibition. As a result, rather than creating flexible spaces that provided for change over time, administrators and trustees chose to reconfigure halls with bricked-in windows, open floors, and predetermined diorama alcoves resembling the radiating chapels of gothic cathedrals.[101] Donors funded new wings designed specially to house these halls, architectural additions that indicated how permanent they believed these exhibits would be.

The diorama halls further testified to the shared faith of museum directors, curators, and exhibit designers in the commonality of human experience in the face of nature—or rather, in the face of its precise replication. According to American Museum president Henry Fairfield Osborn, museum staff created dioramas "to meet modern conditions of life, in which by far the greater number are 'cribbed, cabined and confined,'" far removed from the wild places that curators so loved.[102] Curator Mary Cynthia Dickerson trusted that museumgoers contemplating dioramas would not only see "with unusual vividness and attractiveness the natural history facts involved" but also feel "some part of nature's subtle personal invitation and some reflection of the spiritual response which the original scene might invoke."[103] Confronting these constructed simulations, curators hoped museum visitors would experience the same shock of beauty as travelers did; perhaps, Dickerson wrote in 1918, such exhibits "may succeed in giving one somewhat the rare experience that comes to the person who for the first time visits" new landscapes.[104]

Curators continued to hope this kind of "personal invitation" and "spiritual response" would result in public support for conservation, as well as more awareness of the specter of extinction. Museum workers continued to stand at the forefront of the conservation movement, and they urged their institutions to confront the issue head-on so the broader public might become more active in the preservation of threatened flora and fauna. They used a variety of venues to get their message out. American Museum curator Willard Van Name, for instance, wrote a scalding call to arms in a 1919 issue of *Science*, in which he urged fellow scientists to take up the banner of preservation, for "it is only those with more or less scientific knowledge of animals and plants who can see in advance the need of protective and remedial measures and can direct and carry them out with any hope of success."[105] Exhibit designer Carl Akeley took a more popular route to press for the preservation of African wildlife, publishing books and articles that became widely read and designing the American Museum's enormous African Hall. Others, especially museum administrators and trustees, used politics and policy to push a conservationist agenda.[106] Smithsonian assistant secretary Alexander Wetmore,

Philadelphia Academy of Sciences director Charles Cadwalader and Field Museum director Stanley Field all served on the first advisory committee of the American Committee for International Wild Life Protection, for instance.[107]

As a result of this institutional commitment to conservation, many museum dioramas focused on species on the brink of extermination: the front hall of the California Academy of Sciences, for instance, featured dioramas of the tule elk, the California condor, the California sea lion, and other threatened species, while a diorama of a bird rookery on Farallon Islands attempted to stir public outrage over commercial egg collecting. While it is difficult to measure the impact of dioramas on public opinion, leaders in the conservation movement certainly praised these efforts. As William Hornaday declared in 1922, exhibits like the American Museum's bird dioramas "inspire in the beholder a love of birds and a desire to protect them from slaughter."[108]

Important as they considered the mounts within these displays, in the late 1910s, 1920s, and early 1930s, conservation-oriented curators also began to assign new scientific value to the habitats represented. They had always appreciated backgrounds and accessories as teaching tools for the biological perspective, but in the interwar years, the science of ecology and the notion of preserving the integrity of entire biological communities began to permeate conservation discourse.[109] Staff members responded. Though they continued to design and label dioramas and create supplementary materials to rouse popular interest in saving endangered species, they now added new information that emphasized the importance of habitat protection in accomplishing this goal. Zoologists Charles C. Nutting and Homer Dill's 1914 Laysan Island cyclorama in Iowa, for instance, was a visual indictment of the impact that invasive species could have on a habitat. The exhibit depicted the tiny Hawaiian atoll's albatross rookeries before their devastation—plumage hunters had destroyed hundreds of thousands of birds, and commercial rabbit breeders had allowed their animals to strip the island of vegetation.[110]

The growing presence and power of habitat dioramas in natural history museums significantly improved exhibit makers' institutional status. The exhibit maker should be recognized as "a man of a new calling; a man whose work is as dignified as that of a doctor or a professor," Homer Dill declared at the 1916 meeting of the AAM.[111] Some museums, like the Colorado Museum of Natural History and the Academy of Natural Sciences of Philadelphia, began to provide grander titles and larger offices for preparators, offering curatorships to the best of them.[112] In 1935, for example, the Colorado Museum appointed chief preparator Robert J. Niedrach as its curator of birds. The American Museum, the National Museum, and the Field Museum still reserved the titular honor of curator for professional scientists but did not hesitate to celebrate exhibit makers in other ways. Salaries were in constant negotiation, but leading preparators were now paid nearly as well as curators and, very occasionally, better than them. James Clark, chief of the American Museum's Department of Preparation, Arts, and Installation, took home more than three times the museum's curator of mammalogy in 1925.[113] And the AAM's newly launched *Museum News* devoted the vast majority of its articles to descriptions of exhibit makers' accomplishments and explanations of how even the smallest museums could replicate their techniques.[114]

Exhibit makers took this improvement in status in stride, believing it long overdue, and they used it to highlight their own professional credentials. Preparators constantly augmented their own understandings of animal and plant life, attending evening lectures and classes on zoology and ecology and studying farm and lab animals to learn more about musculature and motion. During his first few years as a museum taxidermist, James Clark later recalled, he "modeled and sketched and sketched and modeled" dead and live animals from local zoos, labs, and aquaria.[115]

Some had spent far more time in the field than curators had. After a childhood of hunting and trapping in Kansas prairieland and Minnesota forest, background painter, bird illustrator, and amateur ornithologist Francis Lee Jaques had traveled from the Arctic to the South Pacific, taking part in more than a dozen expeditions in his twenty-one years at the museum. In an age of scientific specialization and professionalization, most of the taxidermists, painters, and modelers who made up exhibition staff were naturalists in the all-encompassing, nineteenth-century sense of the word, and many felt scientists underestimated their hardearned knowledge, won through decades of close observation, physical interaction, and diligent study.[116] "I suspect that at the time I understood bird flight better than any of them—possibly anyone anywhere," wrote Jaques, recalling his first years at working with the American Museum's ornithologists.[117]

Some exhibit makers went so far as to argue that the dioramas they built would serve as important resources for scientists of the future, capturing soon-to-be extinct animal and plant habitats for future investigation. "Two hundred years from now," exhibit maker and conservationist Carl Akeley predicted, "naturalists and scientists will find in such museum exhibits as African Hall the only existent records of some of the

animals which today we are able to photograph and study in their forest environment."[118] It's likely that curators scoffed at such grandiose
pronouncements behind closed doors, believing expedition notes and
specimen collections far superior resources for the production of knowledge, but few publicly protested such sentiments.[119] Dioramas had become the hands that fed their scientific appetites, after all, and who were
they to bite back?

Museums capitalized on exhibit makers' work with newly aggressive
public relations efforts, and by the 1920s, many exhibit makers became as
well known as the curators with whom they worked.[120] Patrons, visitors,
administrators, and trustees admired their spectacular handiwork and
reporters glorified their exploits in the field.[121] Readers followed taxidermist Carl Akeley's horrifying travails in East Africa as he collected for the
American Museum's planned African Hall and fawned over the bravery
of background painter Zarh Pritchard who, according to the *Los Angeles Times*, routinely worked underwater "in peril of sharks and stinga-
rees [sting rays]," as he sketched fish and coral in a heavy diving suit.[122]
Though such physical courage was impressive, it was exhibit makers'
selfless commitment to public education that truly set them apart, reporters suggested. The public responded enthusiastically to this rhapsodic
rhetoric, interpreting such endurance, fortitude, and self-sacrifice as guar-
antees of scientific credibility.[123] Museums reaped the benefits, for exhibit
makers' carefully crafted public personas both attracted new audiences
and lent dioramas more legitimacy—how could visitors doubt these pictures of nature when they knew exhibit makers had gone through hell
and high water to make them? So institutions publicized their exhibit
makers' exotic adventures to raise money for new displays and halls,
sending out press releases and welcoming reporters into their galleries.

In a few museums, administrators relied more heavily on exhibit
makers' personal charisma to assist in the quest for cash. The American Museum's James Clark and Carl Akeley, for instance, were routinely
given direct access to the museums' wealthiest patrons. Preparators and
philanthropists had not always shared such cozy relationships. On the
first leg of his first expedition to Africa in 1909, for instance, Clark had
spent Christmas at the London Zoo, too lonely and cash strapped to go
elsewhere.[124] But by 1926, after landing in London at the outset of the
Morden-Clark Asiatic Expedition, Clark hired a car to take him directly
to the Ritz so he could drop off his bags. Then he headed over to Saville
Row to get fitted for a tuxedo before meeting the Roosevelts for dinner.[125]
Akeley guided his team of craftsmen and artists into similarly lofty social
circles, insisting in 1925 that taxidermist Robert Rockwell purchase a

tuxedo and new field clothes in London before he sailed to Africa, as he would have to dress properly for dinners while on safari.[126] Throughout the 1920s and early 1930s, taxidermists and artists were popular guests on dinner party circuits in New York and Chicago, where they happily related their adventures in the field to awed audiences and society reporters. "'Everybody' seemed to be going to Africa or elsewhere for a big-game shoot or photography," Clark later recalled. "This was high society's 'thing to do,' for it brought much fame and prestige to the otherwise idle rich and gave them topics for conversation."[127] And museum staff always cited the generosity of expedition sponsors, careful to make sure that reporters wrote down their compliments word for word. The strategy paid off: Clark received so much interest from potential sponsors that he began to hand out brief questionnaires designed to test how much money people were ready to give.[128]

Knowing these exotic expeditions and elaborate dioramas would appeal to the masses quite as much as they did to millionaires, exhibit makers and curators sought to broaden their institutions' donor bases by describing their adventures in the less intimate settings of national lecture tours and popular publications.[129] Local papers and publics responded warmly. The *Chicago Daily News* paid a "substantial advance sum" to fund the Field Museum's 1926–27 expedition to Ethiopia—then Abyssinia—in exchange for exclusive coverage of the adventures of specimen collector and big-game hunter James E. Baum, curator of mammalogy Wilfred Osgood, ornithological painter Louis Agassiz Fuertes, and the wealthy naturalist C. Suydam Cutting.[130] When American Museum curator Roy Chapman Andrews came to speak at the Colorado Museum of Natural History in 1927, Denver audiences raved.[131] "I declare it was the best lecture I ever heard," Colorado entomologist Theodore Cockerell wrote American Museum president Henry Fairfield Osborn. "The judicious combination of scientific information with popular description, the exquisite beauty of the pictures, the touches of humor—never forced, and the personality of the lecturer made a combination which could hardly be excelled."[132] Such direct appeals paid off well beyond mere compliments from scientists: when Andrews announced that he would be leading a new expedition to Asia in order to collect dioramas specimens and fossils, thousands of Americans donated $5 or $10 or $20 a year to finance the expeditions.[133]

Though their fundraising efforts lacked the extravagance or reach of their counterparts in New York and Chicago, museums in smaller cities also cultivated a wide variety of sponsors by capitalizing on the popularity of dioramas and the exoticism of the expeditions preceding their creation. Milwaukee Public Museum director Samuel Barrett, for instance, persuaded local businessmen John Cudahy and Osborne Goodrich to put up the cash for an expedition to East Africa to collect specimens for an African Hall, a trip that became a local cause célèbre. Burt Massee, a Chicago industrialist who had grown up in Milwaukee and cherished memories of the museum there, donated another $16,000 so curators would "get some big animals for 15 year old kids to gaze at and remember when they are forty." Money rolled in from all directions, as local residents eagerly supplied the expedition with everything from fishing tackle and boots and shoes to batteries and malted milk. Director Barrett tried to play down the similarities to an exotic vacation, assuring trustees in June of 1927 that fieldwork was "no picnic." Still, he played the press like a fiddle, describing the exciting nature of safari and returning to Milwaukee with a photogenic lion cub, Simba, which politely posed for pictures on the roof of the museum.[134]

Directors, expedition leaders, and department heads took care to ensure that collecting for dioramas would be the public focal point of expeditions, for many of the sportsmen and members of the public sponsoring dioramas and expeditions in the 1920s had little interest in donating to ventures that didn't yield immediate and immediately comprehensible benefits. Diorama-driven expeditions almost inevitably concluded in tangible success, as the ventures were essentially glorified safaris. Patrons adored the romance of automobile caravans and airplanes, the spectacular scenery, and the fawning press that accompanied such trips. In contrast, "the scientific work and the collections have no special news value; the discovery of a few new species carries no thrill for the public," admitted American Museum curator Clark Wissler.[135] What's more, understanding the significance of expedition data took time, and impressive scientific findings could never be guaranteed.

The resulting institutional dynamics—the elevation of exhibit makers within and beyond the museum, the cultivation of relationships with publicity-oriented patrons, and the prioritization of display above all else—began to unsettle many museum scientists.[136] Curators feared "their institutions are fast becoming mere expositors of elementary facts and embarking upon extravagant schemes of nature-faking," wrote Clark Wissler in 1925. "Often do we hear the term 'kindergarten' applied continuously to present museum policy."[137] Curators admired exhibit makers' capabilities in studio and field but insisted that, in the museum at least, such skills should not command the same kind of prestige as scientific expertise. So while they continued to applaud dioramas in public, curators worried more and more that museums subordinated science to the pleasure of patrons and visitors.

Much against their will, curators were often forced to do the same. Before the 1900s, curators' primary duties were to cultivate, care for, and study specimen collections. Displays were often incidental; indeed, prior to dual arrangement, cataloging specimens and building displays weren't distinct responsibilities. But twentieth-century efforts to reform display had expanded curators' exhibition responsibilities many times over, changing their jobs in ways that weren't always to their liking. Curators now dedicated as much or more time to the development of dioramas as they did to scientific work. They organized collecting expeditions for dioramas and joined preparators in assessing the thousands of resulting specimens, photographs, and drawings to determine what would make compelling displays. Ironically, museums' growing cadres of exhibit makers meant more, not less, work for harried curators. Preparators' own artistic ambitions and ingenuity with materials resulted in ever-more complex exhibits, and curators had the daunting tasks of previewing, verifying, and reviewing the thousands of visual details that found their way into displays. "At best the head of a department or division has not been able to depend on more than a moiety each day for research," Smithsonian geology curator George Perkins Merrill complained. But "in times when exhibits are in preparation, he has been obliged to dispense with even that."[138] "Although we are sending out more expeditions and getting more material than ever before, I have less and less time for study," sighed the Field Museum's Osgood in 1924.[139] Though curators still sat atop the hierarchy of expertise at the museum, rendering final judgment on exhibits' contents and appearance, such powers held less and less appeal for those who just wanted to get back to their research.

Curators also worked alongside preparators to help them reproduce nature as precisely as possible. American Museum curator of herpetology Mary Cynthia Dickerson, for instance, spent long days experimenting with methods that would allow her to render amphibians and their behavior in lifelike fashion. (The preparation of amphibians was particularly daunting, for the animals kept their form "about as well as does a jelly-fish and in truth are just about as satisfactory to cast," Dickerson dryly observed.) She plunged her hands into combinations of kerosene oil, formaldehyde, plaster, and ice for hours each day, trying to persuade the lubricious animals to retain their shape long enough to cast them in wax.[140] In two years, she cast more than twenty of the specimens used in the American Museum's exhibits.[141] Though Dickerson's energy was unusual, her deep involvement in the production of exhibits was not, and by the mid-1910s, administrators considered an artistic eye and mechanical ingenuity critical attributes in a curator.[142]

Of course, diorama building wasn't the only distraction from curators' scientific responsibilities in these years. As museums embraced public education and visitor numbers continued to grow, curators' public obligations ballooned: public speaking, curriculum development, consulting with government agencies and conservation groups, and, of course, tending to visitors. Curators were deluged with earnest questions about the natural world. In the first four days of November in 1920 alone, the Milwaukee Public Museum's curator of botany fielded sixty visits and twenty telephone calls from people hoping to learn to distinguish poisonous mushrooms from edible ones. Sixteen more people requested plant identification, and six others asked for general botanical information. The situation in Milwaukee was not unusual. Administrators at the Buffalo Society of Natural Sciences, for example, estimated that curators spent at least half their time identifying insect pests for local gardeners and farmers.[143] Thousands more inquiries arrived by mail each year, and staff scientists did their best to answer each one or, if they could not, to refer letter writers to relevant specialists or books.

Some curators thrived on the public's shared passion for the natural world, but even the most enthusiastic found these expanded responsibilities a little overwhelming. "With the passage of the years the demands on our time have constantly increased," American Museum curator of ornithology Frank Chapman reflected. "We have become, in truth, a Bureau of Information for everything to do with birds and, to some extent, the countries in which they live."[144] But of all their new responsibilities to the public, many curators, especially those in mammalogy, ornithology, and herpetology, found exhibition work particularly onerous. While museum scientists could shunt schoolchildren onto education staffs, dioramas were more difficult to dodge.

Dioramas and other forms of exhibition consumed not only time but also money. Throughout the late 1910s and 1920s, research and publication budgets languished as museum boards and city councils lavished thousands of dollars on the construction of new displays. "The money supporting this institution comes entirely from the taxpayers, and they are mainly interested in what they personally can get out of it," Milwaukee director Henry L. Ward wrote. As a result, it was "a trifle difficult to justify research work" and nearly impossible to persuade tight-fisted city politicians to fund subventions and scholarly publications when the money could be going to build impressive new exhibits.[145] These funding discrepancies upset curators terribly. Publication was the currency of scientific authority outside of museum walls, and museum scientists worried their institutions would come to be seen as scientific backwaters if museums continued to reduce research subventions and replace scientific journals with popular magazines.[146] Curators feared their own professional reputations would suffer as well. But museum scientists did not control institutions' purse strings, and most boards, stacked with local businessmen, refused to spend money on monographs and lab supplies when there were children to be educated and elephants to be exhibited.

Worsening economic circumstances exacerbated this situation. In the early 1930s, the American Museum continued to direct massive funds toward the construction of new habitat halls but decided to publish its scientific monographs, known as *Bulletins*, only sporadically, and abolished the *Novitates*, short papers describing newly identified specimens. When curators protested, the museum's administration pleaded poverty and refused to provide more money for the scientific journals. Even those curators who believed most devoutly in the power of dioramas were horrified and agreed the displays had been unduly prioritized. "We are told by the Administration that there are no funds for publication," Chapman wrote sarcastically, yet "other activities in the Museum . . . apparently are not so handicapped through lack of funds."[147]

Curators faced a similarly demoralizing disparity in hiring. In many museums, the number of staff members hired to build displays grew exponentially while the number of curatorial staff held steady or declined. The scientific departments of larger museums found themselves continually short-handed in the interwar decades. Even at the moneyed American Museum, where the Department of Arts, Preparation and Installation regularly employed more than a hundred people, some of the scientific departments weren't permitted to hire a single new assistant in the 1920s and early 1930s.[148] Curators begged administrators to provide, "proportionally at least, as good facilities for research as there are for exhibition and other educational activities," but their pleas went unheard.[149] Some scientific departments, through clever accounting, dipped into diorama-building budgets to accomplish related scientific tasks, but geology, entomology, conchology, and botany departments—whose collections did not lend themselves to dioramas—did not have this luxury and found themselves in dire straits.[150] The budget of the American Museum's Department of Invertebrate Zoology, already low to begin with, was not increased proportionately with the general budget, and in 1926, the museum's president, Henry Fairfield Osborn, told its curator that he must expect to finance his department exclusively from sources outside the museum—either donations or foundation grants.[151]

Curators demonstrating ingenuity in display found it easier to garner institutional funding for their departments, some of which could then be used for research. Dickerson's aggressive commitment to display, for instance, ensured that the American Museum's division of herpetology received far more funding than it otherwise would have. When she became a curator at the American Museum in 1908, the department had created a few effective displays—her predecessor, William Morton Wheeler, had placed groups of Texas rattlesnakes, iguanas, and mud turtles on exhibition, as well as a plastic cast of leatherback turtles that visitors could marvel at—but most of the herpetological displays did not draw crowds, and so the department's budget suffered considerably compared to ornithology or mammalogy. Dickerson was both competitive and energetic, and in 1910, she announced her intention to move the exhibition of reptiles beyond glass jars into more attractive, realistic realms. Her involvement in the development of revolutionary exhibits allowed Dickerson to hire her own staff members—one of the most tangible forms of administrative backing. Even though Dickerson was a woman, Osborn was so impressed with her exhibition work that they rewarded her with her own department in 1920.[152] By leveraging dioramas, Dickerson established a healthy program of herpetological research in only a decade, taking the American Museum's collection from a small array of alcohol-filled bottles and dried reptiles to the fourth largest in the nation. By 1923, the collection included nearly fifty thousand specimens.[153] Had her displays not put her in such good standing with the museum's administration, she probably would not have been able to do so.

But as budgets tightened further in the early 1930s, even those departments that had previously been successful in piggybacking their research on diorama work began to come up short. The American Museum's herpetological department was forced to depend exclusively on volunteers to care for their ever-growing collection of specimens. And while mammalogy was one of that museum's favorite sons, the scientific staff of the mammalogy department was so skeletal after the first few years of the 1930s, that it could barely "cope with routine," reported curator of mammalogy Harold Anthony. If the department was not permitted to hire more staff, "research will be reduced to zero," he threatened, resulting in "deterioration of equipment and collections" and, even worse from the administration's point of view, "slowing up efficiency in exhibition."[154]

Museum scientists complained that discrepancies in hiring and funding undermined one of their most fundamental tasks: the identification, description, and classification of specimens. The increase in expeditions collecting for displays resulted in an embarrassment of specimens, but with so many boxes coming in and so many field teams going out, curators had almost no time to tend to new acquisitions. Field Museum curator Wilfred Osgood hardly had time to assess the contents of the specimen crates that had arrived from a recent expedition to the Himalayas, as he had his "hands full getting [specimens] on exhibition to please the patrons who furnished the money."[155] Such dynamics infuriated scientific staff. "We have an enormous amount of material absolutely new to science which calls for description and publication" but too few curators to confront it, director and entomologist William Jacob Holland complained in an annual report to the board of the Carnegie Museum of Natural History.[156] "Merely to gather up vast scientific treasures and to leave them boxed and buried in the warehouses of the Museum is not to fulfill the functions of an institution intended for the advancement of science," he angrily concluded.[157]

Constraints on time and money shrank opportunities for fieldwork, and it became increasingly difficult to reap the scientific spoils of expeditions. In the 1910s, it had been standard practice to send curators, taxidermists, and artists into the field to observe and collect specimens.[158] But as museums launched more and more expeditions, curators were often too busy organizing the details to go afield themselves, leaving exhibit makers and professional specimen collectors in charge.[159] "You have my full and deep sympathy as an organizer of expeditions," Frank Chapman wrote Osgood in 1928, "for that has been my chief occupation for some time and under circumstances breeding many complications. . . . All these affairs arranged chiefly in the interests of others."[160] As the Depression descended, some donors became more penurious, restricting expedition membership to include only themselves and the necessary artists, leaving scientists out in the cold.[161] American Museum curator of mammalogy Harold Anthony begged administrators to guarantee that at least one member of the scientific staff would accompany each collecting trip, reminding them that expeditions were necessary to advance scientific research as well as exhibit making. "It is quite practical to collect not only group material but study specimens as well," Anthony protested. "The acquisition of the latter would place no great additional expense upon the man paying for the expedition."[162] Administrators drowning in red ink had no intention of offending donors with such a request and ignored Anthony's plea.

Scientists at ambitious midsized museums suffered the most from new institutional imperatives for display. Though city officials, board members, and the public continued to clamor for more expansive educational programs and more spectacular exhibitions, these institutions could not afford armies of assistants, lecturers, and preparators to ease at least some of curators' exhibition and educational duties. Nor were their boards prepared to fund the kind of expansive collecting expeditions that addressed both scientific and exhibition needs. Though administrators struggled to accommodate curators' desires for time in the field or lab, they had trouble protecting the time of their scientific staff. "If the amount of work, which we have done along [display] lines, is to be *increased* under present circumstances, the inevitable result will be the further falling into arrears in our strictly scientific work," the Carnegie Museum's Holland warned in 1921.[163] More and more, curators argued, scientific work was falling prey to the very educational reforms they had enthusiastically embraced only a few years earlier. Frustrated, scientists began to push back. By the mid-1920s, curators in a variety of institutions began to balk at the expansion of responsibilities that resulted from the new prominence of exhibition and public education in museums—values literally embodied in dioramas.[164]

Investing in Institutions or Individuals?

Museums' growing investment in dioramas and other educational efforts became increasingly divisive, stirring furious, staff-wide debates over institutional priorities. The popular exhibits may have disappointed a growing cabal of research-oriented curators but not the nation's aging museum men, who insisted that dioramas' pedagogical and public value far outweighed their inconveniences. This divergence in opinion signaled an unpleasant coming of age for natural history museums. As there was never enough money or time to accomplish everything, by the late 1910s and 1920s, staff members were forced to consider what activities and investments institutions should postpone—or even sacrifice. The resulting discussion of museums' relative obligations to scientific research and public education struck at the heart of what a museum should be and do and provoked violent argument and huffy resignations. Tensions rose to the point that many curators dedicated to research felt the need to declare their differences from museum men, who remained museums' most vocal advocates for display reform.

Those redoubtable champions of visitor-friendly displays, museum men, firmly maintained that institutions should direct limited resources toward exhibition rather than research, a conviction that had an enormous impact on American natural history museums. By the late 1910s and 1920s, self-identified museum men enjoyed considerable power as directors and department chairs. They enjoyed chummy relations with trustees and wealthy donors. They represented museums to the popular press, helping to shape public understanding of the institutions. They oversaw the publications and programming of the American Association of Museums and attempted to steer professional discussion toward issues of display, education, and administration. Though they continued to value scientific research and believed it a crucial aspect of museum work, museum men were, at heart, Progressive reformers, fully invested in the idea that society thrived when well-trained experts provided educational opportunities for broad publics. Research, they argued, could come later—museums, and the nation they served, needed displays now. American Museum president Henry Fairfield Osborn readily admitted that the museum's curators, especially the museum's mammalogists, had not had a chance to publish much but insisted that displays had to come first. "When the halls are complete, we can push them toward research again," he concluded.[165]

This didn't sit well with scientists, inside the museum or out, who complained that museums had lost sight of one of their guiding missions: the production and publication of scientific knowledge. Biologists working outside the museum routinely sneered at the disparity between funding for display and funding for research. Joseph Grinnell complained in 1921 that attendees at the meetings of the American Association of Museums no longer spoke of science, only of dioramas and related "'big' things like getting more endowments by improved methods of impressing the public and legislatures."[166] As museums issued scientific journals less and less frequently, choosing instead to fund more popular publications, build more habitat halls, and circulate more loan exhibits, museum scientists admitted stomach-churning anxiety.[167] And rightly so, for curators, citing the overwhelming demands of expedition and exhibition planning, had begun to get a reputation in scientific circles for intellectual passivity. "They have access to a superb collection and they ought to publish more," a reproachful Glover Allen, a mammalogist at Harvard's Museum of Comparative Zoology, wrote Osborn in 1931.[168]

Museum directors acknowledged that prioritizing display over research demoralized individual curators and could devalue museums' scientific reputations, but most were unrelenting in supporting outcomes that best served their institutions' public mission over any larger scientific mission. Though he recognized that the elimination of research opportunities resulted in a "soul sickening stagnation that saps the enthusiasm" of curators, Milwaukee Public Museum director Henry L. Ward wrote in 1920, it was hard for him to sympathize with curators begging for narrower responsibilities.[169] After all, Ward had willingly sacrificed his own interests in paleontology for the sake of institution building.[170] Other museum men had made the same choice. "You're too young a man to think about ceasing scientific work. Don't do it!" a friend wrote the American Museum's director Frederic Lucas after hearing he was going to abandon research altogether.[171] It stung to give up hard-won expertise— "I really feel that I know less in natural history than I did when I came here," Ward sadly wrote a friend in 1915—but museum men like Ward and Lucas believed the education their institutions provided benefited the public more than their own contributions as researchers ever could.[172] So they set their jaws and ignored the scientific colleagues who scolded them for failing to keep up their own scholarship or shifting money from entomology research to education and exhibition departments.

Faith that display should take precedence over research, at least temporarily, clashed with curators' periodic assertions that museums should instead prioritize the production of scientific knowledge. Tempers flared as senior administrators accused younger colleagues of pursuing private glory at the expense of their institutional obligation to public education. In 1922, after one of his curators thanked him for fact checking an article for a scientific journal, Lucas privately expressed his irritation with the curator, who had neglected his exhibition cases while writing his article. "Responsibility!" he scrawled on the thank-you note, underlining the word twice. "He should not allow his private interests to interfere with his duties to the museum and the public."[173] Museum men frequently cast colleagues diligently pursuing research as selfish, concerned more with their own career advancement than with the good of the institutions in which they worked. Frustrated directors complained that, all too often, museum scientists found display work "beneath [their] dignity and of very little interest," as Ward put it.[174]

In the face of such attitudes, many museum scientists fled for the more stable ground of universities, laboratories, or government surveys and field stations. In 1916, the Los Angeles County Museum lost its chief ornithologist and assistant director, Harry Swarth, to the University of California at Berkeley's Museum of Comparative Zoology, "a specialized work more to his liking," according to Director Frank Daggett.[175] In San

Diego, curator A. W. Anthony resigned in late 1923 after the board of trustees instructed him to spend the bulk of his time collecting skunks, rats, and rabbits for school loan dioramas.[176] The following spring he sailed for a field station in Guatemala, where for the next five years, he spent "some of the happiest and most interesting years of his career as a naturalist," according to ornithologist Vernon Bailey.[177]

Changing practical and disciplinary realignments in the biological sciences further complicated the fraught coexistence of museums' efforts in research and display. Field naturalists, who had once mocked cytologists' overreliance on microscopes and lack of taxonomic knowledge, now found themselves disparaged by a new generation of biological researchers, who were winning large foundation grants and developing new laboratory- and technology-driven programs in cellular- and microorganism-intensive fields like genetics and endocrinology. Those biologists who continued to embrace morphology and taxonomy would soon die off and risk being labeled "dead Natural History specimen" by their colleagues, joked entomologist and taxonomist William Morton Wheeler.[178] Many researchers working in museum settings intentionally reoriented their disciplinary self-identification, shifting away from natural history toward emerging fields like animal ecology and conservation biology.[179] This didn't help matters. Biological scientists in both newer university-based fields and older museum-based disciplines, Charles Adams wrote in the *American Museum Journal* in 1917, "were inclined to feel that their group only was concerned with 'fundamental' problems," which led to a fracturing of subspecialties in the 1910s and 1920s life sciences.[180]

In the wake of this splintering, it became nearly impossible for curators charged with display responsibilities simultaneously to keep abreast of all research developments in their field. "Full justice can rarely be done to both these types of museum work by one person at the same time," Chapman later reflected.[181] Chapman admitted he had intentionally divided his career in two parts, devoting the first twelve of his fifty-four years at the American Museum to the development of its exhibition halls. Only after serving exhibits, he later wrote, did he feel he had won the right to concentrate "with equal intensity" on research.[182] But not all scientific curators could compartmentalize as easily and patiently as Chapman; even those who firmly believed in museums' commitment to display reforms resented the erosion of their institution's research culture.

The story of Frederick Charles Lincoln, a young curator at the Colorado

Museum of Natural History, demonstrates how a curator's commitment to scientific research could clash sharply with the new organizational realities in an age of reform. Lincoln had arrived on the steps of the Colorado Museum in 1911, armed with a lot of enthusiasm for local birds and a little experience in preparing their skins for study. Though hired as an assistant to the museum's surly German taxidermist, Lincoln dreamed of becoming a curator of ornithology. A curatorship wouldn't make a person rich, but to Lincoln, it promised the chance to see his name in print, to spend time in field and lab studying the birds he so loved, even to travel to far-off lands. As one of the Colorado Museum's curators gleefully observed, curators got to do for a living what rich men paid to do on their vacations.[183]

Lincoln quickly found a mentor in Colorado's museum director and former exhibition director at the American Museum, Jesse Figgins. Figgins, described by well-respected curators as "the best all-round museum man in the entire country," began advising the young man, providing him with relevant books, and sending him into the field to collect on his own.[184] The director became quite fond of the gangly teenager—so fond, in fact, that within three years Lincoln had become the museum's curator of birds and mammals. If Lincoln worked hard, Figgins wrote a friend, in a few years he was sure to run a museum of his own.

But Lincoln, eager to make a name in ornithological circles, spent his days conducting and writing up field research and avoided preparation and display building. At first, Figgins tried to be accommodating, reassigning some of Lincoln's exhibition responsibilities, even taking over the curatorship of mammalogy himself so Lincoln had time to publish in ornithological journals. "In a sense I very much regret the museum's inability to permit your devoting your entire time to the branches of work you have repeatedly expressed a desire to engage in," he wrote Lincoln. At the same time, Figgins warned, scientific specialization had its perils, for it was "distinctly narrowing and in a sense detrimental" to the museum's mission of general education. Lincoln needed to acknowledge that museum work entailed far more than ornithological research, Figgins admonished, and should turn his attention to the display and preservation of the collections under his care.[185]

Yet the young curator stubbornly continued to avoid exhibition work, which Figgins found intolerable. Lincoln's rapid promotion to curator, the director complained, had "vitalized a form of egotism that was to ultimately take possession of him." "Never for a moment," Figgins wrote, "did he abandon the belief that he should be 'publishing something' and frequently compared his work and position with that of Dr. Chapman" and other well-known ornithologists.[186] Frustrated, Figgins put it frankly to Lincoln: he was no Frank Chapman, and the Colorado Museum was never going to be the American Museum. "The primary object and item of vital importance to this institution is exhibition," he lectured the young curator. "Each of us must be prepared at all times to sacrifice our personal preferences to this."[187]

Exhibition work was no less dignified than research, the Colorado director informed the rebellious Lincoln; in the eyes of the public, it was far more significant. "View the opportunity of assisting in the preparation of groups as a privilege—as I personally do the part I take in them—rather than consider it a hardship," he suggested to the young man. "While I prefer the white collar and carefully brushed clothes I glory in my old coat that is splashed with clay, plaster and paint and am not the least disconcerted when found in it by my personal friends, museum officials or casual visitor."[188] Figgins and Lincoln's relationship buckled under the tension of their competing ideas about the proper goals and responsibilities of a museum curator.[189] In 1919, after considerable conflict, the director tried to replace Lincoln. The young curator fled to the U.S. Biological Survey before he was fired, taking a position that ensured he would never again have to sculpt, plaster, or paint an exhibit.[190] In his place, Figgins hired Alfred Bailey, a naturalist and museum man who believed "fieldwork is the lifeblood of natural history museum" but for whom research and publishing held little appeal in comparison to collecting and display.[191]

Not every museum scientist was a Lincoln, nor was every director a Figgins. Chicago Academy of Sciences director Frank Baker, appalled by his board's insistence on funding and creating elaborate visual display to the exclusion of all other activities, had resigned from his museum the previous year. "The Academy has become a factory," he wrote, with "scientific work all cut out."[192] Yet many curators, Chapman chief among them, continued to celebrate dioramas, describing them as essential contributions to public education. But these same curators also believed wholeheartedly that scientific research should remain the spine of museum work, rather than an adjunct to what American Museum curator Clark Wissler acerbically called "the educational hobby in museums." An excessive fascination with dioramas had upended museums' tenuous equilibrium of education, preservation, and research, curators argued, and proper balance between these missions had to be restored. Staff members who believed natural history museums should remain centers for scientific research feared, as Wissler did, that museums increasingly resembled kindergartens, not universities.[193]

By the late 1920s, display-oriented museum men and research-oriented curators eyed one another with great suspicion. "The term 'Museum Man' [is] by no means synonymous with Curator," Lucas sniffed in 1933, "for not every Museum Man is a Curator and many a Curator is very far from being a Museum Man."[194] Curators with active research programs began to use the phrase "museum man" with as much sarcasm as admiration. In a few museums, wary scientists organized to register their complaints more formally. The Councils of Scientific Staff at both the Philadelphia Academy of Natural Sciences and the American Museum of Natural History established special committees to defend their interests against the encroachment of exhibition responsibilities.[195] Over the next ten years, committee members met regularly to discuss matters of display and bemoan administrators' seeming obsession with dioramas. The presence of such organized dissent on the institutional landscape marked the increasingly combative relationship between research scientists and exhibit makers in natural history museums—a new dynamic that would reshape not only the internal culture of work within museum walls but also museums' broader cultural identity.

Dioramas remained unambiguously popular with visitors, drawing new audiences to museums. But their enormous scale and increasing complexity undermined the sense of the shared—or at least complementary—mission that had allowed them to proliferate in the first place. Though most directors still wholeheartedly believed museums should have superior research, display, and education programs, in an era of dioramas, the realities of limited resources made it impossible to balance these three goals evenly. Though dioramas most frequently raised debate over what museums should value most, the iconic displays were merely one trigger among many in these years. The question of museums' ultimate objectives also arose when allocating funds for school loan programs, institutional publications, or casts of dinosaur bones, when deciding whose needs counted most on collecting expeditions, when determining staff salaries and titles. But the central question remained the same: which mission should museums promote most diligently? Staff opinions varied from person to person, department to department, museum to museum. The question would linger, unanswered, for the next three decades, resulting in increasingly heated arguments, experimentation, and ultimately, major changes in the very idea of the museum.

THREE

Displays in Motion: Experimentation and Stagnation in Exhibition, 1925–1940

For years, curators had listed the ways exhibits could impede museums' contributions to science, but in the late 1920s and early 1930s, they also began to point out that exhibits could hinder museums' efforts in public education. Habitat dioramas and dinosaur skeletons might be popular, these scientists maintained, but they failed to convey the concepts most important to the modern life sciences.[1] A cluster of curators at the American Museum of Natural History became so concerned about the issue that, in May of 1928, they called a meeting of the museum's scientific staff to confront it. "The day has passed when museums should exist primarily for the preservation, study, and exhibition of dead animals," they declared. If the museum was serious about public education, it needed exhibits more in "line with contemporary biological investigations in the laboratories of universities and other foundations for the increase of knowledge."[2] Unsettled by the criticism, the museum's Council of Scientific Staff organized a Biological Education Committee, an Advisory Committee on Biological Aspects of Museum Groups, and a series of other groups to discuss the matter.[3]

The American Museum's curators were not alone in their anxiety, and debates about museum displays of the life sciences swelled in these years. Across the nation, a small but vocal collection of critics suggested that many current exhibits should be retired, or, at the very least, significantly altered, because they were not teaching modern science. Only when a museum committed to modernizing the content and pedagogical approach of its displays, admonished paleontologist and former Colorado Museum curator Harold J. Cook, could it "make a real *educational institution* out of merely an exceptionally effective *show*."[4] Museums should not lean so heavily on specimens and spectacle, scolded AAM director Laurence Vail Coleman in 1931 and should instead "exhibit subjects, not objects."[5] Though museum workers were some of the most vehement detractors of existing displays, educators and businesspeople, journalists and foundation officials also began to reevaluate museums' life science exhibition in this era, offering new perspectives on the old problem of how to educate Americans about nature and science.

At first glance, interwar display critics' motives for improving public understanding of biology varied as widely as their workplaces and worldviews. But most saw the popularization of science as a matter of civic necessity, and they reached similar conclusions about the content of museums' life science displays. Exhibits, critics agreed, should explain modern biological and ecological concepts and should highlight how recent applications of the life sciences affected Americans' daily lives. This vision, loose as it was, guided the clutch of reformers striving to modernize museum display in the 1920s and 1930s. But there was no such consensus about the most effective way to convey this information to visitors. Some critics claimed that independent, appreciative observation remained the best way to learn. Others asserted that more detailed labels and lengthy wall panels would help. A few influential voices suggested more visually dynamic displays, which would beep, whirl, or light up when visitors pressed their buttons. Without a clear pedagogical plan, reformers simply experimented, sometimes trading ideas but more often innovating in isolation. Unlike the museum men, who had used carefully cultivated professional networks to breed widespread support for new approaches to display, most interwar reformers had no interest in developing a national movement to promote particular exhibitionary methods. They just wanted to bring interesting biological displays to their own institutional corners. But as they tested new ways of visualizing the life sciences, they redefined what a museum exhibit—and what a museum—could be.

Consequently, fragmentation and experimentation, rather than consolidation under a single set of new best practices and professional ideals, characterized museum display in the late 1920s and 1930s. The era's dire economy frequently thwarted proposals for reform, and attempts to modernize museum displays often succeeded only thanks to the persistence of an involved patron or a powerful curator. The changes in exhibitions and insitutions that resulted were local or temporary, hardly the national tidal wave of reform that had swept museums only a decade before. But these fifteen years were hardly stagnant, for the fledgling institutions that developed and the exhibitionary experiments conducted during this time would set the agenda for museum display over the next thirty years.

Agitating for Change in Displays of Life

"Science is advancing more rapidly than ever and is more quickly applied to the needs of life. But the scientific habit of mind is not common or commonly respected," Edwin Slosson, director of the newly founded Science Service, wrote in a 1922 issue of the *Century*. "Science, too, needs to be democratized and brought within reach of the many, not as a task forced upon children, but as a lifelong recreation."[6] Sharing Slosson's sentiments, interwar scientists and journalists, broadcasters and film directors pushed stories about science across the pages of popular magazines, through the grill cloths of cathedral radios, and onto the screens of moving pictures.[7] Corporations and organizers of the nation's largest world's fairs presented science as a form of salvation, the blueprint for building a better tomorrow.[8] Philanthropic foundations doled out grants to promote efforts to inform the public about the life sciences—heredity and genetics, more specifically.[9] By the late 1920s, critics of museum display had joined the chorus of voices clamoring for greater science appreciation in America.

Like so many other popularizers, interwar display critics believed that the health of the nation depended on citizens properly educated about science.[10] To be prepared "for the role they have to play as members of a self-governing democracy," wrote Peabody Museum of Natural History director Albert Parr, Americans needed to understand basic biology.[11] Progressive Era museum reformers had voiced similar views, but the biological and chemical horrors of World War I, the mysteries of relativity, the Jazz Age explosion of consumer technology, and the environmental devastation of the 1930s gave old platitudes new power. And like their Progressive forebears, these critics argued that the onus to learn did not rest exclusively, or even primarily, on visitors. Rather, they agreed, museums needed to hold themselves more accountable for educating the public, and museum staff needed to design displays that taught visitors what they did not know about the life sciences. In an age of untrammeled confidence in science and uncertainty about nearly everything else, natural history museums had a special "obligation to furnish up-to-date exhibition and education" about the life sciences, insisted American Museum trustee Perry Osborn.[12] And growing public "interests in science will lead to increasing demands for educational assistance from the natural history and science museums," predicted high school biology educator Benjamin Gruenberg.[13]

Yet beyond this commitment to public science education and their disdain for existing displays, critics had little else in common. Though their backgrounds had varied widely, Progressive Era museum men had been a tight-knit group, sharing duplicate specimens and new taxidermic techniques, swapping stories about fusty colleagues and fussy visitors. Interwar critics, an equally diverse lot, never became a self-conscious group, for their professional paths and personal motives for improving displays rarely intersected. Their ranks ranged from young designers whose work on world's fairs had convinced them that museums' existing exhibits were hopelessly outdated to graying museum reformers, eager for a last chance to realize their rosy vision for American museums. Some worked in museums, but more than a few were already engaged in communicating scientific ideas to the public through philanthropy, journalism, or classroom education; to them, museum display was simply another head of this hydralike challenge.

A few of the nation's leading museum scientists were among the most outspoken of these critics, their interest in display galvanized by public controversy over evolution.[14] Antievolutionists had assailed museum exhibits in the bloody battles over human origins in the early 1920s, demanding they provide more concrete proof of the link between humans and apes.[15] "Is it not more rational to believe in creation of many by separate act of God than to believe in evolution without a particle of evidence?" William Jennings Bryan had asked.[16] Displays in American Museum of Natural History's Hall of Man came under special fire.[17] Though the museum's paleontologists stoutly defended their exhibits' worth, the public's obvious difficulty perceiving displays' underlying scientific concepts, as well as the difference between scientific proof and evidence, got under curators' skin. Many began to wonder if displays they considered quite lucid were actually opaque to visitors, bewildering rather than educating. Creationists' disparagement of the Hall of Man's displays so horrified paleontologist William King Gregory that he asked director George Sherwood for permission to alter the gallery's exhibits, in order to make them "absolutely beyond criticism in clearness and accuracy of arrangement."[18]

Improving displays, believed such curators, could also help to counter criticisms by an altogether different faction: biologists working outside museums. As the scrutiny of specimens in storage rooms had given way to the observation of living animals in the field and the exploration of animal bodies in laboratories, collections work had become—rather unfairly—stigmatized. As a result, early twentieth-century museum taxonomists frequently suffered from their association with what many university and lab-based biologists believed was an obsolete, superficial approach to the life sciences; Charles Adams complained his colleagues often dismissed the research of curators as mere "trifles."[19] Many scientists outside the museum still agreed with geneticist Thomas Hunt Morgan's flat 1910 pronouncement that museum curators "got too much money for not doing very much"; in 1927, American Museum board president and curator Henry Fairfield Osborn confided to a fellow paleontologist that well-known biologists regularly informed him that "what we are doing here at the Museum is an almost criminal waste of money."[20]

But such characterizations of museum-based research were outdated.[21] By the mid-1930s, older forms of basic classification had given way to what biologists called the "new systematics." Museum scientists who once would have limited their work to naming and describing species now used new taxonomic practices to investigate species' evolutionary histories and environmental adaptations. Taxonomy, announced British biologist Julian Huxley, "was no longer a specialized, rather narrow branch of biology, on the whole empirical and lacking in unifying principles." Rather, it had become what he called a "focal point" for new frontiers of inquiry.[22] Systematists eagerly ventured into experimental science and drew more heavily on the tools of field and laboratory biology.[23] Throughout the 1920s and 1930s, curators in museums with active research programs asked their boards to fund experimental field stations and well-equipped laboratories, or, at the very least, in-house aviaries and terraria.[24] They coauthored studies with colleagues in university departments and scientific agencies, collaborations that often resulted from friendships forged through the trials of graduate seminars, lab work, or collecting expeditions.[25] And their contributions to the life sciences were far from peripheral. Some helped to frame important ecological conversations over the distribution and development of species over time and space, while others worked to promote the emerging discipline of ethology by investigating the physiological and psychological mechanisms that unified all animals: locomotion, hormonal changes, social behavior, perception, and psychology.[26] The modern evolutionary synthesis was well underway, and museum curators actively participated in its formation and development.[27]

Yet museum displays implied that curators were trapped in a nineteenth-century state of mind. Whether organized by department, collection, or region, the halls of most interwar natural history museums remained a conceptual and visual hodgepodge. Visitors could wander through mazes of specimen cases, group displays and dioramas, and isolated marvels: a reconstructed *Diplodocus* skeleton, the world's last passenger pigeon, a wax model of a mosquito, its actual size magnified a thousand times.[28] As "it has been the general practice of most natural history museums to strive for a degree of permanency that leaves them in large proportion out of date at all times," complained Albert Parr and other critics, exhibits reflected the pedagogical and biological preoccupations of an earlier generation of curators.[29] And new exhibits tended to be governed by these older paradigms. In the American Museum's Hall of the Birds of the World, which had opened in 1925, ancient stuffed birds, arrayed in red mahogany cases still took up considerable space. Its labels "couldn't be read without a flashlight," recalled background painter Francis Lee Jaques, and stuck to Latin names and dates of collection, providing neither distribution maps nor any explicit information about birds' relationship to the geographical features of their habitats—a missed opportunity, given curator Frank Chapman's considerable contributions to scientists' understanding of the impact of life zones on the evolution of birds.[30]

Impatient to promote their findings to fellow biologists and the broader public, many curators pressed for displays that presented modern museum science. Museums' exhibition and research pursuits had moved too far from one another in recent years, they protested, and the public would be far better served if they were better synthesized. After all, vital museums produced "a constant stream of new discoveries, new scientific material, new exhibits," Gregory explained, so displays should capitalize on museums' ongoing research.[31] "The exhibition value of many recent discoveries in natural history can scarcely be overemphasized," concurred Gregory's colleague, Robert Cushman Murphy. Museum display, Murphy suggested, should feature curators' recent work in ecology, animal behavior and intelligence, genetic mutation, biochemistry, and a dozen other subjects.[32] This kind of self-promotion, wrote American Museum entomologist Frank Lutz, was absolutely crucial to improving the reputations of curators and the institutions in which they worked: though "scientists are notoriously poor advertisers . . . 'it pays to advertise.' " The public would also benefit, he promised. After all, "science depends largely on museums to tell the general public about its wares."[33]

Whether they worked inside museums or far beyond their halls, whether they were driven by public duty or a desire to shore up their professional credibility, interwar display critics shared a vision of what topics museums' displays of life science should address. Modern biological concepts topped the list: germs and genes, cells and structures, hormones and homeostasis.[34] Building on museums' long-standing tradition of exploring "economic nature" and public health, display critics also pushed museum scientists and exhibit designers to create more "exhibits of the moment," in Murphy's words, by focusing more on agriculture, medicine, and other recent applications of the life sciences.[35] (This interest in applied science corresponded with larger trends in science popularization. Interwar educators, cultural critics, and journalists often pushed scientists to explain how biology was "organized and presented in its bearing upon [human] action," as John Dewey put it at an 1934 American Association for the Advancement of Science meeting.)[36]

The biology of humans seemed an especially fertile field for exhibition.[37] Museums had toyed with issues of human health and hygiene in the past but had usually broached the subject obliquely, exploring it through animal behavior and references to human ancestors.[38] But as lingering taboos about public discussion of the human body and its physical functions eased in the 1920s and 1930s, critics urged more detailed and contemporary discussion of the human organism in natural history museum halls. Critics considered displays on the previously forbidden subject of human reproduction especially important. "Progress in this direction," wrote Buffalo museum educator Carlos Cummings, "is essential if museums, devoted to displays illustrating matters connected with the promotion of health and proper living among the people, are to become a recognized and accepted factor in our cultural complex."[39] Spurred on by Scopes, critics also implored museums to expand their exhibits on the history and science of human origins to feature evolution's ugly stepsister, eugenics. Pressure to build or expand such exhibits met with increasing resistance from museum anthropologists and a younger generation of biologists, who resisted the exhibitions' racist overtones.[40]

Interwar display critics pressed for more exhibits on ecology and its applications as well, for the subjects intrigued both scientists and laypeople. Biologists' interest in ecology mushroomed in the 1920s and 1930s: the Ecological Society of America more than doubled its membership during the 1920s, and the term "ecology" appeared in the titles of more than fifty dissertations and three hundred books.[41] Popularizers and policy makers, struggling to explain diseased plants, dead grass, and dust storms to desperate farmers, also stoked interest in ecology.[42] To feed public hunger for the topic, critics argued, museum exhibits needed to explore humans' complex relationship to the environment, rather than assiduously erasing their presence. Critics with a conservationist bent most fervently advocated this approach.[43] Though they firmly believed habitat dioramas had helped build popular support for preservation and conservation efforts, many argued that reproductions of untrammeled wilderness, however beautiful they might be, represented a dated approach. Rather than "distract with the lions of Africa," Parr wrote, museum displays needed to explain "dust bowls and Japanese beetles, Dutch elm disease and fisheries that fail or drown his economy in an unpredictable surplus, a thousand and one other natural hazards to . . . individual comfort, recreations and livelihood, and to the national welfare."[44]

Finally, display critics agreed that museums needed exhibits that gave visitors more insight into modern scientific processes. Specimen displays ordered by Linnaean category had conveyed something about the methods and findings of nineteenth-century taxonomists but hardly represented the diverse processes and practices of modern life scientists. "Men and women can get from the museum something more than a solemn awe or a hushed reverence for the marvels," wrote biology educator Benjamin Gruenberg. If exhibits were to rouse real appreciation for science, he contended, they needed to depict scientific "meanings, sources, workings, developments, and curiosities about how problems have been solved." Only by presenting science in action could the museum "become a democratic educational institution to which people will turn."[45] Critics like Gruenberg insisted that museums had a responsibility—to themselves, to science, to the larger nation—to help the public understand and appreciate the slow methods of research and experimentation. (Curators critical of displays had a personal stake in the matter, believing that such concerted efforts to publicize scientific process were necessary to ensure meaningful public support for their work.)

In the same spirit, critics maintained that museum displays should engage visitors in the unanswered questions with which scientists regularly wrestled. When it came to such messy scientific topics as the mechanics of natural selection or human evolution, "even an imperfect and temporary explanation or hypothesis" would be a better focus for museum exhibits than the minimal but straightforward facts that displays currently imparted, Gregory editorialized in a 1936 issue of *Science*. This, he believed, would at least allow visitors some insight into ongoing scientific debates and evolving methodologies, rather than boring them stiff with carefully edited, frequently obvious certainties. Unfortunately, he lamented, "museums of natural history have been very slow to give this principle a fair trial," a decision that, he implied, resulted in ignorant visitors and frustrated scientists.[46]

Unsurprisingly, critics' ideas about exhibition content overlapped with school science educators' ideas about curricular content.[47] Between the 1880s and the 1920s, educators had dramatically reconceived science education, arguing that its purpose was not only to develop students' intellectual skills but also to develop individuals who could contribute productively to American society.[48] As the secondary school population ballooned in the 1920s and 1930s, educators tried to serve this diverse new group of students by providing socially relevant scientific content, shaping curricula around sciences' applications to social ills, vocational preparation, and "worthy home membership."[49] Interwar biology teachers, aware their subject was often the only science students took during their high school years, focused on topics like nutrition, the prevention of sexually transmitted disease, eugenics, and the biochemistry of yeast— a curriculum that educators dubbed "civic biology."[50] This cluster of subjects, which would become standard in textbooks during the interwar period, reflected a persistently conservative belief that biological science, and especially ideas about evolution, could be used to help guide human physical and cultural progress.[51] Museum critics' interest in applied biology resonated with these curricular trends, as did their commitment to presenting science in action, for interwar educators were likewise working to steep students in the scientific method.[52]

While there was a consensus on the subjects that exhibits should convey, no such agreement existed when it came to deciding exactly how this information should be presented. Opinions of interwar display critics diverged dramatically on the topic. Some critics declared that new content demanded new modes of display and insisted that museums needed to improve or even abandon specimen cases, isolated objects, and habitat dioramas. Stimulated "by reading, by broadcasting, by the movies, by travel," American museumgoers had become more sophisticated, wrote Gruenberg. "Provincial as our vast and heterogeneous population may remain in many respects, most of us no longer . . . stare in amazement at a stuffed ostrich in a glass case."[53] Museum displays, he maintained, should evolve accordingly. Other critics protested that existing display types worked just fine, that only their contents and organization required modernization. Though he was more intent on improving natural history museums' exhibits than most, curator Robert Cushman Murphy countered Gruenberg's point, arguing that it was "the *things*—the sheer mass of evidence of existence" that made museums distinct and special, and museums would be ill-served to abandon them.[54]

A fault line in ideas about the kind of education museum displays should provide lay beneath this debate. A handful of older display critics, many of them curators trained in the 1880s and 1890s, hewed to the long-standing notion that museums should help visitors learn how to look, so displays should encourage the kind of stilled, skilled observation so important to naturalists and field scientists. But educational priorities had shifted since the nineteenth century, and by the late 1920s, fewer people believed museums should continue to focus on developing observational skills. Most display critics were happy to set aside the singular virtues of object-based pedagogy if it meant representing scientific concepts in more explicit, more entertaining terms.

Display critics outside the museum world suggested museums develop more "dynamic" displays, the period's catchall term for displays that invited visitor activation or moved of their own accord. Based in retail, media, and entertainment, unconstrained by institutional tradition or concern for scientific dignity, advocates of dynamic display tended to approach pedagogy as a sales transaction. Museum exhibits, they argued, needed to appeal to distracted customers enough that they would stop to buy the scientific facts offered, and motion offered such an attraction. Evidence of motion's power was all around them, they pointed out—just look at the fascinations of contemporary advertising and moving pictures.[55] Science journalist Waldemar Kaempffert dryly noted that anyone familiar with human behavior could appreciate the importance of motion in a display. "Crowds will watch a machine produce cigarettes by the thousands an hour, but no one will give it more than a passing glance if it stands idle," he wrote. "There is nothing so dramatically instructive as action," he added, and museum exhibits should embrace the era's "tendency to vivify teaching."[56] Others preached the pedagogical power of dynamic displays that required visitor interaction. Julius Rosenwald, a Chicago philanthropist and former president of retail giant Sears, Roebuck & Company, began to tout the approach after a 1911 family trip to Munich, when his eight-year-old son, William, begged his family to return to the visitor-activated exhibits at the Deutsches Museum again and again. The retail magnate also noted the appeal of such displays at world's fairs as well, recalling his own absorbed attention to similar exhibits at Chicago's 1893 exposition. Intrigued, he urged interwar museums to create more exhibits along these lines.[57]

Rosenwald had his finger on the pedagogical zeitgeist, for recent research revealed that both children and adults retained information better when they did more than look. In response, classroom teachers and curriculum designers experimented with hands-on learning, and by the late 1920s and 1930s, the educational establishment had reached a consensus that, for students of all ages, group projects, class discussion, and tactile exploration were just as important as, and often more effective than, more traditional techniques of lectures and quiet observation.[58] A few curators and museum educators slowly reached similar conclusions. "The case for tactile education in museums grows stronger every day," *Museum News* reported in 1926. "It becomes increasingly evident that what people touch becomes a part of their personal experience more completely than anything they merely look at, especially through glass. Even the chance to press a button to start a piece of mechanism . . . relieves the sense of frustration due to the checking of Nature's most active means of individual research."[59] Contemporary psychologists, wrote museum educator Carlos Cummings, "tell us when an observer voluntarily participates in a performance and becomes a part of it, he is to a very much greater degree interested in the outcome, which he unconsciously assumes he personally is responsible for and consequently is proud of."[60]

In retrospect, the era's ongoing debates over life science display centered on museums' ability and responsibility to respond to dramatic changes in interwar science and public education. In the 1900s and 1910s, many displays, some costly and labor-intensive habitat dioramas among them, had been installed with the assumption of that they would permanently occupy museums' halls. But these displays, lauded at the time for their modern approaches to biology and education, seemed outdated only a decade later, compelling critics to confront the problem of rapid obsolescence. How could museum displays keep up with the rapid output of knowledge generated by researchers in laboratories, universities, and even their own campuses? How could they command interest in a world increasingly filled with joyfully blinking, spinning, and shifting amusements? Should they? Such questions were further complicated by the inability of museums to confirm what or whether visitors learned from displays—as *Museum News* asked in 1924, "who can say what they think, or whether . . . they think at all?"[61]

Most display critics remained just that—critics—but a few took action and attempted to modernize museums' stolid halls even as the nation slid into the financial quicksand of the Depression. Unlike the earlier generation of museum reformers, these reformers did not coalesce into a proudly self-conscious group, for they never developed or agreed on a uniform set of practices that would allow institutions to realize their ideas. Instead, would-be display reformers began to experiment in isolation, relabeling, tinkering with existing exhibits, and, in a few instances, launching entirely new approaches.

Ethology in Action: Experimental Displays at the American Museum

At the American Museum, an institution renowned for its commitment to state-of-the-art display, a small group of reformers conducted some of the interwar era's most notable experiments in natural history museum exhibition. Though small-scale experiments in display reform could be found throughout the American Museum in the 1920s and 1930s, two curators interested in animal behavior and psychology—entomologist Frank Lutz and herpetologist Gladwyn Kingsley Noble—helped pioneer these efforts. Seeking what Gregory had described as "perfect cooperation between the producers of scientific values and the teachers thereof," Lutz and Noble designed and developed dynamic exhibits based on their own research agendas and a shared conviction that motion was integral to life science education.[62] Despite public praise for their innovations, their experiments failed to thrive in the harsh financial environs of the 1930s, and their reforms had little immediate impact on the hushed halls of the nation's natural history museums.

Throughout the 1920s and 1930s, display critics at the American Museum believed the biological study of animal behavior was a perfect subject for exhibition. Most of the museum's scientists had grown up roaming field and forest, quietly watching birds, bugs, and beasts going about their business, and they knew well the powerful interest animals could arouse. Visitors, they agreed, would be far more likely to pay attention to complex biological concepts if they could be related through the exploits of animals. What's more, museum scientists' ongoing research often involved animals' life histories, and they believed their displays to be the perfect template for presenting new ethological findings.

But critics bickered over how best to display this content. Some continued to assert that habitat dioramas were the answer, maintaining that the protested exhibits vividly illustrated important biological, ecological, and ethological concepts.[63] Mammalogist Harold Anthony pointed out in 1935 that dioramas were "filled with expressions of biological principles; that the biology was there if the visitor cared to see it; that labels devoted specifically to biology could point out this fact"[64] "Each one," ornithologist Frank Chapman concurred, "might serve as a text for a paper on the home-life of its species."[65] Other curators disagreed mightily, arguing that dioramas were a terrible way to present up-to-date ethological findings, or even basic biological principles. Dioramas concealed scientific concepts "under a vast welter of accurate details," declared Gregory, and these "beautiful but scientifically innocuous . . . centerpieces" provided neither clear facts nor directed educational messages.[66] Lutz split the difference, maintaining that dioramas could be useful if supplemented with other kinds of visual experience. Dioramas offered visitors valuable information about insects' "environment, about the plants on which they feed, and about the animals that feed upon them," the entomologist noted, but they needed to be accompanied by exposure to living specimens in the field if they were to be truly effective teaching tools. "To understand and thoroughly appreciate the lives that insects live," he wrote, visitors had to watch them creeping, buzzing, and veering. Just as important, visitors needed to "sense the light, the shade, the moisture, and the temperature that go to make up insects' habitat."[67]

Noble reached a similar conclusion but based his opinion and his proposals for supplementary materials on entirely different grounds. "No teacher ever taught biology merely by showing his students beautiful scenes in Nature, so how can we expect to teach from the habitat diorama alone?" he asked in 1929.[68] Among both naturalists and experimentalists in the life sciences, he wrote, "there is a growing tendency to look beneath the surface . . . [to] seek out basic causation with the aid of advanced biological research methods." So while "we can enjoy Nature just as we can enjoy our 1930 model without bothering to look 'under the hood,' in the progressive museum, just as in the progressive automobile show," exhibits should not be "confined to exteriors."[69] Instead, he suggested, curators should build exhibits that allowed the public to stand over their shoulders, peering, as the scientists did, into Nature's hidden operations. To accomplish this, dioramas required additional illustrations and diagrams, large explanatory panels, and perhaps even technological wizardry in order to "dissect and analyze Nature in such a way that the public will understand the principles controlling the life of the creatures portrayed."[70] With some imagination and a lot of tinkering, Noble promised, dioramas had the potential to capture "the dynamic activity of the organism while elucidating underlying biological processes and concepts."[71]

Ultimately, the two curators used the diorama form as a springboard for new kinds of display. Lutz began to improvise on dioramas in the mid-1920s. The leading American entomologist of the first half of the twentieth century, Lutz had cut his scientific teeth researching *Drosophilia* genetics at Cold Spring Harbor in the early 1900s. He was a naturalist at heart, however, and firm in his belief that insects were best studied in their natural habitats, so he left the laboratory and its "pesky little beasts" in 1906 for the greener fields of the American Museum of Natural History.[72] At the museum, the hard-driving entomologist was both fox and hedgehog when it came to studying insects. He investigated the flight, sounds, visual perceptions, and geographical distribution of a wide variety of insects, though he always maintained a special fondness for crickets and bees. Lutz published widely on the behavior of honey and stingless bees, the causes of insect diurnality, and bacterial diseases harbored by insects, all subjects of great interest to agriculturists. Though he published foundational research in what would become the field of ultrasonics, he was perhaps best known for his contributions to the larger field of pollination ecology, proving that insects were less interested in flower color than in petals' ultraviolet bull's-eye patterns, which were imperceptible to the human eye.[73] Where his early lab colleagues had described him as "proud and haughty," years at the museum and a happy home life had softened him, so by the 1920s, the bespectacled entomologist had become downright affable. And he was determined to excite the museum's visitors about insect biology and behavior—his own research included.[74]

This interest in popularizing entomology led to ongoing tinkering with the museum's insect exhibits and, eventually, its Hall of Insect Life. Lutz had little patience for taxonomy—he often bragged that he had never named an insect—and he had long pushed to incorporate concepts central to systematics and ecology into exhibit organization. As early as 1916, he had reorganized the department's neatly ordered display collections on the third floor into "bionomic exhibits . . . which illustrate biological principles" and had announced his intentions to create a hall that explored insect behavior, his particular passion.[75] When the Hall of Insect Life finally opened in 1925, it was not a visibly striking departure from older halls, containing mostly specimen cases and dioramas. But its organization offered a very different conception of how museums should present entomology and the broader study of animal life. "Instead of presenting a mere Noah's Ark assemblage of species, two by two, male and female, such as had been the custom of the past," recalled colleague Herbert Schwartz, the hall's exhibits were designed to illustrate the biogeography of species' habitats and other crucial ecological concepts. These, Lutz believed, would help visitors understand the broader significance of insect life.[76]

Lutz also used the Hall of Insect Life to experiment with new exhibition pedagogy. Specimen cases and dioramas could not fully capture insects' "interesting habits or novel solutions to the problems of living," he wrote, and they failed to fire visitors' imagination, for "it is mighty difficult to make dead insects look happy on or under a sheet of celluloid water."[77] Lutz believed he could give visitors a more meaningful knowledge of animal behavior, and a more entertaining experience to boot, if at least some of his displays moved—preferably because they featured live insects.[78] (True to his training as a biometrician, he relied on quantitative evidence to reach this conclusion. He roamed the museum's galleries with a stopwatch to determine which displays commanded the most interest and counted up the crowds crossing the otherwise neglected entomology hall to see its Children's Nature Corner, a menagerie of live snakes and small mammals supervised by nature enthusiast and Babbit Soap heir Benjamin "Uncle Bennie" Hyde.)[79]

Lutz knew he could not rely on the museum to sponsor such experimentation—when it came to funding, the department was, he wryly observed, "a Cinderella waiting for her prince"—so he improvised.[80] He began by re-creating insects' habitats outside of dioramas, putting water and plants into bowls, then dumped dozens of live aquatic insects inside them, and encouraged visitors to inspect the ways the bugs made use of surface tension to scoot about. Pleased by museumgoers' response, Lutz next exhibited a wire cage of "trim, up-on-their-toes cockroaches" and, he recalled, "even New Yorkers stopped to gaze."[81] He received small external grants to continue his display experiments, introducing a beehive and an ant colony into the hall to illustrate aspects of insects' "exceedingly active life," as he put it in 1929.[82] He miked the hall's case of crickets so visitors could listen in, and persuaded administrators to broadcast the insects' musical talents over the museum's public address system. (The puckish curator also pointed out that the crickets were sopranos, chirping a full two octaves above high C and that the frequency of their chirps could be used to ascertain the temperature.)[83] Despite his willingness to experiment with display pedagogy, he retained a measure of scientific conservatism when it came to content. Lutz refused, for instance, to accede to Osborn's request that he revise the labels of the hall's few dioramas to address evolutionary themes on the grounds that too much popular discussion about natural selection and other details of evolutionary adaptation was based on "human imagination," rather than verifiable fact.[84]

Lutz's other attempts to turn dioramas inside out, placing visitors inside the natural habitats of insects instead of just beyond the glass, were more radical. In October of 1924, after a careful survey of the region, Lutz asked museum administrators to help him establish a ten-acre field station for the exhibition and study of insects on a well-traveled route near Tuxedo, New York, about ten miles north of Manhattan. Once Lutz informed them that ardent butterfly collectors and wealthy locals would float the site's purchase and maintenance, Osborn, Sherwood, and Murphy threw their support behind the plan. The museum's Station for the Study of Insects opened the following summer. The station merged Lutz's research and display agenda, providing an extensive insectary for the museum's entomologists and visiting researchers to conduct research and offering visitors the chance to visit what he called a Trailside Museum, a nature trail adorned with exhibits of insects living in situ.[85] Lutz had selected the site strategically: the patchwork of summer resorts and Boy Scout camps blanketing the area supplied youngsters keen to collect insects and clean their cages, and its range of habitats allowed curators to collect study and exhibition material easily and cheaply.[86] Both were essential given the department's tiny budget.

Lutz's Trailside Museum made heavy use of his personal experience as a leader of Boy Scout camping trips and his memories of Uncle Bennie's nature corner.[87] "I wanted [visitors] to enjoy themselves, just as a radio advertiser wants his audiences to enjoy themselves," Lutz later explained, and he relied on his own sense of fun to keep visitors engaged.[88] The so-called Training Trail of exhibits began in a canvas tent filled with displays of household insects like cockroaches and bedbugs, whose biological wonders were extolled alternately by student docents and by limericks authored by Lutz. The trail then wound through various field, forest, and aquatic habitats, where Lutz and his assistants had placed small screen-wire cages around insects busily consuming their preferred food plants. Friendly labels explained the biological significance and behavioral quirks of each species. Once visitors had completed the Training Trail, they could enter the Testing Trail, where fifty numbered tags quizzed walkers on what they had just seen, their answers to which they could then take to be scored by a student docent.

Renowed for their thoughtful pedagogy, Lutz's display experiments proved extraordinarily popular. The live animal exhibits in the Hall of Insect Life attracted so many visitors that other curators attempted to follow suit, mapping out plans to introduce aviaries, terraria, and aquaria into the American Museum's halls. And the Station for the Study of Insects became so crowded that, only a few weeks after it opened, Lutz had to ask the chief engineer of the adjacent Palisades Interstate Park for help. "No one likes to be over-run," he wrote, "however, we are here to help the public. . . . Now, what is the best way to get as many of these visitors as possible and still be able to stand up under the strain?"[89] The nature trail idea spread beyond Tuxedo almost immediately, springing up in hundreds of summer camps, city parks, and empty lots adjacent to schoolgrounds that year. Lutz's close friend, former American Museum director Hermon Bumpus, developed the idea nationally, introducing dozens of trail museums on land administered by the National Park Service.

Lutz was the first to acknowledge that his reforms in and outside the museum were not especially new; conceptually, he admitted, they were quite conservative. "About the only merit we can claim is combining old ideas into a unified scheme and applying them," he wrote Bumpus in 1925.[90] Rather than introducing museumgoers to his experimental methods—explaining how he had devised a tiny treadmill from an old typewriter and tobacco cans to study insect locomotion and its relation to thermoregulation, for instance, or allowing visitors to watch bees navigate their way through mazes to bowls of sugar water Lutz had marked with ultraviolet paint—he simply presented visitors with chummier versions of older modes of display.[91] Ever the naturalist, Lutz remained convinced of the importance of getting museumgoers back to nature in order to teach them to look closely enough that they would, as he confided to Bumpus, actually "*see* what they see."[92]

Noble took a decidedly different tack when it came to display reform, enthusiastically using technology to marry research and exhibition.[93] Unlike Lutz, who never formally studied entomology, Noble had been a keen student of herpetology since his freshman year at Harvard and had written his dissertation at Columbia on the phylogeny of frogs, mapping out the evolutionary relationships between species. An accomplished scientist and an energetic, if abrasive, administrator, the herpetologist became a curator in 1924 and promptly began to push adminsitrators to provide more support for experimental biology.[94] While herpetologists in most other museums were still analyzing species' differences by comparing skeletal and other visible characteristics, Noble drew on neurology and endocrinology to explore the evolution of animal behavior and psychology. Laboratory experimentation was crucial to unlocking the mysteries of animals' sexual and social behavior, he confidently wrote in 1931, and "the modern naturalist armed with hormones and an adequate live room may look forward to working out in a short time the details of habits which have baffled the naturalists for several generations."[95]

Noble also cared deeply about exhibition, and in the early 1930s, he began to add dynamic displays to the museum's Hall of Reptile and Amphibian Life. Developed under the watchful eye of his predecessor, Mary

Cynthia Dickerson, the hall was already relatively modern in its scientific content. It featured not only the museum's famous herpetological dioramas but also displays explaining natural selection, adaptive radiation, the growth factor in evolution, parental care in animals, and the physiology of attraction, all subjects of interest to contemporary zoologists. But Noble was keen to depict recent research on the mechanisms of animal behavior with more verve. Why not display those very mechanisms, Noble wrote in 1930, and ask visitors to help instigate their movement?[96]

Ever confident, he designed and built dozens of moving and interactive displays. In one, a beady-eyed model of a rattlesnake curled in a glass-topped box rattled its tail in fearsome warning when museumgoers pressed a button on the box's front panel (fig. 7). The accompanying label, which explained the biological relationship between behavior, form, and function, satisfied the museum's display critics; so did the deliciously shivery sight and sound of the snake preparing to strike at the touch of a button. "The success of this exhibit clearly shows the importance of dynamic demonstrations and working models in holding the attention of visitors," he wrote in the museum's 1931 annual report.[97] Like Lutz, Noble also introduced live animals into his hall, and because the museum administration had given him state-of-the-art facilities for their breeding and care, he had far greater success keeping his displays alive and thriving in their museum quarters. To emphasize animals' constant movement, he inserted reflective projection equipment into exhibits of frog hatchlings and embryonic snapping turtles, allowing visitors to contemplate the tiny animals' development in large mirrors. Noble made full use of more traditional techniques as well, updating labels to help museumgoers identify the behavioral and physical differences between the lizards and salamanders crawling slowly around their cages.[98]

Noble also began to sketch out a rough but ambitious design for other halls. "Many recent discoveries in the natural sciences including the fields of genetics and physiology receive at present no demonstration in the Museum," he wrote. "It is hoped that a new Hall of Biology will be made available to fill this need."[99] The Depression had sunk its teeth into the museum's finances, so Noble's original plan for a hall exploring the myriad intersections between strands of the life sciences never made it further than his spiral-bound notebooks. The museum opened a small Hall of the Biology of Birds in 1939 and a much larger Biology of Man Hall in 1961.

But another of his proposals, for a Hall of Animal Behavior, became haphazard reality. Drawing on research generated by the Department of Experimental Biology he had convinced administrators to establish, No-

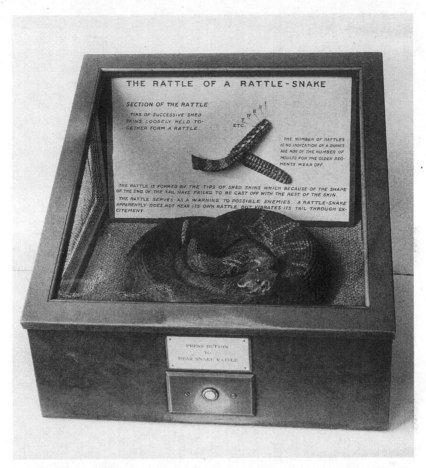

THE RATTLE OF A RATTLE-SNAKE

SECTION OF THE RATTLE

TIPS OF SUCCESSIVE SHED
SKINS LOOSELY HELD TO-
GETHER FORM A RATTLE.

ETC.

THE NUMBER OF RATTLES
IS NO INDICATION OF A SNAKE'S
AGE NOR OF THE NUMBER OF
MOULTS FOR THE OLDER SEG-
MENTS WEAR OFF.

THE RATTLE IS FORMED BY THE TIPS OF SHED SKINS WHICH BECAUSE OF THE SHAPE
OF THE END OF THE TAIL HAVE FAILED TO BE CAST OFF WITH THE REST OF THE SKIN.
THE RATTLE SERVES AS A WARNING TO POSSIBLE ENEMIES. A RATTLE-SNAKE
APPARENTLY DOES NOT HEAR ITS OWN RATTLE, BUT VIBRATES ITS TAIL THROUGH EX-
CITEMENT.

PRESS BUTTON
TO
HEAR SNAKE RATTLE

Figure 7   The rattlesnake snake exhibit in the American Museum of Natural History's Hall of
Reptile and Amphibian Life, ca. 1931, was one of several displays curator Gladwyn
Kingsley Noble used to introduce more dynamism into diorama-heavy museum halls.
When visitors pressed the button, the snake model rattled its tail as if it were about
to strike. The accompanying label explained the biological relationship among the
animal's behavior, form, and function. (Image # 00117728, American Museum of
Natural History Special Collections.)

ble collaborated with exhibit designers to create for an annual member's
day reception a hall filled with displays even livelier than the mechanical
rattlesnake. Additional funding from the Works Progress Administration
in 1935 allowed Noble to develop further the first of his proposed exhibits: a large model of a chameleon that changed its color when visitors
pressed a button, flanked by two diagrams explaining the mechanism of
color change. Noble and his team next created a visitor-operated exhibit

to demonstrate how fish follow moving strips of color. Other exhibits helped visitors see through animals' eyes, giving them a "turtle's eye view of the world," based on curators' conclusion that a turtle's ability to find its way to water was based on its susceptibility to light, and allowing them to glimpse the foreshortened, colorless world dogs experienced.[100]

Noble frequently improvised on the diorama paradigm, tricking it out with technology in order to overcome its limited ability to represent animals' social behavior. In one instance, he supplemented the museum's magnificent habitat diorama of Komodo dragons with a film of the reptiles in action, shot on a 1926 expedition to Indonesia.[101] He also relied on technology to give visitors a window into animal psychology. To help museumgoers visualize the concept of a dominance hierarchy, he built a display in which visitors stood behind a picket fence, looking at a painting of hens and a rooster in a barnyard while a story was narrated over a loudspeaker. At the sentence, "but to the hen every other bird in the yard is a personality," light shone from behind the painting to reveal a new scene in which the hens grew or shrank according to their social status, and the rooster took on enormous proportions relative to the hens.[102]

The public and the press glowed with enthusiasm over Lutz and Noble's displays, yet their exhibitionary experiments soon languished. Though the public continued to stream through his ramshackle field station, Lutz had trouble attracting accomplished scientists and student volunteers to Tuxedo, for most preferred to spend their summers at better-funded, more commodious research sites. So in October of 1928 he loaded that station's lab apparatus into his dusty Model T and closed the site permanently.[103] Noble's Hall of Animal Behavior met a similarly anticlimatic end. The herpetologist's unexpected death in 1940 left the still-unfinished hall without a supervisor committed to keeping its exhibits current, and the dozens of Works Progress Administration workers who maintained the complicated mechanical and cinematic exhibits streamed out of the museum in 1942. Lacking leadership and labor, Noble's hall was eventually dismantled.[104]

A coordinated national movement to reform display might have led to wider and more permanent adoption of Lutz and Noble's techniques, but there was no such effort, and would-be reformers found it difficult to experiment with new kinds of exhibition in the face of the Depression. Throughout the early 1930s, museum work slowed on all fronts as clench-jawed trustees made brutal budget cuts to keep the institutions open. Directors closed halls and reduced visiting hours, suspended ongoing research projects, and canceled long-planned survey and collecting expeditions. They eliminated staff members or cut their hours, and study collections and thriving educational programs slowly deteriorated as a result. The annual budget of the Milwaukee Public Museum—which was $315,000 in 1931—was slashed by 15 percent the next year, and it continued to plunge in the years after. Frank J. Meyer, a curator at the Academy of Natural Sciences in Philadelphia, became so worried about his institution's budget cuts that he sent a thousand specimens from the academy's impressive research collection of rotifers to the Smithsonian and the American Museum for safekeeping during this lean decade.[105] But even the wealthy American Museum suffered in the 1930s, with trustees temporarily halting all collecting expeditions.[106] In this inhospitable fiscal environment, administrators found it difficult to raise funds for anything other than proven forms of display, and they struggled even to find those. Consequently, Noble and Lutz's ventures had little impact on the content and design of interwar natural history museum exhibits.[107]

Specimen cases and dioramas continued to dominate museums' halls; if anything, institutions' commitment to dioramas deepened as administrators used the displays as crutches to hobble over stony economic ground.[108] Dioramas remained wildly popular with wealthy museum patrons, who continued to donate money for the hunting expeditions required to secure the displays' zoological stars and the halls that would ultimately house them. And museums won thousands of dollars in Works Progress Administration money with shovel-ready plans to insert electric lighting, recess cases, and curve backgrounds of dioramas they already possessed. In 1937, at the Colorado Museum of Natural History alone, Works Progress Administration workers molded, assembled and painted 56,031 leaves and 5,200 blades of grass.[109] Dioramas helped museums keep their doors open, so directors continued to tout the displays and pointedly ignored their critics. (Would-be reformers publicly deplored this attitude, pointing out that dioramas' ability to attract funding had tied museums "almost irrevocably . . . to the exposition of one particular aspect of science and made it almost impossible for [them] to respond to changing scientific interests.")[110]

Financial anxiety was not the only source of institutional resistance to new forms of display, however. A powerful contingent of senior natural history curators and directors continued to insist that impressive objects and dioramas remained the most powerful weapons in museums' educational arsenals. Colorado Museum director Jesse Figgins, for instance, swore that one spectacular mount or specimen was "worth far more, from an educational standpoint and as a means of stimulating the right kind of interest in natural history" than hundreds of less impressive displays or circulating school exhibits.[111] Those who admired new approaches also defended the worth of traditional modes of display. "Mechanical contrivances and diagrams to explain how animals react to the cycle of their endocrine glands are excellent," wrote American Museum ornithologist Robert Cushman Murphy, "but they can't completely take the place of a handsome mount of the extinct quagga, a form of African zebra, which the museum never managed to collect and which is a key figure in the story of equine evolution."[112] The combination of economic pressure and this "if-it-ain't-broke" attitude toward exhibition led most natural history museums to maintain a tenuous status quo.

But even the most conservative administrators found it hard to oppose modest attempts to update the scientific content of their halls, so throughout the 1930s, reformers discreetly altered displays to illustrate ecological and biological concepts in more explicit terms. In 1938, the American Museum's Trustees' Committee on the Allocation of Space proposed rearranging existing exhibitions in the museum's forestry hall to show the relation of forests to recreation, soil conservation, water supply, and human use of forest products, while the staff of the Colorado Museum reordered and relabeled their dioramas to illustrate life zones and other ecological lessons.[113]

Though most museum workers were willing to go along with such changes, a handful of curators and many exhibit makers expressed hesitation, arguing that the introduction of labels, charts, and diagrammatic displays were visually disruptive. Patrons had given unstintingly in order to fund display, observed the American Museum's chief preparator James Clark, and their gifts "depended very largely upon the beauty and grandeur of a large hall with spectacular habitat groups." He worried donors would be put off if these halls were suddenly cluttered with smaller exhibits without their consultation. "Anything which detracted from the perfection of these groups as interpreted by the donors would result in invidious comparisons and dissatisfaction on the part of the men who had paid for the group," agreed Harold Anthony, which would not bode well for the museum. Both men warned that "public support had come with the development of the picture groups and interest had been stimulated by beautiful habitat groups," and they intimated that, if these groups were altered in any way, the museum's funding would soon dry up.[114] Concerned, administrators left most exhibits untouched.

Natural history museums' popularity suffered little from this stagnation. Though display critics bemoaned their educational ineffectiveness, the halls of natural history museums filled with hundreds of thousands of visitors, attracted by the promise of free entertainment and education. These "great centers of research and educational influence" had "flourished" in a dark decade, reported Laurence Vail Coleman, and attendance at natural history museums rose steadily over the course of the late 1920s and 1930s.[115] Indeed, in 1938, more Americans attended natural history museums than ever before.[116]

Experimenting with Bodies and Minds: Dynamic Displays at Museums of Science and Industry

Reformers hailing from the worlds of retail, entertainment, and journalism had an easier time building new kinds of life science displays in these years. Outsiders all, they had no allegiance to the responsibilities, customs, and standards of natural history museums, so experimented more freely. The wealthiest of them established museums of science and industry to house these creations. Such museums took European industrial museums and world's fairs as their institutional models, and their exhibits explored biomedicine, industrialized agriculture, and other applications of the life sciences in ways that seemed breathtakingly new to American museumgoers. Rather than strolling past silent specimens, visitors threaded their way through mechanized exhibits that blared, blinked, and spun in order to demonstrate the wonders of modern science. Determined to make education their institutions' exclusive focus, founders of these institutions refused to support curatorial research and collecting, instead relying on corporations to provide them with information and objects.[117]

Chicago's Museum of Science and Industry was the most influential of these museums, and its life science displays the most widely imitated, even though they had been something of an afterthought. Its founder, retail magnate and philanthropist Julius Rosenwald, had actually intended the museum as a place to celebrate the evolution of industry and manufacturing. "I would like every young growing mind in Chicago to be able to see working models, visualizing developments in machines and processes which have been built by the greatest industrial nation in the world," he told a *Tribune* correspondent in 1926.[118] He had considered founding an institution similar to Munich's famous Deutsches Museum for more than a decade after visiting Munich and had even floated the idea to members of the Commercial Club of Chicago in 1921.[119] By December of 1925, he had persuaded the club's members to create a committee to study the matter; a year later, the plan was in action.

After a two-year search, Rosenwald and the trustees of the proposed museum settled on a director: *New York Times* science editor Waldemar

Kaempffert, who had written the copy advertising the position in the first place.[120] Kaempffert's terms were exacting—a $20,000 salary, absolute control over curatorial hiring, and a paid trip to Europe to study the museums of London, Vienna, and Munich—but he seemed to Rosenwald the right man for the job. "I like the enthusiasm he manifests for the job and his evident grasp of the needs," Rosenwald wrote a friend.[121] Once hired, Kaempffert and a formidable crew of curators promptly departed for a four-month tour of European institutions in an attempt to immerse themselves in the latest museum techniques.[122]

Encouraged by what they saw abroad, Kaempffert and his staff decided that Chicago's new museum would focus exclusively on scientific principles, recent discoveries, and emerging scientific applications—exactly the sort of content that critics of natural history museum display, including Rosenwald, had advocated. The team proposed organizing the museum around ten thematic "sequences," beginning with what Kaempffert called "the Fundamental Sciences" of physics and chemistry and ending with the social consequences of science and technology deriving from them. This was a marked departure from the halls of most American natural history museums, which tended to be arranged according to taxonomically defined collections—birds, lizards, mammals, minerals—or in the case of certain diorama halls, geography. This organization went a long way toward satisfying critics' plea to create a museum that revolved around contemporary scientific subjects and ideas, not physical objects. Kaempffert also insisted the museum display applications of the life sciences, especially developments in agriculture and public health. After all, he explained, "the bacteriologist and the sanitary engineer are doctors of civilization," prescribing medicine that allowed modern industrial society "to work and clothe and feed itself without committing suicide."[123]

Kaempffert insisted that the new museum's educational approach be defined as much by innovative exhibition design as by the content of its galleries. Chicago's Museum of Science and Industry would have "no 'hands-off' signs and glass cases," he vowed.[124] He believed the museum needed to improve on traditional exhibition methods, allowing visitors to interact with displays in new ways. Museumgoers "will walk through timberland and saw-mills, see in a diorama what looks for all the world like a forest fire and learn how it is controlled and extinguished," he promised. "There are models of the eye before which you stand and from which you can learn whether you are near-sighted or far-sighted." Visitors, he argued, needed to get up close to the tools of scientific research—objects that natural history museums tended to keep behind storage room and laboratory doors—so he stocked the museum with slides and microscopes. "See for yourself such wonders as the tongue of a fly, the kit of lances and pumps carried by a mosquito, the structure of some deadly germ," he urged.[125] The director also aspired to install as many mechanized, visitor-operated exhibits as possible; exhibits that sprung into action when visitors turned their cranks or pushed their buttons were, he wrote, the only way to make Chicago's museum a " 'live' institution."[126] To help the hesitant, floor "demonstrators" would stand by, providing assistance and commentary on the scientific principles displayed.[127]

Science-themed world's fairs routinely used dynamic displays to make their medicine go down, and museums of science and industry borrowed liberally from these expositions. Though natural history museums had long trafficked in displays from world's fairs, in the 1930s they purchased fewer and fewer exhibits from exposition pavilions. Museums of science and industry, in contrast, snapped them up. Their broader scope of scientific subjects, tolerance for corporate involvement, and enchantment with the new and the mechanized made the museums natural partners with interwar world's fairs; in Chicago, Rosenwald and the 1933 world's fair organizers collaborated frequently in order to make their ventures simultaneously possible, for there was enormous crossover between the two entities.[128] The fair's theme, "Science Finds, Industry Applies, Man Conforms," coincided almost exactly with Rosenwald's original vision for the institution, and the museum's 1933 opening was considered a part of the city's world's fair celebration. Its exhibit designers took direct inspiration from the fair's displays, and the museum's halls became the ultimate repository for the fair's most popular exhibits.[129] (According to the *New York Times*, Chicagoans were especially pleased to hear that the museum would be taking "the medical exhibits that aroused a greater public interest than any of the wonders in the [Fair's] Hall of Science.")[130] Though New York City's Museum of Science and Industry was already open by the city's 1939 World of Tomorrow Exposition, fair organizers and museum administrators had a similarly productive relationship.[131]

Museums of science and industry paid close attention to fairs' corporate experiments in dynamic display, asking companies to loan them out or permanently install them into museum halls. In stark contrast to staff members at natural history museums, who continued to fear that corporations would seek to exercise inappropriate influence over exhibits, the folks at museums of science and industry did not hesitate to reach out to companies for donations or expert assistance.[132] Indeed, they actively sought to develop relationships with American business, declaring that corporations' scientific research and exhibition techniques were just as cutting edge as what could be found in universities or existing museums.

In 1939, for instance, founder of the New York City Museum of Science and Industry Frank B. Jewett invited corporations participating in the 1939 fair to donate exhibits "done in an educational manner" to his museum, enticing them to do so by encouraging them to consider their gifts to be "part of a long-range public relations activity."[133] Corporations happily complied.

As a result, the halls of museums of science and industry vibrated with motion. "Enter the main doorway and pass into the first court. You see—what? Animation everywhere," wrote Kaempffert.[134] In Chicago's Museum of Science and Industry, observed one *New York Times* reporter, a visitor could "do his own exploring and experimenting with the animated exhibits." He could "push a button or throw a switch . . . as long as he will, or until the machine wears out. The attitude of the officials toward the wearing out is: 'When they give out, we'll get others,'" reported the *Times*.[135] The *Times'* report was somewhat exaggerated— though the museum certainly stocked its floor with interactive exhibits, many of the displays were operated by guards or demonstrators, or moved of their own accord, and museum staff certainly cared when they broke down. Still, knowing they could rely on a steady stream of exhibits from major corporations allowed curators in museums of science and industry to take a more relaxed attitude toward interactive displays than natural history museum staff, who spent years building their own.

Such willingness to experiment with display indicated those museums' single-minded commitment to public education through exhibition. Rosenwald was so serious about this mission that he forbade curators from collecting or conducting in-house scientific research, so they would focus exclusively on developing displays. The decision represented a dramatic rupture with museum tradition. Though natural history museum administrators often scolded curators for sloughing off display and collection responsibilities to spend more time on research, most directors and patrons still considered research an important aspect of museum work and expected their curators to pursue it in some way. This was especially true for museums as sizable, ambitious, and well-funded as Rosenwald's. As a result, even neophytes to the museum world, Kaempffert among them, struggled to conceptualize how curators could credibly function in such a setting. But Rosenwald and the founders of other museums of science and industry plowed ahead, scolding administrators who encouraged research.[136]

Within this new institutional structure, exhibit designers had considerably more power than they did at natural history museums, where they worked closely with curators but were generally considered junior partners in the venture. Without scientific collections or research, exhibit designers had the opportunity to ascend to directorships and other high levels of administration, positions that natural history museums still most often reserved for scientists. At the New York City Museum of Science and Industry, for instance, exhibit designer Robert P. Shaw quickly took the helm of the museum, which described itself as a "laboratory of modern exhibition practice," and prided itself on its theatrical exhibits.[137]

To cement this staff structure, Chicago administrators often hired well-established scientists whose primary professional commitments lay elsewhere, ensuring they did not have time to do much more than supervise or serve as figureheads, while the museum's administrators, educators, and exhibit designers ran day-to-day operations. The board agreed in 1933 to hire Evan Carey, dean of the Marquette University School of Medicine, to direct its medical science exhibits, for instance. The arrangement worked out precisely the way the museum's administrators had hoped: Carey helped design and supervise the construction of displays intended for both the larger public and medical practitioners, and lent his name to publicize the exhibits in medical journals, but didn't demand much of a say in larger museum affairs.[138] (The partnership also worked nicely for Carey. By 1937, he had won more than a dozen gold and silver medals from the American Medical Association for his work at the museum.)

Though the purview of museums of science and industry was far broader than natural history museums, the new institutions stressed the life sciences as heavily as their older counterparts did—albeit through the exploration of their applications to human life. When Chicago's museum opened, fifteen out of its forty-four galleries explored biology topics. These displays included fields traditionally associated with natural history, such as classification and comparative anatomy, but mostly focused on experimental and applied biology, ranging from cell biology and microscopic organisms to animal behavior, genetics, and "the wonders of hormone chemistry."[139] The hall also housed an exhibit on eugenics, which the visitor guide cheerfully described as "evolution experimentally produced" and "economic biology."[140]

The biology of humans featured especially prominently in these halls, for displays on the human body held enduring appeal to audiences, and the subject of medical miracles was hard to resist. One of the most popular exhibits featured at museums of science and industry, and soon after, science museums, was the *Transparent Man*, a German-made model of the human body built to allow visitors to view its internal organs.[141]

117

An early version of interactive display, visitors could press buttons to make the model's organs light up. Networked with veins and arteries, the model stood upright, looking upward, arms raised, as if he were reaching up to God. "The organs of this superb figure are separately illuminated, enabling the observer to visualize human anatomy as though possessed of an x-ray eye," announced one gallery guide. That X-ray was rather Victorian, however, for the model was presented sans genitalia.[142]

First displayed in the United States at the Chicago World's Fair in 1933, these transparent models soon found a home at Chicago's Museum of Science and Industry and, soon after, in many more American museums. Four months after it opened its doors to the public, the New York City Museum of Science and Industry put its own *Transparent Woman* on display. Museum publicity agents announced her unveiling with a press release designed to titillate, calling her a "woman without mystery," but also sought to counter any criticism that such an exhibit would be lewd or inappropriate. The museum's Transparent Woman would, the text reassured concerned citizens, be "introduced" by a qualified "doctor-lecturer" to serious visitors only, namely, scientists, public health officials, and members of the medical profession.[143] The exhibit proved so enticing, however, that Shaw estimated that, on most days, more than two hundred laypeople asked to see it. "We had to find some other way to satisfy the demand for a spectacular physiological exhibit of this character," he wrote. Consequently, the museum mounted another exhibit they called *Miss Anatomy*, a medical school skeleton lit by sixteen concealed projectors that illuminated various opaque organs inserted in anatomically correct places in the skeleton's scaffolding.[144]

The *Transparent Man* and *Transparent Woman* were merely the grand marshals to a parade of interactive biology displays in American museums of science and industry in the 1930s. Fascinated with the pedagogical potential of interactivity, organizers of the New York City Museum of Science and Industry's 1937 Story of Man Hall made certain "that visitors can push buttons and pull levers then watch the wheels go round" as they perused the hall's exhibits (fig. 8).[145] Visitors began by contemplating a model of a human egg, enlarged two hundred times, and were then guided by displays through the development of the body and all its organs and their various functions. In order to learn more about the genetic science of blood groups and the medical consequences of cell clotting, visitors could "do a little blood mixing" with the help of a demonstrator, introducing serum into incompatible blood from animal sources. Visitors could also touch a wired model of a human hand, which lit up electronic "nerve corpuscles" to help illustrate the biology of touch. Curators added a model of the human brain, each of its different centers adorned with a button, which, when pressed, would light up a large photograph depicting the activities that corresponded with that particular region of the brain.[146]

Though committed to interactivity, science and industry museums did not fully abandon traditional life science displays, and most continued to rely at least partly on the pedagogy still associated with natural history museums. Chicago's Museum of Science and Industry's exhibit on prenatal development, for example, consisted of a series of preserved human fetuses in various stages of development, a throwback to the era of pickled specimens in jars, even though the exhibit had attracted enormous attention at the Century of Progress Exposition.[147] (By the early 1940s, to the relief of the squeamish, the museum had switched from the specimen jars to wax models, and incorporated a pregnant female mannequin, in order to emphasize the mother's role in birth and childcare.)[148] And its farm exhibits, sponsored by International Harvester, the Delaval Separator Company, and a handful of other companies, featured mounted cows prepared by Julius Friesser, the Field Museum's chief of taxidermy.[149] Still, the institutions primarily emphasized mechanized motion, and new exhibits routinely featured moving parts and flashing lights. A modern vacuum milking machine chugged alongside Chicago's silent cows, and portions of the museums' dioramas lit up when visitors pushed buttons.[150]

The public and media responses to these dynamic biological and medical exhibits echoed the delighted reaction habitat dioramas had once evoked. The displays would provide visitors with much-needed explanations of the mysteries of modern civilization, as well as the contents of Americans' refrigerators, garden sheds, and medicine cabinets, declared the *New York Herald Tribune* in 1935. "The possession by everyone of hundreds of new scientific tools . . . has thrown new curiosities into millions of minds still fitted with the intellectual furniture of the age of . . . the wheelbarrow," wrote the editors.[151] Exhibits in museums of science and industry, they maintained, could help Americans understand the science that permeated their own lives through entertaining illustrations of basic concepts and their applications. Such public accolades translated into financial support. In 1935, and then again in 1939, the Rockefeller Foundation gave major grants to the New York City Museum of Science and Industry to build dynamic displays that would, in Shaw's words, "sugar-coat the educational pill of science."[152]

Despite overwhelming critical enthusiasm, there was little hard evidence that dynamic displays educated rather than fascinated. Though

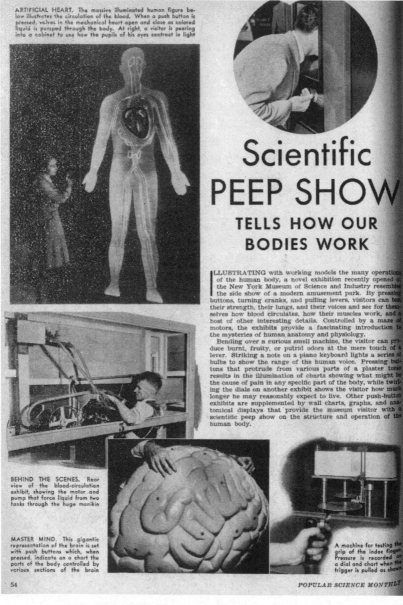

Figure 8 A February 1938 *Popular Science Monthly* article gushed that the New York Museum of Science and Industry's new *Story of Man* exhibit "resembles the side show of a modern amusement park," praising the interactive elements of the displays. "By pressing buttons, turning cranks, and pulling levers, visitors can test their strength, their lungs, and their voices and see for themselves how blood circulates, how their muscles work, and a host of other interesting details." (Image provided by the Boston Public Library.)

SOURCES OF PAIN Buttons in various parts of the figure at the left are pressed to light up a chart that shows what may cause pain in each area. The dummy is opened to reveal the switch system

Nearest approach to the familiar chamber of horrors is the "book man," below, made in vertical sections that open up to show the internal structure of the human body

See your own pulse! A finger nail placed in contact with a rod in this device causes an indicator to swing in time with the heartbeat

STRONG-ARM ACT. A rubber bulb, pumped full of air by means of a foot pedal, moves the arm bones in this exhibit to demonstrate how muscles work. The knob helps control the movements

SMELLS MADE TO ORDER "Phew," this visitor says, as he gets a whiff from the "smell machine." Pressing a lever releases odors from bottles seen in the picture below. By pushing two or more levers at once, additional smells are easily produced

The compass of the human voice at various ages is shown by the exhibit seen at the left. When a visitor strikes one of the notes on the keyboard, the corresponding note lights up on the musical scale, and the chart above it indicates the ages at which the tone is produced

Figure 8   *continued*

Kaempffert maintained that his museum was "no technical Coney Island" and used dynamism only "as a means to an educational end," visitor experiences could belie such convictions.[153] "Great gobs of flame drop and drop in a big glass barrel. Something about phosphorus. Isn't it pretty?" gushed one dazzled young woman in 1933. "Oh, and look at those big globes of wire, with little twinkling lights, like a new-fangled chandelier. Golly, salt molecules or something, a lot of big words, and it's swell to look at."[154] Still, such displays certainly attracted the eye, and the era's museum workers felt that this kind of visual appeal, accompanied by a clear and accessible description of the phenomena depicted, was all that they could do.

Interwar museums of science and industry put forth an alternative model for museum education. The idea of a museum forgoing collections and research altogether was still radical, as was the notion that such a museum could successfully educate the public about scientific topics with only minimal input from scientific staff. Some museum reformers cheered, while others looked askance at the whole project, anxious about the institutions' unabashed willingness to substitute corporate involvement for curatorial input and their obvious commitment to building exhibitions that mesmerized onlookers. As Buffalo's Cummings wryly observed, "Few genuine, dyed-in-the wool curators have any great sympathy" for institutions that catered frankly to visitors' interest in "plain *entertainment*," even if some education was snuck into the fun.[155] But even the most suspicious critics could not deny the extent of popular interest in the life science displays of museums of science and industry, nor could they ignore the institutions' surprising influence. Though few and far between, these fledgling enterprises served as polestars in the era's emerging constellation of exhibition experiments.

From Natural History to Science: The Birth of the Science Museum

Institutional change was afoot in the professional museum community as well. Leaders of the flagging museums in Buffalo and Boston openly criticized existing natural history museums, arguing that the institutions still privileged curatorial specialization over the general public's needs, tastes, and desires. " 'Knowing more and more about less and less' does not of necessity carry with it the requisite characteristics and point of view which are most to be desired in the individual who prepares exhibits and labels for public consumption," declared Buffalo's director of visual education, Carlos Cummings. "Frankly, what has been more or less jok-ingly referred to as the 'Ph.D. complex' can become a serious matter if allowed to run wild among popular educational exhibits."[156] Natural his-tory museums needed to be revolutionized, they concluded; retrofitting a few displays was not enough.

So throughout the 1920s and 1930s, these reformers took bold steps, hiring new directors who reorganized and reopened antiquated societies of natural science into aggressively forward-thinking science museums. Although these two museums maintained their ties to natural history museums, they also reached out to museums of science and industry, developing relationships with staff members and borrowing displays from these new institutions. And like museums of science and indus-try, science museums quietly allowed their research and collecting pro-grams—programs that had defined their older incarnations—to expire. Ultimately, the science museum offered another institutional model, one that selectively embraced the up-to-date organization, content, and dis-play techniques of museums of science and industry while consciously preserving some, though not all, natural history museum traditions, standards, and ethics.

The leaders of these new science museums tended to be scientific gen-eralists, with backgrounds in education and popular science rather than research.[157] Though they themselves were not pushing the boundaries of scientific knowledge, they were determined to familiarize the public with those frontiers and excited to use the latest display techniques to do so. In short, they were museum men, some of them aging members of that first self-declared generation of museum reformers, some of them much younger and newer members of this club. At the Buffalo Society of Natural Sciences, for instance, Cummings elbowed aside William L. Bry-ant, a taxonomist more committed to the study of fossil fish than to dis-play reform. Cummings, by way of contrast, was an institutional mover and shaker, always open to change. A biologist with medical training and personal passions for photography, mycology, and snow crystals, he cultivated relationships with amateur naturalists and professional scien-tists alike.[158] Though he continued to believe that systematic display and habitat dioramas could be useful, he also maintained an active interest in the latest advances in the life sciences and an intent curiosity about new methods of display, eventually authoring a book on the subject.

Cummings's partner-in-crime was Hermon Bumpus, a well-known museum man and invertebrate zoologist who served as the museum's interim head before Cummings assumed its directorship. Both men enthused about creating an institution that would "cover not only the range of activity of the members of the scientific staff, but the entire range of human knowledge so far as the branches of natural history are concerned," an aim that Bumpus had not been able to achieve as the director of the American Museum of Natural History in the 1900s.[159] There, the pugnacious zoologist had clashed repeatedly with its curators when he insisted they present collections along popular, rather than scientific, lines by rewriting exhibit labels and reorganizing exhibits to appeal to lay visitors.[160] His attempts to introduce more microbiology and laboratory-based life science displays and research had alienated him from other curators more interested in a more traditional natural history focus on whole organisms and fieldwork-based exhibits, and he was forced from the museum in 1911. Buffalo provided an opportunity to realize his original vision for display and restore his reputation in the museum world to boot.[161]

Under the direction of Cummings and Bumpus, staff members transformed the Buffalo Society of Natural Sciences into the Buffalo Science Museum, expanding and reorganizing the museum's subject matter to fit its sleek new name and building. They conceived of a completely new organization for the museum, one that would provide more up-to-date life science education by virtue of explaining and connecting all the sciences, rather than focusing on what knowledge could be derived from the close observation of individual objects. "A general study of the science of biology would seem to be much more useful to the general run of museum visitors than an attempt to present a complete collection of birds or the largest number of shells of any institution in the world," Cummings explained. So instead of dedicating its rooms to displaying the prehistoric crustacean and freshwater conchological collections that had so excited Bryant, Cummings and Bumpus broke with natural history tradition and devoted Buffalo's halls to illustrating subjects: in the words of the director, "informative generalizations, well chosen, carefully elaborated, and dedicated to making plain the fundamental laws that underlie and control all things on earth."[162]

Museum staff identified and arranged exhibits according to subjects of scientific and popular inquiry, rather than by specimen collection—one hall explored questions of evolution, rather than dividing the space into cases for invertebrate and vertebrate paleontology, for instance. Throughout the 1930s, the museum opened a succession of rooms devoted to subjects in physics and chemistry, astronomy, earth science, biology (or, as the museum dubbed it, "Life"), evolution, heredity, and physiology.[163] Each hall, Cummings wrote in 1929, was intended to tell a connected story, which, "taken all together, chronicle man's present knowledge of science and natural history."[164]

The reorganization was also guided by Cummings and board president Chauncey Hamlin's firm belief that current events and immediate personal relevance were two of the best ways of exciting public interest in scientific concepts. As a result, curators filled the museum's lobby with special exhibits and rotating shows intended to keep visitors abreast of recent discoveries in the applied sciences.[165] Its subject halls also attempted to combine contemporary scientific findings and ideas with information that visitors would find pertinent. The physiology hall, for instance, featured larger than life-sized graphic displays explaining new discoveries about functions of the human body. Properly appreciated, such exhibits would "point the way to longer life and greater efficiency," promised the museum's guide.[166] Yet the museum remained, at heart, an institution devoted to the natural sciences, and its staff considered displays of scientific applications a means to the end of conveying fundamental scientific concepts.

The institution retained other characteristics of natural history museums as well. Cummings and his staff maintained botany, invertebrate zoology, and vertebrate zoology halls, the most popular older categories of natural history, featuring habitat groups and specimens from the museums' vaunted collections as they did so. Furthermore, it did not introduce the kind of industrial displays and corporation-sponsored exhibits that were staples of museums of science and industry. Buffalo's staff members remained closely allied to the professional museum community and balked at the idea of allowing contemporary corporations to shape exhibits' contents. But there is no doubt that Buffalo's reorganization resembled the layouts of museums of science and industry more than those of natural history museums, and its staff members frequently declared their intention to present the public with a more sweeping view of science than natural history museums did. "What we are doing," Hamlin explained to Rockefeller Foundation officials in 1937, "is to try to write and illustrate the whole fascinating story of modern science in our document—our museum—chapter by chapter."[167]

Display critics in and outside the museum world praised the Buffalo Museum's new organization, asserting, as Coleman did in 1930, that it successfully explained "the principal concepts of each science" far better than natural history museums' specimen halls. "Both the plan and its execution are very different from the traditional mode of exhibition which might call for a room label reading 'Plants' but would never justify one reading 'Botany: the Science of Plants,'" he wrote.[168] Frank Lutz was impressed enough to suggest, in 1929, that the American Museum reorganize its own halls along the lines of Buffalo's.[169] And officials from the Rockefeller Foundation agreed that the museum's redesign had resulted in "the vivid illustration of ideas rather than the display of a multitude of objects."[170]

Boston's archaic New England Museum of Natural History followed Buffalo's lead just a few years later. Though the New England Museum of Natural History had officially devoted itself to education since the 1890s, this mission was never properly funded or fulfilled, and by 1930, the museum was stuffed with "a fearful hodge-podge of ill-cared and often repulsive exhibits," recalled one patron.[171] So museum trustees decided that they would simply start fresh, as had Buffalo. They sent the museum's head of education, Francis A. Young, to intern under Cummings for a year and hired mountaineer and nature photographer Bradford Washburn to help them create a new, more modern institution.[172] Both the press and the museum's trustees believed Washburn's dual careers had prepared him well for the challenge of transforming a small, cluttered semiprivate collection into a popular institution with a more dynamic approach to education. Such a man, wrote one Boston reporter, "could never be happy in a dusty old-fashioned museum," and the furious pace of Washburn's reforms proved the reporter right. Washburn and the trustees promptly agreed that the institution should abandon the vestigial title of natural history museum, becoming the Boston Museum of Science, and "broaden the scope and appeal" of its displays, reorganizing them accordingly.[173]

The gradual introduction of dynamic displays requiring visitor participation also marked science museums' intentional departure from natural history museums' more static pedagogy. While educators and curators at museums of all types had begun to discuss the pedagogical potential of interactivity in the early 1920s, only children's museums had initially taken this call for change seriously.[174] In natural history museums, visitor participation had generally remained a topic of debate, rather than a goal curators and designers actively worked to realize. But the Buffalo Museum of Science—whose curators took to heart contemporary educational and communication research and the success of displays at expositions and museums of science and industry—began to explore the possibilities of participation almost immediately.[175]

When its Hall of Man opened in 1935, for example, visitors and critics exulted over its dynamic displays. Museumgoers could press buttons that would turn model wrist bones or force blood to course through a mechanical human heart. "While you watched the light, a blood corpuscle in your own body traveled the selfsame course and in exactly the same length of time," crowed an article in *Popular Science*. Two "Body Books," detailed wooden models of the human body that were realistically rendered, were sliced lengthwise and crosswise so visitors could page through sections of the head and torso (fig. 9). Visitors could even examine themselves with the help of the museum's radiograph, a version of an X-ray machine attended by a staff member. Such displays were considered tremendously forward thinking for their interactive elements, certainly, but were also commended for the ways in which display techniques embodied basic biological concepts. The press, for instance, celebrated the museum's use of an electrical machine to capture the body's "electrically energized movements" and "exhibit and explain the machine that is man."[176] As the hall's pièce de résistance, the Buffalo Museum of Science acquired Chicago's famous *Transparent Man* in 1935 and exhibited a *Transparent Woman* for six weeks in 1937.[177] Though such exhibits were still the exception rather than the rule at the museum, Buffalo's exhibit designers and curators were thrilled by their reception.

Foundation officials hurried to fund similar displays of health and human biology, pushing science museum staff to incorporate ever-more contemporary—and controversial—content into their exhibits. After lavishing praise on Buffalo's Hall of Man, Rockefeller Foundation official Leonard Outhwaite urged its curators to address the study of evolution and the "sciences of heredity and genetics" in simpler, more direct terms. Few museums "really understand their responsibility in this regard or have faced the issues with which they are dealing," he declared.[178] But the museum's staff was more attuned to the social climate than Outhwaite admitted, so, wary of rousing the kind of ire natural history museums had faced in the early 1920s, sidestepped the subject of evolution. As Cummings reminded his peers: "There can arise no situation more unfortunate in connection with museum exhibits than to have them translated in such a manner as apparently to oppose accepted religious beliefs. . . . Many of us will remember an edict of the Church not many years ago; it took exception to certain exhibits in one of our larger museums and even went so far as to discourage donations to this institution." To maintain visitor confidence and avoid controversy, Cummings concluded, "the wise curator will conscientiously avoid any positive statements along such topics as early man or evolution, for example, and will confine himself purely to a presentation of the proven facts from which the theories themselves were evolved."[179]

Cummings and his staff had no such qualms about displaying more

Figure 9   During the 1930s and 1940s, the popular press eagerly covered new science
museum displays, such as the *Body Book* exhibit from the Buffalo Museum of Science's
Hall of Man, which made the cover of *Science Newsletter* in April 1935. For this same
hall, Buffalo purchased an early *Transparent Man* from the Museum of Science and
Industry in Chicago, evidence of the circulation of exhibits among science and
natural history museums during this same period. (Reprinted with permission of
Science News.)

contemporary applications of evolutionary thought, however, and blithely featured eugenics exhibits. When the Carnegie Institute in Washington offered the museum four large panels detailing the importance of race hygiene, "blood," and "breed" in human reproduction, its curators gratefully accepted. (The panels had originated in Dresden and bore all the ideological hallmarks of Nazi-sponsored biology. The third and fourth panels were particularly ugly, comparing the "defective" ancestral tree of the fictitious Ishmael family, "shiftless, begging wanderers, sound enough in body, their hereditary equipment lacked the basic qualities of intelligence and character on which opportunities could work," to the superior "Roosevelt family-stock," which had produced two presidents.)[180] Similar charts, graphs, and labels surrounded some of the more dynamic displays in the museum's Hall of Man.

While they paid closer attention than ever before to display, leaders of the nation's few science museums intentionally neglected scientific research and collecting. Curators at the old Buffalo Society of Natural Sciences, for instance, had long prided themselves on their research in ichthyological, conchological, and botanical taxonomy, and curatorial publications persisted for a few years after the new museum opened its doors in 1928. But the museum's bulletin appeared only sporadically under Cummings, who was far more invested in improving exhibition than in expanding research or collecting. As the Depression began to eat into museum finances, the museum officially suspended all scientific research to save money, though Cummings declared his determination to continue Buffalo's heralded exhibition and educational programs "at full efficiency throughout the lean years of this trying period." Neither Hamlin nor Cummings gave any indication of wanting to revive research once the crisis had passed.[181] Having been trained to consider "caring for large study collections . . . almost a religion," Cummings did not go so far as to deaccession the museum's holdings, but he did confess that he envied museums of science and industry for their "freedom from all the problems involved in this storage business."[182] Boston's trustees pushed for a similar approach. Though the museum's curators had conducted no meaningful research for decades, in the late 1930s, the board forbade them from adding to the museum's already sparse specimen collections.

In little more than a decade, the museums in Buffalo and Boston had managed to bridge the institutional paradigms of fast changing, commercially influenced museums of science and industry and more traditional natural history museums. The innovations of these two museums proved profoundly influential, yet that influence wouldn't be fully felt for another decade or two. Most of the era's natural history museums were too paralyzed by the Depression to attempt similar metamorphoses. Only an unusual combination of timing, funding, and charismatic personnel had allowed the Buffalo Museum of Science and the Boston Science Museum to muster the political will and financial wherewithal necessary to over-haul their institutions so completely. Still, the example of the science museum and its interactive displays twinkled brightly, providing another institutional model for museum reform.

Measuring Success

A decade and a half of fragmented opinion, bold experimentation, and quiet stagnation in life science displays concluded with some consensus about the future of museums' exhibits—and their institutional identi-ties. "As the full range of science becomes the usual thing, *natural history* is less and less adequate as a descriptive term, *science* is more and more suitable," Coleman reflected in 1939.[183] Museums of science and industry and science museums had begun to address scientific principles more explicitly in their halls, using sweeping storylines to explain how chemi-cal and physical laws set the conditions for all forms of life on earth, chronicling how scientists' recent findings related to broader biological concepts and applied directly to visitors' own lives. Natural history mu-seums slowly followed suit.

Coleman could easily have added the term "dynamic" to his assess-ment of the era's museum displays. By the end of the 1930s, interactive exhibits had roused so much public curiosity and critical praise that me-chanical and interactive displays had become the exhibitionary standard, intriguing even the stodgiest natural history museums. *Museum News*, scientific and educational journals, and the popular press devoted pages to the phenomenon. The dynamic approach, declared a 1935 article in the *Scientific Monthly*, had done "wonders in interpreting science to the public," citing numerous examples of how museum professionals, edu-cators, and foundation officials embraced it with growing enthusiasm.[184] "Everywhere movement has been called upon to make the unseen visible and the visible twice as conspicuous," Cummings echoed in 1940. "The static and immovable exhibit is becoming more and more a thing of the past. . . . Some form of motion seems to be regarded as an essential of all well-rounded exhibits, even in the most unexpected places."[185]

Natural history museums staffed by charismatic display reformers slowly responded to this shift as well and began to borrow ideas and exhibits from science museums, museums of science and industry, and world's fairs. "The trend in exhibits is fortunately leading us," observed Murphy, "into other and even more dynamic roles of expression."[186] Only six years after the American Museum had installed several large blown-glass rotifers, for instance, Noble replaced the models with a "microvivarium," which projected enlarged images of the living unicellular animals—species of amoeba, euglena, and paramecium—found in a drop of water onto screens.[187] The Chicago Century of Progress Exposition and the 1939 New York World's Fair had also boasted microvivaria, but the exhibit was the first of its kind to be displayed in an American natural history museum.[188] Almost ten years after he had announced his hope to present the museum's visitors with more insight into experimental biologists' methods, Noble continued to show the public how lab scientists saw the very small.

Buoyed by the nation's recovering economy, the staff and trustees of natural history museums once again began to discuss what the halls of the future should look like, envisioning more interactive displays of civic biology and grand scientific narrative. Some dreamed big and wild. In 1939, for instance, members of the American Museum's Department of Invertebrate Zoology suggested constructing an immersive diorama of the kelp forests off California's Santa Catalina Island so visitors could wander through thick strands of kelp to see sea stars and spiny lobsters, bright garibaldi and two-spotted octopi up close (fig. 10).

Kaempffert, now back at the *New York Times*, and American Museum of Natural History trustee William Douglas Burden went even further, proposing that museums employ introductory exhibits resembling the General Motors' *Futurama* exhibit at the 1939 world's fair, where visitors had ridden past three-dimensional displays in moving chairs. Everyone who entered the museum, they agreed, should first pass through a multimedia exhibit that dramatized both the underlying mechanisms of life and contemporary scientists' inquiry into these mechanisms.[189] Such an exhibit would, Burden promised, "offer the means of stimulating the appetite for further information so that the visitor will wish to browse in natural history museum halls with greater purpose."[190] Most visions were tamer. Murphy, for instance, urged natural history museums to devote more space to "interpretive exhibits" that explained the latest scientific discoveries and public preoccupations, relying on "mechanical devices, such as the photo-electric cell, sound photography, 'robot-voices'—perhaps even television" to help them do so.[191] And he made sure that the opening of the Whitney Memorial Hall of Pacific Bird Life, a habitat hall he described as the department's "show place," was accompanied by the opening of a Hall of the Biology of Birds.[192] The era's experiments in

131

Figure 10 Though this proposed "kelp forest" was never built, this 1939 sketch by an American Museum of Natural History artist captured the growing fascination with the pedagogical potential of immersive exhibits, as well as natural history museums' renewed display ambitions at the end of the 1930s. ("Santa Catalina Kelp Gardens" by George H. Childs from DR138, Box 1, American Museum of Natural History Library.)

displays might have inspired visible changes in museum blueprints, but they also signaled less tangible shifts in staff attitudes toward the public and displays. With greater frequency, Cummings observed in 1940, curators in all three types of museums accepted "the necessity of knowing 'visitor reaction,'" formally gauging behavior and even consulting visitors to determine just what they were thinking. This attitude wasn't universal, of course. Many staff members believed, as many Progressive Era museum men had, that they were perfectly capable of building displays that would—or should—appeal to visitors and needed no guidance beyond their own convictions.[193] And some continued to take a public-be-damned attitude toward exhibition, Cummings reported. "In the event that the public does not crowd their halls in rapt admiration it merely proves that the average man is not of a high type of mind and really not worth bothering with."[194] Still, he added, growing numbers of curators acknowledged the magnitude of the challenge of teaching science through exhibition and admitted their own relative ignorance of display pedagogy. "The greatest need in museum displays today is a much better understanding on the part of the curators of all the problems properly included under this general heading," he concluded.[195]

Display critics and reformers had urged more scientific investigation of display pedagogy as early as 1925, pleading "for a study of the museum exhibition problem on the level of true research," as curator Clark Wissler put it.[196] They were delighted when, in the mid-1920s, Yale psychologist Edward S. Robinson and his colleagues launched a thirteen-year study of the effectiveness of museum exhibits at a wide range of American museums of history, science, and art.[197] Robinson's Carnegie Corporation–funded team and a similar set of Rockefeller-sponsored studies concluded what many display critics had already assumed: that exhibits were not, in fact, educating visitors as well as curators believed they were. Too often, researcher Arthur Melton observed in one 1935 report, curators overreached in their displays, organizing exhibits and writing labels to "express conceptualized relationships which are thought to be within the mental grasp of everyone" while omitting "essential information which is assumed to be the knowledge of everyone."[198] This approach, Melton pointed out, alienated and confused most visitors, leaving them with memories of objects but no meaningful understanding of the ideas those objects were intended to illustrate. And the numbingly static nature of current displays, found researchers, resulted in what was popularly called "museum fatigue": "aching muscles, tired neck and eyes, and . . . the vague but insistent desire to escape."[199] Visitors were so put off by the motionless halls of New Haven's Peabody Museum of

Natural History that they raced through the museum, taking, on average, only twenty-one minutes to review the history of life on earth during the past 500 million years.[200]

The resulting studies, published throughout the late 1920s and 1930s, came as no surprise, for they reinforced the conclusions many museum workers had themselves drawn after a decade of experimentation and stagnation. The reports helped temper earlier celebration and criticism of dioramas, for instance, noting the ongoing public appeal and educational potential of these brightly lit displays while acknowledging their current limitations.[201] Correspondingly, the national infatuation with— and critics' disdain for—these displays ebbed as the nation stood on the brink of a new decade. Though museums continued to build dioramas and diorama halls, dioramas settled into a more flexible position within museums, remaining an important pedagogical tool, but one that could be supplemented or altered to accommodate new messages and convey different concepts. The age of the great diorama halls, those grand cathedrals of romantic nature, was at an end.

Psychologists' reports also helped moderate hype over dynamic displays, a stance echoed by museum staff. Curators and directors continued to tout dynamic displays, but more than ten years of experimenting with such exhibits, particularly the whiz-bang spectacles poached directly from world's fairs, taught them that "no matter how or where presented, interesting physical phenomena become nothing more than amusement features if some satisfactory type of explanatory label is not properly shown in connection with them," mused Cummings.[202] Mechanical and interactive exhibits also had practical drawbacks, frequently breaking and requiring ongoing maintenance. Visitors themselves proved a constant menace; Cummings wryly observed that many a visitor "achieves his greatest joy and self-expression in wrecking something that somebody else has spent long hours in perfecting."[203] "When such situations arise," he concluded, "about the only thing we can do is hire more guards."[204]

Though museums of all types paid lip service to dynamic displays conveying important biological concepts and applications by the end of the 1930s, fragmentation persisted. Thomas R. Adam, who surveyed museums for the American Association of Adult Education, argued in a 1939 report, *The Museum and Popular Culture*, that this fragmentation was ultimately productive, for "there is no cut and dried story of science that could be unfolded through a visual panorama enlightening the casual observer." Natural history and science museums, he wrote, "are engaged in exploring fields largely neglected by institutions of formal education and misleadingly charted by many selfish interests in politics and commerce," namely, "the *effects* of scientific advances on the lives of men and women."[205] Adam's observations were prescient: science and natural history museums took on the shared responsibility of conveying science's role in Americans daily lives in the 1940s and 1950s, even as diverging missions led different institutions to adopt radically different alignments of research, display, and education. But before any of this could take place, museums had to survive the war.

FOUR

Diversifying Displays, Diverging Museums: Postwar Life Science Education, 1941–1956

As the United States plunged into World War II, staff members of American natural history and science museums set aside preoccupations with display and worked instead to contribute to the Allies' efforts. Younger curators stowed well-thumbed field guides in olive drab duffle bags and prepared to ship out. Older curators found their esoteric expertise—patterns of pigeon flight, the geology of volcanic islands in the South Pacific, flea-borne diseases—had become useful, even indispensable, military knowledge.[1] In 1941 alone, the Smithsonian received over seven hundred requests for information on topics ranging from Eskimo clothing, which the Army wanted in order to help them design cold-weather gear for soldiers based around the Arctic, to maps of the migration patterns of snapping shrimp, which the Navy used to obscure Allied ships from Japanese sonar.[2] A year later, the American Museum of Natural History reported that more than one-third of those staffers who had left to serve in the war had "been called upon to apply special knowledge and skills which the Museum helped bring into existence."[3] Government agencies sent museums' entomologists, botanists, and mammalogists across the world to identify uses for local flora and fauna, to locate rubber, strategic minerals, and quinine, and to study the impact of human para-sites on troop health (see, e.g., fig. 11).[4] Curators made the most of such missions, shipping back crates of specimens to enrich collections.[5] "The story of the Museum's participation in this war is one in which we can all take great pride," declared Assistant Secretary Alexander Wetmore in the National Museum's 1943 *Annual Report.*[6]

But science and natural history museums contributed to the war ef-fort in another important way: by teaching troops and citizens about science. Though institutions shortened public hours and closed off halls to conserve oil and gas, museums established new programs for crew-cut soldiers and sailors, who regularly trooped in to listen to curators with years of field experience explain how to survive on the distant islands of the Pacific front. The Colorado Museum of Natural History screened film series about the locations to which U.S. troops were sent.[7] The American Museum used its planetarium to offer courses in navigation and star identification to more than forty-five thousand servicemen.[8] In Chicago, staff members at the Museum of Science and Industry devel-oped new exhibits and demonstrations to introduce basic physical sci-ence principles to Army Signal Corps recruits.[9] For those on the home front, museums built temporary displays and published pamphlets ex-plaining blackouts and other civil defense procedures.[10] Such programs, explained Colorado Museum director Alfred Bailey in 1941, provided "an outlet for our worries of the present and an education of value for the future."[11]

Throughout the wars of 1940s and 1950s, both hot and cold, museum staffs cast themselves as steadfast guardians of civilization in a time of chaos, a stance that forced them to reconsider their current approaches to science and education and transformed displays more completely than interwar reform efforts. Field Museum director Clifford Gregg, himself on active war duty, explained in a 1942 annual report that his museum would maintain "as far as is possible all of the many services available to the public in order that the influence of this institution may still be felt at a time when normal educational and cultural influences are most necessary."[12] The American Museum of Natural History's new director, Albert Parr, agreed that providing Americans with "an understanding and appreciation of the contents of the civilization for which we are fight-ing, and of which our museum is both a part and an exponent is more important today than ever before."[13] "In the new world," Parr wrote, "Museums of Natural Science may well find their rightful place of lead-ership in the thinking of our people and in educational plans for their

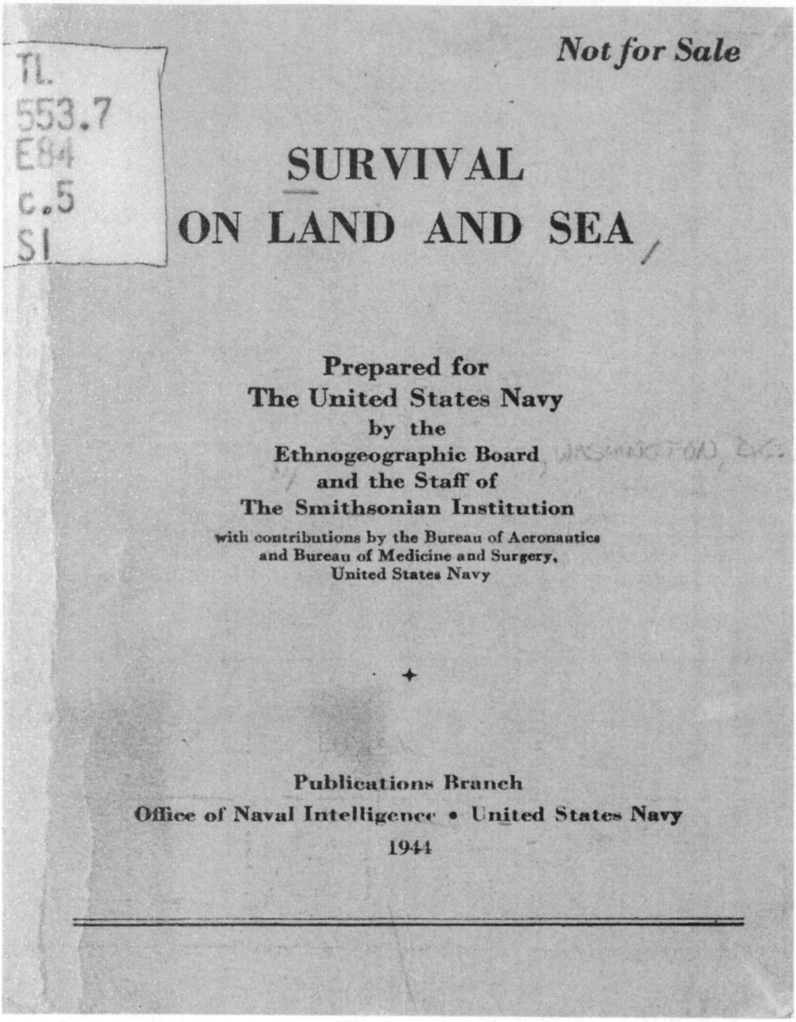

Not for Sale

SURVIVAL
ON LAND AND SEA

Prepared for
The United States Navy
by the
Ethnogeographic Board
and the Staff of
The Smithsonian Institution

with contributions by the Bureau of Aeronautics
and Bureau of Medicine and Surgery,
United States Navy

✦

Publications Branch
Office of Naval Intelligence • United States Navy
1944

Figure 11   Natural history and science museums eagerly contributed to the war effort, providing both the public and servicemen with scientific information. This 1944 edition of the U.S. Navy's widely distributed *Survival on Land and Sea* pamphlet contained information compiled by Smithsonian natural history curators on everything from how to find water to how to identify edible and dangerous plants. (Smithsonian Institution Archives, Image 2009–0434.)

enlightenment."[14] Such optimism about the museum's potential educational stature persisted after war's end. Both science and natural history museum leaders envisioned an expanded role for their institutions in postwar education, describing them as places that could engage citizens of all ages in the newly urgent study of the life sciences. Educators outside museums also began to discuss how the informal educational opportunities museums offered could enhance and supplement school biology teaching, some of which focused on "life adjustment," the educational term du jour for curriculum designed to be directly relevant to citizens' daily lives.

But these shared visions of a more prominent place for museums in national science education gave way to dramatically differentiated institutional missions in the decade after the war. Science museums—a category that now encompassed museums of science and industry—staked out hands-on science education as their territory and continued to rely on interactive display methods as their preferred tools for cultivating it. With the help of newly flush corporate patrons, staff members of these museums developed ever more postwar exhibits featuring experimental biology and medicine. These new exhibits were more closely aligned with the content of life-adjustment biology than dioramas or taxonomic specimen displays, but they failed to realize in full the dramatic changes in pedagogy and content that Depression-era display reformers had envisioned. Science museums retained older, more naturalistic life science exhibits—vestiges of their institutional creations—even as they introduced livelier efforts in biology education through the introduction of live animal displays.

Natural history museums took a different path in the postwar years, electing to reinvigorate their research programs instead as new federal funding for systematics became available in the early 1950s. These institutions developed some new educational and public programs, but straightforward object-based displays and dioramas with minimal updates continued to dominate their halls. Natural history museums without significant research programs had, in theory, a greater impetus to embrace pedagogical change, but attendance was steady and they saw no reason to depart from the standard that had been set by their larger counterparts, so they too left older approaches intact. By the mid-1950s, science and natural history museums had pulled further apart from one another. While they still shared some displays and educational programs, their priorities and pedagogical approaches had become distinctly different.

Museum Learning in the Era of Life Adjustment

Americans had long seen science education as a way to promote social order and economic growth; World War II simply strengthened these beliefs.[15] "The free search for truth by the methods of science has power to rebuild the world and will prevail," geneticist Albert Blakeslee declared to members of the American Association for the Advancement of Science just three weeks after the bombing of Pearl Harbor. Scientists, educators, and government officials echoed his sentiment throughout the war, and so did museum staffs.[16] In September 1943, when the Field Museum of Natural History celebrated its fiftieth anniversary with a day-long symposium on the role of the museum in society, speakers encouraged museums to join the effort to prepare American citizens for a democratic postwar world by educating them about new scientific findings.[17] "The old aspects of truth have become insufficient," museum director Albert Parr told the Chicago audience, echoing the refrain of interwar display critics. Museums could no longer content themselves with dioramas and nature study, he reiterated, but should intensify their efforts to educate Americans about biological concepts, especially those they might encounter in their daily lives. Rather than leaving visitors to draw their own conclusions about the significance of science, exhibits needed to continue to build on decades-long attempts by museums to make explicit science's connections to "individual comfort, recreations and livelihood, and to the national welfare," Parr declared.[18]

Building on foundations laid by Progressive reformers in the 1910s and proponents of civic biology in the 1930s, Parr and other museum leaders continued to argue that exhibits needed to provide visitors—both children and adults—with immediately relevant information about science and nature. If displays could make clear how exhibits about the natural world connected to visitors' own lives, avowed a 1941 American Museum committee report on the future of the institution's human biology and public health displays, then "the meaning and significance of our exhibits on animal biology and natural history would be deepened and extended in a way now impossible." After all, the report's authors admitted, "the life of a frog or a moose can never compete with the human organism in significant appeal to the public except insofar as the frog or the moose contribute to an understanding of the human animal."[19] New exhibits, recommended the committee, should strive to both "transmute into visual expression the findings of those sciences which relate to man" and be oriented to reflect lay, rather than scientific, experience to encourage "the better adjustment of the individual to the world around him."[20] Curator George Vailliant reduced the proposed innovations to a simple slogan: abandon the natural history museum's traditional "God's eye view" of the world, with its emphasis on taxonomic complexity and evolutionary time, for a "man's eye view," which focused on science that related to visitors' own experiences.[21]

This call for relevance echoed the values underlying the emerging life-adjustment curriculum movement, a new effort to reach American students of all backgrounds and capabilities by shaping postwar high school curriculum around older Progressive aims.[22] Though conservative educators denigrated life adjustment for its "well-intentioned but incidental concern with the personal problems of adolescents," as critic Arthur Bestor put it, more liberal teachers and administrators argued that its attempt to link science and other subjects to students' life experience remained a worthy goal. Life adjustment and similar curricular reforms were, declared a 1951 U.S. Office of Education report, essential in helping develop "knowledge as a tool of civic competence."[23] They were also a staple of effective pedagogy, observed classroom teachers. "When a teacher in the rural midwest postpones difficult definitions and starts biology class with an ear of corn from the family crop, he is merely using a psychological approach to understanding basic principles and shows no disrespect for logical sequence," Catholic school teacher Sister Mary Janet wrote in a 1954 issue of *Educational Leadership*.[24]

The approach resonated with a surprising number of university-based biologists and educators. "Whether you like it or not," University of Pittsburgh botanist Robert Griggs explained to National Research Council officer Detlev Bronk in 1946, "there is in the lay mind a very distinct category labeled biology. . . . The subjects of the biological sciences, in effect, comprise the basic attributes of all living things. . . . Upon [this] foundation . . . the new social relationships of man must be adjusted."[25] Education professors also championed this curricular turn toward practical application and urged teachers to abandon traditional disciplinary-based boundaries in biology, and instead, as the University of Colorado's Harl Douglass wrote in 1949, teach "the science . . . called for if one is to live his life as it must be lived today."[26]

Designed to prod students into acquaintance with fundamental scientific principles and improved understanding of the wartime explosion of scientific applications, the life-adjustment approach to biology education also promised to further political and scientific goals.[27] As a result of government newsreels and media coverage of wartime biological and medical advances—especially penicillin, malaria control, and scientific agriculture—most American citizens were aware that science had tremendous power to affect their daily lives. This fact took on a darker cast after the devastation of Hiroshima and Nagasaki. By war's end, public understanding of basic scientific principles seemed more important to the nation's future than ever before. "The survival of modern civilization depends upon an understanding and control of scientific techniques whose power for good and evil dazes human imagination," declared a 1950 report sponsored by the American Association for the Advancement of Science. "The time is short. The task is nothing less than to lift a whole generation of American citizens to a level of knowledge and human goodness which has hitherto been attained by only a small fraction of our people."[28] Scientists and policy makers, worried that the frightening military applications of science would undermine public support for scientific research, hoped that life-adjustment science education would give American youth a more optimistic perspective about science's potential to transform society in positive ways. American citizens needed to understand better the role that science could play in postwar peace, wrote geologist Howard Meyerhoff in 1956, in order to shake off what he called "the anthropological hangover that thwarts the unity *science* and technology could bring to society were the processes of education and communication more effective."[29] "If science is to develop vigorously and to serve the world as it ought to," Yale biologist Edmund Sinnott concurred, "it should not be wrapped in mystery but must be understood, at least as to its spirit and methods."[30] Recent medical advances, physician Dean Clark told Boston Museum of Science director Bradford Washburn, had made life science and public health education "of paramount importance," but "the rate of change of knowledge in these fields is so rapid [that] what is really necessary is continual public re-education in health matters."[31]

Postwar scientists and science educators consequently sought to impress American students with better living through chemistry and physics, but they were especially diligent about applying the life-adjustment angle to biology curricula.[32] "Biology, for one reason or another, is the only science course a great percentage of our high school youth takes," explained California high school biology teacher Barbara McEwan in 1950. "Considering the overwhelming importance and influence of science in our world today, the thought of the vital role which biology teachers play in the development of our future voters is at times terrifying."[33] To this end, biology educators like Benjamin Gruenberg suggested that teachers ask students to investigate the manifold effects of science on their own lives and the surrounding natural world. Rather than relying on research biologists' more specialized categories of knowledge, he encouraged teachers to cut material into units related to "daily affairs of people, the common problems, the common curiosities, the common concerns" that faced all Americans.[34]

Although only a few states formally adopted life-adjustment curricula in their schools, educators in a wide variety of institutions agreed with its general principle of increased accessibility and emphasis on daily encounters with applied science.[35] Staff members in natural history museums explicitly embraced the idea. Life adjustment's human-centered, comparative emphasis fit nicely with natural history museums' contemporary view of their educational mission, articulated in a 1946 American Museum annual report: "Why does man act as he does? The more secrets we can wring from Mother Nature the better we shall understand ourselves and be able to carve out our own destiny with greater intelligence."[36] Staff members' knowledge that they catered to adults and schoolchildren deepened their interest in this pedagogical approach. Museums could step in where high schools and universities left off, offering citizens lifelong opportunities to educate themselves informally about science, University of Chicago president Robert Maynard Hutchins reminded staff members, and as a result, they had a special responsibility to teach "men how to live human lives and how to live them together on a worldwide basis.[37]

Museum staffs signaled their awareness of the life-adjustment trend by building displays that tried to educate visitors of all ages about biology and medicine. Through direct appeal to familiar experience, science museums and museums of science and industry made this shift more gracefully than natural history museums. In an attempt to present pristine landscapes, interwar natural history museums had removed traces of the human experience from habitat diorama halls, so they had to hustle in the 1940s and 1950s to make their most iconic exhibits on nature seem relevant to visitors' lives. They introduced life adjustment into new halls but did so gradually, to the great frustration of curators intent on thrusting their institutions into the frontlines of biology education. In contrast, museums of science and of science and industry rapidly and easily adopted the approach, constructing dozens of exhibits on human biology and laboratory-based biomedicine, two topics that neatly linked visitors' lives to the life sciences.

As early as 1941, for instance, Chicago's Museum of Science and Industry opened a thirty-five thousand square foot Hall of Health illustrating aspects of medical and veterinary science, public health, dentistry, and pharmaceuticals. To the delight of its corporate sponsors, the hall celebrated recent strides made by scientific researchers in collaboration with industry, aiming, in the words of curator Evan Carey, "to present to the public the story of their contributions to the progress of civilization and human welfare."[38] Flashing lights and voiceovers instructed visitors about "the intricacy of the living human machine," Carey proudly explained, and revealed the drama of "the scientific story behind the physician's services and opinions regarding health and illness."[39] The displays in the Hall of Health included a model operating room ("equipped with combinations of lighting for illumination of the room and maximum light directly on the patient's wound") and an interactive exhibit on the effects of hormones on the female body, where visitors could illuminate glands and secretions activated by the monthly menstrual cycle. Likewise, in 1955, the Boston Museum of Science began soliciting local doctors and surgeons for help constructing a "hall of man and medicine." Director Bradford Washburn had one "very lively discussion" with a physician from the U.S. surgeon general's office, who suggested "putting one of the most modern artificial limbs on exhibit—perhaps push-button controlled." The point of such an exhibit, the two men agreed, would be to display both "the remarkable specialization of the human arm" and "how well man is learning to reproduce the motion of the arm and hand mechanically."[40]

One biomedical display of the mid-1950s ultimately became one of American science museums' most iconic exhibits: the walk-through heart. First designed and fabricated in 1952 for the Museum of Science and Industry in Chicago by the medical illustration studios of the University of Illinois at Chicago, the heart model presented up-close, anatomically correct views of the muscle and valve structures, while also making use of a technique that curators had long relied on—the appeal of oversized, immersive environments. Visitors entering the heart heard the thick thud of its beat and walked through glowing red plastic tunnels, eventually leaving through its arteries. "The exhibit is designed," the museum reported to its members in a 1952 newsletter, "to give the public a complete understanding of how the heart works so the problems involved in combating the present greatest 'killer' malady will be readily understood."[41] On exiting the heart, visitors could also see supporting exhibits attempting to educate visitors about the impact of lifestyle choices on heart health. One exhibit instructed visitors to squeeze a lever seventy times a minute, mimicking the heart's continuous exertion. Another push-button panel provided visitors with scientific answers to common health questions involving the heart. Museumgoers could watch blood cells journey around the body, via a model with lights that traveled from

Figure 12   Champions of life-adjustment education believed that children and adults would be inspired to lead healthier lives if they understood more about the human body and the findings of biological science. Chicago's Museum of Science and Industry's *Cardiac Kitchen* exhibit, pictured here in 1952, epitomized postwar museums' embrace of this pedagogical model. Built with the sponsorship of the Chicago Heart Association, the display depicted how cooking, ironing, and other household chores offered women new opportunities for heart health. (Photo courtesy archives, Museum of Science and Industry Chicago.)

the heart to the toe, the finger, and the head in the time it took a normal adult's circulation system to function. Impressed by the heart's success, Philadelphia's Franklin Institute (in 1954), the Boston Museum of Science (in 1960), and the Cleveland Health Museum (by 1965) installed their own versions of the iconic display.

In Chicago, curators developed other exhibits to connect the giant heart to the life-adjustment impulse. Displays surrounding the giant model featured the heart's reaction to emotional and physical activity and compared the human heart to those of other animals—the pencil-eraser sized heart of a hummingbird, for instance, or the sixty-pound heart of an elephant. To appeal to housewives, Dr. Margaret H. Austin worked with the Women and Children's Hospital of Chicago to create an interactive *Cardiac Kitchen* exhibit (fig. 12) to stand alongside the giant heart. The display instructed patients with heart conditions on how to accomplish housework while making necessary adjustments to limit physical exertion. The display advised homemakers to replace metal kitchen utensils and appliances with lightweight plastic and to equip chairs with wheels so they could be pushed rather than carried or dragged.[42]

Enthusiasm for biomedical exhibits at both science and natural history museums intensified as the 1950s progressed, and the most successful exhibits were directed not at students but at adult visitors. In Boston, Washburn ordered the installation of several unusual medical exhibits, designed, he explained, "to acquaint the layman with medical procedures, [and] remove fear of the unknown." The new halls included an exhibit on appendectomies ("ten realistic plastic models show how an appendix operation is performed," bragged the hall's guide) and an exhibit on the biomedical uses of atomic research, on tour from the U.S. government's Atomic Energy Commission.[43] One radioisotopes display featured an innovative hands-on component: by waving a wandlike Geiger counter over four plants, one of which was "hot," visitors could determine their radioactivity. A 1950 *Museum of Science Newsletter* reported that its Geiger counter had once gone off when waved near a visitor who stood nearby. It was later revealed that there were medically administered radioisotopes in that visitor's bloodstream from a procedure performed earlier in the day, resulting in what Washburn proudly described as a "teachable moment" on the medical uses of radioisotopes.[44] Other science and natural history museums hosted variations of the same display. In 1948, the American Museum had hosted the same exhibit, and in 1951, Chicago's Museum of Science and Industry replaced the plants with frogs, instructing visitors to find the irradiated creatures in a small living pond ecosystem.[45]

While museums' embrace of life adjustment resulted in new content, it did not always translate into fresh pedagogical approaches. Sometimes curators literally could not think outside of the box: in a misguided attempt to make the science of animal behavior seem more relevant to visitors, Smithsonian curator Herbert Friedmann arranged mounted chickens in a barnyard setting, then inserted the whole thing into one of the museum's old mahogany cases (fig. 13).[46] Other new displays differed from older exhibits only in the personalized tone of their explanatory panels.[47] The Boston Science Museum's merrily titled *How Your Life Began* exhibit, for instance, used models, embryo specimens, and color slide projections to tell a visitor-centered story "of the beginnings of human life from the moment of conception until 20 minutes after birth." Though its language hammered home the message that knowledge of biology would clearly illuminate visitors' understanding of their own

Figure 13   This 1956 image of ornithologist and curator Herbert Friedmann walking through the
Smithsonian's new Bird Hall illustrates both the vision and the limits of initial postwar
display reforms. To simulate a more natural and more compelling barnyard setting,
designers inserted new background fencing, hay, and chicken mounts in old museum
mahogany cases. The exhibit was then placed against the walls of the cavernous old
gallery, framed only by support pillars and natural lighting "controlled" through
sheet-covered windows. (Smithsonian Institution Archives, Image MNH-11111A.)

lives and bodies, its organization and presentation was almost identical
to Chicago's Museum of Science and Industry's 1930s prenatal display.[48]
And while the Field Museum trumpeted the redesign of its eleven hundred traveling miniature dioramas, boasting that the renovated exhibits
would "answer questions about things within a young person's immediate experience" and "fit the new needs of science teachers" in thrall to
life-adjustment biology, the much heralded overhaul was usually limited
to rewording accompanying labels.[49]

Informal Education:
Collaborations between Schools and Museums

At the behest of worried educators, postwar museums also attempted
to augment school science education with new forms of educational

programming. Discussions of museums' potential to supplement school science had intensified after the war as scientists and educators sought to address what they perceived as a deepening crisis in science education. Debates about whether teachers should emphasize life adjustment or should instruct students in the finer points of disciplinary knowledge were ongoing, but after 1945, scientists and educators concluded that both approaches were failing, and badly—largely because information about science was conveyed to students in such a dry way.[50] Few students understood the scientific facts they were required to memorize, and as Chicago high school principal Butler Laughlin warned, only a few Americans knew enough scientific methodology "to react intelligently to the problems they meet every day.[51] Neurophysiologist R. W. Gerard sounded an equally alarmist note about general biological education in 1956, declaring that "most layman are dimly, if at all, aware of the meaning of the word 'biology' . . . despite the experience in the last war that, whenever one had to deal with a complex multivariable problem, the chances were that it would be a biologist who solved it."[52]

Professional groups, government agencies, and foundations all began to study how to improve science teaching, especially in biology courses, which was the last formal science class most students took.[53] In 1947, the interdisciplinary professional group, the National Society for the Study of Education, devoted its annual publication to the topic "Science Education in American Schools," and the American Association for the Advancement of Science (AAAS) formed a special committee called the Cooperative Committee on School Science and Mathematics, whose membership drew representatives from all the major scientific societies as well as the professional organizations of science teachers.[54] By 1952, scientists, policy makers, and foundations agreed that biology education needed to be revamped entirely, and the National Academy of Sciences Committee on Educational Policies, the National Science Foundation (NSF), and the Rockefeller Foundation all began to fund nationwide efforts to do so.[55] In 1955, the Southeastern Joint Conference of the Science Teaching Societies met in Atlanta to assess the state of high school biology teaching and develop a series of recommendations based on their findings. Their report, promptly published in the *American Biology Teacher*, was not entirely pessimistic. Mississippi's delegate R. W. Caylor noted that the high school biology teacher was finally being taken seriously, "made to feel that his position is more than an adjunct to the coaching of football." Several institutions reported an expansion of teacher training in biology, and Virginia even reported support from nonbiologists: NSF's Scientific Manpower Commission and University of Virginia chemistry faculty member James W. Cole Jr. advocated for the group's plans before the Virginia State Academy of Sciences. "The outlook was never better," Longwood University's George W. Jeffers concluded, "for this body of public school science teachers becoming an 'action group.'"[56]

Classroom educators, who only a few decades earlier had avoided field trips to museums, were among the first to see collaboration with museums as a natural place to bridge the divide between life-adjustment and disciplinary-based science pedagogy. California biology teacher Barbara McEwan, for example, suggested the best way for teachers to "put life into life science" and "recruit the energies" of their students was to arrange field trips to museums, zoos, and botanical gardens, as well as to sites of scientific research.[57] McEwan was not alone: as early as 1942, a survey of the nation's high school biology clubs indicated that their faculty leaders frequently sponsored field trips to museums and encouraged clubs to prepare their own "museum specimens" for school biology classrooms.[58] By 1951, Battle Creek, Michigan, was heavily invested in developing its own "Public School Museum" of natural history, but its director relied heavily on advice from larger national science museums about educational exhibits.[59]

Natural history and science museums responded to renewed teacher interest by cultivating close relationships with local schools.[60] The Colorado Museum of Natural History—which, to keep municipal funding flowing, had become the city-sponsored Denver Museum of Natural History in 1948—reestablished its defunct educational programs in 1954, working with Denver Public School teachers to create museum tours that would correspond directly with the school system's science courses.[61] Chicago's Museum of Science and Industry did the same, asking exhibit demonstrators to plan tours around school curricula. The effort worked well at Chicago: attendance at the museum leaped 33 percent between 1947 and 1949 alone. Between September of 1949 and June of 1950, more than three thousand groups from inside and outside of the country visited the Museum of Science and Industry, most of which were elementary and high school students. According to the museum's in-house magazine, *Progress*, teachers lauded the program. Cooed one local educator: "The Museum has created an educational climate that makes it a pleasure for the teacher as well as the pupil to be a guest."[62] And in 1953, the Massachusetts State Legislature approved one of Bradford Washburn's signature initiatives: a five-year, $50,000 contract that permitted all Boston-area students to attend the newly reopened and resited Museum of Science free during school hours with their teachers. Teachers also received training and supplementary material to plan their visits to the museum and subsequent classes based on what they had seen there.[63] Three years later, Washburn obtained funds from the Lilly Foundation to build a lecture demonstration hall and announced that the hall would be "the nerve-center of our program of intensive science training for elementary teachers and camp counselors."[64]

During the late 1940s and early 1950s, both natural history and science museums worked to strengthen relationships with children through less formal educational opportunities, and scientists and educators cheered the results.[65] Looking at museum displays might rouse students' curiosity, but it didn't necessarily result in sustained interest, agreed museum staff. To elicit a real passion for science in children, they concluded, museums needed to support their ongoing scientific inquiry. As a result, museums embraced the science fairs that had become a positive mania in the United States—hundreds of colleges, science clubs, state academies of science, and corporations joined together to sponsor these contests— and institutions jumped to exhibit competitors' hand-crafted posters of frog reproduction and carefully constructed pipe rockets.[66] Postwar museums revived their own "junior naturalist" and collecting clubs and played host to the science clubs that had sprung up in high schools during the war and were proliferating in peacetime.[67] These clubs appealed to children—the monthly meetings of the Boston Museum of Science's Saturday Morning Science Explorer's Club regularly attracted several hundred children.[68] Clubs often evolved into youth radio and television shows. In the late 1940s, for instance, the Milwaukee Public Museum initiated a series of radio broadcasts based on the museum's Explorer and Junior Explorer Clubs—the show was titled the *Explorer's Club of the Air*.[69] The Boston Museum of Science offered *Ask Science Park*, a radio show emceed by Washburn himself, and *Wonder World*, a weekly television program that had an estimated audience of a hundred thousand local homes.[70] In 1948, the Buffalo Museum of Science also experimented with the new medium of television and began to host *Your Museum of Science*, which, in 1951, became a weekly program. Topics included evolution of the elephant, casting of animal tracks, the identification of various mushrooms, flowers, and birds, and how-to demonstrations, including a show devoted to the construction of birdhouses.[71]

By the early 1950s, scientists and educators believed such informal museum efforts, when supported by teachers and parents, had played an essential role in improving science education. "What grass roots are to agriculture, science clubs are to science education," Science Service founder Watson Davis wrote in 1952. "These bands of boys and girls are a perpetual youth movement, constantly renewed by the innate and undulled curiosity and exploratory spirit of those who are discovering, through doing, the world about them."[72] Davis also complimented museums' venture into popular programming, crediting such efforts with promoting a "rise in science understanding" among Americans.[73] Yale biologist Edmund Sinnott, a prominent postwar advocate of a humanist approach to biology education, also toasted museums' capacity to educate in pleasurable, informal ways. Perhaps more than other educational institutions, "the modern museum serves more and more as a center to awaken interest in science and to disseminate knowledge about it," Sinnott wrote in 1950.[74]

Not everyone saw museums' expansion of educational outreach in such rosy terms, however, and museums' partnerships with schools were frequently criticized on pedagogical and scientific grounds. Alma Wittlin, an educator whose study of museums was funded by the Carnegie Corporation, published a scathing critique of "state of the art" approaches in museum education in 1949. Quoting the foundation's 1937 annual report, Wittlin argued that there was indeed "a need to upset conventions in order to close the gap between what museums are doing and what the world expects of them." "As a rule the main functions of a museum are defined as providing instruction and enjoyment, and serving for the preservation of objects," she wrote. "Yet this definition . . . by its width suffers from vagueness and does not really outline ways of action. . . . Instruction—in what and for whom?"[75] Many curators and trustees at natural history museums also objected, believing outreach programs had reduced their institutions to serving as mere adjuncts to public school systems. The Los Angeles Natural History Museum's chairman, A. W. Bell, published a 1941 editorial in *Science* detailing what he saw as the injustices suffered by "those scientists who are striving to create . . . a respect-worthy museum" in a city that had consigned the institution to a site for field trips. Over the course of the previous three years, Bell noted, governance of the institution had been slowly transferred to Los Angeles County supervisors, while scientists had been demoted to an advisory role. This had resulted in a situation "at wide variance with the plan of the [museum's] founders," for the museum now served "principally as an instructing agency for the schools and for circulating study materials." These shifts, he concluded, had worked "to the detriment of science in Southern California."[76]

Despite the fanfare that accompanied collaborations between museums and schools, they fell far short of the decisive pedagogical revolution that educators had hoped they would produce. Institutions could barely cope with the thousands of young children scuffing their floors and smudging their exhibits, much less meet the continuous demand for innovative yet scalable educational programs. Many museums built new wings or moved into new buildings as a result, but such investments took time, and they struggled to educate students about the life sciences in temporary settings. When the Boston Museum of Science closed its doors between 1947 and 1950, for instance, it resorted to a traveling nature tour to accommodate teachers. But the station wagon they turned to was no permanent home for the museum's squawking, smelly live animals, and museums' life science programs flagged without a more expansive, physical infrastructure.[77] The problems weren't limited to physical plants. Educators hoping to use museums to help update schools' science programs were quickly disillusioned by the mismatch between museums' grand talk and their actual efforts. In 1956, intent on improving his district's science programs and intrigued by Washburn's boosterism, Boston's commissioner of education wrote to request assistance from the Boston Museum of Science; Washburn directed him to the museum's Education Department. Teachers protested that the resulting partnership was a waste, complaining that the museum did not supply teachers with sufficiently detailed content and rarely provided teachers with enough information to "precondition" their students about what they were likely to see and do at the museum on field trips.[78]

Science and natural history museums themselves admitted that educational programming, however important, still played second fiddle to the work of exhibition. More than ever before, museums described exhibition as the embodiment of their institutional missions. Displays reflected curators' research agendas and staff ideas about pedagogy, and they reflected staff members' understanding of museums' responsibilities to visitors. As early as 1941, the American Museum's Committee on the Coordination of Exhibits had concluded that "the vitality of a Museum is measured in its exhibition," an opinion the museum's incoming director, Albert Parr, shared. New exhibits, he reiterated in 1947, were the foundation "on which we must base our continued existence as an effective scientific educational force."[79] Educational programs were important, agreed Boston's director Bradford Washburn, but they could not exist without exhibits. It was exhibition that attracted people to the Boston Museum of Science, he wrote in 1950, persuading them of "the Museum's contention that 'Science is Fun.'" "More than a thousand visitors a week are now paying admission to see the exhibits," he added, and public response to "the informality [of exhibition], the interested attention of staff, the varied demonstrations and activities offered" indicated the potential of displays—and museums—to assist in nationwide efforts to improve science education.[80]

Pushing Buttons and Petting Owls: Corporate Sponsorship and Interactivity at Postwar Science Museums

Science and natural history museums evolved in distinctly different directions in the late 1940s and 1950s, a deliberate divergence that resulted in a realignment of the nation's museum community. Where science museums had once seen themselves as hybrids of natural history museums and museums of science and industry, sharing qualities of both kinds of museums, the institutions aggressively repositioned themselves in the postwar years. In contrast to natural history museum directors, whose official line was that research and education missions complemented one another, directors of science museums announced that museums needed to choose between research and public education if they were to thrive. Research programs retarded the development of exhibition and education, argued Boston Museum of Science director Bradford Washburn, "by diverting attention and dollars from the basic issues at stake."[81] "Our Museum of Science is primarily a teaching institution," he explained in a 1956 annual report, proudly noting that science teachers, not research specialists, dominated his staff. Staff members were hired not on their records of research but for their "ability to create and sustain an enthusiasm for learning."[82] Washburn put his money where his mouth was: when one wealthy patron offered to fund the salary of an ecologist who would devote his time exclusively to research, Washburn turned him down. Though members of the museum community had carefully distinguished science museums from museums of science and industry in the 1920s and 1930s, by the 1950s, the boundaries between the two had collapsed. The institutions became almost interchangeable in the minds of museum professionals, and they began to use "science museum" as a one-size-fits-all moniker.

Rather than relying on their scientific staffs to generate displays, then, postwar science museums increasingly turned to corporations for ideas and assistance. The historical involvement of corporations in museums of science and industry made them logical sponsors for postwar exhibits in science museums, and exhibit directors, keen to develop new displays that distinguished their institutions and their own careers, forcefully pursued this sponsorship throughout the 1950s.[83] In the economic turmoil immediately after the war, corporations' donations to science museums declined, a problem for which museum administrators were quick to blame themselves. "Industry and business as a group has become educated to contribute to Community Chest and Health drives, but in comparison has contributed little to science museums," Bruno Gebhard, director of the Cleveland Health Museum, wrote in 1946. "This is largely the fault of museum people themselves. In this country money is available, but museum people have not 'sold' the advantages of their museums."[84] By the 1950s, however, science museums and corporations courted one another more aggressively. Corporations' long history in science museums and their inextricable relationship with science and government in the postwar period made them logical sponsors for postwar exhibits in science museums. Indeed, throughout the 1950s, corporations sponsored most of the biological and medical exhibits at Chicago's Museum of Science and Industry.

Corporations' interest in sponsoring interactive life science displays pleased science museum directors, who believed interactivity the most effective way to educate museumgoers about the biological sciences. Natural history museums might boast authentic objects, but dynamic celluloid models and lights that flashed when visitors flipped a switch captured far more attention than bone fragments or stuffed animals under glass ever could. As Lenox Lohr, the director of Chicago's Museum of Science and Industry told *Time*, "The average man is not interested in things over his head." Interactivity, not taxonomy or spectacle, was the key to educational display, he maintained, for "people are most attracted by exhibits they may operate themselves."[85] Most curators at the Chicago museum shared Lohr's perspective, and the director fired those who did not.[86] Thomas Hull, formerly the chief diagnostic laboratory scientist of the Chicago Public Department of Health and now Lohr's director of medical exhibits, happily asked corporations to help build biological and medical exhibits at the Museum of Science and Industry.[87] Companies or trade organizations proposed some exhibits to Hull, but most of the museum's biomedical displays resulted from the curator's solicitation of industry sponsors. Push-button displays on new biomedical topics—nutrition, disease, anesthesia—rapidly replaced dioramas on historical medical practices and bas-reliefs of Hippocrates.[88] By 1953, visitor surveys indicated the medical section was one of the museum's most popular attractions.[89]

Corporate sponsors found none too subtle ways to exhibit new developments in health sciences, all the while pushing corresponding products. Abbott Laboratory, for example, funded an exhibit on the effects of vitamin deficiencies, which allowed visitors to illuminate images of the fourteen then-known vitamins—many of them sold by Abbott. Each time visitors lit up a vitamin, corresponding organs of a three-dimensional transparent human body also lit up, to illustrate those parts of the body affected by specific vitamin deficiencies.[90] A few years later, Abbott sponsored an exhibit called *The Conquest of Pain*, intended to educate visitors about contemporary scientific understandings of pain.[91] The exhibit, its designers argued, should not attempt to cover the wide history of anesthesia but should instead provide "confident assurance that anesthesia and the doctors who are in this science are to be given trust in life's crises of the present and future."[92] At the introduction to the exhibit stood an eight-foot-tall figure of a human male composed of nerves, "a matching piece for the . . . heart which stands across the balcony," gushed the museum's newsletter. Push buttons activated flashing lights that represented how nerve endings received pain messages, then sent those messages on to the spinal column and finally to sections of the brain. The exhibit exploited visitors' inner masochists; by pushing buttons, museumgoers could inflict various forms of pain on the body and watch its neurological and physical reactions. To watch the body react to a burn, visitors could push a button on the hand, which activated lights from the hand to the brain, then lit up the hand muscles, which retracted in response. The exhibit also featured an anatomical model to which visitors could "administer" different anesthetics and observe the consequences for a patient's well-being. This display suited Abbott's objectives perfectly. Anesthesia was the company's largest and most profitable pharmaceutical product, and the Chicago display premiered just as the company was bringing a new topical anesthesia to market.[93]

Corporate funding, however, was not always such an easy fit: even when science museums and industrial giants could agree that collaborating on exhibits and educational programming might be desirable, they could not always agree on the medium or the message. In 1955 at the Boston Museum of Science, for instance, an official from Davis and Geck, a Connecticut surgical supply subsidy of American Cyanamid, contacted Washburn about showing the company films in the museum's auditorium. "We know that you have an ideal location to achieve a medical audience for postgraduate teaching programs," the corporate official argued. But Washburn declined to show the films, noting that few, if any, physicians would go out of their way to see them. Less than a year later, when Washburn wrote to pharmaceutical firm Smith, Kline and French requesting funding for an exhibit on "medical facilities and services," the company deemed the planned exhibit "a public service" but declined to ·

offer any assistance. The corporate vice president who responded told Washburn that, while they had checked with their Public Relations Department, the head of which was "distinctly exhibits minded," the project nevertheless "just seems a little out of our line."[94]

The chick hatchery display at the Museum of Science and Industry in Chicago represented another case in which corporate funding enabled the creation of an interactive exhibit with a clear marketing message. Hull had toyed with the idea of creating a live animal exhibit as early as 1952, proposing a "research zoo" of laboratory animals that would educate visitors about new advances in biomedical sciences.[95] Less than a year later, the museum instead mounted a temporary live animal display to attract young visitors during the Easter season: a large incubator in the International Harvester exhibit on farming. The incubator's glass top, curators believed, would give visitors the chance to watch developmental biology in action.[96] The chick hatchery was so popular that, when Swift & Company approached the museum later that summer about building a permanent exhibit on the nutrition, director Lenox Lohr suggested the hatchery become part of the new installation. The resulting Food for Life Hall was designed to display the "complete story of man's food: where it comes from, how it is adapted to fit his needs, how the various elements entering into the production of food affect man," as reported the museum's house publication, *Progress*.[97] Though the permanent exhibit included some displays of plant ecology and photosynthesis, the crowd-pleasing chick hatchery remained its centerpiece. By the end of the decade, the success of the chick hatchery had resulted in a full-scale animal nursery, where visitors could feed trembling piglets, lambs, calves, rabbit kits, guinea pig pups, ducklings, turkey poults, and, in a disturbing twist on the hall's agricultural food theme, puppies, kittens, and baby hamsters.[98] Those who just couldn't get enough of baby animals could follow the activities of this large and motley flock in the museum's annual reports.[99] The addition of living animal exhibits reflected popular and academic interest in animal behavior, as well as administrators' shrewd belief that live animals at museums could draw the same kinds of sizable crowds they drew at contemporary zoological parks and children's petting zoos.[100]

Devotees of a discipline-based approach to science education questioned the value of petting and feeding animals, but science museum staffs insisted live animal displays were a sound form of science pedagogy and introduced them to their own institutions. The resulting displays were designed to afford visitors more engagement with living specimens than they had gotten from natural history museums' aquariums, terrari-

Figure 14   A Boston Museum of Science education curator displaying Herkemiah, a porcupine with deactivated quills, in the museum's main gallery in 1950. Such live animal demonstrations, reminiscent of contemporary zoo exhibits and popular with young visitors, were considered pedagogically interactive. (Museum of Science, Boston.)

ums, and microscope displays, usually by offering visitors to the chance to interact with a single "star" specimen.[101] Trained demonstrators encouraged a lucky few to touch, feed, and interact with the specimen in question while other visitors posed questions about the animal's habits and behaviors. Many of these exhibits installed animals in spaces designers intended to be visitor centered; at his own Boston Museum of Science, Washburn was especially fond of what he described as "a miraculous, revolving, push-button exhibit with thirteen cages for live New England snakes."[102] The museum also featured what exhibit designers dubbed "animal demonstrations," which encouraged carefully monitored, hands-on contact between visitors and animals. Museum staff invited trepidatious children to touch animals like Black Beauty, a seven-foot-long indigo snake with unique camouflage markings, or the slightly more approachable Herkemiah and Cuddles, two pet porcupines whose quill release mechanisms had been disabled (fig. 14).[103]

Live animals were worthwhile entrees to science education, according to science museum staffs, for they demonstrated, explained a member of

Figure 15   "Science in Action," read the caption of this picture of the Boston Museum of
             Science's "Junior Assistants in the Animal Room" published in a 1956 *Member
             Newsletter*. The assistants' back-of-the-house work not only made live animal
             exhibits possible, argued director Bradford Washburn, but also offered assistants
             invaluable lessons about animal biology, behavior, and husbandry. (Museum of
             Science, Boston.)

the Boston Museum of Science Education Department, "the wonders of
the world we live in" and were important for "motivating the young to
learn more about their environment." Washburn and his staff members
considered their own live animal exhibits to be an excellent marriage of
the museum's dual commitments to interactivity and exposure to scientific research.[104] Whereas natural history museums had shunted care of
museums' animals onto lowly laboratory assistants and Works Progress
Administration workers in the 1930s and 1940s, postwar science museums declared that animal care afforded students valuable hands-on scientific experience. By handling the animals on a routine basis and learning
what it took to keep them healthy and happy for museum use, teen-
aged assistants were participating in what biology educators had deemed
"science in action," explained the Boston Museum of Science newsletter
(fig. 15).[105]

The Boston Museum of Science's most famous live animal demonstration embodied the integration of exhibits and school partnerships
by postwar science museums, as well as the limited interactivity these
institutions claimed as their hallmark. After finding a fledging owl and
reviving the fluffy ball with a medicine dropper, two Milton, Massachusetts, residents had called Washburn to see if the museum wanted the
tiny bird. Christened Spooky, the young owl was featured in exhibit halls
after only a few weeks at the museum, and to museum staff, the owl took
on near-human status as a charismatic educator. Chauffeured by assistant
director of education Gilbert Merrill and gawked at by nearby drivers
amazed to see a great horned owl perched on the front seat of a sedan,
Spooky frequently visited New England schools and community groups,
chalking up nearly fifty-two hundred miles of road travel in a single year
(fig. 16). By the mid-1950s, Spooky had become a celebrity outside of the

Figure 16   This 1959 photograph pictures Spooky the Owl accompanying Boston Museum of
Science education curator Gib Merrill on one of his many tours to local schools and
community centers. Though the museum described Spooky as a model for interactive
education, younger visitors didn't get much of a chance to interact with the owl and
sometimes questioned whether the animal was live or stuffed. (Museum of Science,
Boston.)

museum's walls, appearing in *Look* magazine and on *The Today Show*.[106] Merrill claimed that Spooky encouraged younger visitors' engagement with science. According to Merrill, Spooky's initial stillness forced children to ask themselves if the owl was a live animal or a stuffed specimen, and then wonder, if it was alive, why it was not acting like the fierce predator they knew an owl to be. This may have been true for some small students, but others did little more than stare at the animal. For those students and visitors who really wanted to learn about owls, Spooky was a revelation; for others, the owl was merely a matter of passing interest, a feathered form of entertainment.

Insistence by science museums that such exhibits were exceptional blatantly ignored the historical and ongoing use of live animals in display and educational programming by natural history museums. Philadelphia's Academy of Natural Science's chief of exhibition Harold T. Green, for instance, built a live animal house in the academy's courtyard in 1956, placed lovebirds in the Bird Hall and pythons in the Asia Hall, and in 1958, began to offer a weekly demonstrator-conducted "Wild Animal Show." The shows were so popular that they became daily events and moved from an exhibit hall into the lecture hall. (Parents of small children must have been relieved at the decision to shift the location to a more structured venue, as the show featured, among other animals, a red wolf, a coyote, and, before it became too hard to handle, a puma.)[107] Eventually museum administrators appointed Green to the position of chairman of live natural history. Natural history museums across the country took a similar tack, though usually on a smaller scale: the Denver Museum of Natural History put a live great horned owl on display, while at the American Museum, Alice Gray, an educator and entomological expert at the American Museum, presided over a menagerie of living insects, and educators at the museum's Natural Science Center conducted their own live animal shows.[108]

Yet science museums maintained that such work was their particular province and insisted that interactive displays remained a defining characteristic of science museums and their unique contribution to national science education. "Our exhibits," declared Boston Museum of Science director Bradford Washburn in 1956, "demand the *active participation of the visitor* in the learning process" and lent "a new dimension to the word Museum."[109] In the early 1950s, science museum staffs made grand plans to develop more and more interactive life science exhibits, declaring that more interactive displays would finally achieve the kind of science education that biologists and museum reformers had envisioned as early as the 1930s.[110] Under the banner of modern pedagogy, museums installed many of these displays by the end of the decade, but the ability of institutions to demonstrate that students learned additional things from the newly interactive displays remained, in this period, elusive.

A handful of science museum staff members began to think more introspectively about the kind of experiences they were providing visitors, designing studies and surveys to measure visitor reactions and educational outcomes from exhibits.[111] Lucy Nielsen, an associate curator at the Museum of Science and Industry in Chicago, was at the forefront of this effort. As a youngster Nielsen was a precocious student: admitted to the University of Chicago at the age of fourteen, she excelled at both science and playing the violin.[112] When the war stripped the museum of its male curators, administrators increasingly turned to young women like Nielsen to fill in. The museum had hired Nielsen to rewrite exhibit labels but soon asked the twenty-year-old to introduce basic principles of physical science to young recruits. In this role, Nielsen observed firsthand the difficulties of informal education. Soldiers were impatient, more eager to finish their museum tours than to learn from them. "The ideas were complicated and it seemed like these groups were just racing through the Museum," she recalled.[113] Nielsen admired the early work of constructivist educators on kinesthetic learning, and, intrigued by museums' potential in this arena, she approached the education school at the University of Chicago and proposed a dissertation to be titled "The Motivation and Education of the General Public through Museum Experiences."[114]

Nielsen's resulting research seemed to support the assertion that interactive displays that required physical activity were, in fact, more effective with visitors than those that limited visitors to looking. Her visitor exit surveys showed that those visitors who pushed buttons really did come away with more scientific knowledge and reported they were more engaged by their time in the museum. Visitors learned better at their own paces, Nielsen observed, so science museums had to take care to "seek a balance between directing visitors to the lessons and letting them discover them on their own."[115] But she strongly cautioned science museums about relying too heavily on corporations to provide such displays, expressing concern that industry-sponsored exhibits would subvert the promise of interactive museum-based education. Corporate sponsorship could reduce the effectiveness of museum display, she warned, because education departments would have little or no control over content or the pedagogy of presentation. If corporations were given complete freedom with the displays they funded, Nielsen concluded, science museum exhibits would become little more than advertising.[116] Most science museums, however, seem to have believed the golden eggs of sponsorship were well worth the burden of the corporate goose and ignored Nielsen's predictions.

In spite of the popularity and confirmed effectiveness of interactive displays, science museums installed them more gradually than their publicity indicated, echoing the slow construction of dioramas by natural history museums a generation prior. Alongside Spooky and Herkemiah, the giant heart, and radioactive plants, Boston installed an observation beehive, replicating earlier displays at the Smithsonian and the American Museum, which had cast the insects as industrious honey producers. In addition, a few habitat dioramas were built as well.[117] A display of black widow spiders gave the shivers to visiting children, and curators also installed a few iconic examples of New England wildlife and an exhibit of American bison. But the science museum invested most of its budget in push-button exhibits and other displays designed to engage children's hands as well as their eyes. The museum's new multimedia show of the human body was far more representative of its postwar mission and effort. The show starred the museum's *Transparent Woman* (fig. 17) and featured a female voiceover explaining the location and function of each of her organs. A score written especially for the display by Boston Pops' conductor Arthur Fiedler accompanied each explanation.[118] Local critics lauded the spectacle as a model of how contemporary scientific understanding of human anatomy and physiology could be incorporated into interactive display.

Investment in interactivity by science museums established these institutions as active contributors to science education in subsequent decades, a reputation directors traded on for funding and goodwill. With the help of Herman Serieka in the Department of Education at Chicago's Museum of Science and Industry, director Lenox Lohr raised a bundle of cash to expand the museum's Electric Theatre, a hall that began with a simple chemical demonstration of how a firefly produced light and concluded with exhibits on nuclear medicine.[119] At the Boston Museum of Science, Washburn launched a $2.5 million campaign to expand exhibit space to accommodate new school groups. In 1955, he reorganized the museum's staff so they could more easily integrate pedagogical goals and principles into the exhibit-making process rather than tacking science education on as an afterthought. On the museum's new organization charts "Exhibits" and "Education" departments were equally important, and as a result, education department workers frequently suggested possible exhibits, which were then translated into prototypes and tested on visitors. Science museums' forays into display diverged sharply from the postwar path taken by the nation's natural history museums,

Figure 17   Boston Museum of Science *Member Newsletter* photo, showing crowds of visitors
of all ages coming to the newly renovated East Wing during the spring of 1951.
Throughout the 1950s, adult and family group attendance at natural history and
science museums grew; one internal note cited "10,000 in the museum" during one
weekend when a *Transparent Women* show was first offered. (Museum of Science,
Boston.)

many of which subordinated innovations in display and education to
research.

Postwar Natural History Museums:
Reviving Research, Downplaying Display

In the years immediately after the war, natural history museum staffs
gestured to the importance of science education, freshening exhibition
halls as they did so within established parameters, but curators had little
inclination to disrupt well-worn organizational patterns. Instead, natural history museums' leaders of this period demonstrated their scientific
relevance by reserving their most importance resources for research, not
display. Efforts to restore research's crown in the nation's natural history
museums intensified in the early 1950s, as funding from the National

163

Science Foundation began to trickle into the nation's scientific institutions. Research's ascendance dominated many museums in this era, and broader interest in innovating through exhibition began to fade somewhat, resulting in a period of relative stasis in display.

Even before the war ended, the American Museum's new director, Albert Parr, had begun to push for a new focus on scientific research.[120] Born in Norway, the tall, blue-eyed oceanographer had the kind of checkered background and adventurous field experience typical of an older generation of museum workers. After earning high honors at the Royal University of Oslo, Parr had traveled the world as a member of the Norwegian merchant marine and inspected Norwegian fisheries, then had turned his oceanographic expertise toward museum work, directing the Peabody Museum before assuming leadership of the New York City natural history museum.[121] Intent on restoring luster to the reputation of natural history museums as research institutions, Parr reassured his own museum's suspicious scientists that they now had an administrative champion. He set a floor on scientific departments' budgets, expanded eviscerated research and publication budgets, and raised curators' salaries in line with those of scientists at leading universities and research institutes.[122] "With these measures, the tendency during a period of forced economy to sacrifice the scientific departments for the administrative and public work of the Museum was checked, and the importance of the scientific work properly recognized," the museum's relieved Council of Scientific Staff reported in 1942. "In the future, the Museum should be able to enlist the services of the most promising young scientists."

Curators at the American were less pleased with Parr's assertions that natural history museums would recover once formidable scientific and educational reputations only if they abandoned collections-based research, and they were horrified when he implemented policies reflecting these beliefs. Systematics was naturalists' play, not scientists' work, Parr declared in 1943, and rather than continuing to classify and categorize, he maintained, "we must subordinate our search for more information about the variety of nature to the study of the natural laws."[123] "If we do not do this," he warned, "we shall merely go on accumulating an ever more intricate inventory of nature, beyond anybody's needs or capacity to use, and the gratitude we can expect will be as slight as the service we render. But if we do it, we shall find that the road to regained scientific prestige is also the road to recaptured public interest."[124] He had particularly harsh words for fossil collections. Though dinosaurs were crowd pleasers, the director coolly declared, evolution was "a finished issue." The study of fossils, and the broader scientific discipline of paleontology, could "contribute little further knowledge of any importance to the world," he maintained, and those who studied prehistory should be united with those who studied the living forms of today.[125]

To the museum's curators, Parr's statements must have seemed incomprehensibly ignorant. Parr neither seemed aware of the "new systematics" movement in evolutionary biology nor did he acknowledge the crucial role the American Museum's paleontology collections were playing in advancing this synthesis of Mendelian genetics and neo-Darwinism. As Parr was planning to dismantle the department, curator George Gaylord Simpson was using its collections as the grounds for *Tempo and Mode in Evolution* (1944), a theoretical cornerstone of the evolutionary synthesis and a powerful argument for the usefulness of paleontology in evolutionary studies—though a more traditional examination of taxonomic principles, Simpson's 1945 *The Principles of Classification and a Classification of the Mammals*, also argued convincingly that paleontology had much to offer new thought in evolutionary systematics and biology.[126] So when the director dissolved the museum's Department of Vertebrate Paleontology, he faced open revolt. "To tell a paleontologist that this department is to be discontinued is almost like telling him that the end of the world has come," Simpson angrily wrote. Curators would have been even more furious had they been aware that Parr had recommended to trustees that the museum divest itself of the vast majority of its collections.[127]

After the war, Parr relented on vertebrate paleontology, and the department was reconstituted, but the director applied his considerable fundraising talents to other ends, paying little mind to the budget requests of the museum's systematists. The director tirelessly raised money to promote experimental work in ethology and ecology. As early as 1946, the museum established a marine laboratory in the Bahamas and built a comparable facility for work on desert reptiles in 1954.[128] In the postwar years, the museum also ran several small field stations in New Jersey, obtained an islet in Long Island Sound to use as a plein air lab, and established stations in Arizona, south-central Florida, and the barrier islands of the Gulf coast.[129]

Older curators' steadfast defense of systematics proved prescient, however, for postwar federal science policy makers invested heavily in collections-based research and the expansion of its institutional infrastructure. The architects of the new National Science Foundation, founded in 1951, announced that systematic biology and museum collections would stand alongside the physical sciences as early and consistent funding priorities for the agency. Systematics research deserved renewed support, wrote NSF director Alan Waterman in a 1952 annual report, because of its potential contributions to the biological sciences and to postwar society as a whole: it would "serve as the basis for the assessment of natural resources and hasten the introduction of new and economically important groups."[130]

Though all kinds of life scientists fought fiercely to win postwar funding, museum-based systematists emerged from these battles with more money than they had seen in the past thirty years. From 1952 through 1956, the NSF funneled hundreds of thousands of dollars in grants to systematists at the American Museum and the Smithsonian, as well as to smaller regional institutions like Hawaii's Bishop Museum.[131] Although NSF funding accounted for only 4.8 percent of all basic biological research support in 1954, fully 61.3 percent of federal funding for systematic biology came from the agency.[132] Initial awards in systematics were small, often less than $10,000, and the money was used to develop checklists, inventories, and monographs updating the morphological studies of specimens, sometimes in connection with new expeditions.[133] But by the later 1950s and early 1960s, these grants ballooned; in 1967, the NSF Division of Biological and Medical Sciences spent $6.6 million on basic research and curatorial support for systematic biology, nearly 80 percent of total federal support in this field.

There was never enough money, space, or staff for museums to accomplish everything they aspired to do, and NSF support for systematics forced institutions to choose whether to invest in display or in the considerable scientific infrastructure and staff required to win additional federal funding. By the mid-1950s, many larger and older natural history museums had decided to subordinate, at least temporarily, exhibition to research; some museums literally shoved exhibition to the side to make more room. At the Smithsonian, for instance, "there were galleries in which cases were deliberately pushed around and out of sequence to make cubby holes . . . for the storage of collections or even work spaces for scientists to work in," National Museum administrator Frank Taylor recalled.[134] The space crunch at the Smithsonian in the wake of increased postwar research funding got so bad that the Smithsonian eventually granted the Museum of Natural History administrative independence and its own building, decisions that made plain the museum's investment in expanded systematics research throughout the 1950s.[135]

The commitment to research required to win NSF funding was substantial, and those smaller and newer institutions that had invested in display rather than scientific infrastructure subsequently found it difficult to attract the agency's eye. Throughout the 1940s and until 1957, for instance, only one curator at the Milwaukee Public Museum possessed a PhD, a situation that became a self-fulfilling prophecy. Without PhDs, the museum could not apply for research funding from government organizations or private foundations.[136] But without research funding, it could not attract PhDs. Frustrated with its inability to build the kind of scientific program it now desired, the museum returned its attention to traditional strengths—public education and anthropology—and left biological research to larger museums. Some of these institutions held out hope that the NSF's secondary mission in science education would lead to federal grants for exhibits or educational programming, but they put those hopes aside almost immediately. Though the agency generously funded graduate education, by the mid-1950s, its scientists had relegated primary, secondary, and adult education to the bottom of their pedagogical priority list.[137] Nonclassroom science education efforts, including museum exhibitions, received no funding at all, and those museums that had devoted themselves to public education since the 1910s found themselves without any hope of winning federal money.

Natural history museums rarely used federal grants to advance display in the 1950s, and even when they did, they did not use them to revamp exhibition techniques in any meaningful way. At the Florida State Museum, biologist-director Arnold Grobman hired an unprecedented number of curators after the war and began sorting through the collections backlog at the museum. This strategy paid off handsomely: Grobman was able to win a 1.1 million–dollar NSF grant to house both the scientific collections and new public galleries in a new building on the University of Florida's Gainesville campus.[138] But with relatively luxurious newfound postwar support for research at hand, directors like Grobman appear to have had little incentive to funnel resources toward revolutionizing displays or museum pedagogy. Rather than exhibiting and educating about postwar biological "big science," natural history museum curators were now busy producing it, and they no longer depended, as they had in during the 1920s and 1930s, on display building to make their scientific work possible.

So administrators and curators calmly proceeded with the same exhibitionary strategies they had been using for decades. When T. J. Schneirla, a curator in the American Museum's Department of Animal Behavior, returned in 1952 from studying army ants on Barro Colorado Island, he simply inserted a hunk of meat into a case containing a brooding colony of ants, then placed the case in one of the museum's halls. Schneirla's approach differed little from the labeled aquaria and terraria that had punctuated museum galleries for nearly a half century.[139] That same year, Field Museum zoology curator Kurt Schmidt replaced half of the objects in his synoptic cases with relevant photographs and drawings but did nothing more. And museums steadily continued to build dioramas. Between 1942 and 1959, for instance, the Colorado Museum opened thirty-four new dioramas in four new diorama halls; the Academy of Natural Sciences in Philadelphia opened an Audubon Hall of Birds and a Hall of Philadelphia Birds and completed its halls of North American and Asian mammals; and the American Museum continued to add dioramas to its North American Mammals, Pacific Bird Life, and Birds of the World Halls.[140]

Curators' reluctance to deviate from past practices stemmed, in part, from distraction but also resulted from continued confidence in their own abilities to translate scientific knowledge for the public. In contrast to science museums, which welcomed educational research and tried to assimilate findings of which into their displays, postwar curators, boards, and directors of natural history museums assumed that their own scientific authority and the hours they had put in walking museum halls would lead them to intuit the proper reforms for exhibition—not, they maintained, that any were needed. After all, visitors seemed happy enough. Chicagoans who saw Schmidt's revised displays at the Field, for instance, gushed over the small improvements, calling them "excellent examples of the 'new' style of museum display of selectivity and interpretation."[141]

Natural history museums periodically freshened their displays, but the cautious nature of these changes made evident the considerable distance between educational aspirations and institutional realities. At the American Museum, for instance, Parr boasted throughout the 1950s about his commitment to the improvement of display; a 1951 report by the board of trustees noted that he considered his greatest contributions to be in exhibit planning.[142] It's true that the museum looked different than it had under his leadership, for Parr quickly replaced the indifferent background colors of halls and exhibits—drab hues the director described as "the color of dirty fingermarks"—with bold blocks of contrasting colors and updated graphics. He insisted on painting the Whitney Memorial Hall of Pacific Bird Life a deep red, in order to highlight the green tones of the hall's habitat dioramas, a decision that raised almost as much uproar as his decision to close the Department of Vertebrate Paleontology.[143] But the halls built during his tenure as director reveal just how limited Parr's definition of revolution really was. Though Parr announced these spaces would "teach ideas, not merely exhibit objects" and inserted graphic explanations of various scientific functions into the halls, the museum largely stuck to display techniques that were half a century old.[144]

The museum's new Hall of North American Forests epitomized Parr's approach. Though board president Alexander White hailed it as "testimony to the Museum's special competence in the presentation of the natural sciences," and Parr declared it affirmed the museum's "position of leadership in the presentation of nature as it exists on its own, undisturbed by man," it was hardly a departure from past practice.[145] Its labels were a bit longer, its conservation message a bit sharper, and its biological lessons a bit more explicitly described, but its dioramas were much the same as they had been four decades prior.[146] To Parr, this continuity needed no defense: "The traditional exhibits of undisturbed nature and of human life . . . have a great function to perform in our educational system," he argued in 1958, though he did acknowledge the validity of earlier complaints about the educational opacity of dioramas. "The habitat group, when offered by itself alone, places too great demands upon the public," he agreed, as "the educational method of the habitat group is indirect. . . . [It] offers abundant opportunities for those not versed in scientific deduction to overlook or misunderstand the messages implicit in the exhibits." Consequently, curators had integrated smaller educational exhibits that depicted particular natural laws or phenomena into the halls. "The aim is to have exhibits explained by exhibits in interrelated series, rather than exhibits explained by labels," he explained. Curators had been requesting space and resources to make more such hybrid displays, and yet, to Parr, such exhibits could never rival well-constructed dioramas, "the dramatic impact of which carries its message implicitly in a 'total recall' of nature."[147]

One museum leader openly worried about the widening divide between natural history museums' public and scientific missions; and the resistance he faced as he struggled to bridge this chasm further illustrates why display inertia at these institutions persisted throughout the 1950s. Frank Taylor became increasingly concerned about the deplorable state of exhibits at the Smithsonian's National Museum and spent several decades of his career trying to improve them. Having worked his way up from the bottom of the museum ladder, Taylor was a populist and a popularizer at heart and believed that education and entertainment were central to the museum's public mission.[148] Though much of Taylor's early rhetoric mimicked that of other natural history museum curators interested in modernizing museum display, he distinguished himself by embracing the hands-on exhibition and insisted that displays should elucidate scientific or technical processes—an approach more typical of the era's science museums. At an institution in which overcrowded displays were now nearly thirty years out-of-date, such ideas were radical, to say the least. Taylor wasn't entirely sure what kinds of displays he did want to see in the museum's halls, but he knew that he did not want to see what currently existed.

In the ten years immediately following the war, Frank Taylor single-handedly battled the curators who ruled the Smithsonian's decidedly out-of-date exhibits. Though the Smithsonian had led efforts to reform displays seventy years earlier, by the 1950s, it was literally a mausoleum: mahogany cases crowded with dead specimens filled its exhibit halls. Disorganization and inattention to contemporary educational trends were the rule—the Smithsonian's living beehive, for instance, one of its few displays that appealed to contemporary pedagogical sensibilities, was banished to the Comparative Technology section of the museum well into the 1960s.[149] In contrast to George Brown Goode, who had believed the Smithsonian should serve as a vehicle for the diffusion of scientific knowledge to the general public, scientist Remington Kellogg, the National Museum's current director, firmly believed research should come before exhibits. He and the museum's curators regularly disavowed the barest mention of display reform as a result. Anthropologist Herbert Kriger, for instance, flat out insisted that there was no need for new exhibits, just new paint. Curators should just "polish up the Old Rolls Royce," he told one colleague, then return their attention to research.[150]

As a result, when Taylor began his efforts to modernize the National Museum's natural history exhibits in 1948, he struggled mightily. It took him nearly six months to persuade Kellogg to convene a committee on display and appoint Taylor chair.[151] Taylor requested that committee members come to their first meeting with a typewritten response to a "check list of inquiries" about their department's exhibit experiences.[152] Though responses to Taylor's Smithsonian-wide survey revealed a wide range of opinions about the value and meaning of displays, natural history curators concurred that exhibits' exclusive purpose was to promote the museum's collections and curators' own research. The response of entomologist Waldo Schmitt, head curator of zoology, typified this attitude. He impatiently swatted away Taylor's questions, declaring that the exhibits they had were fine, for the displays had no real function but to advertise the scientific research of the Smithsonian's various scientific departments. "Our exhibits are our show windows for displaying our wares and accomplishments; our advertisements are the publicity given those exhibits and the publications describing the material that they represent," Schmitt declared. Schmitt had other things on his mind: he and some like-minded colleagues were leading a coordinated international effort to revive systematic biology and to bring the declining numbers of systematists to the attention of the broader scientific community.[153] While the field zoologist was all in favor of cleaner glass and new paint, he had little use for interactivity or educational improvements. Pandering to the public, he implied, was a waste of curators' valuable time. "In working with and on exhibits, should we not give first consideration to our own self–not merely selfish—interest . . . as one of government's leading research establishments?" he asked in a pointed memorandum. Schmitt went so far as to challenge Taylor's underlying premise, rejecting the idea that there could be a "common ground of exhibition policy and theory for the whole Institution."[154] Other scientists responded more briefly and more bluntly. "We firmly believe that scientific work comes before exhibition," botanist Ellsworth Killip wrote in response to Taylor's survey.[155]

Smithsonian curators' indifferent responses failed to deter Taylor. In October of that year, he generated a document titled "Preliminary Report on Exhibits" from the committee's initial work. Its findings were both grim and impracticable. "The exhibition element of the Institution's work has reached a low level of accomplishment," Taylor wrote. "No policy on exhibition exists. A negative policy that 'scientific work comes before exhibition work' is so widely accepted that it is practically official." If the Smithsonian were serious about improving its education program—and along with it, its display—it would take considerable resources and would require the National Museum to divert funding from research to display. Taylor estimated "four million dollars as a reasonable guess for the total job." But Kellogg, a midwestern paleontologist known informally by his staff as "the Abominable No-Man," had little patience for Taylor's grand plans. Kellogg told Taylor that he had read the report with interest but wryly noted that, at current funding levels, the renovations he proposed would not be finished for forty years. Kellogg flatly rejected Taylor's initial plea to launch a museum-wide exhibits modernization program, instead approving very small changes. Rather than creating new exhibits or renovating entire halls, Kellogg provided money to prepare and install more legible and attractive labels for selected exhibits and to freshen up the cases by hanging monk's cloth in back of specimens.[156] Museum men and research-oriented curators at the Smithsonian had battled for decades over the mission of natural history museums, but in the years immediately after World War II, scientists seemed to have the upper hand.

Postwar natural history museums without costly research programs didn't have to deal with these tensions, but their staff members rarely favored the visitor-centered, interactive exhibitions proposed by natural history museum reformers and educators in the 1930s and early 1940s and promoted by science museums in the 1950s. Instead, they revived a model of exhibition more typical of the 1910s and 1920s.[157] In 1940, for instance, the Denver Museum of Natural History's director Alfred M. Bailey took his staff to review displays at what he described as "practically all museums in the east," then proceeded to build the kinds of exhibits that had characterized natural history museums for the past fifty years, emphasizing "ecological displays," the postwar descriptor for habitat dioramas.[158] In Milwaukee, new director Will McKern also continued to build habitat dioramas and displays of specimen collections, many of which were based on the kind of well-publicized safaris that had stood museums in such good stead some twenty years earlier.[159]

The nation's natural history museums, large and small, did little to alter their exhibits or promote cross-institutional dialogue about display in this period, relying instead on the strategies winning their institutional cousins popularity and public accolades. Natural history museums in the 1950s expanded their work with live animals and edged slowly toward interactivity and multimedia. In 1956, for example, the American Museum introduced the Guide-a-Phone, a trim attaché case that housed a tape recorder and attached to a set of headphones, so visitors could listen to lectures on various exhibits as they traveled throughout the museum.[160] Even the Smithsonian, whose displays continued to lag decades behind those of most other museums, purchased a *Transparent Woman* complete with voiceover and flashing lights for its new Hall of Health in 1957, though the display was relegated to the museum's Arts and Industries building, far away from the natural history collections.[161]

In short, when it came to exhibition, throughout the late 1940s and early 1950s natural history museums talked radically and acted conservatively. Whereas habitat dioramas had been determined by a clear aesthetic vision but didn't always have obvious contributions to science education and public engagement, displays in postwar natural history museum had clearer educational goals but no corresponding clarity or commitment to determining what changes should be made to displays to make them more effective pedagogically. Although, as Smithsonian secretary Alexander Wetmore put it in 1949, curators were still nominally committed "to pursu[ing a] program of activities in the 'increase and diffusion of knowledge,'" just a few years later these assertions rang false.[162] Indeed, if "the value of our contributions to the community and the nation càn, in a sense, be measured by the use of our halls," as Parr argued in 1948, then the American and other natural history museums committed to scientific research lagged behind science museums when it came to public engagement.[163]

However different their halls may have appeared, by the late 1950s, the educational programs and displays of both natural history and science museums revealed a shared faith that providing visitors with the opportunity to contemplate biological principles, practices, and discoveries would eventually improve public understanding of the life sciences. It was not always clear what or how museumgoers learned, but confident staff members felt certain that their exhibits had an osmotic and ultimately positive effect on visitors. Museums should abandon exhibits on the history of science and technology, argued advocates of this pedagogical paradigm—contemporaries called it "exposure"—and instead expose visitors to the latest developments in these fields. Fascinating exhibits and programs would do their educational work not didactically but indirectly, inspiring children to pursue careers in scientific research or at least more formal training. Educators and the press agreed that the informal exposure to science offered by museum exhibits was one of the best ways to ensure that both adults and children developed a genuine, and potentially lifelong, interest in the field. "The aroused interest in science and technology resulting from these visits cannot be measured, but if the shortage of scientists and engineers is ever to be met the interest of elementary students in these subjects must be stimulated," and museums were able to stimulate this interest more effectively than most other institutions, affirmed the *Boston Post* in 1955.[164] After a 1957 tour of European science museums, Bradford Washburn was able to report that the premise of exposure had been embraced abroad as well and happily noted that Karl Bässler, the director of Germany's esteemed Deutsches Museum, had lauded the Boston Museum of Science's approach to education via exhibition. Like Washburn, Bässler believed visitors should be allowed to guide themselves through museums and that administrators and curators shouldn't worry too much about what museumgoers were taking away from the experience. "Exposure is the important thing!" avowed Washburn's trip notes.

Although natural history and science museums shared educational programs and a few displays by the mid-1950s, postwar science museums' more consistent emphasis on the relationship between education and display would prove consequential over the next two decades. Natural history museums' strategic investments in research moved them further from contemporary conversations about science pedagogy; science museums, by contrast, moved toward the center of these dialogues as their experiments in interactivity began to change national conversations about science pedagogy. In the wake of Sputnik's launch, these differentiated missions helped determine institutions' respective relevance and irrelevance to national debates over Cold War science education policy.

FIVE

"An Investment in the Future of America": Competing Pedagogies in Post-Sputnik Museums, 1957–1969

A week after Boston Museum of Science director Bradford Washburn returned from his 1957 tour of European museums, news anchors reported that the Russian government had launched Sputnik, the world's first earth-orbiting space satellite. Despairing that they had fallen behind their Soviet enemies in scientific and technological development, the American public and the U.S. government began to reconsider the relationship between research, education, and public understanding of science. A mere six weeks after the launch of the tiny silver sphere, federal officials announced their intention to overhaul American science education. The improvement of science education was a more important priority than producing more missiles or expanding the Armed Forces, Eisenhower soberly declared from the White House lawn. "My scientific advisers place this problem above all other immediate tasks," the president reassured an anxious public.[1]

Science and natural history museum directors seized this moment to insist that the government should invest more heavily in the informal education that museums offered. Museums were crucial to science education because they sparked student's interest outside the classroom, proclaimed a 1957 editorial in the Boston-based *Christian Science Monitor*, but museums needed larger buildings and more federal financial support if they were to promote to Americans more broadly the kind of scientific know-how that would defeat the Soviets. "With too many Sputniks to Soviet Russia's credit," the editors argued, "this is no time to run out of space for teaching science."[2] "High school and college science enthusiasts don't just HAPPEN—they are the result of continued systematic EXPOSURE to science at an early age," Washburn wrote in a follow-up letter to the editor. "America had better wake-up and start doing something significant in this area." Reiterating the importance of what he would come to call the "exposure model" of science education, he concluded by declaring that funding for science museums "must not be considered an *appropriation*. It's an investment in the future of America and will pay off far better in the next 20 years than all the superhighways and bridges."[3] Smithsonian secretary Leonard Carmichael had a similar response, insisting that natural history museums had an important role to play in post-Sputnik science education, for they allowed visitors to learn "as they cannot elsewhere about the natural resources of America, the natural history of the world, and special aspects . . . of their own United States. Many leave better informed and are truly more patriotic Americans than when they came."[4]

But post-Sputnik concord over the role of museums in science education was short-lived. Agreement over pedagogical ends quickly dissolved into conflicts over practical means, and exhibits once again became the terrain on which battles over what constituted effective life science education were waged. Beyond museum walls, teachers, science policy makers, and scientists argued broadly over the relative merits of discipline-based science pedagogy and science literacy in the late 1950s, conflicts that found their way into museum exhibits. Within museums, the ongoing professionalization of exhibition work continued to improve the appearance of displays, and evolution as a biological concept slowly found its way into exhibits. Though continued internal disputes over who should determine the shape, nature, and final form of exhibits hindered the development of meaningful pedagogical innovations.[5] At natural history museums, decades-old tensions between scientific and exhibits staffs reignited, and museum workers fought over models, narratives, and multimedia the way they had once fought over habitat dioramas. Science museums, ostensibly more open to experimentation with exhibition, continued to invest in interactive, corporate-sponsored displays. But their new exhibits, which focused on contemporary successes in scientific research, did little to challenge what had become questionable educational assumptions about the diffusion of knowledge as a simple trickle-down process.[6]

As the 1960s wore on, federal educational leadership jump-started more productive dialogues about display pedagogy, and these dialogues attracted renewed attention to museums' potential contributions to science education. In 1966, the U.S. government sponsored a meeting on museums and education, bringing together policy makers, museum staffs, educators, and scholars to discuss the future of museums. There, conference participants caught a glimpse of an alternative vision of museum exhibition, one that would redefine science and science education in museums over the course of the coming two decades. Museums had big possibilities for small Americans, maintained physicist-turned-elementary science educator Frank Oppenheimer, but it needed to capitalize on its distance from the more formal demands of schoolroom study, rather than attempting to cater to established educational traditions. Embracing Oppenheimer's vision would require museum staffs to do more than renovate their displays or update their educational programming. It would compel them to redefine, once again, their institutions in entirely new ways.

How to Teach Science after Sputnik: From Classroom Debates to World's Fair Displays

After Sputnik, educators and scientific experts revived their call to expand Americans' knowledge of science, borrowing language used during the depths of World War II. Without the critical thinking skills that general science education provided, the American public would "have the illusion of full self-determination" but end up "willing victims of [their] own subjugation," argued University of Chicago education professor Joseph Schwab.[7] "No longer is it an intellectual luxury to know a little about this great new tool of the mind called science. It has become a simple and plain necessity that people in general have some understanding of this, one of the greatest of the forces that shape our modern lives," agreed Warren Weaver, a former director of the Division of Natural Sciences at the Rockefeller Foundation, in a 1957 issue of *Science*. Without understanding science, Weaver warned, "we simply cannot be intelligent citizens of a modern free democracy, served and protected by science. Without this we will not know how to face the modern problems of our home, our school, our village, state or nation."[8] Such rhetoric echoed the polity-minded emphasis on relevance and life adjustment that had shaped science education in the 1930s, 1940s, and early 1950s, but with a looming difference. As Weaver pointed out, science not only served the nation, it protected it as well. Post-Sputnik science education was no longer just a tool for achieving a utopian standard of living in a prosperous postwar world. Rather, it was now a weapon in the home front's arsenal, a way of standing firm against the impending Soviet threat.

Before 1958, federal support for science education was piecemeal, lacking consistency and coordination. The National Science Foundation, for instance, had adopted a "let a hundred flowers bloom" approach, cultivating programs intended to enhance existing school science and promote the work of government scientists in schools. In 1955, the agency-funded traveling libraries of 150 books, selected by a combination of scientists and teachers, were intended to "broaden the science background of high school students, and to assist students in . . . choosing a career."[9] Whether students actually read, enjoyed, or learned from the books was outside the agency's purview. That same year, NSF officials supported a series known as the Traveling Science Demonstration-Lecture Program in conjunction with the Atomic Energy Commission. Scientists from Oak Ridge and other national laboratories toured the country with laboratory equipment designed to illustrate basic physics principles. According to the NSF's 1957 annual report, demonstrators' enthusiastic performances would "stimulate interest and competence in science education" and provide students with "a deeper appreciation of science."[10] Students were expected to absorb scientific knowledge by watching silently and were left to take the initiative to pursue more information once the school day had ended.

After Sputnik, educational policy makers got more aggressive. The NSF no longer left students' scientific interest to chance, instead supporting more ambitious efforts to increase the flow of precollege students into collegiate and graduate science and engineering programs. At first, the agency attempted to shepherd the nation's highest achievers into advanced graduate study but soon expanded its efforts with a sweeping attempt to improve classroom science education from elementary school all the way through high school graduation.[11]

Classroom science educators came under attack for their lack of rigor—prominent historian and educational critic Arthur Bestor chided them for giving students "only the most fleeting glimpse of the great world of science"—but initially, they held their ground.[12] Teachers continued to argue that student enthusiasm for science mattered and that such enthusiasm depended on science's relevance to current events or their own experience. But rather than continuing to call this kind of appeal to student interests "life adjustment," which suggested that individuals needed science in order to adapt and create a healthier social ecosystem, some educators reframed it as "science literacy." Influential Stanford education professor Paul DeHart Hurd pioneered the latter term in 1958, asserting that "more than a casual acquaintance with scientific forces and phenomena is essential for effective citizenship today. Science instruction can no longer be regarded as an intellectual luxury for the select few."[13] Despite its immediate popularity, the parameters of scientific literacy remained vague. Some educators defined it as greater content knowledge in a broad range of scientific fields. Others described it as the knowledge and skills that allowed one to read and understand science discussed by mass media.[14] Still others defined it as the ability to evaluate scientific claims made by the news media or the lay public, and the ability to determine the potential social relevance and importance of scientific developments.[15] Its malleability was likely what made the concept so appealing.[16]

The idea of science literacy faced greater challenges from a competing post-Sputnik model of science education based on the ideal of rigorous disciplinary training embodied in the various NSF-sponsored curriculum projects of the 1950s and 1960s. Critics of the NSF began to voice their opposition to the life-adjustment approach internally as early as 1956, when a member of the Educational Division argued that schools' attention to the personal and social needs of children sometimes militated against the development of meaningful scientific achievement, which mandated the mastery of a particular body of knowledge about nature. Indeed, by the late 1950s, many scientists and NSF officials came to see discipline-based science education, where the intellectual study of science was valued for its own sake, as the antithesis to science literacy and earlier Progressive approaches.[17] The scientist-architects of the NSF's first curriculum project, the Physical Science Study Committee, explicitly sought to develop students who could reason from evidence and who understood the practice of scientific research: its primary tools were demonstration films and low-cost laboratory equipment. But biology was trickier educational terrain, admitted NSF program leaders, for educators rarely agreed on how, when, and even what to teach students. "Biology faces important challenges, for it is at a crossroads," NSF assistant division director for biology, Arnold Grobman, wrote to his boss. "There is so much disagreement as to what biology is." The NSF's second curriculum project, the Biological Science Curriculum Study (BSCS), alternatively, focused on promoting both the importance of their field and what its participants argued was a newly coherent story about nature that the evolutionary synthesis had made possible.[18]

Yet biology remained the heart of American secondary science education, and biologists successfully pushed policy makers to recognize the primacy of their field at all levels of classroom study. It was through the study of biology that Americans would best be able to "to meet Russia's scientific advances," declared zoologist Walter Taylor. "After all, no appropriation of money, no investment of time, no . . . effort in the fields of physical science is of the slightest concern or meaning to us as human beings . . . except in relation to the living matter of which we are composed."[19] Advocates of both approaches agreed that curricula should explore "the remarkable congruence of function and form in living things" but diverged over how exactly this should be accomplished.[20] Teachers in the high school trenches continued to insist that students responded best when they were interested in the topic at hand and could position facts and theories within a meaningful context. Many scientists dismissed this pedagogical approach as soft, giving, one scoffed, "the impression that biological science was only important to man."[21] "We may lose a potential biology major in a beginning class criticized as being overloaded with details," argued American Institute of Biological Sciences' Education Committee chair Oswald Tippo, but such losses could not be allowed to retard the effort to prepare the nation for international scientific competition. "To bring these courses up to date," he concluded, curriculum reform instead required a forceful "focus on fundamental biological principles."[22] The BSCS tried to settle the issue by devising three different textbooks for biology classrooms: one basic introductory overview and two more specialized books, one focusing on cell biology and chemistry, the other on molecular biology and evolution. In many ways, BSCS revisions were a throwback to the sociobiology promoted by oceanographer William Ritter and other Progressive Era reformers, for they too presented the science in the context of human social progression. Yet they offered classroom teachers a well-coordinated approach to the subject that circumvented both the critics of science literacy and critics of discipline-based pedagogy.

Resolving the competition between science literacy and discipline-based approaches beyond the classroom proved more challenging, as the diverging approaches of two federally sponsored sets of educational displays at the 1958 Brussels World's Fair made clear. The U.S. Public Health Service approached its few displays at Brussels with general science literacy in mind, featuring the applications of biomedical research in exhibits designed to capture public interest. With the biomedical exhibits, private industry had jumped in to help, funding appealing displays that featured the Salk polio vaccine, antibiotics, and other success stories of scientific research that had benefited broader society.[23] The NSF-sponsored exhibits, in contrast, relied on discipline-based pedagogy, tacking up wall charts featuring the latest American developments in physics, chemistry, solid-state physics, and biology. In accordance with the discipline-based model, these charts didn't give an inch to casual visitors, an approach that suited the fair's European organizers to a tee. "While it is desirable that simple minds should be given a less complicated idea of the significance of various phenomena," the Belgian government's commissioner-general had explained, "it is essential that specialists, scientists and scholars who visit the International Science Pavilion shall find there the means to enrich their knowledge."[24] But visitors found the NSF-sponsored exhibits downright dull, and many press outlets ignored them entirely. Though Cold War fears likely stifled outright criticism of the displays, scientists later agreed among themselves that the exhibits had adhered too closely to the discussions of specialists and had failed to communicate scientific information effectively as a result.[25]

Though the NSF never officially sanctioned one method over the other, federal funding directed at specialized training in the sciences after Sputnik reinforced the discipline-based approach and had the unintentional effect of limiting access to science education.[26] After Congress passed the National Defense Education Act, the NSF shoveled money into specialized graduate training in biology, physics, and engineering and funded dozens of institutes designed to provide teachers with advanced knowledge of their fields. By 1960, more than half of the agency's nearly $176 million dollar education budget was dedicated to teacher training, most of it going toward institutes where secondary school teachers could learn about the latest developments in scientific research.[27] The foundation's funding priorities had a cascading effect: rather than expanding attempts to address the needs of a broad middle that had little intention of pursuing higher education, instead many of the nation's colleges, secondary schools, and even elementary schools diligently began preparing ambitious students for the kind of advanced coursework that graduate programs and scientific careers required.[28] Students without college plans were left out in the cold.[29]

The discipline-oriented scientists who dictated NSF grant making routinely overlooked museums as potential contributors to science education. The agency was not averse to the idea of what it would later call "informal education" or its venues—the agency had established a "public understanding of science" program in 1960, for instance, and the fledging effort directed money toward mass media explorations of science—but program officers ignored museums. They did not consider museums as possible candidates for collaboration with NSF-sponsored teacher training institutes, nor did they make an effort to establish conversations about public education through museum display. Boston Museum of Science director Bradford Washburn repeatedly failed to win NSF funding for his museum in the late 1950s and 1960s, for instance.[30] In 1962, after the agency refused to grant him funds for a Summer Teachers' Institute in the museum's galleries, NSF program officers disdainfully explained that "we are fundamentally interested in *instruction rather than exposure* to science. We are interested in the real stuff, not the kind of thing that you appear to be doing now."[31] While visitors seemed to like the Boston Museum's exhibits, educators and policy makers did not view them as sufficiently innovative or sensitive to new pedagogical ideas. To NSF program officers, pushing buttons and petting owls was not science education. The model of learning that Washburn and others had called "exposure" seemed old-fashioned to them, a passive method that ran counter to the emphasis on scientific practices and standardized curricula that were hallmarks of the new discipline-based approach to science education. As the NSF set the terms of debate for post-Sputnik science education and research, institutions that failed to win its recognition were marginalized, and museums received almost no support for displays or educational programming.[32] Despite strong staff interest in participating in the developing national infrastructure of education, museum staffs found themselves left out, abandoned by policy makers eager to fund institutions that embraced pedagogical methods and ideals more in line with their own.

Battles over Exhibit Design: Themes and Narratives in Museum Education

Post-Sputnik museums, struggling to regain educational credibility, increasingly turned to professional exhibit designers to help them build exhibits that would at once improve Americans' science literacy and satisfy those who advocated the new disciplinary model of science education. Consequently, museum directors hired a new generation of designers, charging these experts with the responsibility of creating displays that would satisfy diverse and discriminating audiences. To carry out this mandate, designers leaned hard on narrative, an approach they believed would persuade audiences of the power of science and offer a new outlet for curatorial research.[33] Still, various staff factions objected to designers' new power and used "narrative display" as a political football as they sought to reclaim authority over museum halls.

Exhibits were a unique contribution to science education, museum staffs believed, and they had to withstand new scrutiny from scientists, policy makers, and educators. Exhibitions distinguished museums from formal schooling and the other educational venues slowly beginning to attract funding from government agencies, and post-Sputnik museum leaders were keen to build displays that not only communicated scientific content in clear and appealing fashion but would also allow museums to elbow their way into national debates about science pedagogy. Half a century after the museum men had reformed exhibition for educational ends, exhibition continued to seem the most efficient way to educate museumgoers—or, as Washburn insisted, "expose" them to science— even if museum staff weren't entirely sure how exhibition fit into the debates over science education raging in educational and policy circles.

The increasing professionalization of display in science and natural history museums gained further traction from the expansion and rising prestige of exhibition design in the United States.[34] Midcentury American artists, influenced by constructivism and the Bauhaus, demonstrated new respect for those whose creations weren't limited to the canvas or marble block. Those at the vanguard of the nation's art scene regularly sought out commissions for such work. More and more, corporations employed well-regarded artists and design professionals to create extravagant exhibitions for the trade shows that filled the nation's largest hotels and brand-new convention centers throughout the late 1950s and 1960s. The government, busily waging a global propaganda campaign to promote the American way of life, also invested heavily in exhibition design. Agencies ranging from the U.S. Information Agency to the U.S. Atomic Energy Commission hired accomplished artists and industrial designers to design traveling shows and vast pavilions for the era's international expositions.[35] After Sputnik, pressures to create more compelling, more educational exhibitions about science were especially intense, even resulting in the establishment of federal administrative units devoted to exhibition design. In 1963, for example, the National Science Foundation hired German-born exhibit designer Leonard W. Nederkorn to head its newly founded Office of Science Exhibits. A seasoned designer who had worked with Frank Lloyd Wright and Chicago's New Bauhaus movement, Nederkorn coordinated educators and scientists in the design and construction of the Science Pavilion at the Seattle World's Fair in 1961.[36]

And so, throughout the late 1950s and 1960s, museums across the United States began to invest more heavily in exhibition staff, hiring full-time professionals to help them create displays. While science and natural history museums had long employed professional taxidermists, artists, and model makers to create their habitat dioramas and transparent women, most exhibit makers had worked on a contract basis. The Carnegie Museum of Natural History, the Field Museum, the Philadelphia Academy of Natural Sciences, the American Museum of Natural History, and a few other influential museums had maintained independent exhibition departments for decades, and the specialists who staffed them were renowned for their skill. But most institutions could not afford such a luxury and, well into the late 1950s, continued to rely on subcontractors or one or two in-house, jack-of-all-trade exhibit makers—what Smithsonian exhibit designer Joseph Shannon later called "semi-amateurs."[37] "There wasn't a Charles Eames or a George Nelson among them," he wrote.[38] Ladies' auxiliary associations often assumed exhibition responsibilities, volunteering their time and taste so curators could devote themselves to managing collections and conducting scientific research. At the Buffalo Museum of Science, for instance, members of the Women's Committee oversaw the construction and organization of both permanent displays and traveling exhibitions.[39] But in the post-Sputnik era, science and natural history museums across the nation began to hire more professional input.

Museum administrators expressed their new interest in exhibit design in the clearest way they knew how: through cash and real estate. In 1959, for example, the Denver Museum of Natural History established a department of graphic design to presketch, construct, and install exhibits in fields outside the habitat group range, to replace old signs and add new ones, and to provide miscellaneous art work for projects like motion picture maps, titles, book illustration, and book cover designs.[40] By the early 1960s, the Boston Museum of Science regularly consulted with Verner Johnson, a Boston architect who also specialized in museum exhibit design. Like many designers, Johnson expressed some frustration at how poorly museums integrated professionals into the process of exhibit design. He, for example, was called in to work on the museum's heart exhibit only after members of the Education Department realized they could not construct it on their own, and he continually bemoaned how he received only "second hand information and instructions" when creating displays.[41] But the fact that he was working in the museum at all represented a "quantum leap from amateur night to the shot-up sophistication of the freaky 1960s" for such institutions, as another contemporary designer put it.[42]

Even the Smithsonian, long a dinosaur on display, came around to the idea of professionalizing the design of their exhibits. Tired of Frank Taylor's constant complaining, Secretary Leonard Carmichael authorized the creation of an institution-wide Office of Exhibits in 1955.[43] The new entity provided exhibit makers with considerably more prestige than they had ever before enjoyed, and its now official Exhibits Modernization Program, one Smithsonian curator later reflected, served to "dramatically alter the status quo of 100 years."[44] By the 1960s, the institution's hesitant support of exhibit design had become a firm embrace. The Smithsonian hired well-established professional designers to help them recast their images and rethink their halls.[45] Some of these relationships continued to occur on a contractual basis: in 1965, for instance, Carmichael's successor, Dillon Ripley, used corporate funding from IBM to hire renowned designers Charles and Ray Eames to direct a film celebrating the museum's hundred twenty-fifth anniversary.[46] But the institution's investment in design extended far beyond temporary celebrations. Impatient to renovate the aging galleries of the National Museum of Natural History, the Smithsonian's Office of Exhibits also offered full-time positions to artists and corporate exhibit designers.

Few of these new exhibit designers had the kind of field experience that had so profoundly shaped the aesthetics of the taxidermists and artists responsible for the habitat dioramas of the first decades of the twentieth century. Instead, their backgrounds lay in art, architecture, industrial design, and mass communications, and the contemporary sensibility of these fields influenced their displays. John R. Clendening, for example, trained in painting and drawing at the Corcoran School of Art while working full time as an exhibit designer at the Smithsonian during the early 1960s, and his Hall of Medical Science for the Museum of History and Technology reflected aesthetics he had picked up at the Corcoran.[47] James Mahoney had formal training in industrial design and had cut his teeth producing traveling displays for the U.S. Information Agency and the Brussels World's Fair, before the Smithsonian snapped him up for its Office of Exhibits in the late 1960s. Mahoney's own approach to exhibition reflected his interest in contemporary communication theory, as well as his modernist devotion to simplicity. "I usually cut straight through to a couple of very important points," he later reflected. "Number one is what's the message? Number two, what's the audience?"[48] Mahoney was appalled by what he called the "frighteningly permanent" clench of habitat dioramas and heavy, Biedermeier-style specimen cases on natural history museums. The halls of the American Museum of Natural History and the Smithsonian were in desperate need of overhaul, he confided to Frank Taylor, for they themselves had become museum pieces. Indeed, Mahoney added, some of the habitat dioramas "should be saved strictly as examples of a disappearing art form."[49]

Exhibit designers like Mahoney frequently relied on what contemporaries called "themes and narratives," clear story lines intended to interest visitors and appeal to science educators; the approach was both flexible and familiar. Taking cues from communication scholars like Marshall McLuhan, designers pushed to create exhibitions with more narrative thrust and more clearly defined themes: museums were the medium and stories about science were the message.[50] The notion of narrative as a pedagogical strategy for museum display was hardly new. Diorama creators had prided themselves on inserting stories into their displays, and museum reformers ranging from Carlos Cummings to Gladwyn Kingsley Noble had pressed for more narrative panels to explain objects' scientific significance. In the postwar decades, however, exhibit designers increasingly supported the approach, arguing that specimens' value lay in their unique ability to illustrate scientific ideas but that they were only effective in doing so if embedded in broader stories. In short, the movement toward narratives and themes among museum designers was a more thoughtful, visitor-centered approach to what museums had always attempted to do: to present and contextualize objects in such a way that they became meaningful to visitors. But this approach resonated with advocates for science literacy, who also plugged narrative as an effective means to an educational end. And the professional designers of the postwar era proved far more pragmatic about creating narrative- and theme-based displays than the curators and taxidermists who had preceded them. Rather than counting on visitors to deduce narratives from halls' organization, new exhibit designers relied on explicit explanation.

More and more, these designers began to script out exhibit halls, creating stories for visitors to follow as they moved through the hall. Though such narratives fit easily with the trend toward science literacy, they were a more awkward fit for the facts-and-practices emphasis of discipline-based science education, so designers also employed a thematic approach to scientific content, believing a montage of objects and images organized around a single scientific concept would give visitors a deeper understanding. Curators interested in education and exhibitions were delighted. "For the first time," recalled one designer, "some curators became interested in the possibility of developing *their* disciplines into similar 'story' exhibits where formerly they were consumed only by their research."[51] Throughout the late 1950s and early 1960s, voluble champions of this approach included the American Museum's Albert Parr, recently retired from his directorship but still a productive oceanographer and a prolific museum critic, and ornithologist Herbert Friedmann, the Smithsonian's chief curator of zoology and eventual director of Los Angeles's natural history museum. But curators accustomed to straightforward cases of taxonomically arranged specimens vehemently resisted designers' efforts to incorporate narratives and themes into their exhibitions. Field Museum curator Karl Schmidt deplored the shift. "There is a conspicuous modern trend to attempt, by means of thoughtful arrangement and labeling, to set forth abstract concepts and principles rather than to merely show objects, however intrinsically fine these may be," he wrote in 1958. "It seems evident that this shift of emphasis from the particular to the general is a pervasive one, found or to be expected in all museums everywhere. Not all of the efforts in this direction have been successful," he warned. "Efforts to transform exhibition halls in this direction are sometimes over-zealous, and there is danger that the museum baby may be thrown out with the bath."[52] Schmidt's resentment of an emerging "narrative approach" to natural history display was hardly isolated. Scientists and educators who favored discipline-based science education opposed narratives on the grounds that they left out too much information. They argued that shoehorning objects into a strict story line led to distortion of facts, or worse yet, outright inaccuracies.[53]

As Schmidt pointed out, exhibits developed by professional designers "demand far closer cooperation between the scientific and the educational staffs than the 'old style,' take it or leave it exhibits," a situation that rarely sat well with curators.[54] To create these exhibits, designers worked closely with staff scientists and educators over months and sometimes years to develop a story line and determine which themes should be highlighted. Throughout this process, designers, educators, and curators frequently functioned as equals; while curators continued to guard their veto privilege over the scientific knowledge presented, the collaborative nature of the work eroded their ability to control how science would be defined and presented in exhibit halls. Scientists griped about the way narrative-and-theme exhibits ate away at their time and traditional powers. Curators already had to do research, train students, and look after collections, protested American Museum paleontology curator Edwin Colbert in 1958.[55] Planning and supervising the installation of exhibits added yet another responsibility to a plate already heaped with them. "How is the curator to manage his life and his time among these several and in some ways quite separate duties?" he asked.[56]

Tensions over these issues flared dramatically at the Smithsonian in the late 1950s and early 1960s when Taylor consolidated all the institution's display planning and construction into one organization-wide Office of Exhibits.[57] Most natural history museums with vibrant research programs expected their curators to participate at least tacitly in the planning and production of exhibits. At the Smithsonian, however, curators in the division of natural history had never been expected to engage in these kinds of collaborative activities to the same extent. Their idea of public outreach was largely limited to identifying specimens for members of the public.[58] Ironically, in the first years of its existence, the Office of Exhibits had only contributed to this expectation, as professional designers took on the responsibility of creating displays, leaving curators free to conduct research without interference.

The ongoing renovation of the museum's Hall of Marine Life served as a flashpoint for conflict between professional designers (and the administrators who supported them) and the scientists who still believed the Smithsonian's primary responsibility to the public lay in scientific research. In 1956, Frank Taylor and Herbert Friedmann had started to discuss the need to renovate the Hall of Marine Life, a process that began in earnest two years later. Hoping to bring the old hall in line with recent trends in science education and exhibition design, Friedmann, Taylor, and the Smithsonian designers, pushed for a narrative about "life in the sea."[59] Biologist A. C. Smith, who had been appointed director of the Smithsonian's newly independent administrative subdivision of natural history, now called the Smithsonian Museum of Natural History, readily agreed to their plan.[60] But with the exception of Friedmann, who eventually decamped to assume the directorship of Los Angeles's natural history museum, Smithsonian's head of exhibits John Anglim and other exhibit designers found themselves without friends among the museum's head curators. Desperate for allies among the museum's scientific staff, they ended up working closely with Ernest Lachner, a cigar-chomping associate curator from the Division of Fishes.[61] Marginalized within his own department despite his considerable research accomplishments, Lachner hoped to use the renovation to redress what he charitably described as "spotty" displays of his own favorite species of marine life.[62]

The resulting Hall of Ocean Life script and design tactfully assimilated designers' interest in narrative with older traditions of natural history museum exhibition. Lachner and the exhibit designers had tried to create associated displays highlighting the kind of biological principles that dioramas' early proponents had hoped visitors would observe and deduce on their own. Rather than assuming visitors would notice the ecological relationships, food chains, and quests for survival in the scenes they contemplated, the new exhibit script called for more explicit discussion of these themes. Beyond traditional biological and ecological themes, Lachner and the designers also proposed narratives connecting specimens more directly to visitors' experiences, beliefs, and interests, a technique museum reformers had discussed since they had first attempted to rewrite labels in lay terms in the 1910s but had rarely realized. The Smithsonian team proposed developing displays on the themes of ocean exploration and long-standing myths about the sea, suggesting that the hall feature models of bathysubs, gondolas, echosounding, and newer televisual and photographic technologies, as well as displays that illustrated how ideas about sea monsters, mermaids and other legends had developed.[63]

Still, the proposal by the Hall of Ocean Life committee offended discipline-minded curators intent on presenting specialized, contemporary information about marine biology, and they objected to the project's premise and scope. Quite unintentionally, designers had hit curators where they hurt: the interdisciplinary nature of the narratives and themes challenged the museum's traditional departmental divisions, and systematists—already under fire from university biologists for the narrow scope of their research—were sensitive to breaches of the carefully guarded boundaries of their work. Many protested that the miscellany of topics was more appropriate for a nineteenth-century natural history museum than a twentieth-century institution devoted to research in the life sciences. The hall, they argued, should present more focused research findings, not cater to the masses' roaming curiosity and short attention spans. As Curator of Mammals David Johnson complained to Friedmann, "Some of the topics that are listed on the outline indicate a tendency to wander off into related fields rather than sticking to the subject of zoology." Mythological and primitive conceptions of marine life belonged in anthropology, exploration of the sea in history, declared the frustrated curator. Johnson believed that "in the eyes of the visitor each of our halls is a specialized element. . . . If each hall branches out into the field that are more expertly covered by others, we will end up with a patchwork arrangement that is confusing."[64] Whether Johnson's assessment of museum visitors was correct, his protest certainly reveals curators' continued adherence to a rigid, discipline-based understanding of the sciences.

To protest the new design, many members of the scientific staff wrote angry letters, snubbed designers, or simply refused to cooperate with the plan. Throughout a lengthy and expensive expedition to the South Pacific with the exhibition team in 1961, Associate Curator Joseph Morrison went so far as to ignore exhibition designers altogether. Morrison, a specialist in mollusks, had been sent to New Caledonia specifically to help collect for the new hall, but the intransigent curator flatly refused to dive for specimens that designers needed and made a mere three trips to the sites from which designers wanted specimens. John Anglim was simply appalled. "This trip, like everything else associated with the new Ocean Life Hall, has been accorded very little cooperation from the curatorial staff," he wrote Taylor. "They seem to feel that it is a project schemed up by the exhibits office for the sole purpose of harassing them and interrupting their orderly way of life. We are in the untenable position of trying to force them to do a job that has been assigned to them by the Museum."[65] In one of his final acts as the museum's chief curator of zoology, Friedmann, desperate to find a compromise, approved a design that incorporated specialized research as well as some of the broader themes pushed by exhibition staff.

The exhibit that ultimately opened wasn't remarkably innovative by national standards of exhibition design, but at the hidebound Smithsonian, it was revolutionary. At first glance, the resulting hall closely resembled the Hall of Ocean Life that had opened at the American Museum three decades earlier, but a few key differences betrayed how exhibition techniques had evolved over the course of those thirty years. Its narrative and broader themes were crystal clear to critics and visitors. As the *Baltimore Sun* explained a few days before the exhibit formally opened, "*Life in the Sea* attempts to show how some of the world's marine animals—ranging from the largest to the smallest—look in life; how they are adapted to resist the physical forces of their environment, to elude their enemies, and to reproduce; and how their existence benefits or harasses man."[66] And designers' belief that objects were little more than illustrations of these broader ideas, rather than items of scientific value in and of themselves, was impossible to miss. Where the American Museum's designers and curators had hung whale skeletons from the ceiling, Friedmann had agreed to allow exhibit designers to create a massive, carefully detailed model of a blue whale in lieu of an actual specimen (fig. 18).

Designers hoped the whale model would not only serve as a powerful embodiment of marine life but also give museum visitors a sense that they were immersed in a marine environment. Though the American Museum had created vast spectacles designed to evoke a sense of travel to exotic or outdoor climes throughout the 1920s and early 1930s, its immersive designs remained an exception: most museums had limited

Figure 18 The entrance to the 1963 *Life in the Sea* exhibit at the Smithsonian National Museum of Natural History, which the American Museum of Natural History director Albert Parr praised for its combination of scientific accuracy and dramatic design. Painted dark blue to evoke the feeling of immersion in water, the hall featured a life-sized sculpture of a blue whale. Beyond its narrative theme, however, it broke little new exhibitionary ground. (Smithsonian Institution Archives, Image MNH-1115.)

their re-creations of nature to the tightly framed space of the habitat diorama. But in the early 1960s, a new generation of exhibition designers insisted on breaking the proscenium boundaries of dioramas, removing the glass that had traditionally separated specimens from viewers. When the Milwaukee Public Museum moved to a new building in the mid-1960s, for instance, designers of the new African exhibits abandoned cases altogether, taking the opportunity to create what they called "open-air" dioramas of African zoology to force visitors to feel as if they were in a savannah or a mountain bamboo forest. Smithsonian designers and the curators who supported them hoped the whale model would function in a similar way, making visitors to the hall feel as if they were actually in a ship, moving alongside the whale.[67]

The carefully planned narratives of *Life in the Sea* received glowing reviews from museum affiliates interested in display reform. The Smithsonian design team had successfully combined modern narrative exhibit design with accurate understanding of ocean life, wrote the American Museum's Albert Parr, and in the process, combined modern display strategies with educational goals. Parr heaped especially high praise on the modern aesthetic that had guided designers' decisions. The hall's bold asymmetry and curving balcony railings evoked "memories of ships without straining actual credulity by imitation of form and structure," he wrote. "By these, and other gently suggestive devices, the designers have very cleverly managed to imbue the space itself with a subtly nautical air that makes the whale, in defiance of all logic, seem a far more reasonable and attractive sight than it ever did before under a roof."[68]

By 1965, the increasingly popular efforts of designers at the Office of Exhibits and new pressure from administrators brought both curators' attitudes and halls into line—or at least into the same century—with other museums. Anglim breathed a sigh of relief at this gradual change, confessing to Frank Taylor that less curatorial hostility toward exhibition meant some long overdue items on his modernization agenda could finally "move toward a start."[69] Yet the museum remained relatively conservative; even new exhibits, like its 1966 Hall of Medical Sciences, looked old and relied on a nineteenth-century hospital period room and parades of medical objects ranging from Zuni fetishes to modern heart valves.[70] Anglim was thrilled when the federal government budgeted $500,000 to the Smithsonian General Services Administration that year so the museum could modernize the exhibit halls.

Curatorial resistance to professional designers persisted, a situation well illustrated by conflict over the museum's new Hall of Insects. At Anglim's request, Taylor allotted nearly a quarter of the display-modernization budget to renovate the natural history galleries' decrepit Hall of Insects. He also asked popular natural history writer and Smithsonian exhibits staffer Peter Farb to create the hall's script, granting Farb broad latitude in selecting subject matter for new exhibits. Catering to the research interests of the museum's own entomologists, Farb planned the hall around the themes of insect ecology and behavior. Its focal point would be an oversized model of a grasshopper with features that would light up at the touch of a visitor's button—an interactive insect equivalent to the blue whale in the marine hall.[71] Not a year into building the fourteen-foot-long model, however, Anglim had to request additional funds for scientific research. Anglim had asked the museum's entomologists to provide him with a more detailed analysis of the arthropod's nervous system, and, to their great surprise, they found themselves unable to do so. Curators, taken aback that display was driving research rather than they other way around, irritably argued that the task would distract them from their own and suggested the Office of Exhibits hire entomologists from outside the museum to conduct the necessary studies.[72]

Building an accurate grasshopper model should have rallied curators and designers around a common cause, if not a common project, but instead the project came to exemplify the ways that ongoing staff conflict served to cripple the museum's development. Though they reluctantly acknowledged the final version of the model augmented scientific understanding of the anthropod nervous system, curators ultimately refused to allow the grasshopper model to be placed in the hall, on the grounds that the exhibit's theme of "insects and their allies" was too scientifically narrow. The exhibit didn't provide a complete enough picture of the extant natural history knowledge of grasshoppers, they argued, and partial information was just as bad as inaccurate information.[73] Exhibit designers saw scientists' protests as illogical intransigence, a childish attempt to shore up their eroding authority even at great cost to the institution's educational mission. Rather than put the model in storage, the Smithsonian sold the grasshopper to the Boston Museum of Science, which promptly made it the centerpiece of its own new insect exhibit. The giant grasshopper remains on display in Boston to this day, a six-legged testament to the historical struggles of natural history museums over exhibit design.

Ultimately, attempts by natural history museums, post-Sputnik, to create exhibits that would please advocates of both science literacy and discipline-based science education did not reinvigorate their role in national science education but instead simply fed existing institutional rivalries. Those who embraced the exposure model found themselves pitted against those who believed the public needed to understand the specific concepts that made up the larger body of scientific knowledge. While curators, exhibit makers, and administrators could be found on both sides of the debate, this contest over pedagogical approaches felt to many like a replay of generations-old tensions, between those who prioritized accessible public education, on the one hand, and those who prioritized research, on the other. The friction had hindered display innovation, and natural history museums remained irrelevant to policy discussions of science education.

Interactivity Again: Corporate Sponsorship, Exhibit Design, and Educational Rationales in Science Museums

Science museums were far more welcoming to professional exhibition designers and to new trends in displays, for they did not have the burden of attempting to balance education and exhibition with research, collecting, and preservation. Nor were they weighed down by the need to maintain a sense of scientific dignity. Sputnik seemed to have validated the decision by science museums to focus exclusively on education, as well as their increasing attention to children—a demographic Bradford Washburn affectionately referred to as "fingerlings."[74] Yet the child-friendly, interactive displays that professional designers created with money from corporate sponsors failed to interest policy makers in exposure and solve the larger question of the role of museums in post-Sputnik science education.

Without the boundaries established by disciplinary departments and collections, staff members at science museums were quicker to collaborate than natural history museum staffs had been, and curators, educators, and exhibit designers worked more closely—and more calmly—than their counterparts in natural history museums. Rather than squaring off with sullen scientists, exhibit designers in science museums faced resistance from a different quarter: staff educators. Science museums had always valued their education staffs more than natural history museums had, and after Sputnik, the importance of the education staff only increased. They began to have more input into exhibits and saw themselves as guardians of accuracy in exhibits, resulting in tension with designers, who were trying to make displays more attractive to the public. Both staff factions genuinely believed they were contributing to the larger project of exposure, and disagreements were usually mild, but the friction could also be destructive. In the late 1950s, for instance, after the closing of

Boston's city aquarium, Bradford Washburn hoped to establish a new aquarium at the Boston Museum of Science's new site. University biologists and the region's museum leaders cheered the project, but the plan was ultimately foiled by disagreements between exhibits and education staff. Educators objected when John Siebanaler, manager of the Florida Gulfarium, a commercial marine tourist attraction, was proposed as the director of the project, arguing that his selection indicated the museum was more interested in entertainment than education. "There might be some bad feelings and misunderstanding if [Siebnaler] came thinking that his ideas of 'the show's the thing' were going to be accepted entirely," educator Ned Pearce tersely informed Washburn. Such conflict killed the project, and ultimately, the Boston Museum ended up with no aquarium, only a tank devoted to "game fish of New England."[75]

More often, however, the groups were able to work out these conflicts, or at least compromise, for the single mission of education espoused by science museums meant they shared goals and professional values, even though they held different perspectives and skills. As exhibit designer Richard Sheffield later recalled, "the tensions involved sometimes were . . . on a particular thing that you wanted to do as a designer—that *I* wanted to do, from my point of view and as a designer, and the education department person would say, 'But that wouldn't explain it very well. You really should do it this way.' Then you really might lock horns for a while. But it didn't bother us too much. A lively discourse going on all the time is what we lived for."[76] Boston's staff members learned to value this conversation, for it kept the museum from making expensive mistakes. In 1963, for instance, the museum commissioned the construction of a walk-through heart that would incorporate electronic, interactive features to demonstrate the organ's sound and motion. Hopes for the costly model were high; Bradford Washburn even hoped to use it to teach medical students, as well as members of the public. But when the heart was unveiled to staff members, the museum's educators balked at the model's inaccuracy—a fact the director's brother, primatologist Sherwood Washburn, had noted when he toured the museum. (The model's proportions were "*way* off," he informed the horrified director.) Sensitive to the designers' financial and professional investment in the heart, educators suggested the model could still have educational value if framed in the right way. Rather than scrapping the display altogether, grateful designers agreed that educators should write and place a large explanatory label near the heart to remedy the situation. Having learned their lesson, educators and exhibition designers worked closely on later additions to the museum's Pierce Hall of Medical Sciences.[77]

Exhibit designers and educators agreed that interactivity continued to be the key to effective display pedagogy—and would help distinguish science museums' contribution to national efforts in children's science education to boot. As a result, when designers advised them to jettison the aging push-button displays of the late 1930s, 1940s, and 1950s, science museums responded, looking to corporations to fund the design and production of new displays. As American industry thrived in the late 1950s and early 1960s, its leaders looked for new venues to advertise their successes, and corporate sponsorship of cultural and educational institutions ballooned.[78] Science museums relied heavily on corporate support to help them replace obsolescent displays or renovate outdated halls.[79] Administrators in museums with long-standing corporate partnerships became bolder, leaning on local businesses, rather than waiting for businesses to come to them. When the Harvester Farm animatronic cow in Chicago's Museum of Science and Industry began to act up in 1957, acrid smoke pouring out of the poor beast's nostrils and ears, administrators had few qualms about hitting up International Harvester for money to replace the exhibit.[80] A transfusion of corporate cash likewise revived the museum's famous but ailing heart hall in 1960.

But the new generation of corporate-sponsored interactive displays science museum designers developed often succeeded in entertaining, rather than educating, younger and younger visitors. One new Chicago Bell Telephone–sponsored exhibit, *Telefun Town*, ostensibly aimed to teach children more about telecommunications and used cartoons and animal sounds to teach children how telephones worked. Bees buzzed to signal that lines were busy, children could hook telephone lines from miniature phones to miniature poles, and speak to Mother Goose once they correctly connected the lines. Exhibit designers reported proudly that "this new exhibit brings into the more serious fields of science and industry, emphasized at the Museum, the same kind of showmanship so successfully used in recent years in the increasing number of children's zoos introduced at zoological institutions about the country."[81] Critics simply rolled their eyes.

The most profound problem that science museum display faced in this period was its ongoing association with the paradigm of exposure, a pedagogical ideal many NSF program officers and other advocates of discipline-based science education increasingly disavowed on the grounds that it was passive and lacked rigor. But science museum leaders maintained their faith in the approach and refused to give ground on its merits. Bradford Washburn, one of exposure's most aggressive promoters, wrote endless editorials, letters, and speeches aimed at persuading

Figure 19   Director Bradford Washburn (*left*), Harvard zoologist George Howard Parker (*right*), and two child visitors meeting in the halls of the Boston Museum of Science. This picture, staged for a 1951 member newsletter, placed Parker, who had visited the museum as a child, "on display" in order to illustrate Washburn's assertion that the museum's work with youngsters helped direct them toward scientific careers (Museum of Science, Boston.)

politicians and the public of the efficacy of his approach. "There is no doubt," he wrote to Massachusetts governor Endicott Peabody in 1962, "that the exposure of youngsters to the whole broad vista of science at an early age is of tremendous importance to our whole national education program today."[82] The director shrewdly collected evidence of the success of his approach and presented this evidence in public venues, to show how museums were playing a critical role in educating the next generation of scientists.[83] In member newsletters, photos of new exhibits featured small children staring reverently at the displays, while the museum's annual reports featured tales of adult scientists who had written letters recounting happy memories of early years spent at the museum (fig. 19). He tracked the numbers and ages of museum attendees and attempted to cross-reference them with national surveys of science enrollments in public schools.[84] In 1959, Washburn sought to quantify the validity of his method, using a grant from the Lilly Foundation for several outside educational experts to conduct a survey of children who had taken part in the museum's educational programs. The Lilly survey resulted in a ringing endorsement of Washburn's philosophy, concluding that a science education program like Boston's, "which creates such enthusiasm among a group of young people, a large proportion of whom have said they intend to prepare realistically for a science career, should certainly be of interest to those concerned with children, education, and science."[85] Yet the NSF remained aloof, suggesting through silence that exposure to science through museum displays was not a valuable pedagogical approach—and, as such, not worth financial investment.

Museums that explicitly announced their attempts to blend science literacy and discipline-based pedagogy in exhibits work achieved greater support from scientists and government officials in these years. In 1962, when Tacoma-born marine biologist Dixy Lee Ray argued that science literacy and disciplinary-based science education could be successfully combined to create what she called a "living science center," the federal government agreed to provide enough funding to allow her to turn the temporary science pavilion of the Seattle World's Fair into a permanent institution that reflected both ideals.[86] Possessing an almost evangelical faith in the beneficial power of scientific rationality, Ray was a science-policy insider: she served on the National Academy of Science's Committee on Oceanography and, in 1960, became the NSF's special consultant in biological oceanography, where she helped establish the National Center for Atmospheric Research. But Ray also sought to demystify science for the masses, hosting a weekly television series called *Animals of the Sea* for Seattle's KTCS and lecturing to local audiences on a wide variety of scientific topics. She had little interest in building a traditional museum, one where "objects of permanent interest in the arts and sciences are preserved"; instead, she suggested, the institution should "make the people, all the people, of the Pacific Northwest scientifically the most literate" of any region in the nation, all the while allowing them to engage in the kind of hands-on scientific practice that advocates of information-based science education adored.[87] "Here, people may learn about science by participating in science," she promised.[88] Willing to take a risk on her vision, the National Science Foundation gave her $100,000 to create and direct the Pacific Science Center.

Though Ray's rhetoric appealed to educators and policy makers across the political spectrum, battle-scarred museum directors noted the vagueness of her conception and questioned whether her experiment represented real pedagogical change. The Pacific Science Center presented a "magnificent challenge," Bradford Washburn acknowledged, but it would "take some very rapid and tough decisions to get it solidly on the track." "How on earth they are going to finance this operation is a mystery to me," he admitted.[89] As Washburn had predicted, funding problems handicapped Ray's grand designs. Despite support from illustrious scientists like geneticist and recent Nobel prize winner Edward Tatum, from established museum leaders like Washburn, and some seed money from the NSF, the Science Center found it difficult to raise enough money to keep its doors open, much less realize Ray's initial vision. Six months after assuming leadership of the Science Center, she directed staff members to abandon earlier plans for a less expensive approach, relabeling and rearranging existing displays along taxonomic lines.[90] In those first few months, the Pacific Science Center presented films and science demonstrations, helped form a local mycological society, and prepared an exhibit on arctic life. Rather than becoming a new kind of institution, one filled with learning equipment for scientific participation, one that provided visitors with new ways of thinking about and doing science, by 1963, Ray's Pacific Science Center had become a place where learning looked an awful lot like it did in a traditional museum.

The Smithsonian Conference and Beyond

By the mid-1960s it had become clear that, despite the Pacific Science Center's failures, many of the nation's scientists, educators, and exhibit makers shared Ray's interest in promoting science through meaningful hands-on participation. In private correspondence and public speeches, museum staffs began to suggest this strategy could reframe post-Sputnik debates about display pedagogy. The act of doing science, rather than simply standing back and viewing it, appealed to advocates of discipline-based science education as well as champions of Progressive, child-centered education, and museum staff on both sides of the debate increasingly joined hands to urge hands-on opportunities for the primary school set. Yet they had no idea how to create displays based on this ideal. Active participation in science was easily accomplished in school biology laboratories and even in museum-sponsored science clubs, but museum staffs could not imagine ways that this kind of learning could translate to museum exhibition. In 1966, however, participants in a government-sponsored conference glimpsed a vision of how this might eventually be accomplished.

Intrigued by science museums' live and interactive displays, reform-minded staff members at natural history museums began to discuss how they might adapt them for their own halls. On returning from a tour of several U.S. science museums, Smithsonian botanist Richard Cowan raved to colleagues about the Boston Museum of Science's interactive displays of microscopic organisms and suggested the National Museum consider introducing similar displays into their own halls. Children seemed to relish the experience of looking at the properties of objects through microscopes at the museum, Cowan reported. "I think this is something we could use to great advantage," he wrote. "A simple thing like this could lead some of the younger viewers into a life in science."[91] This kind of exhibit was a throwback to Progressive era hopes that visitors would engage in the kind of focused looking that would result in lifelong habits and passions, but Cowan found it deeply compelling.

Cowan's suggestions to incorporate more hands-on exhibits in the Smithsonian's halls were also motivated by a desire to promote the life sciences—his own field of study—more aggressively. In post-Sputnik science education efforts, Cowan speculated, biology had gotten short shift from both policy makers and the public. Because policy makers more readily funded the study of atoms than they did animals, Cowan anxiously wrote, "many young people who might naturally be attracted to biology as a career are never acquainted with the breadth of opportunity and . . . skills utilized in this profession." The result, he added, was a "dearth in the biological sciences of scientists . . . and other biologically-oriented personnel."[92] Interactive museum displays, he argued in an internal report, just might improve the status of the life sciences in the post-Sputnik United States. Such exhibits might convey the excitement of life sciences and "something about how people *do* research" to museumgoers and, just as important, to "people on the Hill, who have their hands on the purse strings."[93] Cowan's British colleague John Cannon, a botanist at London's natural history museum, put it more succinctly: "A well-informed public is more likely to support our activities than one that feels [we] have little time to make their subject intelligible to the layman."[94]

The Smithsonian's imperious new secretary, ornithologist S. Dillon Ripley, could not have agreed more with Cowan's points about the pedagogical and political power of interactive exhibits—and the need to expand the presence of such exhibits at American natural history museums more generally. As Ripley wrote in the Smithsonian's 1964 annual report, "Museums and their related laboratories are just entering a new era, and museum resources are being drawn upon as never before for general education."[95] More interactive exhibits promised to impress not only politicians but also the people who put them in office, a group Ripley recognized was becoming more and more vocal. By the mid-1960s, even the thick walls of museums had begun to feel tremors from the ongoing earthquake of civil rights and the gathering storms of antiestablishment political culture. Aware of the nation's shifting cultural mood, reformers in the museum community pressed to adopt exhibition approaches and topics visitors found exciting and accessible, rather than continuing to rely on those that curators found fitting. Hands-on exhibits seemed a first step in this direction.

Ripley made his plans to emphasize his museum's educational mission clear to his exhibits staff and fellow administrators, but he knew that his still-suspicious Smithsonian scientific curators would need more convincing. In 1965, he hired a museum outsider, Charles Blitzer, to become the Smithsonian's new director of education and training. A sweet-tempered administrator with considerable political savvy, Blitzer was the first to admit he had been hired to help Ripley "break down the barriers between museums and the outside world."[96] Blitzer had not cut his teeth in museums the way so many administrators had—his background was in education and policy circles—and he made a few false starts, including recommending that the Smithsonian install a slum exhibit in its newly opened Anacostia neighborhood museum in Washington, DC, complete with live rats in the walls, as a way of demonstrating that the museum could be attentive to the interests of its less affluent visitors.[97] Nonetheless, Ripley believed Blitzer's considerable diplomatic and negotiating skills would heal old wounds from battles between curators and exhibits staff and, in the process, help the moribund institution transform its role in society and education during the coming decade.

Outside of museums, educators and policy makers again began to reconsider the value of informal educational venues for popular science education. In 1965, for instance, the American Association for the Advancement of Science hired E. G. Sherburne, a scientist who had worked extensively in educational television, to head up its revived public understanding of science program. In public statements, Sherburne argued that popular science education had received short shrift and that it was time to change this policy. Discrepancies in NSF budgets between funding for education and research indicated "a gap in national thinking and planning," he argued. Though school children had been well served in federal efforts to improve science education, the broader public had been neglected. "There is remarkably little formal assumption of responsibility by government agencies for informing and educating the public," he concluded.[98]

Sherburne's sentiment found a welcome audience among museum administrators, who also deplored how government interest in science education had focused almost exclusively on school science. Frustrated with the lack of federal attention to museums and their efforts in educational research, the Smithsonian's Charles Blitzer proposed that same year that the U.S. Office of Education sponsor a research conference on the subject of museums and education. "In recent years, American museums have dramatically widened the range and increased the extent of their educational activities," Blitzer wrote in his proposal. "This commitment to teaching constitutes a real change in their traditional role," he explained, a change that had created new opportunities but had also raised "a number of problems, both practical and theoretical."[99] The government approved the grant, and in August of 1966, Blitzer assembled a team of scientists, exhibit designers, and educators to discuss these topics at the rural Vermont estate of his society friend, Mrs. Vanderbilt Webb.

As he planned what soon became known as the Smithsonian Conference on Museums and Education, Blitzer went out of his way to insure that attendees included not only museum administrators and curators but also nonpractitioners interested in museums and education. Albert Parr attended and acted as a discussant; so, too, did Eric Larrabee, a magazine editor and college professor of architecture and environmental science at State University of New York–Buffalo who had only recently become interested in museum education. He gathered representatives from the Carnegie Foundation, the Rockefeller Foundation, and the Science Service; museum educators like Alma Wittlin, Edgar Richardson, and Sue Thurman; public school teachers like Ruth Zuelke, an art teachers from Birmingham, Michigan; and Robert Hatt and Michael Butler, exhibit designers from Detroit's Cranbrook Institute of Science. Paper topics were equally expansive, ranging from generic overviews of museum education to more provocative and theoretical diatribes ("To Gawk or Think?" and "Exhibits: Interpretive, Under-Interpretive, and Misinterpretive").

Richard Grove, who had approved the grant, reminded his fellow participants that the federal government did not usually fund conferences on museums, but the recent expansion of federal programs aimed at the institutions had necessitated it. Between 1963 and 1966, the government had established the National Science Foundation's Office of Science Exhibits, the National Endowment for the Arts Museum Program and the National Endowment for the Humanities—the latter two of which intended to fund educational displays—and the Department of Health, Education, and Welfare's Arts and Humanities Program. In just three years, the federal government had shifted from a laissez-faire stance on museums to a level of financial support not seen since the Depression's Works Progress Administration. Museums had promptly formed regional consortiums to share ideas about how to best take advantage of this influx, but government policy makers scorned these organizations as too self-interested and too theoretically minded and, instead, sought out independent, research-based opinions on exhibits and how they worked. The purpose of the conference, Grove reminded them, was to help think about how museum displays could contribute to education in new and meaningful ways.

Grove also urged conference participants to establish "a rapprochement between museums and schools, which had drawn apart as a result of the post-Sputnik battles over pedagogy and federal grantmakers disinterest in the kind of science education museums provided. "Museum men," Groves concluded, "are surprised by what seems to them to be lawmakers' lack of perspicacity. Are museums not an unparalleled and mighty educational resource? Why are they not directly eligible for federal aid? [But] school administrators are suspicious and ask what kind of educational institutions are museums anyhow? They are quite outside of the educational establishment." "What we ask for," Grove pleaded to conference participants, is help in synthesizing modes of formal and informal education "widely and well. . . . From what I see, I do not think this is being done as well as it could be."[100]

The ensuing discussion represented some of the clearest articulations of the ideological conflicts and practical challenges facing those who genuinely sought to unite museum work and education in the mid-1960s. Larrabee, who later became provost of the State University of New York–Buffalo, reported that the Smithsonian conference "managed to be a source of frustration for all participants" because "speaker after speaker seemed confidently to be handling ideas with which the Conference as a whole was unable to cope." Several attendees agreed that museums needed to conduct more research on their educational failures and successes but admitted that such self-examination would require institutions to change their attitudes entirely. "Too many museums are scared of finding out how wrong they have been," curator and critic Albert Parr explained. [101] Educator Sue Thurman appealed to participants directly to reach across institutional barriers and embrace their conflicts about museum pedagogy: "What I think is we need to be is a good deal gutsier . . . what I am knocking is the over-abundance of educational friendliness. . . . Whatever we find, let us not just review what we already know: that in both the school field and the museum field, 'sweetness and light' can cut out a great deal of illumination."[102]

The toughest issue that participants faced, Larrabee later wrote, was deciding "whether museums should make a deliberate effort to accommodate themselves to the formal curricula of schools, or to intervene in the schools' process of arriving at curriculum." Answering this question, Larabee concluded, was inexorably dependent on "alternative approaches to the question of what the museum is and does, and the more we talked, the more it became apparent that idea and object are so bound together that none of us could successfully disentangle them."[103] Working groups formed on the second day, and combinations of museum and education workers offered some pragmatic answers geared toward greater collaboration, suggesting that every school teacher should be required to study museum education as part of their training and that museum educators should be appointed to school curriculum revision committees.

Professor Frank Oppenheimer piqued the interest of other participants when he rejected the school-museum dichotomy altogether. Nervously lighting and tapping a seemingly inexhaustible supply of cigarettes, the lanky physicist insisted that, for science at least, choosing between formal and informal education was unnecessary. The two processes did not need to be exclusive but could, instead, mutually reinforce each other. "In order to produce real understanding, in order to have science come to something—so that it really affects the way one thinks about the world—[science curricula] leaves things out," Oppenheimer mused. "Some of this can be picked up through reading but I think most has to be picked up through props, and that the museum's role is to fill in these holes. The school provides the narrow channel and the museum is the thing that broadens it. Without this broadening, I don't think one can get a decent education."[104]

Though he remained something of an outsider throughout the conference, few participants at the Smithsonian were better suited to speak to the question of how museum exhibits might transform education. Oppenheimer had turned to science at the urging of his brother, Robert, who freely admitted his younger brother's superior talent for "getting his hands dirty in the laboratory."[105] After obtaining a PhD in experimental physics from Caltech, he had taken a research position at Stanford, making frequent trips north to Berkeley to visit his more-famous sibling. Throughout the war, the wiry young physicist worked with the Manhattan Project, and in 1947, he took his first regular academic appointment in physics at the University of Minnesota. In 1949, three months from being awarded tenure, he was forced to resign from Minnesota under harassment from the Federal Bureau of Investigation for his involvement with the youth wing of Berkeley's Communist Party.[106] Deeply disillusioned, unable to find work as a physicist, he left science entirely. He sold one of the Van Gogh paintings he had inherited from his wealthy family, purchased land in southwestern Colorado, and then began, in his own words, "eking out a living as a working rancher."[107] In 1957, however, he was drawn back into science when the local high school, in the midst of a teacher shortage, begged him to step in; that year he taught biology, chemistry, physics, and general science.

While teaching at Pagosa Springs High School, Oppenheimer had developed a deep commitment to participatory learning through hands-on experimental demonstration. At a Parent Teacher Association meeting that first year, Oppenheimer told local parents of his desire to move the school's science curriculum beyond memorizing facts to what he called "multiple proficiencies." Rather than limiting students to textbooks, he explained, he wanted to make learning "satisfying and fun" by emphasizing how students could master some or all of the full range of skills required in doing science: "the manual dexterity of setting up and performing experiments, the mental dexterity of solving numerical problems, the technique of observing the results of an experiment. . . ." Taught this way, Oppenheimer argued, high school science would not only "enrich the student's individual life" but would also "make more useful scientists."[108]

By 1959, with McCarthyism on the wane and attention to science education on the rise, the political climate had shifted enough for Oppenheimer to return to higher education. He took an appointment in the physics department at the University of Colorado, but instead of returning to research, he continued his pedagogical work, revamping the university's physics laboratory curriculum in order to, he explained, give students "a sense of power to actually do something" in science and in life.[109] In Boulder, he created what he called a "library of experiments" in order to teach physics first principles with hands-on, student-paced experimental methods.[110] In 1958, he was invited to participate in the Physical Science Study Committee, which aimed, in the words of one participant, to revise high school physics curricula by "examin[ing] the fundamental processes involved in imparting to young students a sense of the substance and method of science."[111] In 1964, though Oppenheimer had only two years of high school teaching experience, the well-regarded Princeton Day School invited him to direct its science program.

Intrigued by the potential of the long-heralded but rarely realized hands-on approach to science pedagogy, Oppenheimer began to wonder how to expand the reach of these new educational methods outside of the classroom. "There is an increasing need to develop public understanding of science and technology . . . to bridge the gap between the experts and the layman," he declared.[112] By the mid-1960s, he had become convinced that museums, even more than schools, offered a place where this could occur. Inspired, in part, by a United Nations Educational, Scientific, and Cultural Organization resolution "concerning the most effective means of rendering museums accessible to everyone," in 1965 Oppenheimer applied and received a Guggenheim Fellowship that took him to England.[113] Though the fellowship was officially intended to fund the study of physics and its history at University College London, Oppenheimer took the year to explore various European science museums.[114]

Throughout that fellowship year, he later reported, he began to visualize a new kind of museum, one that would re-create a "laboratory atmosphere," one where all kinds of people could "become familiar with science and technology and gain understanding by controlling its props."[115] To Oppenheimer, the physical materials and manual labor associated with science were both crucial to public understanding of the field, for "explaining science and technology without props can resemble an attempt to tell what it is like to swim without ever letting a person near the water."[116]

Though admirers have suggested Oppenheimer's methods were original, the product of isolated genius, midcentury trends in science education and museum exhibit reform shaped his approach.[117] In historical context, his methods were a long-awaited realization of desires first voiced by the Progressive educators who advocated visitor participation and, later on, by those biology teachers in the early 1950s who promoted science in action. But as his contributions to the Smithsonian conference and his publications make clear, his ideas were also framed by post-Sputnik educational debates, for Oppenheimer advocated hands-on pedagogy as a way of fulfilling demands for both general science literacy and more rigorous discipline-based science education. In keeping with his own Progressive educational ideals, Oppenheimer's pedagogical views were at once traditional and revolutionary. He advocated mandated teacher training to ensure that elementary and secondary school teachers kept up with the latest advances in their field. At the same time, he believed, as his Physical Science Study Committee colleagues Jerome Bruner and Jerrold Zacharias put it, that "the intellectual work of a research scientist and an elementary school pupil are essentially identical"—an unusual stance for a university professor to assume—and he pushed for as much hands-on experimentation as possible.[118] Educators in classrooms and museums swooned over his appealing combination of practicality, social idealism, and playfulness.[119]

Oppenheimer was at the forefront of conversations about hands-on science education methods that would be viable on a broad scale, and his ideas enthralled long-time museum and educational reformers. Just after the Smithsonian conference, Oppenheimer devised what he called a "Proposal for a Museum of Science and Technology at the Palace of Fine Arts." In it, he clearly articulated his vision for a new kind of museum, one that would provide "a flexible framework for achieving broader educational purposes," one filled with well-designed opportunities for hands on exploration of scientific principles and methods. He also wanted, he explained, to create a place that rewarded teachers and exhibit designers for employing innovative strategies, where "creative pedagogy [would] be raised to the same status and be granted the same rewards as creative research."[120] *Curator* editor Albert Parr promptly published Oppenheimer's proposal, and Smithsonian administrators invited him to found and direct a new federal museum based on the idea.[121] Oppenheimer turned them down, a rejection that resulted partly from his independent tendencies and partly from his belief—one shared by many other scientists and policy makers—that existing museums represented a barrier, rather than a tool, for educational reform. Instead, Oppenheimer announced his intention to build a place in San Francisco for the city's diverse public to come in and do science. The museum he had in mind would form the basis for a decade of museum educational experimentation.

SIX

The Exploratorium Effect: Redefining Relevance and Interactive Display, 1969–1980

For nearly two years, Frank Oppenheimer worked to realize the vision of the new museum he would describe again at the second NSF-sponsored Belmont conference. By 1969, he had scraped together an improvised institution he planned to call the Exploratorium. The Exploratorium, he declared, would provide a "humanistic atmosphere" where visitors devised their own museum experience, a place where "even novices can ask and answer" questions about science.[1] The idea raised eyebrows. After *San Francisco Chronicle* science writer David Perlman interviewed Oppenheimer in 1968, he expressed doubt about the physicist's open-ended approach to science education. "I thought he was pretty far out," Perlman later recalled. "He talked about things I had never thought about in the context of a museum. I thought: 'This is never going to fly at all.'"[2]

The pragmatic hurdles Oppenheimer confronted were as daunting as critics' skepticism. Short on money, materials, and time, he resorted to scrounging old lab equipment from Stanford to build exhibits. He relied on family to help him install new wiring and sweep the floors of the city's old Palace of Fine Arts, a rebuilt remnant of the 1915 Panama-Pacific International Exposition. Neighbors vehemently objected to the Exploratorium's presence, fearing the insti-

Figure 20   The Exploratorium's "open" floor, around 1975, with some hands-on *Perception* exhibits and visitors. Director Frank Oppenheimer, an adherent of Progressive education, argued that pedagogically effective museum spaces did not compel visitors to follow preordained paths through exhibits but, rather, created opportunities for them to experiment and pursue their own interests. (© Exploratorium.)

tution would attract San Francisco's unrestrained youth counterculture to the otherwise quiet Marina district.[3] They needn't have worried. When the Exploratorium opened in August of 1969, it barely registered with the public. No renowned scientists spoke, no ribbons were cut, no press releases were sent.[4] Instead, one morning, Oppenheimer and the few members of the staff just opened the doors, and kept them open.[5] Few people entered that first day. Two runners jogged in and "never slowed up," Oppenheimer recalled. "They just went back and forth, and finally went through the curling path and out the door again."[6]

Despite its unassuming beginnings, the Exploratorium had an enormous impact on both the nation's museum and science education communities. Its attempts to democratize the museum experience and make scientific knowledge accessible appealed equally to renowned scientists and long-haired social activists, to policy makers and peaceniks, to museum professionals and museum protesters (fig. 20). Although the idea of a "science center" was not originally his, Oppenheimer's fresh take on museum-based science education became the hallmark of the science-center movement that swept the United States in the 1970s and early 1980s.[7] Indeed, the lanky physicist was so closely identified with this movement that his famous statement—"no one ever flunked a museum"—became its rallying cry.[8] Museum studies scholars now mark 1969 as the beginning of what might be called the "Exploratorium effect." The phrase captures how the Exploratorium's twinned values of informal learning and hands-on experience became the foundation on which science and natural history museums built new relationships among content, pedagogy, and display. As they did, these institutions adopted the rhetoric and exhibition techniques that the Exploratorium pioneered.[9]

Yet the Exploratorium's much-publicized success obscured the challenges the institution encountered as its staff members attempted to use new kinds of interactivity to transform museum education. Nowhere were these challenges more evident than in its life science displays. Displaying the content and methods of biology according to new interactive principles was tricky. Hands-on live animal exhibits required more scientific expertise and material support than mechanical displays did, and the Exploratorium's exhibit designers had trouble capturing—and controlling—biological phenomena in ways that were reliably transferrable and scalable to the museum's displays. And while the populist pedagogy of the Exploratorium offered an enticing alternative to more traditional museum displays, its assumptions about visitor response actually reinforced older ideas about proper scientific method. Furthermore, the science center did not resolve the question of where and how museum-based science education fit into a politically electrified culture. Throughout the 1970s, critics and activists assailed natural history and science museums as oppressively out of touch. Even the Exploratorium's doggedly apolitical life science exhibits had led to public protests, forcing Oppenheimer to defend what seemed to critics to be an almost naive faith in the experimental method and the value of participatory learning.[10]

Nevertheless, the Exploratorium had a lasting effect on museums' life science displays—and on museums more broadly. As the Exploratorium gleefully toppled established museological traditions, natural history and science museums began to reflect on their own roles—and their own relevance. They strained to make themselves more germane to contemporary science and education but also tried to capitalize on existing strengths in order to preserve their distinct institutional identities amid the era's kaleidoscopic changes. Some museums overhauled the content of exhibits to reflect a scientific take on contemporary political and social concerns. Other institutions reasserted the importance of staff members' scientific expertise and the primacy of their collections. They pointed out that these, too, were public resources on par with the Exploratorium's exciting educational programming and that time-honored exhibition and research work allowed natural history museums to present broader, more nuanced understandings of biology than interactive science centers ever could. Still other museums continued to install push-button exhibits, which, in the wake of Oppenheimer's work, seemed as retrograde as dioramas had decades earlier. Ultimately, the effect of the Exploratorium on American natural history and science museums mirrored the effect of the late 1960s and early 1970s on America. More people had a say in how museums displayed life, but this expansion of stakeholders multiplied conflicts over how to define and achieve educational success and social relevance in museums.[11]

The Politics of Science Literacy in an Era of Economic Decline, 1968–1975

By the late 1960s and early 1970s, broader American public debates about social justice, public responsibility, and individual rights had found their way into both classrooms and museums, and debates about science education became heavily political as a result. Rather than worrying about Soviets and space races, educators increasingly wondered how to provide science education that would be meaningful to all American children.[12] Educators and some scientists argued that a broader, child-centered form of science education would reach those students who had no intention of becoming scientists and began to emphasize hands-on participation and creative experimentation as a result. The child-centered classroom became the ideal; those who favored teacher-centered pedagogy were charged with dehumanizing students through mindless repetition.[13] Champions of discipline-based education disdainfully described this trend as a revival of Progressive educational reform's worst aspects, but the shift resonated with the public, and national newspapers and magazines from *Newsweek* to *Reader's Digest* to the *Wall Street Journal* all ran optimistic features on open classrooms and alternative schooling. The pedagogy of science classrooms changed to reflect these values. Instead of continuing to train the best and brightest to wage the Cold War through discipline-based science, educators now talked about reaching all students regardless of ability or inclination, echoing the populist rhetoric of the 1930s as they did so.

Consequently, high-school science teachers began to retreat from the rigorous disciplinary training embraced by NSF-sponsored curriculum projects, adopting broader, more student-centered explorations of scientific processes and principles instead. Interest in science literacy sparked after Sputnik's launch caught fire in the 1960s, fueled by new anxieties about clean air and water and the publication of books like Rachel Carson's *Silent Spring* (1962) and Paul Erlich's *The Population Bomb* (1968). "With nuclear holocaust just outside the window and the polluted atmosphere already seeping in, we simply cannot afford to train a generation of students who know the *how* and *why* of scientific phenomena, but do not have a process for inquiring into the values issues raised by the topics they study," declared a 1970 article in the *Science Teacher*.[14] By the 1970s, science literacy advocates regularly argued that science education, and especially biology classes—the one science course that the majority of American students took—should train students to confront the personal and social consequences of scientific knowledge, rather than just teaching them formulas and facts.[15] In its 1971 position statement, "Science Education for the '70s," the National Science Teachers Association identified scientific literacy as its single most important objective: "The major goal of science education is to develop scientifically literate and personally concerned individuals with a high competence for rational thought and action."[16] Many scientists also began to campaign for value-centered science curricula. "Citizens should acquire the kind of general understanding that facilitates recognition and evaluation of the social consequences of science and technology," microbiologist René Dubos argued in 1967. "For lack of this understanding, the citizen will have to submit to the tyranny of the expert, who will then become a decision-maker without being answerable to the community. In contrast, if the public can share in a more enlightened manner in the decision-making process involving scientific problems, democratic societies may regain the social coherence which is the condition of their survival."[17]

Influential museum educators welcomed this new approach and urged museums to rethink their own pedagogy in turn. In 1970, for instance, Alma Wittlin revised and reissued her 1949 book on the subject, now titled *Museums: In Search of a Usable Future*. Wittlin updated her survey of museum education, but now also encouraged her peers to reevaluate the meaning of education. "Is it our sole desire to feed information into people—or do we understand education as a process of wider scope: a tuning of people to best think judiciously?" she asked. She reserved her harshest pedagogical criticism for the pedagogical ideal of exposure. "Exposure is not enough," she stated flatly. Merely showing and telling, she concluded, was an outdated approach. Museums needed to engage more directly in conversations with educators, she concluded, if they hoped to remain relevant to public science education.[18]

Wittlin's warning was a prescient one, for educators and policy makers often overlooked museums as potential participants in the general science literacy movement of the late 1960s and early 1970s. (The NSF was a prominent exception to this).[19] When the first "Report on the Public Understanding of Science" appeared in 1970 from the American Association for the Advancement of Science, for instance, it heralded the importance of science literacy in a wide variety of formal and informal educational settings but made no mention of museums. Later that same year, when the AAAS Commission on Science Education recommended possible sites for new programs, it again ignored museums.[20]

The exclusion of museums from contemporary conversations about life science literacy resulted in part from the public's growing conviction that the institutions were out of touch, unable or unwilling to engage in the controversial political debate that science literacy could inspire. Americans increasingly dismissed museums as conservative, even reactionary. Throughout the late 1960s and 1970s, natural history museums' once-serene halls rang with angry demands that institutions confront the racist, colonialist implications of their research, collecting, and exhibition traditions.[21] Some of these protests embodied the era's free-floating hope and antiestablishment rage, while others were aimed at specific institutions.[22] One particularly damaging episode for the Smithsonian came to light in 1969, when NBC news reported that the institution's largest grant for natural history research—the multimillion dollar Pacific Ocean Biological Survey Program initiated by Philip S. Humphrey, curator of birds and chairman of the Department of Vertebrate Zoology— existed not only to survey Pacific migratory birds but also to provide the U.S. Army and Navy with information about the effectiveness of these animals as potential vectors for biological weapons. National Museum of Natural History director Richard Cowan was blindsided by the report. He told his colleagues that, although he himself had never had full access to survey program's records, it was prudent for the project to be removed from the museum and resited at a university or even at the National Science Foundation.[23]

But the damage was already done. Museum visitors saw natural history museums as indifferent to the politics of their research and display and envsioned their curators as too siloed in their own research enterprises to tend to public concerns. "There's *that* American Museum of Natural History, of course," wrote *New York Times* reporter Robert Stock, "the public institution. The bedlam of screams and scampering shoe leather. Thirty-eight exhibition halls in 19 buildings on a 23-acre plot of land beside Central Park. . . . And then there is another museum, a private world peculiar to those who administer, and maintain, and pay for, and create the public institution."[24]

A few natural history museum directors pushed back against these political critiques, arguing, as Cleveland art museum director Sherman Lee did in 1972, that museums needed to "resist actively the chaotic demands forced upon them equally from the swinging set to moralizing Maoists."[25] But most found it difficult to respond directly to public ire and did so only under the cover of the collective professional voice of the American Association of Museums. At the 1970 AAM annual meeting, for instance, shaken curators pledged to "attend to the problems of racism, sexism and repression by implementing new programs which will further their solution."[26] That same year, the AAM issued six more mea culpas in print, each time pledging to do more to address the social issues swirling around them.[27] But natural history museums didn't change quickly, and criticism didn't quickly abate.

Individual museum workers were so wary of jeopardizing federal research funding and so worried about provoking violent public reaction that they often backed away from public debates, even those that touched on their most cherished areas of expertise. In 1970, for instance, the NSF-backed social studies course for fifth graders, *Man, a Course of Study* (MACOS), generated a firestorm among Christian fundamentalists. Protesters denounced the curriculum for its implicit validation of evolution and its supposed cultural and moral relativism and leveled their sights at *Man, a Course of Study*, the Biological Sciences Curriculum Study textbooks, and other curricular materials.[28] Parents, educators, and professional organizations of scientists and science educators rallied in support of NSF-backed curricula, but curators remained on the sidelines.[29] A handful of university-based curators defended the federally sponsored science and social science curricula against creationists' attacks, curators in the nation's municipal natural history museums tended to stay mum.[30] The avoidance of the bloodiest aspects of this debate by museum biologists was unsurprising; the agency, under pressure from conservative Congressional representatives, ultimately terminated its support for *Man, a Course of Study*. Curators, given their almost total dependence on NSF funding for research, may not have wanted to risk a similar financial fate.[31]

The increasingly troubled finances of institutions gave curators another good reason to steer clear of the debate over evolution. As inflation climbed and the stock market plunged in the late 1960s and the early 1970s, both government-funded and nonprofit museums found themselves without enough money to cover their operating costs. Eroding urban tax bases forced cities to slash long-standing allocations to local museums, at the same time that museum operating budgets grew, in large part because the oil shocks of the 1970s made it more expensive to heat their large, drafty buildings. The experience of the Buffalo Science Museum was typical. Under tremendous financial strain in the late 1970s, the Rust Belt capital cut its contributions to the city's Museum of Science until, by 1978, the city only provided the museum with the money required to prevent its grounds and building from total decay; the museum was forced, in 1981, to begin charging for admission.[32] The nation's wealthiest institutions also suffered: in 1970, despondent administrators at the American Museum of Natural History considered charging for admission—a reversal of a century-long tradition of free public access.[33]

Desperate, museums began to cast about for new sources of funding. Many turned to volunteer labor to make up the shortfall. By 1974—when the National Endowment of the Arts published *Museums USA*, another comprehensive survey of U.S. museum activities and finances—there were two volunteers for every full-time paid employee at museums in the United States, when taken in their entirety. Museums also pressed government officials for help. By 1974, nearly 40 percent of science and natural history museums in the United States reported that they depended almost entirely on federal, state, or municipal funds for their operating budgets. Another 45 percent of museums were nonprofit entities, which relied on a combination of endowment yields, donations, foundation funding, and public subsidies for support.[34] Curators, knowing just how precarious their institutions' finances were, could hardly afford to jeopardize public or political good will with an aggressive defense of school science curricula—at least not if they hoped for continued employment.

Paralyzed by their alienation from the nation's educational debates and frantic to qualify for new sources of funding, the American Association of Museums announced the creation of a special committee to define the shared purposes and future directions of American museums. A clearer institutional mission, they hoped, would allow rudderless institutions to chart a new course through unfamiliar political and economic landscapes. And, savvy lobbyists pointed out, if museums could persuade the government to recognize them as institutions with official educational capacities, they might be able to qualify for direct public funding under the Elementary and Secondary Education Act.[35] Committee members met at Belmont House, a country estate in Maryland, throughout 1967 and, in 1969, issued the *Belmont Report*, which analyzed the results of the first AAM comprehensive survey to the federal government.

Though the *Belmont Report* forcefully argued that museums were educational institutions and, as such, deserving of more federal support, the report's contents also made clear that museum professionals were almost crippled by their desire to appeal to all groups or, at the very least, not to alienate potential institutional allies and wide swaths of the public.[36] Excruciatingly conscious of the diversity of their newly vocal constituents' desires and the increasing variety of institutions for which they claimed to speak, the committee couldn't even bring itself to define the term "museum." "To construct an air-tight precise definition to cover all American museums," the report admitted, "may be impossible." Those common denominators they did identify were comically general: according to the authors, museums were nonprofit institutions "concerned in some way or another with education. And they perform a function, which no other institution does."[37] What exactly that function was remained anyone's guess.

As the AAM and the Belmont committee members examined the educational capacities of museums, they concluded that museums needed to reprioritize exhibition, education, and community relations. Natural history museums would find this especially challenging, a 1972 AAM report predicted, for postwar administrators and curators regarded education and exhibition departments "as fairly unimportant, concessions to modern times, distractions from the main thrust of the museums, which has been thought to be scholarship oriented toward the enhancement of the collections." "But," the authors warned, "in the crisis now confronting museums and in the opportunities which are theirs in working with the new constituencies of the inner city, such traditional academic snobbery is more than ever out of place."[38] The AAM assessment was breathtakingly ahistorical, ignoring decades of museum reform, but it wasn't inaccurate. And staff members at natural history museums knew it.

Oppenheimer's Synthesis: Hands-On Scientific Exploration at the Museum

To a museum community desperate to educate and entertain visitors while also pleasing policy makers and social protesters, the Exploratorium's synthesis of hands-on pedagogy and rigorous scientific inquiry must have appeared to be a silver bullet. Oppenheimer's exhibits circumvented long-running debates because they looked nothing like the elderly collections of arrowheads or the cheerfully fawning, industry-sponsored exhibits of biotechnology that raised such ire among youthful protesters of natural history and science museums. One of Oppenheimer's few rules for the new museum was that it should avoid "any reliance on the types of diorama displays."[39] They were too expensive and too static, he explained. Instead, Exploratorium exhibits were works in progress, experiments intended to be broken, fixed, and improved by the fixing.

The Exploratorium's first few exhibits, which explored visual perception, were a remarkably succinct embodiment of Oppenheimer's vision for what a museum should be.[40] *Perception* was a typical Exploratorium venture, reflecting Oppenheimer and his staff's interdisciplinary, idiosyncratic interests in art and philosophy while sitting at the intersection of contemporary research in physics and biology.[41] In a 1967 letter, Oppenheimer explained that exhibits should not respect traditional academic boundaries: the Exploratorium, he noted, planned to feature exhibits on the biology of vision alongside demonstrations of the physics and mechanisms of neurosensory perception.[42] This exhibit also capitalized on the expertise of a broad base of local scientists working outside the museum, a routine practice for the Exploratorium. While planning the exhibits, Oppenheimer leaned on board member Arthur Jamplosky, a biologist who directed the Institute for Medical Sciences at the Smith-Kettlewell Institute of Visual Sciences, to recruit an advisory committee of Bay Area researchers interested in vision, including optometrists, doctors, and marine biologists, to recommend relevant scientific topics.[43] He also contributed his own ideas about the ways resonance and interference could explain a wide range of visual phenomena.[44]

Just as the Exploratorium's exhibits departed from existing museum taxonomies of knowledge and hierarchies of expertise, so, too, did its educational programs dispense with traditional practices. Rather than providing docent-led tours, supplying miniature exhibits to schools, or setting aside a gallery to display student science projects, the museum instituted a "student science project program." This program opened up the museum's exhibits-building shop to high school students working on projects for science fairs and encouraged them to work alongside the museum's own exhibit development teams. The museum hired interested students to be "explainers," offering them training and a combination of academic credit and an hourly wage. Clad in trademark red vests, shaggy-haired explainers circulated among Exploratorium visitors, enthusiastically answering questions and suggesting the best ways to interact with exhibits.[45] Explainers also built and repaired exhibits, using the manual and mechanical skills that Oppenheimer believed so crucial to scientific discovery.[46]

Students found these "drop-in jobs for drop-outs," as Oppenheimer

cheekily described them, far more rewarding than the limited participation afforded by most museum internships or high-school programs.
"The more I built, sweated, measured, cleaned, watched, and collected,
the more completely enriching of an educational experience I have received," recalled one of them later.[47] While the museum clearly needed
the cheap labor students provided, Oppenheimer also argued that the opportunity to teach material would help these students learn more about
science; though the explainers "initially need instruction concerning the
material . . . by immediately using their knowledge, their learning becomes more profound and precise," Oppenheimer reported.[48] The NSF
was convinced and agreed to fund the program, which became one of the
agency's first and longest post-Sputnik investments in informal science
education.[49]

The Exploratorium and its fledgling educational programs also satisfied San Francisco science teachers, and idling yellow buses soon lined
the curb of Palace Drive. As the Exploratorium's budget swelled, Oppenheimer expanded the museum's mission to include teacher training,
which local educators adored. "Your attempt to bridge the gap between
the layman and the expert is one of the most exciting ideas in education today," one delighted teacher wrote in 1973.[50] Early on, however,
the museum lacked formal educational programming for students. Each
month, science teachers herded nearly thirteen hundred students—a
number equivalent to the crowds the Exploratorium attracted on a weekend day—into the hangarlike space, only to find that the museum had
no organized programs for students or school teachers.[51] By 1972, the
staff had remedied this, using money from the Rockefeller Brothers fund
to develop a pilot program for the "School at the Exploratorium," also
known as SITE.

The School at the Exploratorium synthesized formal and informal
science education through a combination of extended demonstrations
of the museum's exhibits for students and take-home materials for teachers. One of the most popular SITE segments was the dissection of a cow's
eye, as demonstrated by an explainer. Teachers followed up on this experience back at school, using Exploratorium handouts that illustrated
the eye and its neural pathways of perception.[52] The School at the Exploratorium was so successful that it required the museum to renovate
its building to accommodate public demand: by 1974, the museum had
drawn up detailed plans to construct four demonstration rooms where
classes could meet away from the noise and activity of the exhibit hall.[53]
By May 1978, SITE was serving over fifteen hundred students and their
fifty teachers with onsite classes each month, often sending them back

to school with items from the museum's "lending library" of props, kits, and exhibits.[54]

Oppenheimer's vision of exhibit designers, visitors, and their teachers as collaborative learners fit nicely with existing imperatives to make both science education and museums more socially relevant and publically accountable. Educators and policy makers, museum reformers and foundations adored his ideas, even though many found themselves hamstrung by existing commitments to more formal educational venues. After Fritz Mulhauser of the National Institute of Education visited the Exploratorium, he lamented it was a "great pity" that his agency had "no relation to the 'education' going on informally by the whole rest of the population, in private and in public places such as yours." He would have loved to provide more support, he explained, but "we are tied by tradition, political and scarce dollars, and dealing solely with formal schooling and only at the lower levels."[55]

By emphasizing the visitor's interaction process rather than standardized school curriculum or pre-scripted curator-created content, Exploratorium exhibits resonated with the countercultural trends of the late 1960s—among them, the celebration of individual choice and the challenge to authoritative knowledge.[56] Yet the Exploratorium also promised visitors and museum workers a shared experience of something quite traditional: learning about scientific methods and concepts through predefined paths of experimentation and interpretation. Every exhibit, for instance, featured explanatory panels guiding visitors about "what to do and notice" and "what is going on," a far cry from the messy uncertainties of actual scientific research.[57] When discussing this approach with donors, Oppenheimer skillfully shifted attention away from visitors' prescribed interactions to exhibit designers' investigations of them. "Although we are not engaged in research with a capital R," Oppenheimer wrote to the Sloan Foundation, "it had been my experience that any attempt to develop nontrivial pedagogical material has inevitably involved an important element of questioning and subsequent discovery. I am certain that our efforts here will inevitably make some contribution to new knowledge."[58]

Ironically, given Oppenheimer's efforts to associate this approach with museum modernization, the interactive exhibits that lay at the core of the Exploratorium's mission harkened back to the educational promises and problems of the chaotic collections of late nineteenth-century natural history museums. Oppenheimer insisted that visitors interact with exhibits and draw their own conclusions from what they encountered on the museum's floor, so he refused to provide the kind of pedantic narrative explanations that twentieth-century Americans had come to expect from natural history museum displays. Neither would he organize exhibits in the taxonomic progressions that had characterized Progressive Era displays, nor would he present biology as a matter of civics or "life adjustment," as museums had before and after World War II. In an unwitting echo of the pedagogical approach so popular among nineteenth-century American life scientists, Oppenheimer contended that lack of obviously didactic explanations led visitors to make what he called "scandalous revelations" through close observation and, in the process, provided a balance between skill acquisition and broader content overviews of the subject.[59] "If we do not tell people what they are supposed to find, many will leave with a sense of frustration, but a few will have become addicted to finding more than anybody knew was there," Oppenheimer reflected in 1976. This approach worked brilliantly for some visitors, but others, uncertain of what to make of this kind of interactive imperative, found the museum's exhibits maddeningly opaque. Oppenheimer refused to accommodate those who preferred their science entirely predigested. "How many frustrated people is one addict worth? Since there is no going back if one gives away too much, we tend to lean toward [one] . . . answer to this arguable question," he explained. "And we do have a large number of addicts who come back for more."[60]

Oppenheimer's decision to cater to the "addicts" resonated most with university-based scientists and school science educators—people so addicted to scientific experimentation they had consciously devoted their careers to its pursuit and promotion—and they praised his approach effusively. They wrote to Oppenheimer, describing how his museum had inspired them to explore new ways of teaching and learning, and providing anecdotal confirmation that hands-on methods and Exploratorium-style displays not only taught scientific concepts but, in addition, as one teacher raved, "made young people enthusiastic about science and mathematics."[61] "It is *not* a museum in the static, almost aseptic, sense that the latter term implies," enthused University of Washington physics professor Arnold Arons. "You have an active, on-going, vibrant operation in which people . . . were 'doing' things."[62] "I think that the best measure of success was the noise of the dozens of young people that were having a happy science experience while I was there," concluded physicist Albert Bartlett, one of Oppenheimer's former colleagues at the University of Colorado.[63] "In a society that is increasingly anti-science, there were a large number of people who were clearly learning that science could be fun," Temple University physicist Robert Weinberg wrote to the San Francisco Chamber of Commerce after his visits to the Exploratorium.

Echoing museum directors' post-Sputnik rhetoric, Weinberg assured its members that "You are gaining much value for your investment."[64]

Science museum educators quickly came to think of the Exploratorium as an informal clearinghouse of educational strategies and poached its most successful experiments. Staff members at the Boston Museum of Science, for instance, wrote to Oppenheimer inquiring about how to train explainers to guide visitors through a museum.[65] The Exploratorium's success also spawned imitation on a larger scale: in the decade between 1968 and 1978, the United States experienced what museum policy-maker Lee Kimche described as science museums' "greatest growth in ten years," as measured by the number of new institutions and the expansion of existing facilities. To mark their departure from traditional museum practices, many of these organizations came to adopt the moniker Dixy Lee Ray had coined in the late 1950s: "Science Center." Between 1973 and 1975, visits to science centers more than doubled, rising from 14.4 million visits per year to 36.5 million.[66] "There is no sign," Kimche observed, "that the trend will slow down soon." Kimche was hopeful that the influence of science centers would ripple into schools, suggesting that teaching science through hands-on exploration might be "adapted to enhance more conventional methods of teaching."[67]

Education policy makers and foundation officers also expressed unwavering enthusiasm for the Exploratorium, for it represented to them an unprecedented educational achievement. As an institution of informal learning that seemed to realize long-held dreams of museum accessibility and student-centered learning while reconciling post–Cold War visions of content-based science education and general science literacy, the Exploratorium could have been an unwieldy mess of contradictory undertakings. Instead, it was an elegant solution and could be used to accomplish the aims of any number of educational institutions. "A number of sources suggested that the Exploratorium represented a 'must-see' [for those] interested in viewing education in a broader setting than simply the schoolhouse or traditional classroom," a special assistant to the U.S. commissioner of education wrote to Oppenheimer in 1975. "All reports indicate the efforts that you and your staff engage in are imaginative, creative, and productive." It was "one of the most unique museums I have ever visited," later concluded Joseph Califano, the U.S. secretary of Health, Education, and Welfare.[68] In 1970, largely as a result of its perceived successes with its initial Exploratorium grants, the NSF arranged a conference at the Smithsonian on the role of museums in precollege science education.[69]

Foundations were similarly eager to support the replication of the

Exploratorium's most successful programs. In the late 1970s, the Exploratorium helped convince the Kellogg Foundation that museums were too important to ignore. As a result, the foundation supported fellowships at the Exploratorium for exhibit designers and also gave Oppenheimer money to conduct studies of why the museum's nonlinear interactive exhibits worked for teaching science.[70] And while the Sloan Foundation, which devoted funds to the public understanding of science and technology, did not give Oppenheimer anywhere near the million dollars he had requested for the *Perception* exhibition, they awarded the museum a seed grant of $1,500 in 1970, supplementing it with $125,000 the following year. Edwin Land, the cofounder of the Polaroid Corporation, was so inspired by *Perception* that he made annual donations of $50,000.[71]

Shrimp Ballets and Watchful Grasshoppers: The Challenges of Living Science Exhibits

Within the first decade of the Exploratorium's existence, federal and foundation excitement about the museum made it possible to extend Oppenheimer's interactive display methods to biology, a shift that posed both logistical challenges and political quandaries for the museum. Though Oppenheimer was eager to demonstrate commonalities in the way plants, animals, and human beings perceived stimuli in the Exploratorium's *Perception* exhibition, he hesitated before jumping into life science display, worrying that live plants and animals would transform the Exploratorium into a zoo, a botanical garden, or, what he feared most, a traditional museum that housed specimen collections.[72] Oppenheimer knew that life science specimens—especially the breathing, moving kind—would be expensive to acquire and even more expensive to maintain. They would demand ongoing care and would require the Exploratorium to hire trained biologists, whose interests and skills would likely be entirely different from those of the freewheeling exhibit builders he had begun to court.[73] It was NSF support that ultimately pulled the physicist into the life sciences, for, in 1970, education program officers strongly encouraged the Exploratorium to include biology along with physics in the exhibits on perception and offered to put up the money to do it.[74] In 1972, the NSF awarded the museum its second grant, allowing it to hire staff members with expertise in biology and to construct a handful of biological exhibits.[75]

To head up his life science staff, Oppenheimer hired a scientist with experience in experimental biology as well as experimental museum display design. Evelyn Shaw, the daughter of working-class Polish immigrants, earned a doctorate in biology at New York University before joining the American Museum of Natural History as a curator in the Department of Animal Behavior (see chap. 3). There, she curated the new Hall of Man exhibit in 1960 and published extensively on the behavior of schools of fish.[76] When Shaw moved to San Francisco in 1970, Oppenheimer offered her a chance to continue her scientific work and to publicize her methods and results through exhibits in the Exploratorium.[77]

Soon after, Shaw hired a young Berkeley alumnus, Charlie Carlson. An experimental biologist and graphic designer who was, conveniently enough, dating Shaw's daughter, Carlson was a perfect fit for the Exploratorium, temperamentally speaking. He was comfortable with interdisciplinarity, experimentation, and idealism—he had double majored in zoology and communications and embraced the creatively berserk culture of 1960s Berkeley—but he also retained a hardheaded common sense when it came to making scientific concepts meaningful. Plus, he owned a truck. When Oppenheimer heard that last detail, he hired Carlson on the spot, taking for granted, correctly, as it turned out, that the young man would allow the Exploratorium to take frequent advantage of his pickup.[78] Throughout the next several years, Shaw and Carlson worked together closely to create a series of displays themed around animal behavior.

Shaw and Carlson's life science exhibits, like the perception exhibits that Oppenheimer himself designed, were educationally innovative museum experiences because they demanded more from visitors than observing, button pushing, or animal petting. Instead, they placed visitors in roles of experimental scientists. The displays' designs were flexible enough to allow visitors to relate to exhibits in multiple ways but constrained enough that visitors could find them meaningful even if they had only the most rudimentary knowledge of science. As such, they were a radical departure from the binary design of earlier interactives, where visitors might make a light go on or off, and from animal demonstrations, where the educational demonstrator controlled the exhibit and told visitors what was meaningful about it.

One of the earliest Exploratorium biological exhibits, dubbed the *Brine Shrimp Ballet*, epitomized this philosophy in practice. Better known as sea monkeys, *Artemia* or brine shrimp had been marketed as novelty aquarium pets since the early 1960s. But the feathery crustaceans appealed to scientific researchers as much as they did to bell-bottom wearing children. Shaw knew these tiny marine creatures well from her experimental biology days at the American Museum and believed them ideal for the

Exploratorium's purposes. At a centimeter long, brine shrimp were man-ageably sized. They thrived in salt water, consumed fish food, and were cheap to acquire. Visitors could observe and handle them without harm-ing either human or shrimp. In a cylindrical tank filled with salt water and illuminated at the top, thousands of brine shrimp fluttered upward toward the light. When a visitor pushed a switch so that the light shone from the bottom of the tank, the tiny creatures turned in unison, mov-ing downward toward the new light source. Exhibit designers put the biological phenomenon of phototropism on display by having visitors interact with it: they could observe the initial movements of the shrimp and watch how they spiraled in different directions when the light source was flipped.

What visitors were less likely to realize, however, is that the life sci-ence displays were as experimental for the exhibit designers as they were for the visitors. Shaw and Carlson's work on the brine shrimp exhibit demonstrated the Exploratorium's commitment to exhibit design as itself a kind of scientific and educational exploration: broken exhibits, unexpected or complex animal behaviors, and even the most irritating visitor behaviors generated no failures, only "teachable moments.[79] At first, Shaw and Carlson had believed that brine shrimp reacted to light wavelengths rather than light intensity, so they installed alternating flashing red and blue lights in the tank. But this psychedelic light show had resulted in a profoundly boring display, generating little movement and leaving nothing for visitors to do, so they replaced the colored lights with white bulbs that changed direction when visitors pushed a button.[80] Once visitors could choreograph the direction of the shrimp, however, other challenges presented themselves. Younger or irrepressibly fidgety visitors flipped the light switch back and forth so quickly that the brine shrimp in the tank became habituated to the light's changing direction and no longer swirled toward it. On realizing this, Carlson and Shaw began to rotate new shrimp into the display regularly.[81]

Though visitor interaction was the governing purpose of the Explor-atorium, it could be tough on living specimens. Staff members found it a real challenge to design interactive biological exhibits that were interest-ing, educational, and could withstand the daily onslaught of thousands of small hands and shrill voices. Shaw's second exhibit, *Feather Worm*, featured *Spirobranchus*, a lovely, reef-living marine creature whose spiral-ing fronds gently swayed back and forth with water movement. Sensitive to light and touch, when visitors cast shadows or tapped the tank's glass, the worms immediately recoiled into their protective tubes. Shaw had high hopes for the exhibit but removed it from the Exploratorium's floor after only a few weeks. Her attempts to create an exhibit demonstrating the biological effects of environmental stimulation had succeeded to the point of boring visitors. Like the brine shrimp, the feather worms had become unresponsive, numbed by the barrage of noise and visitors' inquisitive tank tapping. Intent on making the larger scientific point, exhibit designers replaced the worms with mimosa plants, whose leaves folded inward when touched or shaken then reopened moments later. When Oppenheimer touched a cigarette to the tip of one plant and watched its leaves collapse, the staff realized that greater heat caused a greater reaction, and Carlson promptly installed a heat wand near the exhibit so visitors could see the phenomenon for themselves.[82]

Unsurprisingly, this exhibit design and monitoring process was slow and open-ended, and after only a few years it became clear that the Exploratorium's biology division was becoming the costly, labor-intensive burden that Oppenheimer had dreaded. In addition to the demanding work of exhibit development, staff members found themselves burdened with deadening daily tasks: cleaning tanks and pots, skimming out dead animals, replacing stressed specimens with fresh samples. Overwhelmed, Shaw left to teach biology at Stanford University, write a book on the behavior of female animals, and open a catering company.[83] Carlson stayed on but begged Oppenheimer for more help. Oppenheimer was sympathetic but short on staff and money and so encouraged the young man to reach out to research universities and scientific foundations.

The suggestion proved remarkably fruitful for generating new life science exhibit ideas and new funding streams. Carlson went hat in hand to one of his former professors at the University of California, neurobiologist Gunther Stent, and asked Stent if he knew of graduate students and postdoctoral researchers who might be willing to donate time or ideas to the Exploratorium. Stent recommended biologist Richard Greene, who was interested in the intersection between artificial intelligence and brain cell research, as well as in nontraditional science pedagogy. Greene knew how to implant electrodes into brain cells, a method that served to enhance the neurophysiology exhibits and eventually became the starting point for a new *Nerve Cell* section of the life sciences.[84]

But the most productive result of Carlson's ongoing relationship with Stent was his eventual introduction to the Grass Foundation, a small, Boston-based foundation devoted to neuroscience research and education. Stent referred Carlson to postdoc Bill Kristan, who, in turn, introduced Carlson to the foundation's directors. Kristan was a good matchmaker. The foundation's trustees were impressed by the museum's attempts to allow visitors to engage in the experimental methods, and they were intrigued by staff members' inventive use of simple machinery. Albert and Ellen Grass had made their fortune developing instruments used by neurophysiologists, including electroencephalography machines that measured electrical activity in the brains of epileptics, and the unabashed pleasure that Exploratorium staff took in developing tools to measure, record, and amplify biological forces for visitors appealed to them. "We have an exercycle set up with the handlebars as electrodes, displaying cardiac impulses," Oppenheimer enthusiastically wrote the foundation in 1973. "We demonstrate, using both audio and an oscilloscope, the nerve impulses to the flight muscles in the thorax of a house fly, from the stretch receptors in the abdomen of a crayfish, and from touch receptors on the leg of a cockroach."[85] Such innovations charmed Grass family members, whose foundation donated laboratory equipment and supported documentation for the Exploratorium's biology exhibits. They also gave money to Carlson to attend national neurophysiology meetings, where he demonstrated the Exploratorium's exhibits for college biology professors interested in using the exhibits in their own classrooms.

As a result of the Exploratorium's relationship with the Grass Foundation, visitors to the old Palace of Fine Arts building played with an astonishing number of neurological experiments throughout the 1970s and 1980s. Perhaps the most popular was *The Watchful Grasshopper*, an exhibit that married the Grass family's interests in the nervous system and the Exploratorium's lively approach to life science display.[86] To build a new display that would clearly illustrate the relationship between neurological stimuli and behavior, Carlson and other staff members started with an old exhibit concept: helping visitors to see through the eyes of the animal. They implanted wire electrodes into the ventral nerve cord of *Schistocerca nitens*, popularly known as the gray bird grasshopper, in order to record its impulses. This arrangement, they explained, allowed visitors to explore the grasshopper's visual field, to determine what triggered impulses in the insect, and to watch how it became habituated to repeated stimulation. When visitors moved in front of the animal's visual field, they could watch and hear the neural effects of the grasshopper watching them (fig. 21). To work properly, *The Watchful Grasshopper* exhibit required sensitive instrumentation: an oscilloscope, to record and amplify extracellular signals in the grasshopper's brain, and amplifying speakers, so visitors could hear, not merely see, the grasshopper's nervous response to stimuli. The Grass Foundation was more than happy to provide the latest machines for their displays; in an inversion of the usual corporate exhibit relationship, the Grass Instrument Company received

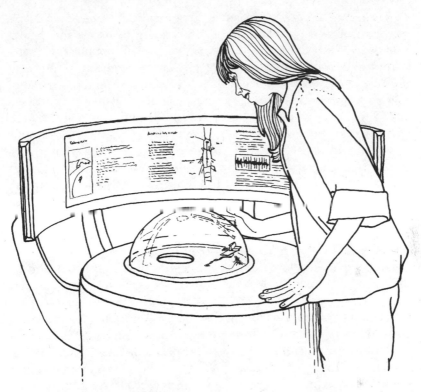

Figure 21   A line drawing of the Exploratorium's *The Watchful Grasshopper* exhibit, taken from
Ron Hipschman's *Exploratorium Cookbook 2* (1983). The series of physiological
displays, of which this exhibit was a part, put the visitor in the role of experimenter.
A grasshopper, mounted to the wire platform shown under the Plexiglas dome, had
wires surgically inserted into its central notochord. When a visitor approached the
exhibit, the insect's visual field would be activated and "displayed," through the
motion of an oscilloscope and the clicking sounds. (© Exploratorium)

prototypes from the Exploratorium for trade show displays of its latest
wares in return.[87]

But scaling up the life science displays was even more challenging
than Oppenheimer had predicted. The relative technical sophistication
of *The Watchful Grasshopper* exhibit presented practical challenges even
greater than those posed by brine shrimp. Obtaining grasshoppers was
the first challenge. *Schistocerca nitens*'s status as an agricultural pest in
California meant there was existing research on them, but it also mean
that they were nearly impossible to acquire through commercial venues,
forcing staff members into temporary careers in grasshopper husbandry.
The extensive preparation the exhibition required was another obstacle.

To be effective, Carlson's interactive grasshopper exhibit required detail-oriented, labor-intensive exhibition preparation of a scale and quality not unlike that required by the elaborate natural history museum dioramas of the 1930s. The exhibit demanded expertise in the frustrating field of grasshopper surgery, which forced Carlson to take on responsibilities of both preparator and researcher. Though enough scientific research existed on the animals' nervous systems to pique exhibit designers' interest, only a few scientific specialists understood the species' basic anatomy. So Carlson read what he could on *S. nitens*, conducted his own intensive observations and experiments, learned how to insert electrodes into their ventral nerve cords, and then taught exhibit support staff, many of whom were students or temporary volunteers, to do the same. This specialized surgery required considerable patience and practice. The grasshoppers needed to be immobilized with dental wax—especially hoppy ones were anesthetized through ten minutes of refrigeration—then placed under a dissecting microscope, and the electrode implanted and tested. Finally, the preparator needed to use a mixture of beeswax and rosin to cement the electrodes inside the animals. "If this preparation is carefully made without too much trauma to the animal," Carlson wrote, "it will last one week or more, up to a month, without noticeably affecting the behavior or health of the animal," a period of time his harried staff members must have found irritatingly short.[88]

Though museum preparators had long engaged in scientific observation in order to build exhibits, whether for habitat dioramas or large models, the resulting knowledge was usually played down or dismissed as lay knowledge; the Exploratorium's experimental displays and intentionally democratic culture, however, afforded its exhibits staff a platform on which they could proclaim their own findings. Carlson created a supplementary display explaining the anatomy and life cycle of grasshoppers, featuring the museum's grasshopper colony and the exhibit designers' scientific knowledge of behavior and husbandry. Graphics that accompanied *The Watchful Grasshopper* also illustrated the electrode implantation procedure, carefully emphasizing that it did not hurt or permanently injure the insects.[89]

Biologists and science educators rewarded Carlson for his efforts, calling *The Watchful Grasshopper* and the Exploratorium's other new life science displays a "tour de force of exhibit construction."[90] After Carlson and his staff traveled to the 1977 meeting of the American Neurobiological Association to demonstrate the displays to academic scientists, for example, he received numerous requests to create instructions so teachers could replicate them. But because grasshopper surgery was more challenging than they anticipated, local medical and physiology professors instead often brought their classes to use the Exploratorium biology displays as laboratory experiments. Biology graduate students suggested additions that would make the exhibits more useful for their own research agendas.[91] These responses were gratefully received, for in the early 1970s, the opinions of scientific experts about exhibits still counted more for fundraising than hard-to-measure visitor behaviors did.

By the late 1970s, however, foundations, government agencies, and other grant makers required more evidence of visitor interest than effusive praise by scientists and teachers. Increasingly, they demanded more rigorous evaluations of educational effectiveness before granting funds. As early as 1965, the U.S. Atomic Energy Commission and the Smithsonian Institution hired educational psychologist Harris H. Shettel to develop evaluation criteria for determining the effectiveness of museum exhibits.[92] Shettel's belief in the science of exhibit evaluation, much like Paul Rea's earlier belief in the "scientometrics" of museum demographic data, was unfailing. "Exhibit complexity is subject to rational investigation," Shettel wrote in 1966, "and current educational practice in the areas of both measurement and instructional technology hold out the promise of real effectiveness." Ultimately, Shettel's research cautioned that, while exhibit makers and visitors often had different goals, certain universal principles did apply to good exhibits, including "maximum use of relevant, overt responding on the part of the viewer" but also "informing the visitor of the correctness of his response."[93]

Perhaps the decidedly nonconstructivist educational perspective of evaluators like Shettel explain why, in 1974, the Grass Foundation commissioned its own audience-evaluation studies of the life science exhibits they had funded at the Exploratorium. These studies showed that, although popular with scientists, life science displays did not draw the same response from the public: according to an in-house study conducted in the late 1970s only approximately 5 percent of the Exploratorium's 560,000 annual visitors interacted with the displays.[94] The Grass Foundation was not the only Exploratorium donor to demand quantitative evidence of visitor response. Though eloquent letters from fans of the Exploratorium biology exhibits had sufficed for grant applications about the educational efficacy of the displays for most of the 1970s, by the end of the decade, federal agencies and foundations required formal surveys to determine if the museum was making good use of federal funds. Yet these evaluations, while more methodologically sophisticated, clung to older hierarchies, continuing to solicit the opinions of scientists as experts rather than addressing the thorny issue of how best to measure the educational experience of actual visitors.[95]

Had officials inquired more deeply into the individual opinions held by those who visited the Exploratorium's life science exhibits, the findings might have unsettled them. Just as Oppenheimer had hoped, museum-goers did draw independent conclusions from the museum's exhibits, but these conclusions didn't always echo those drawn by the Exploratorium's own staff and the broader scientific community. In 1981, for example, a local teacher lauded the Exploratorium as a "wonderful place for making people become aware of and excited by science" but noted that she found *The Watchful Grasshopper* downright disturbing. "Whatever the instructional value of such an exhibit it represents cruelty to animals and only encourages people to treat animals as playthings without feeling," she wrote to Oppenheimer. "I sincerely hope you will remove that torture chamber. I did not go through the animal behavior section after seeing the grasshopper in fear of seeing more of such disturbing sights."[96] Perhaps the museum could show a short animated film with sound to illustrate this phenomenon instead, she suggested. Another visitor, Lee-lane Hines, also expressed discomfort with the exhibit, describing it as "counterproductive" and reminding Oppenheimer that "what we learn is not always what people think they are teaching." Rather than conveying information about animal behavior and neurological pathways, Hines argued, the exhibit propagated a reprehensible lesson: "When creatures are less than human, we, as superior more knowledgeable beings, need not treat them with respect or kindness. Lesser beings may freely be used for our own (scientific) (genetics) (self-defense) purposes."[97]

Flooded with letters like this, Oppenheimer was forced to defend the Exploratorium's pedagogical embrace of the scientific method and participatory learning. As he tried to explain the methods and motivations behind *The Watchful Grasshopper*, Oppenheimer's correspondence makes clear that the Exploratorium's interactive approach to the life sciences had created new and entirely unanticipated pedagogical and ideological problems. In 1981, for instance, he attempted to reassure an angry visitor that the grasshoppers in the exhibit weren't receiving visitor-administered electrical shocks, as the visitor had assumed. "The live grasshopper that you saw was not hooked up to a shock stimulator!" he protested. "Fine, flexible wires were connected to the back of the grasshopper who lives very happily and can move around" and measurement was very different from torture, he explained. Still, he agreed, the exhibit could probably use some clarification. Going forward, he promised, "we will do everything we can to make it clear that we are showing that animals, as well as people, transmit information by generating their own electricity and that this process is going on all the time in all of us.[98]

Museumgoers committed to the expansion of animal rights found Oppenheimer's defense of scientific practice and his ethical reasoning unconvincing. Visitor Muhaima Startt insisted that the museum would never conduct a similar experiment on a human being and, as a result, that the Exploratorium's life science exhibits were fundamentally incompatible with a broader respect for life in all its forms. "My simple point of view is that animals have their own things to do in their natural environment, and that should be respected," Startt wrote. If visitors wanted to experiment on something, she concluded, they would do better to experiment on themselves. "It may be of more learning to folks to have more tools for self-exploration available—the bicycle reading one's heartbeat and the simple EMG register are well thought through in this respect."[99] Ultimately, many Exploratorium visitors concluded from the life science displays that the museum prioritized scientific discovery over human ethics.

The Age of Introspection: Seeking Pedagogical and Political Relevance, 1968–1980

Science museums and natural history museums enjoyed increasing popularity with the American public in the late 1960s and 1970s, but the emergence of science centers and mounting political turmoil threw their institutional identities into doubt, forcing them to examine long-standing pedagogical practices and define their missions more precisely. Between 1973 and 1975, visits to science centers more than doubled, rising from 14.4 million visits per year to 36.5 million.[100] Natural history and science museums remained popular as well and continued to be stolid staples of elementary school trips and rainy-day family excursions. Even though their core constituencies had decamped to the suburbs, these museums continued to post significant visitor numbers. In 1972, for instance, science and natural history museums combined attracted 38 percent of the nation's museum visitors, even though they made up only 16 percent of the nation's museums.[101] "We outdraw everything in Boston but the Red Sox," boasted the aging director of Boston's Museum of Science, Bradford Washburn.[102] Yet these older museums found they no longer commanded the widespread cultural approval they once had. Even the most sympathetic educators and museum critics, those who believed in the continued value of research and collecting at museums, increasingly disparaged the continuing reliance of science and natural history museums on exhibits in wooden cases and boring educational programming.[103]

Some of the most pointed criticisms of natural history and older science museums now came from the heads of the nation's brand-new science centers and science museums. Recently established institutions declared their approach to education was a modern alternative to the dysfunctional, disdainful, and dusty pedagogy of older museums. In 1971, for instance, Helmuth Naumer, director of the recently established Fort Worth Museum of Science and History, criticized what he described as museums' "traditional priorities." "The object comes first, the pet interest of the director or curator comes second, the social image of the institution comes third, the donors' special interests come fourth, and the public's involvement comes ninth or somewhere down the line," he complained.[104] "Museums should provide programs and activities— including exhibits—that arouse the participant's curiosity and test his power of observation, thus teaching him to teach himself," Naumer argued.[105] Even in collections-based museums, looking should be replaced by touching and exhibits should be accessible to all. "Realistically, how many hands does it take to wear out a mammoth tusk or a Ghiberti bronze?" he asked. "Proximity to an object is not enough for long lasting influential education; personal involvement is the key."[106]

Many older museums tried to reassert their preeminence by publicly asserting their differences from these unnervingly hip and messy homes for scientific exploration. The Exploratorium and other science centers didn't even officially qualify as museums, protested longstanding members of the professional museum establishment because they lacked collections—the characteristic that defined museums. In 1972, AAM defined a museum as "an organized and permanent nonprofit institution, essentially educational or aesthetic in purpose, with professional staff, which owns and utilizes tangible objects, cares for them, and exhibits them to the public on some regular schedule."[107] The definition excluded museums like the Exploratorium, which did not own, care for, or exhibit collections of objects but jubilantly combined, destroyed, or discarded objects in the name of scientific discovery. This official designation was more important than it seemed: many foundations and government agencies refused to accept grant applications from institutions that weren't accredited by the AAM, but the AAM would only accredit those institutions that fit its official definition of "museum."

Science-center leaders pushed back, determined to claim what they believed to be their hard-won educational place alongside older natural history museums. In 1972 Oppenheimer and other directors of the nation's most prominent science centers teamed up with the older science museum leaders, and with funding from the National Science Foundation, they founded the Association of Science-Technology Centers.[108] One of the association's first crusades was to lobby the AAM to expand its definition of a museum. "We feel that science and health museums are being discriminated against," wrote the association's president and director of Chicago's Museum of Science and Industry, Victor Danilov. "The proposed alignment treats non-collection museums as second class citizens (regardless of how it is worded)." By July of that year, the Association of Science-Technology Centers and AAM had hammered out an agreement that allowed science and technology centers to be eligible for accreditation "even if they do not own collections of artifacts of intrinsic value—assuming they meet other professional standards." By the mid-1970s, science centers, admitted AAM officials, had become "numerous enough and possess sufficient maturity of concept and breadth of experience to qualify as successful museums."[109]

Museums or not, science centers were a gadfly to the museum community, and their pedagogical innovations stung many institutions to accommodate the Exploratorium's methods. Field Museum officials, intrigued by the buzz, even tried to poach Oppenheimer, suggesting in 1971 that he join their Department of Education. "Someone has suggested that you might have exactly the qualities we are looking for," curator Robert Inger explained.[110] By the late 1970s, many older science and natural history museums were following the Exploratorium's example in more agile, creative ways.[111] Soon after the Exploratorium founded an artist-in-residence program to formalize its ongoing relations with artists who performed or contributed scientific art to Oppenheimer's halls, the Boston Museum of Science submitted an application to the National Endowment for the Arts for funding for a similar program of its own.[112] Other museums relied on new forms of media to make their displays as playful and appealing as those in Exploratorium-style science centers. As the 1970s wore on, the halls of both science and natural history museums increasingly featured computer screens and digitally activated videodisc displays, as museum educators sought ways to create opportunities for interaction. Visitors were amazed when the computers called them by the names they had earlier typed into the screen and asked them questions "according to age and sophistication," explained educators at the Cranbrook Institute of Science.[113] "If a computer is doing the asking," they wrote brightly, "visitors love to think."[114] Some exhibit designers predicted that such interactions would soon become the norm: museum displays whose authentic objects had once provided the "physical convincers" of the natural world they represented were now being replaced by television and high-quality color printing, wrote B. N. Lewis, a designer at London's Natural History Museum.[115]

Not all museum workers were persuaded by the Exploratorium effect, however, and many complained that interactivity was often sound and fury. "Now when I go to the Museum of Natural History or the Museum of History and Technology, an anxiety and revulsion consumes me, when faced with the throngs among the noise and flash in these department stores of educational bombast and trickery," wrote a frustrated Joseph Shannon, former Smithsonian exhibit designer. "I can't get out too soon. Shopping in a giant commercial mall is less stressful and in many ways more rewarding."[116] Shannon took special offense at the onslaught of technology that visitors now faced when they entered museum halls. "By taking aboard the 'concept' exhibition and consequently the audio visual as a primary technique to achieve 'relevance,' museums—particularly the so-called teaching museums of history and natural history—are going to hell on a psychedelic choo-choo."[117] It was time, Shannon maintained, for museums to remember they "are only supplementary educational instruments, and, at their best, teach only what the objects in their collections can teach, no more, no less."[118]

Critics quoted educational psychologists who expressed reservations about this latest generation of interactive displays, pointing to articles by researchers like Harris Shettel, who argued in *Museum News* in 1973 that such exhibits too often asked visitors to participate for participation's sake. Physical interaction got people moving, but the movements rarely related to the objectives of the exhibit. "Anyone who doubts this should spend a Saturday afternoon in a 'modern' museum watching children racing from one exhibit to the other, pushing all the buttons, flipping all the switches, and turning all the cranks without waiting to see what happens as a result of these activities," Shettel added.[119] Educational studies conducted at Cranbrook in the late 1970s revealed that Shettel's concerns about such developments might have been justified: museum visitors, the researchers concluded, were far more engaged with interactive exhibits if a teacher was present to ask questions and stimulate a discussion.

Two old lions of the American Museum, Albert Parr and Margaret Mead, proved formidably thoughtful critics of the Exploratorium effect. Though Parr had been an early champion of the Exploratorium, he now deplored its larger influence on the museum world. "Instead of playing up the qualities that are uniquely their own, it often seems as if museums have been more bent upon 'me too' in the media in which other establishments have all the advantages," he wrote in 1973. "The bravura performance of fashionable instruments and methods [are] probably more appreciated in public relations than by the public."[120] Mead took a similar stance, lamenting both the merry mayhem of science centers and the turn by museums to multimedia and clearly mapped narratives. Museums' current efforts to "produce miles of illustrated print about matters better read in books, batter the ears with recorded lectures, insert continuous films which gather small crowds instead of the large crowds of the movie theater, introduce sub-objects to be manipulated or experimented with which belong in classroom and laboratory" would only make the institutions obsolete.[121]

Convinced that newer, louder, and livelier was not necessarily better, curators, educators, and exhibit designers defended older pedagogical approaches, sometimes crossing party lines as they did so. Both Mead and Parr, for instance, agreed that natural history museums needed to return to a less programmed model of learning. No longer could visitors find in museums "that atmosphere of quietly enjoyable personal contemplation, which used to be one of the greatest appeals of a museum as a setting for the display of the things that attracted you," Parr wrote sadly in 1978.[122] Rather than attempting to compete with science centers like the Exploratorium or commercial venues like world's fairs, wrote Mead, natural history museums needed to enhance their own distinct brand of nonclassroom education, welcoming people with quiet, clean, beautiful spaces and with challenging exhibits that prompted discussion of the present and future.[123] Likewise, many science museums continued to develop the kinds of displays that had made them famous in the 1940s and 1950s, but with period riffs on the usual institutional approach. In 1974, for instance, the Museum of Science and Industry in Chicago bragged about its new Medical Balcony, twenty-five exhibits on various topics in health and medicine whose designers promised "Public Education with a Pushbutton." The Medical Balcony featured a new transparent women, renamed Tam. Tam was exactly the same as she had been since the 1930s, with only one exception—her designers had tipped their hats to the ongoing sexual revolution. Rather than standing simply, the model was now posed seductively, with hands flared out at her sides and her hips pushed forward.[124] Meanwhile, back at the Boston Museum of Science, director Washburn solicited funds for a new wing by driving across the Massachusetts Turnpike in a truck with the plaster cast head of a twenty-foot high *T. rex* model in the back. The head was eventually installed at the museum.[125]

The Exploratorium's displays had forced museums to resolve, or at least reflect on, the problem of how museums could stay pedagogically pertinent, but they had not answered the insistent question of how museums could remain socially and politically relevant to a nation now deeply suspicious of institutions. Natural history and science museums had long enjoyed a sort of permissive neglect from the public, a laissez-faire attitude borne of the vague conviction that mission of museums, however the institutions chose to define it and carry it out, was in the best interest of science and society. But in the late 1960s and 1970s, museums of all stripes were forced to confront the fact that, while they had been cultivating good will from foundations, donors, corporations, and various government agencies, they had neglected to begin conversations with their publics. Americans were no longer hesitant to protest such cavalier neglect. And so science centers, science museums, and natural history museums alike grappled with the difficulties of being accountable to a public at war with itself.

Public attacks on museums shocked and troubled staff members accustomed to thinking of themselves as the most liberal-minded members of the professional American museum community, and they struggled to decide how to respond. According to a 1974 study by the National Endowment for the Arts, some 98 percent of natural history and science museum directors argued that their institutions' primary purpose was to provide educational experiences for the public. Eighty-one percent believed providing instruction to children was more important than any other objective. Most tellingly, 26 percent of natural history and science museum directors believed that "encouraging positive social change is very important"; only 6 percent of history museum directors in the United States shared this opinion.[126] "The quest for 'relevance' nearly approached paranoia" in the mid-1970s, one exhibit designer recalled. "It was a time when everything a museum was, or could be, was being questioned and attacked. The museum, like all institutions, was sinking into a miasma of doubt. We had to deal with this; we had to be 'relevant.' "[127]

Museums' quest for relevance proved a double-edged sword: no museum could deny its importance and still remain credible with their publics, yet if museums pursued relevance, they could not escape the era's harrowing political turmoil, not even the nation's determinedly apolitical science centers. Oppenheimer had fully intended to sidestep the problem of relevance, but he, too, found his institution sucked into escalating debates about whether museums should address social and political questions nonetheless.[128] Oppenheimer himself actively disavowed exhibits whose lessons could be easily applied to current events and had intentionally excluded specific discussions of broader political and cultural discussions of life science and its methods from the Exploratorium exhibits. Because he believed that the educational mission of his reformed museum "was not to give people the right answers . . . but to help them gain the confidence to make discoveries for themselves," he resisted calls by staff and visitors to develop exhibits promoting popular environmental or cultural causes.[129] By connecting its visitors to science, Oppenheimer declared, the Exploratorium gave them a way to connect to social problems as well: if visitors "give up" on "the physical world around them," he wrote, "they give it up with the social and political world as well, and then . . . they don't know how to act on any of the problems that come up."[130]

Scientists who believed their work to be largely apolitical admired Oppenheimer's approach, as did those who believed the political left had co-opted the mission of science education. Too much emphasis on "polluting and ugly, tawdry uses of technology" led to "hostility towards science," argued *New York Times* reporter Walter Sullivan. The Exploratorium was right to maintain a careful innocence, Sullivan wrote, for only in this way could it demonstrate that, "in science, there is also beauty and joy."[131] Many visitors, and an increasing number of science educators and politicians, however, declared this stance to be naive, even misguided. Science, and especially life science, was necessarily political, they argued, and as such, science education should place scientific developments and discoveries in social and cultural context.[132] But because of their emphasis on doing rather than representing, and because of their sweetheart status with both countercultural and establishment figures, science centers like the Exploratorium were able to evade many of these criticisms.

Older science museums were not always so lucky, and they struggled to respond to the angry politics of the 1970s. Though they didn't have the same skeletons in the closets—literally—as natural history museums, science museums' cheerfully unexamined support of science didn't play well with a public newly attuned to the darker consequences of scientific discoveries. After a decade of intense discussion about atomic fallout and pesticide use, Americans no longer held unfailing confidence in the scientific-industrial complex. Exhibits devoted to human biology and other aspects of the life sciences were especially fraught. Critics routinely skewered the Boston Science Museum for failing to confront the social implications of scientific discoveries, arguing that the museum's biology displays skirted topics that might trouble the city's many Irish- and Italian-Catholic residents. One display maintained that curbing population growth was the long-term answer to starvation but failed to discuss either birth control or the ethics of imposing the values of the industrialized West on developing countries. Another noted only that there might be "implications of tampering with human genetics," rather than delving into the complicated political issues that DNA research and engineering raised. According to the August 1970 issue of *Science for the People Newsletter*, the bimonthly publication of a left-leaning group of Cambridge scientists and engineers, the Boston Science Museum typified "institutional misrepresentation of science and is a disservice to the community it purports to serve." High admission fees ("four bucks for an Inner City Family of Four") yielded visitors access to "a plastic middle class institution" and the inherent hypocrisy of a museum "that squats over the stinking Charles River and has no actual exhibits on pollution." Most displays, the anonymous author noted, "are nothing more than hypes for the companies contributing to them."[133] This critique was resonant enough that *Science* reprinted the article nearly in its entirety, "minus obscenities and other extraneous matter," editors noted.[134]

Efforts by older museums both to accommodate and to redefine what it meant for their institutions to be relevant in the contemporary cultural and political climate were exemplified by the American Museum of Natural History's lopsided responses to new public demands. On the one hand, the museum proudly retained vestiges of a resolutely hierarchical past. Administrators remained aloof from calls to alter its research programs, for instance, citing their determination to retain the museum's scientific independence in the face of populist rumblings. This stance provoked a variety of cultural clashes, culminating in a collision over animal rights during the winter of 1976 and 1977. Over the course of fifteen years, respected biologist Dr. Lester Aronson, chairman and curator of the Department of Animal Behavior, and his assistant, Madeline Cooper, had studied cats' sexual behavior, removing the glands, nerves, or brain tissue of hundreds of live cats in the process. New York City animal rights activist Henry Spira learned about the research via a Freedom of Information Act request for National Institutes of Health records and approached the sixty-six-year-old Aronson privately, requesting that he end his work. Aronson refused to discuss his research, waving Spira off as an antiscience nut. In contrast to the ad hoc complaints of the animal welfare activists who objected to the Exploratorium's *Watchful Grasshopper* exhibit, Spira and his colleagues organized and went public with the information, picketing the museum and burying administrators, trustees, and donors with angry letters. Though Aronson and Cooper's research didn't exactly endear the museum to the public, it was museum director Thomas Nicholson's perfunctory response that ultimately damned the institution. Nicholson had previously defended such experiments on the grounds of scientific relevance and intellectual independence: "If anything has distinguished this Museum, it has been its freedom to study whatever it chooses, without regard to its demonstrable practical value. We intend to maintain this freedom." But by the mid-1970s, visitors, taxpayers, and government officials found this attitude high-handed at best. Protesters and editorialists bandied about Nicholson's phrase, citing it as evidence of the institution's sense of entitlement and its insensitivity to the public will. The damage to the museum's reputation (and to the National Institutes of Health, whose unusual funding to Aronson's enterprise had supported the study throughout the 1960s and 1970s) was enough to force administrators to end the project.[135]

While Nicholson and other defenders of the museum's scientific independence may have been tone deaf, they were far from insensitive to the institution's public mission, and they worked hard to create exhibition and educational programs that would engage disaffected activists and traditionally disenfranchised groups.[136] To create exhibits that would interest previously overlooked audiences and appeal to—or at least appease—the museum's critics, the American Museum played to its strengths, plumbing its collections for specimens from previously colonized or exploited regions. Curators and educators used these objects to create temporary exhibits, community lecture series, art and photography shows, and even dance programs that celebrated the very cultures it had, only a few decades earlier, dismissed as primitive or romanticized as extinct.[137] In the late 1970s, for instance, hoping to attract visitors from the swelling Puerto Rican neighborhoods along the subway line that stopped at the museum, the museum hosted a series of exhibitions on Puerto Rican nature and culture.[138] The museum also reached out to the city's roiling young, expanding its Natural Science Center for Young People and hiring new educators to develop some events for new audiences—rehabilitation centers for drug addicts, day care centers, and other community organizations.[139]

The blossoming of environmentalism in the late 1960s and early 1970s provided science and natural history museums with another way to engage contemporary public concerns and redefine their relevance. With the Cuyahoga River aflame and oil lapping at Santa Barbara's wide beaches, environmentalism seemed to have little of the divisiveness of the antiwar movement or other controversial social causes, and museum directors felt they could champion environmentalism without putting themselves at risk of political controversy or straying far from their traditional mission. As New York Botanical Garden president and former AAM chief William Steere assured his colleagues in 1969, "Museums are admirably equipped, through their diverse collections, their expertise in display, their long experience in conservation, and their vast clientele, to play an important role in making the public aware of environmental problems, as well as of the means for their solution. Museums can bring vividly and unforgettably to the American public, especially to its youth, the basic facts of the human environment, how it deteriorates and how it can be improved."[140]

As a result, many science and natural history museums built displays designed to educate visitors about the scientific dimensions of social problems, exploring urban pollution as a civil rights issue, for instance. Indeed, increasing public concern over water and air pollution, the effects of DDT, and habitat destruction fit easily with museums' longstanding tradition of romantic nature education. Environmentalist exhibits had the added convenience of allowing museums to plunder older displays and revise them with an eye toward the present. Museums made new use of megafauna collected in earlier eras, using new labels and smaller explanatory exhibits to convert hunting trophies into icons of environmentalism.[141]

Natural history museum staffs had always been ardent environmentalists, and in the late 1960s and early 1970s, they leapt at the chance to help prepare new exhibits addressing urgent ecological problems.[142] In 1969, the American Museum of Natural History, for example, launched the nation's first major exhibition along these lines, with the reassuring title *Can Man Survive?* Artifacts, audiovisual displays, and disturbing lighting made the argument that continuing to indulge in wasteful energy consumption and thoughtless but ecologically devastating practices would lead to collective suicide. The public wasn't quite ready for the exhibition's apocalyptic message and objected to the sense of "demonic dread"—to use the words of critic Robert Smithson—that it called forth. (In his review of the exhibit, Smithson also observed that the exhibit had been "generated through the courtesy of the enemy: General Electric, General Motors, I.B.M., Bell Systems, etc.")[143]

Other environmentalist exhibits at museums were more successful, as museums began to master the balance between advocacy, guilt, and education. Supported by a grant from the U.S. Consumer Protection and Environmental Health Service, then a department within the U.S. Public Health Service, the AAM's Environmental Committee developed a handbook of techniques for creating more effective displays to address environmental concerns or "human ecology," as they called it. The handbook compiled dozens of environmentalist-oriented outlines, themes, techniques, and illustrations for museums to adopt and adapt as they saw fit.[144] Individual museums also promoted their own solutions to the challenge of presenting depressing news in a way that would inspire action. In 1970, for example, the Smithsonian developed a traveling exhibit, *Population: The Problem Is Us*, which was intended to demonstrate the interrelationships of the environment and how people use science and technology to solve environmental problems. Funded by the NSF Office of Public Understanding of Science, the exhibit was a hit and toured both museums and science centers throughout the 1970s.[145] That same year, curators and educators at the American Museum used Earth Day to present a slightly more palatable version of its earlier message of impending doom, wandering the galleries with live insects and snakes for the public to pet and inviting public questions about environmental matters via phone and radio.[146] Exhibit designers also got in on the act, slapping large "ENDANGERED" signs across the glass of some the museum's most beloved habitat dioramas (fig. 22).[147]

Museum staff found human evolution, another of their long-standing institutional preoccupations and a hot-button political issue, to be more difficult to address. Throughout the 1960s and early 1970s, creationists had pressed their case with increasing force, and at the end of the decade, they took a direct swing at natural history museums. In the mid-1970s, building on the success of the National Museum of Natural History's 1974 exhibit *Ice Age Mammals and the Emergence of Man*, Congress quietly allocated nearly $500,000 to the museum to develop an exhibit on evolutionary biology titled *The Emergence of Man: The Dynamics of Evolution*.[148] Creationist and retired missionary Dale Crowley's advocacy group, the National Foundation for Fairness in Education, sued the Smithsonian Institution in 1978 on First Amendment grounds, arguing that these exhibits promoted "secular humanism" over alternative religious interpretations of the origin of life.[149] A federal judge immediately dismissed the lawsuit, and the exhibit opened as planned in 1979, but the results were an object lesson in keeping mum in the face of impending political explosion.[150] Whenever possible, exhibit designers eliminated text, instead arranging specimens in ways that they thought would suggest the story of biological evolution to, as director Porter Kier put it, "a person who went through on roller skates."[151] The results were almost comically minimalist. One display in the hall featured a brown bear mounted on a base designed to look like a forest landscape next to a polar bear

Figure 22   In the 1970s, natural history museum curators and
exhibit designers struggled to make older displays seem
relevant to contemporary visitors. For the museum's
inaugural Earth Day celebration, exhibit designers taped
"endangered" signs to the glass front of the American
Museum of Natural History's Akeley gorilla diorama.
(Image scanned for authors from *AMNHAR* 1970,
p. 13. by the American Museum of Natural History.)

on a simulated ice cap, with a sign in front reading: "Differentiation:
Continents."

Museum director Porter Kier represented the exhibit as revolutionary—
"not just a collection of birds or dinosaurs, rocks or plants, but a visual
interpretation of complex concepts, it may be the most ambitious and
unusual undertaking yet by the Smithsonian," he proclaimed—but critics
thought it was a political and scientific cop-out. Though the display's entrance placard announced that "this exhibit explains how most scientists
think the evolutionary process works," "that simple phrase—'how most
scientists think the evolutionary process works'—belies a major scientific struggle" that the museum had refused to confront, protested *Science
News* reporter Susan West. To ensure audiences did not misinterpret or
reject outright fundamental concepts in evolutionary biology, curators
and exhibit designers had presented straightforward biological principles
to which even creationists subscribed, and had tried to present them
as simply as possible. The approach backfired. In its attempts to avoid

confusion or offense, critics complained, some of the exhibit's content had devolved into bland mistruths.[152] "There are some statements" in the exhibit, admitted entomology curator and head of the exhibit committee John Burns, "that I don't myself subscribe to."[153]

Other museums shared the Smithsonian's fear of negative publicity and exercised similar caution. As a result, American natural history museums in the 1970s waded more tentatively into displaying controversies over human evolution than they would twenty years later. Some museums created displays obliquely connected to the debate but shied away from aggressive scientific proclamations about evolutionary biology. The California Academy of Sciences' 1979 display *The Origins of Culture*, which exhibited Ice Age art artifacts alongside narrative panels, delicately separated the specimens from controversies over human biological evolution. Instead, labels explained, the objects epitomized "common ideas of what makes us human . . . the creative products of emotion and imagination." It was hard to voice political or religious objection to such pleasantly vague truths. Other museums limited their participation in recent debates over human origins to scientific circles and kept their display halls out of it. In 1974, a team of anthropological researchers, including Cleveland Museum of Natural History curator Donald Johanson, uncovered "Lucy," a hominid fossil whose features suggested there were two distinct bipedal human ancestral lines represented in their collections. Curators rushed in to challenge and debate Johanson's findings, but it would take almost a decade for American museums to put the Lucy controversy on public display.[154] And those few museums bold enough to take on the challenge often went it alone. In 1980, for instance, when American Museum curators proposed staging an international exhibition of original human fossil specimens, it quickly became clear that the museum would have to commit itself to underwriting nearly all the costs of the exhibit, as most federal and private sources of funding deemed the project too controversial to support.[155]

Though display halls received far more public attention than museums' laboratories and field stations, the front of the house wasn't the only place where museums could demonstrate their public relevance, agreed staff members. So museums in the 1970s increasingly altered their research agendas to present their work as potentially important contributions to positive social change. "Washington administrators of basic science are getting into step with a political tempo that calls for aiming science toward doing something about the problems that are here and now," observed journalist Daniel Greenberg in a 1970 issue of *Science*.[156] Consequently, throughout the 1970s and early 1980s, curators worked to apply their knowledge of systematics to the struggle to preserve biodiversity, an emerging interest of environmentalists and policy makers worldwide. Botanists at Pittsburgh's Carnegie Museum of Natural History, for instance, began to inventory rapidly disappearing tropical forests across the globe, trying to determine the consequences of deforestation. They worked closely with the Surinamese government to help determine which of the nation's imperiled lands should become national parks and what areas needed most to be saved from impending environmental destruction.[157] Throughout the 1970s museum biologists continued to receive steady support for work on plant and animal systematics from the National Science Foundation.[158]

But museum staff understood well the fleeting nature of such attention, and warned one another that the battle for institutional relevance had not been solved by recent interest in biodiversity. In 1980, Carnegie Museum director and AAM president Craig Black reminded his colleagues to be vigilant about making clear to the broadest possible audience the social and environmental importance of their research. "If we fail to make our case convincingly," he cautioned, "the next decade will see a gradual withering of research budgets in our museums, with reduction in curatorial and support staff and the deterioration of collections[,] . . . and the variety and breadth of research programs will decline."[159]

As they had since the beginning of the twentieth century, museums' scientists struggled to balance responsibilities to the public with their own research—a struggle made more complicated as a result of new laws. In 1969, Congress passed the National Environmental Policy Act, which required real estate developers to submit environmental impact statements before moving forward on projects. Environmental consultants and engineers hired by developers increasingly turned to natural history museum curators for geological, floral, and faunal survey information, asking to inspect existing museum collections and requesting that museums hold the voucher collections the law required developers to create. Initially, many curators were delighted to help, pleased that their collections and expertise were proving so useful. But museum staffs were soon overwhelmed by the onslaught of requests from businesses, municipalities, and private developers. Scientists were appalled to realize that consultants casually circulated the data they had spent years collecting, and in some instances, released or published it before curators were able to prepare their own reports or claim credit for their own research.[160] As publicly funded institutions and "the custodians of the only permanent record of the earth's diversity of plants and animals and, to an extent, of our cultural heritage," museums were accountable to the larger public, wrote museum director Craig Black. But public accountability, in this realm at least, was proving to be a financial hazard and a deterrent to scientific research.[161] To balance their accounts, a few natural history museums, the Academy of Natural Sciences of Philadelphia and the Bishop Museum of Honolulu among them, began to charge for such services. "Our institutions and staffs must participate in this arena of public service," Black concluded, "but we cannot bear the resulting financial burden alone."[162]

As the 1970s drew to a close, older natural history museums and science museums had successfully persuaded various audiences that their exhibits and research programs were now more in step with contemporary interests. In 1977, for instance, fully 70 percent of those listening to a Washington-area radio call-in show argued that the Smithsonian's curators knew what they were doing and should be left alone to continue to make the museum, in the words of one gentleman, "so enjoyable."[163] Smithsonian Office of Personnel director Vincent Doyle said it best when he observed that staff members' public writings did more than confirm that curators were "the most knowledgeable and intimately aware" of the subjects on display in the museum's galleries: they "affirmed our responsibilities in the field of Education."[164] Yet the visitor-as-experimenter approach pioneered by the Exploratorium remained the gold standard of exhibits-based pedagogical success, and on this count, older institutions continued to fall short.

The Exploratorium had created a series of new possibilities in the museum world, upending traditions of staff hierarchy, display, and visitor behavior. Its life science displays made use of trailblazing biological practices, and their designers had pushed scientific and pedagogical knowledge forward even more aggressively than dinosaur mounters, diorama makers, or grasshopper modelers had in earlier years. Most revolutionary, its exhibits had encouraged visitors themselves to explore the meaning of these phenomena by employing the experimental methods of biologists. At the Exploratorium, Shaw, Carlson, and the museum's other exhibit designers had finally achieved the exhibitionary synthesis that earlier museum reformers had envisioned but never realized. The Exploratorium's exhibits and organization had an outsized influence on the museum world as a result, forcing museums to rethink their own displays and redefine their own institutional identities. Arguably, although the institution was itself resolutely apolitical, the Exploratorium effect hastened the transformation of institutions already metamorphizing as a result of the era's political challenges. Though some of their exhibits and agendas remained distinct from one another, by the end of the decade,

American natural history museums, science museums, and science centers were largely unified in their renewed attention to audiences, exhibit pedagogy, and educational programs.[165] The financial crises and culture wars of the late 1970s and into the 1980s would soon pose new challenges to these institutions, drawing these very different establishments even closer together.

SEVEN

From Diversity to Standardization: Edutainment and Engagement in Museums at the End of the Century, 1976–2005

Throughout the first seven decades of the twentieth century, natural history museums had struggled to define and preserve those elements that distinguished them from other educational institutions or places of leisure: specimen collections, scientific research, casual learning through the contemplation of displays. As science museums were established throughout the 1930s, 1940s, and 1950s and as science centers developed in the 1960s and 1970s, each new form of institution aggressively asserted its differences from its older cousins. Too often, protested Thomas Nicholson, the director of the American Museum of Natural History, Americans concluded that museums, like schools, were part of a broader system of education, one with relatively standardized practices, constituents, and purposes. This, he proclaimed, was precisely the opposite of reality. "Diversity is the rule, not similarity," he wrote in 1981. "There is little unanimity among us as to what a museum is or ought to be." This multiplicity of approaches and aims, he asserted, was one of the great strengths of the museum community in the United States.[1]

But museums' exuberant experimentalism subsided as the 1970s came to an end. Where their approaches to life science exhibitions and their definitions of science and science education had once diverged profoundly, by the 1980s and 1990s, these three types of institutions increasingly shared shows, sources of funding, and pedagogical strategies. The ongoing professionalization of museum education reinforced these similarities, standardizing once ad hoc methods of educational programming and evaluation. Reagan-era budget cuts and market forces also steamrolled variety—in their haste to capture dwindling audience, many museums came to look more and more alike. Though museums had long traded individual exhibits among themselves, their halls—and institutions—had retained a largely local feel, reflecting staff interests or regional preoccupations. But as the twentieth century waned, the nation's varied institutions remade themselves into streamlined sites of family "edutainment," a term contemporaries used to describe a potent, often commercially motivated blend of education and entertainment. Little more than a decade after Nicholson's declaration that diversity was the rule, natural history museums, science museums, and science centers had begun to resemble each other to the point that their halls were frequently indistinguishable.

By the late 1980s, most American natural history museums, science museums, and science centers maintained some combination of the same life science displays: interactive halls of health, discovery zones, and live animal exhibits. They possessed IMAX theaters, well-stocked gift shops, and eating areas crowded with toddlers and schoolchildren. They featured the loud, colorful, and heavily publicized traveling exhibits known as blockbusters, which gained momentum over the next two decades as designers, administrators, and eventually curators harnessed them to achieve varying scientific and economic aims. Only the stilled, permanent galleries of habitat dioramas of natural history museums, the yellowing transparent women of science museums, and the battered biological displays in the back corners of science centers reminded visitors of these museums' long and curious coevolution. Had the walls been transparent, the offices and storage spaces of natural history museum curators, crammed with collections, would have reminded visitors that some profound differences between the kinds of institutions persisted—though in many museums, back-of-the-house differences were also beginning to erode as funds for research and collection maintenance tapered off. Although tensions between the public and scientific missions of natural history and science museums persisted, at the outset of the twenty-first century, more Americans had come to see displays, not collections, as the primary defining characteristic of a museum and its work.[2] Museums of science and nature had become culturally important sites for informal science, engaging diverse audiences in scientific ideas and ideas about scientific practice and awakening new investment in museums as institutions. As such, they collectively embodied museum professionals' new vision of effective public science, as well as their shared goal of defending it.

Confronting Financial Crisis: Corporate Sponsorship in the Carter and Reagan Years

American museums found themselves slapped with an especially difficult set of financial choices in the late 1970s and early 1980s. Most had struggled to make ends meet since the late 1960s, and the stagflation, deindustrialization, and energy crises of the 1970s had increased museums' expenses and diminished municipal tax coffers. Annual inflation averaged 6 percent throughout the 1970s, and museums' deficits soared in response, forcing institutions to cancel shows, cut staff, and consider new ways to raise money.[3] Even the Exploratorium, a darling of both the scientific and museum worlds, struggled because of the somnolent economy. In 1981, the million-dollar grant from the MacArthur Foundation that Oppenheimer had won that year covered only an eighth of what the science center required to operate, forcing the institution to charge adults for admission.[4] "We hope that this new policy does not exclude anyone who wants to come here and does add as much to our coffers as we estimate it will," the membership magazine announced apologetically.[5]

Federal funding helped museums make up a portion of the difference between income and expenses, and museums did what they could to make themselves attractive to government grant makers while treading water to keep afloat. The Institute of Museum Services' willingness to provide money for operating expenses was a godsend in a decade when many museums wondered how they could afford to keep their lights on and their lobbies heated. Museums also applied for federal grants in accordance with the 1970 Environmental Education Act and continued to rely on the NSF, which provided funding for museums in the guise of its Public Understanding of Science program. To attract federal grants, museums and science centers reorganized "back of the house" positions, hiring more midlevel scientist curators, as well as in-house accountants and attorneys to help them navigate increasingly complicated tax codes and regulations and professional grant writers to help them win money from federal agencies.

But the tax revolt of the late 1970s and early 1980s threatened to eradicate this funding, leaving museums in a perilous financial position.[6] Once Ronald Reagan was elected to the U.S. presidency, the Institute of Museum Services came under perpetual siege.[7] In 1981, for instance, the Office of Management and Budget proposed eliminating both the Institute of Museum Services and the NSF's science education efforts, including all support for science museums. Lobbyists, aware that local money would never be able to cover institutional expenses, urged directors to build stronger relationships with the National Science Foundation and federal legislators concerned with science education.[8] Hustling to find ways of remaining financially soluble, museums quickly redrew their organizational charts, establishing development departments and placing once-sleepy marketing staffs into positions as powerful as—or more powerful than—the scientific or exhibition staffs.[9]

As in the past, museums and science centers turned to corporations to help them stay in the black, a relationship that some institutions found easier to cultivate than others. Science museums were already used to asking for—and receiving—corporate support. Chicago's Museum of Science and Industry had long featured donated products from International Harvester, General Motors, and the Bell System, for instance, and the Buffalo Museum of Science had relied on the financial assistance of Fisher-Price Toys to open its children's room in 1968.[10] But after the mid-1960s, the explosion of American corporate foundations precipitated an expansion of the relationship between businesses and museums.[11]

Beginning in the late 1970s, for instance, natural history museums joined science museums in more regularly trolling corporate foundations and public relations offices for cash.[12] Such partnerships made sense for industry, observed *Museum News*, because natural history museums tended to attract well-educated, affluent consumers who traveled, owned real estate, and ate out regularly, "the very audiences the corporations want to target."[13] Though corporations had long sponsored exhibits that coincided with their core products or public mission-of-the-moment, they began to expand their support, using family-friendly natural history and science museums to try to counteract reputations of ruthlessness.[14] In 1979, for example, Chevron USA sponsored a massive exhibit, *Creativity: The Human Resource*, which traveled to science museums in twenty-one major cities before finding a permanent home in the Pacific Science Center. Citicorp/Citibank supported educational programming to accompany the traveling *Black Achievers in Science* exhibition, while ExxonMobil donated $2.5 million to modernize the American Museum's dinosaur and fossil vertebrate halls.[15] Life science exhibitions especially appealed to corporations, for animals were a sure hit with visitors of all ages, and explorations of the human body held similar fascination for most museumgoers.

Journalists, watchdog groups, and public agencies questioned this increasingly cozy relationship between American corporations and museums.[16] Corporate-sponsored exhibits "result in glossy, image-building advertisements, at best, and in blatant propaganda masquerading as education, at worst," authors of a white paper published by the Center for Science in the Public Interest argued in 1979. "Some science museums have strayed so far from the path of objectivity that they function as public relations tools of major industries."[17] Though the NSF was still eager to fund science museums, its own officers increasingly recognized potential conflicts of interest. "Science museums are mission-oriented activities that involve getting your hands dirty with science and technology; charitable donors want to keep clean," acknowledged Michael Templeton, program director for the National Science Foundation's new Informal Science Education Program. "This means that far too much leverage is given to industry donors and corporate supporters. This is not a good source of support for projects that are thoughtful and philosophical."[18]

The case of the National Museum of Natural History's interactive *Insect Zoo* illustrates tensions over corporate sponsorship. The exhibit opened in 1977, sponsored by a clutch of companies that made millions each year off the production and sale of insecticides. Though the exhibit received mixed reviews and though educators and corporations complained its message was unclear, Orkin was pleased enough with the results that it upped its commitment, donating half a million dollars to refurbish the exhibit. *The O. Orkin Insect Zoo* is still open at the Smithsonian today (fig. 23).[19] Wary Smithsonian employees had kept sponsoring corporations at arm's length throughout, using in-house staff to develop and update the exhibit, but the exhibit attracted criticism nonetheless.[20] During a 1993 federal appropriation hearing, for instance, Congressional representatives asked Smithsonian secretary Robert McAdams if the zoo was merely an excuse for free corporate advertising, noting that Orkin's corporate logo was conspicuously displayed at the museum's entrance and on the educational materials to be distributed to teachers across the country. McAdams promised committee members that the museum relied exclusively "on the expertise of professional educators in developing these materials. . . . The Orkin Corporation will have no intellectual

Figure 23   A 1977 publicity still from the Smithsonian's *Insect Zoo* opening, taken for the member magazine, *Torch*. Unlike the grasshopper shown here, most of the live insects on display were under glass; a few years later, the National Museum of Natural History got additional corporate funding from Orkin Pest Control to expand the popular exhibit and rename it *The O. Orkin Insect Zoo*. (Smithsonian Institution Archives, Image 94–2867.)

control over the exhibit or related educational materials."[21] When asked if he planned to allow more corporate sponsors to splash their logos across the National Museum's walls, the secretary hedged. "Whether the public can expect to see more corporate insignias in the future will be a function of the Institution's success in raising corporate contributions," Secretary Adams calmly responded.[22]

Smarting museums eventually got wise. By the late 1980s, most had implemented policies to prevent exploitation by corporations. Museum staffs often refused to allow corporate sponsors to participate in exhibit development, except as nonvoting members of broad-based advisory committees, and insisted that curators retain veto power over any proposed content.[23] But museum development officers and public relations advisers figured out how to have it both ways. They satisfied critics by minimizing the visibility of corporate logos on museum walls but pleased corporate sponsors by prominently featuring their logos in splashy new advertising campaigns. Over the course of the decade, the American Museum, for example, spent about $200,000 annually on advertising for exhibits, programs, and services, a good chunk of which was provided by highly visible corporate sponsors. Museum administrators weren't merely acting as corporate shills, of course; the advertising benefited them as well. At the Field Museum, for instance, attendance increased over 13.5 percent between 1983 and 1985, a shift museum president Willard Boyd credited to new corporate-sponsored advertising.[24]

Experiments and Evaluations: The Science of Education in American Museums

In the late 1970s, the kind of "authoritarian condescension" that museums had shown toward their audience throughout so much of the twentieth century had disappeared almost entirely.[25] Desperate to prove their relevance, determined to fulfill their educational responsibilities to their visitors, museum staffs quietly continued the counterculture rebellion against reigning institutional conventions. Even when the nation's politics rushed to the right, continuing concerns about the declining quality of science education perpetuated a spirit of experimentalism, and throughout the 1980s, American museums and science centers used their miscellany of pedagogical methods to involve themselves in educational debates more directly than ever before. Yet the professionalization of museum-based education and evaluation and attempts to replicate successful programs on a national scale soon ironed out the wrinkles of diversity that had resulted from this experimentation.

The content and forms of museum display initially became more varied in the late 1970s and 1980s as museums catered to the mood of what Christopher Lasch has described as the "age of narcissism" and emphasized how science and nature affected the lived experience of museum visitors. This learner-centered approach to science education had roots in Progressive Era biology courses, the life-adjustment curriculum of the 1930s and 1940s, and the science literacy movement of the 1960s and early 1970s, but the ideology underlying this approach was entirely different. Instead of casting their displays as the kind of science visitors *should* know and striving to make those subjects relevant to their audiences, museums of the late 1970s and 1980s increasingly considered what visitors *wanted* to know. "What does it mean to me? It is a question that people ask first, even subconsciously, before opening their minds to learning," wrote Jeffrey Birch, part of the generation of exhibition and programming consultants that thrived in the museum world during these decades. "If museums are to continue to be important places in the years ahead, they must more fully recognize the importance of this question."[26] This visitor-oriented approach to education resulted in part from the ideal of social relevance that had motivated museums in the late 1960s and early 1970s, as well as from staff members' new desire to reach those in the "me generation."

Natural history museums, with their established fields of inquiry and hoary traditions of display, were forced to cast a wider scientific net than they had in the past in order to present information that would, in Birch's words, "promote a personal understanding of our world and ourselves."[27] It could be challenging to convince museumgoers that the life cycles of birds and bats were significant to the modern American experience, outside the fleeting wonder that such beasties elicited. Shows on the environment and the cultural heritage of museum visitors had satisfied public calls for relevance in the 1960s and 1970s, but for visitors exhausted from decades of social activism, these could be difficult sells, and natural history museums struggled to find a focus.

So natural history museums added new technologies to their display repertoire, hoping that their novelty would keep visitors engaged and critics satisfied. As early as 1983, the Milwaukee Public Museum used computers to conduct exhibit-related quizzes, revitalizing a long-standing museum practice as they did so.[28] Museum staff marveled at how visitors willingly waited to use the bulky machines and their squealing modems. The National Museum of Natural History relied on more spectacular technology, borrowing techniques and components from existing aquaculture research projects to build a living coral reef, which opened in 1980. Determined to introduce biodiversity in a way that would resonate with contemporary audiences, curators filled a three thousand–gallon glass tank with three hundred species of animal, fish, and plant life to create an intact, self-sustaining biological community—something like the Soviets' BIOS-3 or the University of Arizona's Biosphere 2, but for sea creatures, not hungry scientists.[29] The tank alone was a technological marvel, powered by solar energy. An artificial wave machine circulated oxygen through valves that opened to a connecting tank that served as a lagoon. A series of documentary shorts depicting the reef's inspiration, construction, and biota played nearby.[30]

But while new technologies dazzled, they didn't always deliver clear scientific or educational messages. An exhaustive visitor survey conducted in 1981 revealed that the Smithsonian's *Coral Reef* left visitors more puzzled than pleased, for example. Few came away with more insight into coral reefs, underwater life, or biodiversity. Some asked if coral was vegetable, animal, or mineral. Others inquired if the fish on display could be bought in pet stores. One visitor explained that the exhibit had taught him that a coral reef was alive but confessed that he didn't understand how it stayed that way.[31] "I know that this exhibit is giving an important scientific message. It's obvious by its size and the drama that the scientific concepts are really significant. But beyond knowing that it is a living system, I really don't know much about it," admitted another. "Are they studying fish behavior in natural environments? If so, wouldn't it be better to do the research in the ocean?" asked one budding biologist.[32] By November of 1980, *Washingtonian* magazine's Hits and Misses column had relegated the *Coral Reef* to the miss category. "Originally an innovative idea, it's quickly becoming just a dirty old fish tank," editors wrote.[33]

Science museums and science centers found it easier than did natural history museums to adjust their exhibits to fit new public preferences, but critics derided their displays as thoughtless, even irresponsible. Science museums had traditionally emphasized science's relevance to society, and science centers rarely worried about the social or personal relevance of exhibit content. For institutions like the Exploratorium and the Pacific Science Center, science's personal applications lay in the immediacy of the hands-on experience, rather than in the information conveyed. Yet their apolitical approach elicited new censure from scholars who believed that museum displays should confront the power structures and social pressures that lay beneath the life sciences. "Is an understanding of current scientific principles and contemporary technical devices all there is to the public understanding of science?" asked historian of technology Melvin Kranzberg in 1984.[34] "What we require are museums that tell us what we once were, and what is wrong with what we are, and what new directions are possible," New York University professor and communications critic Neil Postman agreed in 1989. At the very least, he maintained, museums should provide a narrative "different from the vision put forward by every advertising agency and political speech."[35] Political groups and professional critics continued to complain, as they had throughout the 1960s and 1970s, that museums too often ignored the complicated legacy of scientific discovery and promoted an unequivocally positive view of scientific progress instead.[36]

Museum leaders confessed that many of these criticisms were valid. Too often, science museums attempted to "confirm and restore man's confidence and pride in how he has deliberately altered his world," a stance that capitulated to corporate sponsors and visitors who liked their science simple, admitted Willamette Science and Technology Center director Alice Carnes in a 1986 issue of *Museum News*.[37] Science museums, she concluded, needed to decide if they were going to be showplaces blindly cheering for technological progress or "lively forum[s] for learning, controversy and the search for solutions."[38] But diverging from older paths risked offending government agencies and corporate sponsors intent on promoting science as a keystone of national progress—a risk that museums and science centers of this era found it progressively more difficult to take.

Tired of costly pedagogical and technological gambles, exhausted by defending their educational efforts to political critics, exhibit designers and educators in museums and science centers turned to science to help them develop visitor-centered exhibits and educational programs. In the early 1980s, they began to ground new exhibits and educational programs in the findings of developmental psychologists like Jean Piaget and Erik Erikson, whose scholarship had been embraced wholeheartedly by social scientists and the nation's schools of education by the 1970s and 1980s.[39] Piaget had described learning as an active exchange between the learner and his environment but emphasized that age, intellectual interests, and psychological development affected how individuals experienced their environment. This research bolstered the credibility of science centers' hands-on, carefully tested methods, while further discrediting the dated arguments espoused by Bradford Washburn and other post-Sputnik advocates of "exposure." "Many museum people accept as an article of faith that visitors—especially children—will learn something simply by being exposed to exhibits and educational programs," Robert Matthai, director of the American Exploration Project at the American Museum of Natural History, explained in 1984. This notion, he added, was completely wrongheaded. One could not, "through intuition or practice, teach children successfully or create comprehensible exhibits, without reference to relevant learning research and theory."[40]

Museums' increasing embrace of developmental psychology led them to outsource evaluations of the casual education their institutions supposedly did best and, in the process, tied museums more closely than ever to heated debates within the nation's educational establishment. By the mid-1980s, both museums and science centers relied heavily on the methods and findings of education researchers, whose use of visitor focus groups and statistical modeling promised to measure educational success scientifically. Few staff members of museums had experience with developmental psychology, so institutions hired consultants and education school graduates to help them evaluate exhibitions and educational programs.[41] They needed the help, for Piaget and Erikson's ideas sent education and exhibition departments scurrying to consider how a wider spectrum of visitors reacted to their programs and displays, rather than assuming a one-size-fits-all model of learning or dividing museum audiences into binaries of lay and expert, child and adult, as they had in the past. Designers, educators, and psychologists began to test the audience appeal and educational effectiveness of their exhibits more rigorously, compiling and sharing stacks of data about what engaged visitors' attention, and what simply resulted in children idly splashing water or flipping switches.[42] Experts did not always agree on what or how to evaluate, however; Piaget's constructivism inspired plenty of researchers, but others, like psychologist Harris Shettel and his survey research group who worked to develop a universal set of exhibits evaluation criteria, denounced his assumptions.[43]

The era's newly exacting approach to what the NSF, in 1982, deemed "informal education" vaulted long-lowly museum educators into more respected institutional positions. Where museum educators had earlier felt so marginalized that they had threatened to secede from the AAM in 1973, a decade later, their colleagues regarded them as critical components for institutional success.[44] Their work was no longer described in the ancillary terms of "outreach," but positioned as central to the missions of museums. Indeed, "Museums for a New Century," the AAM's 1984 report on the status of the nation's museums, affirmed that education was a, if not the, "primary" purpose of museums in the United States.[45] Museum educators found themselves newly indispensible to display development and fundraising, for grant writers hoping to win foundation money increasingly relied on their input. By 1990, evaluations by educators had become central to fundraising. To win a grant from the NSF's new informal science education division, museums had to provide the agency with a detailed plan for evaluation and even identify who would be carrying it out.[46] Most educators were quick to seize what they felt was overdue recognition. Mary Munley, an educational psychologist and acting director of George Washington University's museum education program, applauded her colleagues for developing a "stringent set of professional standards and practices for education and audience-related functions." "Museums stand on the brink of a major expansion in their responsibility to society," and recent innovations in museum education, she promised, would "change our notions of museums forever."[47]

Studies sponsored by philanthropic foundations mediated against the potentially divisive effect of educators' conflicts and further contributed to the standardization of museum practices, displays, and educational programs.[48] In the late 1970s, the W. K. Kellogg Foundation had supported studies of exhibition design and educational efficacy at the Exploratorium, and in 1982, it granted money to the Field Museum and the Smithsonian for similar research. By 1987, staff from eighty museums had participated in Kellogg workshops.[49] The Kellogg Foundation also funded the publication of a series of manuals and "Exploratorium Cookbooks" based on these studies. The cookbooks provided institutions across the country with recipes for exhibitions previously tested by the Field Museum, the Smithsonian, or the Exploratorium.[50] These manuals allowed museums to cherry-pick content and methods, and their halls became less predictable as a result—natural history museums increasingly borrowed exhibits from science museums and science centers, and vice versa.

Whereas natural history and science museums had aggressively sought to distinguish themselves from one another, overlapping educational and display practices in natural history museums, science museums, and science centers no longer seemed to bother anyone; to the contrary the wider museum community extolled the combination of professionalization and informality that had come to characterize their shared educational mission. And they lauded the institutions for contributing to the broader science of education. Museums and science centers, noted one Association of Science-Technology Centers–commissioned report, "provide access to educational technologies that schools cannot afford or have not yet installed," but more and more, these institutions allowed "active experimentation and investigation [in education that was] parallel to that of science itself."[51]

By the 1980s, the respective decisions of museums to emphasize their exceptionalism or the inclusiveness of their particular brand of informal science education seemed to be more a matter of strategic positioning than actual methodological difference. After decades of attempting to partner with schools, a handful of natural history museum directors rejected the notion that museums should be considered adjuncts to school education. "Museum education is and should be treated on its own merits, not as a part directly of the vast educational apparatus in this country, but rather as an essential and continuing enrichment in people's lives," wrote former Smithsonian secretary S. Dillon Ripley. Museums, he suggested, had little responsibility to alter themselves for the sake of the nation's formal school system.[52] Other museum leaders, striking a more conciliatory and collaborative tone, reached a different conclusion. While nothing like schools, Frank Oppenheimer noted in a 1982 address to the AAM, museums and science centers "are basically educational institutions. It is a mistake to think that preserving culture is distinct from transmitting it through education. . . . Whether they like it or not, curators are educators."[53] As a result, he maintained, museums and science centers had a responsibility to work with schools to improve science education nationwide.

Ultimately, federal pressure helped push museums and science centers into more active relationships with schools. Though Cold War fears of Russian scientific superiority had waned, ongoing economic woes and the explosion of Japanese technologies had created new anxieties about the state of science education in the United States.[54] A slew of national reports excoriated the nation's science teachers for their lack of preparation in both subject matter and method.[55] Poor training was exacerbated by diminishing resources in the schools themselves—federal money earmarked for science education was largely eliminated in the first years of the Reagan administration, and teachers taught too many classes, faced overcrowded classrooms, were provided with poor facilities, and had little or no time to prepare demonstrations. Their own programs hobbled by funding cuts or abolished altogether, both the NSF and the U.S. Department of Education urged museums to find ways to improve primary and secondary science education directly rather than merely supplementing established courses with programming or exhibits.[56]

Obliged by a sense of national duty and lured by the promise of eventual funding, science centers and museums got involved in the messy business of changing school classroom practice by bringing teachers to them. The Exploratorium led this effort, for it had a track record of success in this field. "Upgrading science teaching is a long-term, long-haul problem," explained Rob Semper, the Exploratorium's associate director and a leader of its School at the Exploratorium program. "The real trick is doing long-term, intensive teacher development."[57] Its staff members fashioned an aggressive program to make over American science pedagogy according to Oppenheimer's ideals, a program that, like their exhibits, was decidedly hands-on and emphasized scientific process and open-ended experiment, and throughout the early 1980s, the Exploratorium offered an intensive, eighty-hour summer program for teachers, followed by activities during the academic year.[58] In-service teacher training, first begun in the 1950s, soon became a staple of the educational programming of science museums and science centers. In 1985, for instance, the Pacific Science Center began weeklong "Science Celebration" seminars to train elementary school teachers in Washington State to teach "hands-on" science.[59] "This program is actually starting to change the way and amount of science that is taught in our district," said Olympia School District resource specialist Judy Porter. "It's magic. We train the teachers enough so they can answer questions, and we show them the principles of interactive learning; how to let the kids discover for themselves. And away they go."[60] By 1988, the science center was training twenty-one hundred teachers each year.

Such efforts paid off handsomely as the economy began to recover, and museums won federal and corporate funds for similar programs in formal science education. In 1984, inspired by the success of the Exploratorium's program, the NSF permitted museums and science centers to compete for science education grants for the very first time. "Both the state and the federal government have been lobbied into funding us," gleefully explained Dennis Schatz, associate director of the Pacific Science Center.[61] In 1986, the Buffalo Science Museum received a grant for $135,063 from the United States Department of Education to educate Buffalo public school teachers about how to teach the sciences in effective ways, calling the grant TEAM—Teacher Education at the Museum. This teacher education venture bore additional fruit for Buffalo—in 1990, the museum received $1,286,195 from the NSF to expand Teacher Education at the Museum and create a partnership between the museum and the Buffalo school system, training teachers to expose children to life science, earth science, and physical science and to teach experimental, hands-on science.[62] Corporations, short on employees with sufficient scientific training, also devoted money to these programs. Throughout the 1980s, the GE Foundation sponsored the Science Teacher Education at Museums program, directing cash to the Association of Science-Technology Centers, which distributed the money to member institutions as mini-grants for teacher-education programs. The Carnegie Corporation of New York, the Danforth Foundation, and the William and Flora Hewlett Foundation also provided funding for museum-based training of primary and secondary school science teachers.[63]

Beginning with their collective embrace of visitor-centered learning and more scientific methods of evaluation, by the late 1980s and early 1990s science centers, science museums, and natural history museums had begun to overlap in mission, mantra, and practice. Once self-consciously separate organizations, their identities had blurred, and it became more difficult to distinguish these institutions from one another. It also became more difficult to distinguish them from other institutions devoted to education and entertainment. The late-century emphasis on relevance and educational access had brought museums closer and closer to schools. And school science classrooms increasingly resembled the halls of museums. Cultural shifts and the consequences of the financial turmoil of the late 1970s, the 1980s, and the early 1990s would bring museums and science centers even closer to one another—and bring both nearer to theme parks, malls, and blockbuster films as well.

Buildings and Blockbusters: The Rise of Edutainment

Beginning in the late 1970s, museums of science and nature jostled for Americans' attention in an increasingly crowded field of venues promising to offer entertainment with an educational thrust, which marketers called edutainment.[64] To entice and impress consumers, a new complex of theme parks and other for-profit ventures began to display impressive specimens or host traveling exhibits.[65] Museum staffs saw this as intrusion into their territory, but they were uncertain how to respond at first. After all, in the early part of the century, professional museum staffs had described their halls as places for uplifting leisure but had vehemently insisted on their difference from commercial entertainments. Natural history staff became more comfortable with comparisons to venues of consumer culture as the twentieth century progressed, arguing that they, too, offered a product—scientific knowledge—and so could legitimately employ advertising, theatrics, or other persuasive strategies in an effort to make their wares more attractive to potential customers.[66] The growth of science museums and science centers certainly encouraged this stance. But curators, directors, and educators remained squeamish about the cheerful commercialism that motivated or, at the very least, accompanied most edutainment. The ongoing financial crisis and eroding audience share made commercialism seem more palatable. By the mid-1980s, museums offered their own version of edutainment, using museum shops and exhibitionary spectacle to help them get visitors—and their much-needed dollars—in the door.

Edutainment gave museums an excuse to usher overt consumerism into the halls of science museums, natural history museums, and science centers in the last decades of the twentieth century.[67] To attract busy tourists and local leisure seekers, museums advertised more heavily than ever before, unfurling banners across their front entrances and adorning city streetlights with advertisements.[68] Some institutions, like New York's American Museum of Natural History and the San Diego Museum of Natural History, opened their halls in the evenings.[69] And they offered the ever-growing audiences that tramped through their halls new ways to spend money. "Museums must improve their services or risk losing their audience to such money-making amusement organizations . . . which are going into object interpretation for the first time," director of the Fort Worth Museum of Science and History Helmuth Naumer cautioned his colleagues in a 1971 issue of *Museum News*.[70] Consequently, institutions placed museum shops in entry halls, hoping visitors would purchase coffee-table books on rainforests, sea monkeys, or plush zebras.[71] Some institutions turned to licensing, lending their names and collections to corporations in exchange for royalties, then selling the resulting jewelry, paperweights, and stuffed animals in their newly built shops. As early as 1972, the Smithsonian signed a major contract with the Tonka Corporation of Minneapolis for lines of playthings, hobbies, crafts, and games based on objects in the Smithsonian's halls, and in the late 1970s and 1980s, other museums attempted to copy the model.[72] The Exploratorium, for instance, sold "home science experiment kits," promising visitors that the kits would allow them to relive their museum visit on their own time and territory; it even considered franchising its store.[73] By 1985, *Time* magazine estimated annual sales for museum shops at $200 million. By 1989, estimates ranged from $500 million to a billion dollars.[74]

Museums also used their physical space in new ways to recruit new audiences and generate new income streams. Natural history museums introduced profitable "experiences" like sleepovers and IMAX/Omnimax theaters. Museums constructing new buildings instructed architects to include well-equipped conference rooms, auditoriums that could double as community theaters, and soaring atria designed to be rented in the evenings for fundraisers, weddings, and bar mitzvahs.[75] This approach contrasted starkly with that of earlier eras. In 1963, when Seattle University asked to rent the Pacific Science Center for its junior prom, the center refused, for the request was "clearly outside the purpose and policy of the Pacific Science Center."[76] By the 1980s and early 1990s, though, major science and natural history museums and science centers regularly rented out newly built wings to local businesses and wealthy revelers. The Exploratorium went so far as to offer prospective renters cost comparisons with other meeting spaces in town. Administrators were not subtle about their motives, hinting in brochures that event participants might "in some way assist and support the Exploratorium somewhere down the road."[77]

Blockbusters and major traveling exhibitions were another dramatic change, one that both embodied consumerism and further ensured that visitors would experience the galleries of science centers, science museums, and natural history museums more similarly. Big shows were not a new phenomenon, nor were traveling exhibits. National traveling shows had started in the 1950s and 1960s, as government agencies and the Smithsonian sponsored dozens of Cold War exhibits promoting the promise of the sciences and the glories of American culture. But the blockbusters of the late 1970s, 1980s, and 1990s elevated these same practices to an entirely different scale and became the engine for most museum merchandising.

Blockbusters were first mounted in art museums in the 1970s and early 1980s, but natural history museums, desperate for cash, quickly acquired versions of their own.[78] Many larger natural history museums had already begun to expand their exhibition space to accommodate public demand; as a result of the blockbuster phenomenon, they hastened and enlarged these expansions. After adding three thousand square feet of temporary exhibition space in 1972, American Museum officials added another cavernous ten thousand–square-foot space four years later, allowing the museum to feature a quick succession of massive shows.[79] Midsized museums coveted blockbusters but struggled to find funding to build the large flexible spaces traveling exhibits required—many exhibits needed ten thousand square feet, and by the early 1990s, some demanded close to twenty thousand square feet. As a result, most museums didn't attract these kinds of exhibits until at least a decade later.[80]

Early on, natural history museums, with their miscellanies of objects and traditions of anthropology and ethnology, found blockbusters an easier fit than science museums and science centers did. When the

Treasures of Tutankhamun exhibits traveled to Chicago in 1977, followed by *The Great Bronze Age of China* in 1980, it was the Field Museum, not the Art Institute, which played host. In the 1970s and early 1980s, most blockbusters featured art or jaw-dropping cultural artifacts, which natural history museums seamlessly incorporated into their own diverse halls. Science museums, in contrast, were left contorting their missions to justify displays of Leonardo paintings and the Egyptian boy king's burial mask. Natural history museums' collections also allowed them to outpace their younger institutional cousins in the development of blockbusters, for their ability to marshal rare biological and paleontological collections for exhibits allowed them to generate their own epic exhibitions. The promise of seeing forty original human fossils, a substantial proportion of the most significant physical evidence of human evolution to date, together in a single exhibition hall brought more than half a million people to the American Museum of Natural History's 1984 *Ancestors: Four Million Years of Humanity* in just five months, for instance.[81] The exhibit buttressed the museum's finances, to be sure, but also its institutional credibility. That the museum was able to bring together and display these fossils, cloistered in institutions scattered all over the world, was not only a formidable administrative feat, *Science* declared, but also a powerful blow to creationism and proof of the ongoing scientific and educational value of museums.[82]

By the mid-1980s, however, science museums and science centers, tired of watching crowds flock to other institutions, "broadened their missions," to use the public relations term, and welcomed such exhibits into their halls. From 1985 through 1988, for example, *Ramses the Great* circulated through the nation's art and natural history museums before stopping at the Boston Museum of Science—a spokesperson described the museum's decision as an attempt to "unite art and science." And in 1984, a six-month show titled *China: 7,000 Years of Discovery* attracted 727,861 people to the Pacific Science Center—almost triple the number of visitors to the Science Center in the same period in 1980 and double the visitors to the Science Center during the same six months in 1982. Though critics complained that the show's "artifacts" were mainly replicas and that the show "was conceived and presented to elicit gee whizzes rather than serious inquiry," it proved phenomenally successful, earning a half a million dollars more than anticipated, allowing the Science Center to launch the first capital drive in its twenty-two-year history.[83]

Robotics and animatronics pulled science museums and science centers further into the center of the blockbuster trend and merged the life sciences with sophisticated technologies of entertainment. The American

Museum had toyed with the idea of automated models of dinosaurs as early as the 1939 world's fair, and Disneyland had introduced animatronic dinosaurs to American audiences in its 1964 *Primeval World* exhibit, but these early prototypes were largely playful novelties, without much emphasis on scientific accuracy. By the 1980s, however, commercial companies like Dinomation and Kokoro had made the most of recent advances in robotics and special effects and began to rent animatronic dinosaurs to museums, theme parks, zoos, and malls eager to profit from the spectacle of these stiffly moving beasts. When the wildly popular *Dinosaurs Alive* exhibition came to the Buffalo Museum of Science in 1987, for instance, the museum more than doubled its admissions fees and began charging families or guests of members for entry.[84] When the exhibition ended early in 1988, the museum reduced these admissions, but they never returned to the pre-1987 levels.[85]

Dedication to blockbusters pressed museum leaders to make what seemed to be counterintuitive managerial decisions, but in the long run, their judgment bore out: creating and hosting blockbusters provided invaluable economic stability. In 1987, the Pacific Science Center persuaded Washington State's Department of Public Instruction to help them purchase several dinosaurs from Dinomation as part of the museum's program of educational outreach. Exhibit designers positioned the animatronics within simulated prehistoric environments, surrounding them with actual artifacts, supplementary exhibition materials, and volunteers stationed behind carts laden with scientific games, then barnstormed through the largest towns in Washington State and Alaska. It was rough-and-ready travel; using forklifts they fondly dubbed "dino-dumpers," staff members would set up the dinosaurs in tents at muddy local fairgrounds and grit their teeth as the extremely expensive models became coated with layers of fine dust. But the exhibition was wildly successful, attracting nearly half a million rural and small-town visitors.[86]

Inspired by this success, the Pacific Science Center took out a million dollar loan and designed one of the decade's most successful biological blockbusters around an equally massive set of animatronic beasts: *Whales: Giants of the Deep*. The eight thousand-square-foot exhibition attracted more than 4.6 million visitors on three different continents.[87] It combined commercial spectacle with the ostensible interactivity of science centers and the artifacts of traditional natural history displays, offering a sound environmentalist message to boot. Life-sized animatronic models of orca, gray, humpback, narwhal, and sperm whales, which whistled, moaned, or waved their flippers, formed the exhibition's centerpiece, but visitors could also explore more "hands-on" displays of cetacean biology.

Visitors could wander from station to station, trying on replicas of whale flippers and fins to get a sense of the animals' scale, moving a brushlike "baleen" through water to collect food pellets, testing their knowledge of marine biology at question-and-answer stations. For ancillary displays of whale physiology and products, exhibit designers borrowed whale oil, whale bones, whale meat, and other confiscated materials from the National Oceanographic and Atmospheric Administration. The center also provided promotional and educational materials, and, of course, a prefabricated gift shop.[88]

Ultimately, the *Whales* exhibit made the center enough money to pay back its initial loan, plow hundreds of thousands of dollars back into operating expenses, and begin building its next traveling show. Commanding a rental fee of $225,000 as a traveling exhibit, *Whales* wasn't wildly expensive compared to an artifact show, which usually required sky-high insurance rates and considerable security, so this same exhibit baptized a number of smaller science and natural history museums into the blockbuster circuit.[89] It was the single largest exhibit hosted by the San Diego Natural History Museum in terms of cost, size, and number of displays up to that point, and it helped the museum reduce its deficit from $450,000 to $50,000.[90] Small and midsized museums found themselves hooked.

Blockbusters were expensive to create, but their benefits were distributed: once an institution invested in an exhibit, it had the potential to pay longer-term financial and demographic dividends across the entire science and natural history community. Most shows demonstrably expanded museum audiences, bringing ever more diverse adult visitors to natural history and science museums. "We don't expect to recover all our costs, but we'll pick up new membership, and that's a long-range benefit," director of the American Museum Thomas Nicholson explained in 1979. Courtesy of its blockbuster schedule that year, the museum's membership rolls almost doubled, expanding from eight to fifteen thousand.[91] "We've increased our exhibition audience from 11 to 40 percent," the satisfied Nicholson told the *New York Times*.[92] That directors now envisioned their accountability for display investments much like corporate chief executive officers envisioned their accountability to shareholders provided museums with the credibility they needed to garner more support from government and industrial sponsors.

Like dioramas had decades earlier, blockbusters expanded public interest in museum halls. As a result of highly publicized "premieres" and limited runs, visitors eagerly anticipated these temporary exhibits and made a diligent effort to catch them before they left town.[93] As a result of their status as events, visitors discussed them as they did professional sporting events or new films. Marketing campaigns persuaded new visitors to venture past once-forbidding pillars and cajoled them into purchasing souvenirs of their experiences. Though museum stores only averaged, in 1989, $1.28 per customer in mean revenue, that tiny per-person figure could become a tidy sum if hundreds of thousands of visitors tramped through an exhibit.[94] By the early 1990s, sales from museum stores, stimulated by blockbuster audiences, were often responsible for a quarter or more of museums' annual revenues.[95]

This fervent cultural and material consumption of blockbusters, their advocates maintained, ultimately served the larger cause of science education and the broader mission of the museum, encouraging museumgoers to explore halls they might not otherwise have visited, piquing interests that resonated long after visitors had left the museum. "We may agree that there's a vast difference between 'consuming' culture and studying it seriously, but that base level of consumerism is where the blockbuster trend draws its energy," New York State's commissioner of education Gordon Ambach told AAM members at their annual meeting in 1986.[96] Indeed, the consumerist orientation of these new exhibitions ultimately reinforced museums' public-service rhetoric, serving as evidence of involvement by museums with the larger community. "It was difficult to charge museums with standoffishness and withdrawal when hundreds of thousands of visitors would make their way, in just a few weeks, to particular cities simply because of their interest in a museum exhibition," noted museum scholar Neil Harris.[97] "As long as tens of thousands of museum visitors are willing to spend an hour looking at the backs of three rows of other visitors heads instead of looking at artifacts, these popular exhibitions may save a handful of larger museums," Bob Flasher, an educator at the Oakland Museum wrote in 1983.[98]

Despite their popularity, these new permanent and blockbuster exhibits faced sharp criticism from some members of the museum community and cultural critics at large, who argued they were distracting museums from other aspects of their mission. Many museum professionals saw the popularity of such exhibitions "as the beginning of the inevitable compromise of scholarship . . . prelude to a headlong sprint down the garden path to the lowest common denominator," wrote exhibit designer Stuart Silver.[99] Some staff members got so fed up with the blockbuster obsession that they fled the public museum world for the quieter realms of university museums. "I voted with my feet," explained Ken Miller, an astronomer who left the Pacific Science Center for the University of Nevada's planetarium after the Science Center's massive *China: 7,000 Years of Discovery* show. Exhibits developer Steve Brugger also left, protesting that the hoopla surrounding the massive exhibits had obscured the Pacific Science Center's focus on science education. "There was a lot of frustration among the staff," he recalled. "More and more, the issue became not whether a project was valid as science, but whether it was 'viable' to the Science Center."[100] The traveling *Art and Technology of the Muppets* show that was scheduled to arrive at the Pacific Science Center in 1985 seemed to prove his point.[101]

Echoing the pedagogical complaint that Frank Oppenheimer had made in the late 1960s about natural history museum displays, Albert Parr, the elderly director emeritus of the American Museum, opposed blockbusters on the grounds that they eliminated the need for independent thought or activity. While Parr acknowledged the innovations science centers had made in encouraging visitor engagement, he firmly believed that discovery didn't require hands-on activity. Some of his favorite exhibits, he wistfully told the AAM in 1982, remained the dioramas of the Denver Museum of Natural History. "I keep finding new things in them—a lizard here, an arrowhead there or a high-altitude heliotrope behind some grass," he explained. "In a recent visit, the people around me were constantly pointing out to each other, 'Oh look! Over there behind that rock.' The dioramas reproduce the sense of discovery that occurs when one is wandering in the high country."[102] Older display paradigms may have been static, he suggested, but they offered an intellectual freedom and imaginative challenge that newer ones shouted down. In modern life, "children and adults have few opportunities for exploration," Parr wrote in 1982. "They usually know ahead of time where they are going and why they are going there. They rarely just wander on the off chance that they might find something interesting. . . . As a consequence, the whole tradition of exploration is being lost for entire generations." Museums were one of the last places where it was possible to explore, he declared, and the managed experiences of blockbusters were "too rigid, too structured or too channeled" to allow genuine intellectual and emotional discovery.[103]

In their efforts to satisfy both the loose tastes of visitors and the tight fiscal checklists of in-house comptrollers, museums had become more like malls or theme parks than ever before, a shift that raised serious questions about the broader social role of museums. In the past, museum staffs had been determined to keep their reputations and appearances separate from those of commercial establishments—in the 1900s and 1910s, museum men like Jesse Figgins and Henry L. Ward had fired staff members for putting price tags on exhibited objects, professing "repugnance for the appearance of commercialism in an institution intended for Natural

History."[104] Not so in the go-go 1980s and 1990s. As a result, staff members increasingly asked themselves who should set the agenda for exhibition and what that agenda should be. Was popularity synonymous with accountability and accessibility, or was it a different beast altogether? For two decades or so, the professional trend had been to disparage those museums that were too curator-centric and to look to institutions that gave visitors what they wanted for guidance. "Museum exhibits are too often creator oriented rather than visitor oriented," Charles Alan Watkins complained in a 1991 issue of *Museum News*. "Theme park exhibits answer the questions people want answered, not the ones museum people necessarily want to answer."[105]

But as the blockbuster juggernaut rolled on, the museum world began to wonder what it portended. Keeping the doors open was important, curators and educators knew, but would their blockbuster efforts to do so ultimately become a Pyrrhic victory? Throughout the 1980s and early 1990s, staff members quietly asked themselves if museums should continue to allow the public to dictate the terms of display or if they should return to the museum traditions of the past—if that was even possible. Formerly, the public and museums themselves had placed great value on museums' scientific expertise and allowed them to set the educational and social agenda of their exhibits as a result, often tackling topics or questions that the public would not have known to ask. Some museum staff murmured that museums should return to this tradition. Others sought to balance old and new, expert and lay, collection and experience. Regardless of their personal misgivings, most continued to march forward, eager to capitalize on the excitement blockbusters generated.

Designing Experiences and Reclaiming Research in the Edutainment Age

The aesthetics of blockbuster exhibitions and commercial entertainments soon found their way into museums' permanent halls, and exhibition designers found themselves at the center of a turn toward edutainment by museums. Throughout the 1980s and early 1990s, museums overhauled galleries that hadn't been touched in decades and engineered elaborate, immersive "experiences" in their permanent halls.[106] Sensitive to museums' increasing competition with for-profit entertainment venues, designers faithfully imported lively commercial aesthetics, using set design, video displays, special effects, animatronics, and plenty of black lighting to create thrilling simulations, often of the natural world. But it remained unclear if such spectacles advanced anything other than the bottom line for museums. Their educational benefit was ambiguous, and the damage they did to museum research programs was considerable. Alternately frustrated and furious, curators resisted these changes at first. But by the mid-1990s, they had learned to exploit edutainment for their own ends.

Professional exhibition design thrived as a result of blockbuster shows and new interest in edutainment, and exhibition designers commanded new power in the museum world as a result. With some prominent exceptions, for most of the twentieth century designers of museum exhibits had not gotten much credit.[107] By the late 1970s, however, exhibition design had planted a professional flag, and designers attending the AAM meetings formed a rump group, the National Association of Museum Exhibition. Museums expanded their in-house exhibit departments, which looked to the methods of display and evaluation promoted by exhibition firms founded to serve commercial venues like shopping centers, trade fairs, and exposition pavilions. Increasingly, museums called on these firms to assist with museum exhibitions. By the 1980s and 1990s, what *Museum News* called a " 'high-design' installation imperative" had become just as important as the contents and educational intentions of displays. Museums routinely relied on full-service exhibition and interpretive design companies, whose other clients included malls, hotels, and theme parks, to help them design every step of their elaborate exhibitions, from hall renovation to marketing blitzes to affiliated merchandising. Though museums had long borrowed from popular culture's most effective exhibition techniques, the systematization and sheer scale of the overlap was new—and hotly contested.[108]

Many curators at natural history museums resisted museums' turn toward professional exhibit design firms, anxious that they would inadvertently undermine the reputations and scientific authority of these institutions.[109] Eager to entertain the public through interactive opportunities, designers didn't always respect the requirements of science educators or the expectations of research scientists. During the development of the Smithsonian's *Insect Zoo* in the late 1970s, for instance, curator Richard Cowan successfully prevented Charles Eames from designing the exhibit, arguing that industrial designers like Eames were "not knowledgeable enough about biology generally, or entomology specifically, to take on the position."[110] So the exhibit proceeded in-house, with the museum's own designers taking the lead. In-house designers worked to develop an exhibit that would entertain while educating and promote the kind of interactivity that was then winning the Exploratorium so much praise.

But the label scripts they presented seemed to realize Cowan's worst fears about using designers outside the museum. The exhibit, a series of open terraria, was designed to encourage visitors to interact with live insects.[111] "The entire presentation," wrote Smithsonian entomologist Richard Froeschner, "seems to be a disconnected series of selected examples with an effort to tie them together with generalized statements that are often incomplete, obsolete, contradictory between themselves, or sometimes downright wrong." The exhibit featured tarantula spiders, for instance, which weren't even insects. "Dead insects properly arranged" and labeled, Froeschner argued, "can show infinitely more than live ones lying exhausted or cowering in a corner."[112]

Yet edutainment's visual requirements ultimately overwhelmed many curators, who readily admitted their ignorance of AutoCAD, Adobe, and audience research. In the new blockbuster economy, wrote one critic in 1990, "the most erudite museum professionals suddenly, disarmingly, find themselves functionally illiterate."[113] Some curators embraced their sudden obsolescence; after a century of often grudging service on exhibition planning committees, they gratefully limited their role to consultation or fact checking. Others, like Cowan and Froeschner, went out kicking and screaming.

Throughout the 1980s and early 1990s, design firms pushed museums away from older display formats, urging them to adopt more marketable aesthetics of contemporary businesses. The success of edutainment in theme parks and other commercial spaces should prompt "the guardians of the glass case, indirect lighting and low noise levels to look outside their ornate doors and iron gates to the success of the world's most visited education/entertainment centers," wrote one consumer researcher in 1990.[114] Given the past relations with industry and the era's current economic realities, many designers and consultants argued that museums and science centers should not disparage Disney's "imagineers" but should, instead, seek to mimic their methods and what they increasingly referred to as "experiences."[115] In 1989, the Association of Science-Technology Centers held its annual meeting in Orlando: the conference's theme—"making science memorable"—explicitly examined the relationship between education and entertainment.

To create memorable experiences for visitors, exhibit designers increasingly used museum halls to simulate total environments.[116] The new simulations that designers created were no mere diorama halls or hanging models; instead, they realized the full-body immersions in science and nature dreamed up by earlier world's fair exhibit designers and refined by Disney's imagineers. Postwar exhibition trends had made simulacra ever more acceptable—no longer would displays need to be tethered to the objects museums had in their collections. Exhibition designers, less encumbered than ever by loyalty to "the real thing," worked to create immersions so grand that they would evoke awe in the face of nature. In the late 1980s, the Milwaukee Public Museum created a Costa Rican rainforest in one hall, building a floor-to-ceiling waterfall in the middle of the gallery in order to display groupings of tropical plants and animals and to lure visitors with the sound of falling water—the rock structure that propped the waterfall up doubled as an amphitheater for videos on tropical biology.[117] In 1994, for instance, visitors entered the new Pacific Northwest Museum of Natural History through a lava tube complete with flowing magma, blasts of hot air, and robotic bats and butterflies. Once inside the museum, visitors wandered through a series of ecosystems of the Northwest, surrounded by the smell of coastal seaweed, the mossy floor of an old-growth forest, and castings from volcanic formations to convey the texture of the high desert's rimrock. To help them re-create the outdoors indoors, the museum had hired Academy Studios, a Marin County–based exhibit fabrication firm established by the former California Academy of Sciences Exhibits chairman Dean Weldon.[118] The ironic result of attempts by designers to create unique experiences through simulated environments was that many museum halls came to resemble not only one another but also hotels, shopping malls, and other commercial ventures.

Though such overlaps made curators squirm, designers argued that they offered more compelling opportunities for learning than drier displays did. While building the exhibits for the Pacific Northwest Museum of Natural History, for instance, Academy Studios was also in the midst of doing research and development on a project for the Hilton in Reno, Nevada. "They actually want to reconstruct a Northwest forest right down the center of the casino, a Northwest forest with a huge waterfall, a big special-effects sky with projected images, lightning bolts that split trees, and about 50 robotic pieces that go into that," explained Weldon. "Museums are being affected by entertainment and attractions, and, inversely, attractions are being influenced by museums."[119] Designers argued this overlap enhanced, rather than betrayed the educational missions of museums. As one industrial designer reflected, education theory and entertainment theory intersected naturally: the museum's "permanent exhibits were designed to get away from the traditional linear approach to learning and to adopt something more spatial. . . . We're working on creating exhibits that almost act like opera sets. You really have a few moments within a large space where you need to create a mood, a feeling, an environment that sets the stage for the information you are going to portray."[120]

The informality and the inquiry-driven graphics of these museum experiences appealed to media-literate children, parents, and teachers. In 1998, for instance, the National Science Teachers Association praised the informal, participatory environments of science centers and museums with immersive displays, noting they were "especially well-suited to learners of mixed-language backgrounds and varied learning styles."[121] Such displays also afforded visitors a new perspective on the natural world. Designers pointed out that immersion into simulated habitats eliminated the psychological sense of superiority over nature that dioramas and neatly arrayed collections of objects tended to elicit and, instead, could make museumgoers appreciate the enormity of the natural world.[122]

In natural history museums, however, immersion remained an educational strategy fraught with a century of tensions over exhibition design. Some curators and educators protested that simulation, no matter how detailed, eliminated the messiness of nature, transforming difficult realities into unrealistic Disneyesque fantasies.[123] Others wondered if wandering through a simulated environment counted as learning. Some suggested immersive displays could be made more meaningful with the intentional insertion of ambiguity, an educational technique that discomfited those who wished to present the museum as a definitive source of knowledge. (As one senior curator planning the Field Museum's *Traveling the Pacific* exhibit declared, "if you can't say it clearly and with conviction, then don't say it at all.")[124]

Whether blockbuster shows and exhibitionary experiences educated visitors properly remained a matter of debate during the 1980s; the toll they took on natural history museums' research programs did not. The impact of blockbusters on research budgets had been blunted somewhat in the mid-1980s as scientific researchers, members of the public, and patrons turned their attention to extinction and DNA-based discoveries. Flush with donations, a number of museums had expanded their scientific research programs as a result—Philadelphia's Academy of Natural Science, the American Museum of Natural History, and the Smithsonian's National Museum of Natural History all built biochemistry labs in the 1980s, and even smaller museums, like the Santa Barbara Museum of Natural History, built new facilities for their scientific staff.[125] But as the economy turned sour again at the end of the decade, administrators at natural history museums confronted a familiar but unpleasant choice: research or exhibition?

As dioramas had in the interwar years, blockbusters and renovated permanent halls presented museums with hard trade-offs. They were surefire ways of raising money, but their heavy costs forced other areas to curtail spending and, in so doing, compromise their scientific mission. Between 1990 and 1993, for instance, the San Diego Museum of Natural History unveiled a new hall of mineralogy, launched an all-out marketing campaign, and introduced a dizzying number of temporary exhibits to its halls.[126] The museum also cut the annual budget of the museum's library to $6,000 and laid off one-third of its scientific staff, including internationally respected life science researchers like Frederick Schram, a paleontologist known for his research on the evolution of crustaceans and Amadeo Rea Jr., an expert on endangered coast bird species.[127] The scientific community was outraged. Though the museum's finances soon stabilized, research remained in decline, and administrators decided to restore only one of the eliminated research jobs after the museum's fortunes had reversed themselves. In 1992, the new director, Michael Hager, agreed that the number of scientists remained "below what is probably critical mass for a scientific culture within the institution" and promised he would increase the scientific staff when the museum could afford to pay their salaries.[128] This scenario repeated itself at museums all over the country. "This museum started with science at the core," Smithsonian botanist Vicki A. Funk told *Science* in 1992. "Everything else was based on the collections and on the science. Education and exhibits came second. And that's not true anymore; exhibits and education have taken on a life of their own."[129]

Recession-strapped government agencies and foundation officers ignored curators' cries for help. Under increasing pressure from taxpayers and board members, they began to fund primarily those projects that would yield visible results almost immediately.[130] "There is less sympathy for fundamental research and more emphasis on societal benefits, short-term benefits, technology transfers," Smithsonian assistant secretary for science Robert Hoffmann noted in 1992.[131] Throughout the 1980s and 1990s, federal agencies and foundations continued to provide museums with grant money but increasingly limited their awards to exhibits and education programs. The NSF, for instance, provided some money to natural history museums for collections, but its contributions to educational programs far outpaced those to scientific research. Between 1988 and 1992, the NSF's informal science education program increased by 130 percent, rising from $15 million in 1989 to $35 million in 1992. The agency's support for collections increased by 66 percent in those same years, from $4.4 to $7.3 million. But the funding for museum research projects in systematic biology increased by only 21 percent, rising from $12.7 million in 1989 to $15.4 million in 1992, leaving American museums wondering how to fund already reduced research agendas.[132] Beginning in the 1970s, in fact, the NSF had slowly moved away from funding research in systematic biology, arguing that there were few obvious direct applications for knowledge. The National Institutes of Health and the Department of Education also made a few grants to researchers available in these years but provided money only "where clinical applications of the work can readily be seen" or in "mission-oriented studies that emphasize the environmental impacts of human activities."[133]

Municipalities further aggravated the financial headaches of museum scientists as they pared museums' operating budgets and pressured the institutions to eliminate collections and the staff who cared for them. In 1994, for instance, Los Angeles politicians floated a plan to force the city's natural history museum to donate the bulk of its fossil collections to universities, including the skull of a sixty-two-inch, half-ton *Purussaurus*, the largest species of crocodile that had ever lived.[134] The museum's scientists were horrified. "There are lessons to be learned by identifying and studying collections," Daniel M. Cohen, deputy director of research and collections for the Los Angeles County Museum of Natural History, protested. "Our exhibits are based on collections, our educational programs are based on collections."[135] City officials were unmoved. "Our concern is how to keep (the museums) open seven days a week," explained city spokesperson Robert Alaniz in 1994. "We are certainly not suggesting that the museum get rid of all its collections . . . but there can be a look at what the priorities should be."[136] Los Angeles was not an isolated case. In 1992, the New York State Museum considered eliminating most of its scientific positions in order to carve $500,000 from a $5.7 million budget. Honolulu's Bishop Museum of Cultural and Natural History hired collection managers, exempting curators from the long-held duties of caring for specimens but also paving the way for the museum to rid itself of more expensive, better-educated scientists. The Smithsonian National Museum of Natural History used attrition to trim staff—as of 1992, administrators had managed to eliminate about 8 percent of staff positions that way.[137]

Scientists within and beyond museums angrily protested the decision by museums to fund exhibition at the expense of research, arguing that starving museum science not only would damage individual institutions but would also retard the progress of socially relevant scientific knowledge.[138] To prove this point, they cited instances of the ongoing relevance of collections to contemporary public health concerns. When a new hantavirus was discovered in 1993, for instance, scientists were quick to note that frozen tissue collections in museums verified the virus's existence at least twenty years earlier and documented its geographical and taxonomical range.[139] Others pointed to the use of biological tissues collected around now-heavily polluted sites like Chernobyl, which could be tested and compared to determine the impact of radiation and heavy metals on ecosystems across time and place. Scientists also defended museum collections as essential tools for modern evolutionary biologists.[140] "As insect-net-wielding curators of a natural history collection, we resent the implication that museum-based research is a dust-laden activity irrelevant to the study of evolution today," wrote Andrew Brower and Darlene Judd in a 1998 response to an article in *Science* suggesting that museums had become scientifically obsolete.[141] "The value of these collections to the science of evolution and to society is immeasurable," wrote museum curators Robert Baker and Terry L. Yates.[142]

Rather than watch their research programs slowly die, curators decided to adapt to this new world order and use it to support their own work. In 1993, for instance, when science centers and natural history and science museums found themselves flooded with more dinosaur-hungry visitors than usual because of the popularity of Steven Spielberg's film *Jurassic Park*, savvy curators scrambled to make the most of the moment. The Dinosaur Society—a group of university and amateur enthusiasts and professional paleontologists working in museums that was determined to expand the tiny pool of funding available for paleontological research—welcomed onto their board of trustees Michael Crichton, who had written both the screenplay and the novel on which the film was based.[143] Members of the Dinosaur Society persuaded Spielberg to lend the society the film's props and fiberglass and latex models of dinosaurs to use in a museum exhibition, *The Dinosaurs of "Jurassic Park."*[144] When the exhibition opened at the American Museum of Natural History, forty thousand people visited it in the first ten days. If attendance were to hold up, predicted the *New York Times*, "it [would] rank among the most popular special exhibitions the museum has mounted, right up there with 'Ancestors' and 'Pompeii A.D. 79.'"[145] (Sales at the museum's shop that year topped $3 million dollars, an increase of 11 percent over the previous year's gross.)[146] The popular show zigzagged across the nation for the next several years, making stops at the San Diego Museum of Natural History, the Pacific Science Center, the Cleveland Museum of Natural History, Chicago's Museum of Science and Industry, and other institutions, while an alternate version of the exhibition toured internationally. The Dinosaur Society split the considerable proceeds with host museums, all of which earned an additional bundle through dinosaur-themed concessions and gifts.

On its face, the *Jurassic Park* exhibit had more to do with props than paleontology, but curators believed they could resist the relentless force of edutainment even as they exploited it. Using a crushed Jeep, an enormous chunk of fake amber, a Barbasol can, an embryo storage tank, storyboards, film clips, computer graphics, and the massive molds of the film's reptilian stars, curators staged a behind-the-scenes look at the film and its extensive special-effects technology, then sprinkled in dinosaur factoids and biological clarifications.[147] Yet throughout the jungle setting of the exhibition, small placards rebutted nearly every scientific premise of the movie. Curator authored labels briskly reminded visitors that there was no direct evidence that ornithomimid dinosaurs traveled in herds, that the head of the film's *Brachiosaurus* was far larger than the fossil evidence indicated it should be, and that it was, as yet, impossible to clone dinosaurs from fragments of DNA lurking in blood consumed by mosquitoes preserved in amber. By the exit, the exhibit's final plaque read: "There would be no Jurassic Park, no recreated dinosaurs for our entertainment, without science."[148] "This is an exhibition that shows signs of awesome struggle, and it's not between flesh-eating giants," wrote *New York Times* critic William Grimes, "Rather, it's the struggle between real science and the juiced-up version presented in the film." American Museum paleontologist Mark Norell, the exhibit's cocurator and an active Dinosaur Society member, was well aware of these tensions. "This may look like an extended commercial for the film," he admitted to Grimes, "but my view is that we can use this to teach people about science."[148]

Whether or not the show taught people about science, it certainly funded scientific research. By 1996, nearly two million people in five different countries had seen the exhibition. The proceeds of the show had exponentially increased the Dinosaur Society's income, generating more than $2 million in annual revenue and allowing the Dinosaur Society to fund nearly 20 percent of all worldwide research in paleontology. The paleontologists who ran the Dinosaur Society were careful to not only praise Spielberg but also nod to the public tastes that had sustained their work: "These exhibits have allowed us to generate money for scientists, and scientific discoveries made can perhaps allow someone to make another movie," one scientist tactfully told reporters.[150]

Curators and administrators, emboldened by this success, soon after began to search for new ways to fund research via exhibition, developing cross-institutional financial partnerships, educational programming, and display development protocols in the process. In 1997, when the largest

Figure 24   The blockbuster exhibit *T. rex Sue*, took many forms: casts were replicated from the original fossil and circulated beyond Sue's permanent home at Chicago's Field Museum, with the goal of creating a variety of popular and educational experiences that could be fit-to-site. When Sue arrived in Washington, DC, in 2000, she was installed in Union Station, along with simulated crates from the fossil dig, right outside of a family-friendly Pizzeria Uno restaurant. (Mark Wilson for Newsmakers/Getty Images.)

and most complete *Tyrannosaurus rex* skeleton known to science went on the auction bloc, representatives from Chicago's Field Museum partnered with the California State University System, Walt Disney Parks and Resorts, McDonald's, and a number of individual donors to purchase it for a cool $7.6 million.[151] Affectionately dubbed "Sue" after fossil hunter Sue Hendrickson, who discovered the skeleton, the fossil served many masters. McDonald's and Walt Disney used Sue to open the hearts, mouths, and wallets of American families with dinosaur-obsessed offspring. ("We saw unlimited potential," declared chief executive officer Jack Greenberg, "in the education area, which has always been a top priority for McDonalds.")[152] Museum scientists, in turn, used their corporate sponsors to open new research labs—one at the Field Museum, the other at Disney World's newly opened Animal Kingdom—to study Sue.

During Sue's preparation, the scientific process itself went on display, which pleased all parties. The public watched through large windows as paleontologists and preparators painstakingly pried bone from rock, trying to get to a point where they could interpret what Field Museum geology collections manager Bill Simpson called "the Rosetta Stone for anybody who wants to study *Tyrannosaurus rex* and other large thero-pods." To build buzz for Sue's public unveiling at the Field, the museum put annotated footage of this work on the Field Museum's website. When the skeleton was finally displayed in May 2000, the museum clocked ten thousand visitors on opening day alone. "She has drawn tremendous attention to Chicago, and she is not going anywhere," the Chicago De-partment of Tourism director told the media. "She is something you can only experience in Chicago."[153] Chicago boosters discounted the fact that visitors across the globe could experience satisfyingly close approxima-tions of the *T. rex* elsewhere beyond Chicago. A cast of her skeleton was displayed at Disney's Animal Kingdom in Florida, and two others toured the country, sponsored by McDonald's; ultimately, 6.5 million visitors in forty-three U.S. cities and fourteen foreign countries saw copies of Sue's bones. By 2000, a cast of Sue arrived in Washington, DC, and was mounted at Union Station rather than at the Smithsonian (fig. 24).[154]

What Is a Museum? Defining and Defending Public Science

By the end of the 1990s, museums of nature and science seemed more unified in mission and appearance than ever before. This unification was, in many ways, the price and product of complexity. Throughout the twentieth century, new sources of funding and more varied groups of visitors had complicated the aims of museums. Specialized staff members multiplied their ambitions and stakeholders, which now included a diz-zying array of groups: scientists, directors, exhibit designers, educators, evaluators, patrons, government agencies, foundations, corporations, and, of course, visitors. Educational and financial crises, along with the enveloping influence of contemporary design and commerce and ever-changing ideas about what defined the biological and life sciences, had forced museums to reconfigure their appearances and their institutional taxonomies. And museums had responded by developing a shared set of educational practices, values, and goals.

In a 1999 article addressing the blooming, buzzing, confusing his-tory of museums, Neil Harris asked, "What are we to make, then, of the position of museums in modern American life? . . . How are we to measure the impact of the museum experience, or even to talk about so diverse a universe of institutions in any categorical way?"[155] As inter-nally and culturally contradictory institutions, do museums and science centers still have the power to, as Harris put it, "shape anyone's values, validate anyone's identity, impose any lasting sort of order?"[156] In this book, we have tackled a small corner of Harris's query, chronicling the social history of their exhibits to explain how science and natural history museums have conceived and reconceived their institutional roles in relation to one another, as well in relation to life science and society in the twentieth-century United States. As a result, we believe that a few categorical statements can be made about these diverse institutions and their transformation over the past hundred years.

First, if collections were the defining trait of a museum in the nineteenth century, by the end of the twentieth, displays had become their distinguishing characteristic. Though public education through exhibition has long been a declared goal of museums of science and nature in the United States, we have shown how and why it assumed new importance over the last several decades.[157] In the 1890s, curators and directors like Carnegie Museum director William J. Holland could declare that they had no interest in visitors who "seemed to make no serious study" of museums' scientific collections. Neither did they need to bother, Holland and his peers believed, with younger visitors who "came frequently but seemed rather to take a general view than to make a careful inspection."[158] A century later, it was unthinkable that a director—indeed, that any staff member of a natural history museum, science museum, or science center—would take such a stance, much less announce it in a public forum.

Second, as these museums attempted to educate visitors, they worked to balance their wealth of professional expertise—scientific, educational, and aesthetic—with public participation. As the twentieth century progressed, museums tried to elicit new kinds of visitor participation and became more sensitive to public interests. This transformed museums' social and cultural roles. Though George Brown Goode foresaw public museums' potential to become important institutions for the promotion and dissemination of scientific knowledge, neither he nor the museum men he inspired could have imagined the efforts of contemporary museums to redefine, in collaboration with one another and their visitors, what has come to be called informal science.[159] Elements of informal science were first envisioned by Progressive Era educators and museum staff and, later, refined by Oppenheimer and his exhibit designers, but contemporary directors, curators, educators, and designers are once again actively rethinking the meaning of noncompulsory interactive engagement in the museum space. Science centers, for instance, now regularly host large community debates about climate change, fracking, and other controversial scientific issues in their exhibit galleries.[160] They also use new technologies to recruit online users of the museum and create a virtual architecture of participation in which these new visitors generate, share, and curate exhibition content.[161] Natural history museums have relied on a more top-down approach to achieve similar ends. Using both staff interactions and new exhibit arrangement, for example, they have more actively confronted creationist challenges to evolutionary biology in the last few years. Through displays that explore the controversies surrounding the 2004 *Kitzmiller v. Dover Area School District* case about intelligent design, exhibit designer and educator Judy Diamond argues, museums can empower visitors to be "better able to handle it when it comes up." "Museums, as a field, we have recognized, have to take a more proactive role in evolution education," Diamond explained. "Just telling visitors they are wrong is not going to be effective."[162]

Third, the balancing act first introduced by the New Museum Idea continues, and it is through this work, not in spite of it, that science and natural history museums have found their institutional calling in the new millennium.[163] As we have shown, deliberations over what constituted effective science museum displays both reflected and contributed to ongoing political conversations about the role museums would play in shaping relations between sciences and their publics. Progressive Era natural history museum workers and civic leaders alike worked to transform museums into institutions for training a scientifically literate citizenry. During the Cold War, Washburn and Oppenheimer disagreed whether exposure or interaction was the best approach to science education, but they shared a belief that museums could and should shoulder some responsibility for recruiting the next generation of scientists and engineers.[164]

By the end of the twentieth century, museums had largely succeeded in their attempts to reform themselves. This is not to say that museum workers—or anyone else, for that matter—are satisfied with their exhibits or the processes surrounding their construction. As we've demonstrated, museum displays have elicited all manner of debate about the nature of science, the ends of education, and the ethics of display. But this cycle of discussion and dispute offers the museum world recurring opportunities to reflect on these institutions. Through mounted elephants and walk-through hearts, chick hatcheries and grassoppers' nervous systems, museums have allowed Americans to visualize aspects of increasingly specialized categories of knowledge and to grapple with a set of disciplinary and representational practices that shape everyday life. In turn, a diverse group of Americans has allowed American museums to reinvent their missions.

On the whole, visitors have responded to the outcome with enthusiasm. According to the American Alliance of Museums, more Americans visit museums each year than attend all major league sporting events and theme parks combined.[165] Though G. K. Chesterton concluded in 1931 that museums were "not a product of popular imagination, but of what is called popular education; the cold and compulsory culture which is not, and never will be, popular," he got it wrong on both counts.[166] Museums of nature and science will remain enormously popular with Americans, who rightly consider them sites for imagining and reimagining public science.

Notes

ABBREVIATIONS FOR ANNUAL REPORTS, HOUSE PUBLICATIONS,
AND PROCEEDINGS CONSULTED

AMNHAR
> American Museum of Natural History annual report, American Museum of Natural History Library, New York (the actual title of the report has appeared variously as *Annual Report of the American Museum of Natural History*; *American Museum of Natural History Annual Report*; *American Museum of Natural History Annual Report of the President*; and *Annual Report of the Trustees of the American Museum of Natural History*)

BMoSAR
> *Boston Museum of Science* annual report, Lyman Library, Boston

CMAR
> Carnegie Museum annual report, Carnegie Museum of Natural History Library, Pittsburgh (the actual title of the report has appeared variously as *Report of the Carnegie Museum*; *Annual Report of the Carnegie Museum*; and *Report of the Directors of the Carnegie Museum*)

CMNHAR
> Colorado Museum of Natural History annual report (the actual title of the report has appeared variously as *Annual Report of the Director to the Board of Trustees of the Colorado Museum of Natural History*; *Articles of Incorporation, By-Laws and Annual Report of the Director to the Board of Trustees of the Colorado Museum of Natural History*; and *Annual Report of the Colorado Museum of Natural History*)

DMNHAR

Denver Museum of Natural History annual report, Bailey Library and Archives, Denver Museum of Nature and Science, Denver (the actual title of the report has appeared variously as *Annual Report of the Denver Museum of Natural History*; *Annual Report for the Denver Museum of Natural History*; *Denver Museum of Natural History Annual Report*; and *Annual Report, Denver Museum of Natural History*)

FMNHAR

Field Museum of Natural History annual report, Field Museum Library, Chicago (the actual title of the report has appeared variously as *Annual Report of the Director to the Board of Trustees, Field Museum of Natural History*; *Report of the Director to the Board of Trustees, Chicago Natural History Museum*; *Annual Report, Field Museum of Natural History*; *Field Museum of Natural History Annual Report*; *Report, Field Museum of Natural History*; and *Field Museum of Natural History Biennial Report*)

LAMAR

Los Angeles County Museum of Science, History and Art annual report, Research Library of the Natural History Museum of Los Angeles, Los Angeles

PAAM

Proceedings of the American Association of Museums, archive.org

PANS

Proceedings of the Academy of Natural Sciences of Philadelphia, Academy Library and Archives, Academy of Natural Sciences of Drexel University, Philadelphia

PDANS

Proceedings of the Davenport Academy of Natural Sciences, Putnam Museum of History and Natural Science Library, Davenport, IA

PMOSI

Progress, Museum of Science and Industry Archives, Chicago

SIAR

Smithsonian Institution annual report, Smithsonian Institution Archives, Washington, DC (the actual title of the report has appeared variously as *Annual Report of the Board of Regents of the Smithsonian Institution* and *Annual Report of the Smithsonian Institution*)

YBANS

Year Book of the Academy of Natural Sciences of Philadelphia, Academy Library and Archives, Academy of Natural Sciences of Drexel University, Philadelphia

AMNH

American Museum of Natural History Archives, American Museum of Natural History Library, Special Collections, New York

- DM-AMNH

  Department of Mammalogy Administrative Papers, Archival Collections, Department of Mammalogy, Division of Vertebrate Zoology, American Museum of Natural History, New York
- DMAP-AMNH

  Department of Mammalogy Administrative Papers, American Museum of Natural History Archives, American Museum of Natural History Library, Special Collections, New York

BL-UCB

Exploratorium Collection and the Frank Oppenheimer Papers, Bancroft Library, University of California, Berkeley

BMoS

Boston Museum of Science Archives (2005), Boston Museum of Science Library

CMNH

Colorado Museum of Natural History Archives, Bailey Library and Archives, Denver Museum of Nature and Science, Denver

- CH-CMNH

  Charles Hannington Papers
- FT-CMNH

  Frank Taylor Papers
- JDF-CMNH

  Jesse D. Figgins Papers

MSIC

Museum of Science and Industry Archives, Chicago

DANS

Davenport Academy of Natural Sciences Archives, Putnam Museum of History and Natural Science Library, Davenport, IA

DPL

Special Collections, Western History and Genealogy Division, Denver Public Library

FAL

Frederic Augustus Lucas (PP), Archives, Lu Esther T. Mertz Library, The New York Botanical Garden, Bronx, NY

LAM

Files of the Office of the Director, Museum Archives, Natural History Museum of Los Angeles County

MPM

Milwaukee Public Museum Archives, Milwaukee Public Museum Library and Archives

- CF-MPM
  Correspondence Files
- DF-MPM
  Director's Files

PM

Putnam Museum Archives, Putnam Museum, Davenport, IA

PSC

Pacific Science Center Archives, Pacific Science Center Library Services, Seattle

SIA

Historical Records of the National Museum and the National Museum of Natural History, Smithsonian Institution, Smithsonian Institution Archives, Washington, DC

INTRODUCTION

1. G. K. Chesterton, *All Is Grist: A Book of Essays* (Freeport, NY: Books for Libraries Press, 1967), 196–202. On the ways Chesterton's trip to America shaped his views, see Ian Ker, *G. K. Chesterton: A Biography* (New York: Oxford University Press, 2011), 115–58.

2. For a useful disquisition on the etymology of the term museum and a general overview of the early history of the public museum, see Jeffrey Abt, "The Origins of the Public Museum," in *A Companion to Museum Studies*, ed. Sharon Macdonald (Malden, MA: Blackwell Publishing, 2011).

3. Paula Findlen, *Possessing Nature: Museums, Collecting, and Scientific Culture in Early Modern Italy* (Berkeley, CA: University of California Press, 1994); Oliver Impey and MacGregor, eds., *The Origins of Museums* (Oxford: Clarendon, 1985).

4. See, e.g., the essays in sec. 3 ("Discipline, Discovery, and Display") in N. Jardine, J. A. Secord, and E. C. Spary, eds., *Cultures of Natural History* (Cambridge: Cambridge University Press, 1996); E. C. Spary, *Utopia's Garden: French Natural History from Old Regime to Revolution* (Chicago: University of Chicago Press, 2000); R. F. Ovenell, *The Ashmolean Museum, 1683–1894* (Oxford: Clarendon Press, 1986); E. Miller, *That Noble Cabinet: A History of the British Museum* (Athens: Ohio University Press, 1974).

5. American universities and academies of science also maintained museums throughout the nineteenth century, but these institutions hewed to narrower definitions of the term, reserving their specimens and spaces for dedicated students and scholars. Few such museums permitted members of the public to wander through them. Not that hordes of people were clamoring to do so—such museums could be profoundly dull. As an 1883 editorial in the Syracuse University newspaper observed, "over one-half the students regard the Museum as a mysterious, gloomy, dusty affair that is a necessity to a University but of no meaning or use to them" (as cited in Sally Gregory Kohlstedt, "Museums on Campus: A Tradition of Inquiry and Teaching," in *The American Development of Biology*, ed. Ronald Rainger, Keith R. Benson, and Jane Maienschein [Philadelphia: University of Pennsylvania Press, 1988], 30). On museums of universities and academies of science in the nineteenth-century United States, also see Sally Gregory Kohlstedt, "From Learned Society to Public Museum: The Boston Society of Natural History," in *The Organization of Knowledge in Modern America, 1860–1920*, ed. Alexandra Oleson and John Voss (Baltimore, MD: Johns Hopkins University Press, 1979), "Curiosities and Cabinets: Natural History Museums and Education on the Antebellum Campus," *Isis* 79, no. 3 (1988): 405–26, and "Parlors, Primers, and Public Schooling: Education for Science in Nineteenth-Century America," *Isis* 81, no. 3 (1990): 424–45; Mary P. Winsor, *Reading the Shape of Nature: Comparative Zoology at the Agassiz Museum*, ed. David L. Hull, Science and Its Conceptual Foundations (Chicago: University of Chicago Press, 1991); George F. Goodyear, *Society and Museum: A History of the Buffalo Society of Natural History, 1861–1993, and the Buffalo Museum of Science, 1928–1993*, Bulletin of the Buffalo Society of Natural History, 34 (Buffalo, NY: Bulletin of the Buffalo Society of Natural History, 1994); Daniel Goldstein, "Outposts of Science: The Knowledge Trade and the Expansion of Science in Post-Civil War America," *Isis* 99

(2008): 573–99; Victoria Cain, "From Specimens to Stereopticons: The Persistence of the Davenport Academy of Natural Sciences and the Emergence of Scientific Education," *Annals of Iowa* 68 (2009): 1–36.

6. Though scholars have long described Peale and Barnum in opposition to one another, recent scholars have focused on the way their intentions and institutions overlapped. See, e.g., Les Harrison, *The Temple and the Forum: The American Museum and Cultural Authority in Hawthorne, Melville, Stowe, and Whitman* (Tuscaloosa: University of Alabama Press, 2007). For more specific histories of Peale and Barnum's respective museums, see Toby A. Appel, "Science, Popular Culture, and Profit: Peale's Philadelphia Museum," *Journal of the Society for the Bibliography of Natural History* 9 (1980): 619–34; Neil Harris, *Humbug: The Art of P. T. Barnum*, Phoenix ed. (Chicago: University of Chicago Press, 1981); James W. Cook, *The Arts of Deception : Playing with Fraud in the Age of Barnum* (Cambridge, MA: Harvard University Press, 2001); Andrea Stulman Dennett, *Weird and Wonderful: The Dime Museum in America* (New York: New York University Press, 1997). For more general information on the history of American museums before 1870, see Joel J. Orosz, *Curators and Culture: The Museum Movement in America, 1740–1870* (Tuscaloosa: University of Alabama Press, 1990).

7. On the founding of municipal public museums in these decades, see Steven Conn, *Museums and American Intellectual Life, 1876–1926* (Chicago: University of Chicago Press, 1998); Helen Lefkowitz Horowitz, *Culture and the City: Cultural Philanthropy in Chicago* (Chicago: University of Chicago Press, 1976); P. H. Oehser, *The Smithsonian Institution*, 2nd ed. (Boulder, CO: Westview Press, 1983); John Michael Kennedy, "Philanthropy and Science in New York City: The American Museum of Natural History, 1868–1968" (PhD diss., Yale University, 1968); Nancy Oestreich Lurie, *A Special Style: The Milwaukee Public Museum, 1882–1982* (Milwaukee: Milwaukee Public Museum, 1983).

8. Chesterton, *All Is Grist*, 200.

9. Ibid., 202.

10. Duncan F. Cameron, "The Museum: A Temple or a Forum?" *Curator* 14 (1971): 11–24, esp. 17. Les Harrison characterizes, then takes issue with, Cameron's "cultural warfare" formulation, arguing instead that "the distinction between temple and forum has never been as absolute as [this] language would imply." See the preface to Harrison, *The Temple and the Forum*, xv.

11. Museum historian Joel Orosz has since characterized this initial accommodation of multiple missions as the emergence of an "American compromise" (*Curators and Culture*).

12. For more on practice-oriented history of science, especially in relation to the history of biology, see the section titled "History of Biology in the Twentieth Century" in the bibliographic essay at the end of this book.

13. Richard R. John, "Why Institutions Matter," *Common-place* 9, no. 1 (2008), http://www.common-place.org/vol-09/no-01/john/. The foundational text

of the new institutionalism is Theda Skocpol, "Bringing the State Back In: Strategies of Analysis in Current Research," in *Bringing the State Back In*, ed. Theda Skocpol, Peter B. Evans, and Dietrich Rueschemeyer (New York: Cambridge University Press, 1985), 3–37. On the importance of institutions for the history of science, see Sally Gregory Kohlstedt, "Institutional History," in "Historical Writing on American Science," ed. Sally Gregory Kohlstedt and Margaret W. Rossiter, special issue, *Osiris*, 2nd ser., vol. 1, no. 1 (1985): 17–39.

14. Everett Hughes, *The Sociological Eye: Selected Papers* (New Brunswick, NJ: Transaction Books, 1984). See also Richard Langlois, "*The Institutional Revolution*: A Review Essay," *Review of Austrian Economics* 26 (2013): 383–95.

15. For further discussion of literature and historiography on this theme, see the sections titled "Science, Technology, and Nationalism" and "Public Science, Popular Science" in the bibliographic essay at the end of this book.

16. Andrea Witcomb, *Reimagining the Museum: Beyond the Mausoleum* (New York: Routledge, 2003), 89.

17. Tony Bennett, *The Birth of the Museum: History, Theory, Politics* (London: Routledge, 1995).

18. Thomas Bender, "Reforming the Disciplines," *Daedalus* 131, no. 3 (2002): 59–62.

19. Alan Friedman, "The Evolution of Science and Technology Museums," *Informal Science Review* 17, no. 1 (1996): 14–17.

20. Sharon Macdonald, "Exhibitions of Power and Powers of Exhibition," in *The Politics of Display: Museums, Science, Culture*, ed. Sharon MacDonald (New York: Routledge, 1998), 9–15.

21. Orosz, *Curators and Culture*.

22. See, e.g., Nina Simon, *The Participatory Museum* (Santa Cruz, CA: Museum 2.0, 2010).

23. Sally Gregory Kohlstedt, "Thoughts in Things: Modernity, History, and North American Museums," *Isis* 96, no. 4 (2005): 587–95.

CHAPTER ONE

1. Mary Antin, *The Promised Land* (1912; reprint, New York: Penguin Books, 1997), 252. Barbour is quoted in Mary Desmond Rock, *Museum of Science, Boston: The Founding and Formative Years: The Washburn Era, 1939–1980* (Boston: The Museum, 1989), 5.

2. Robert Ridgway, letter to Witmer Stone, June 29, 1893, as cited in Robert E. Kohler, *All Creatures: Naturalists, Collectors and Biodiversity, 1850–1950* (Princeton, NJ: Princeton University Press, 2006), 108. On Davenport, see W. H. Pratt, letter to Mary Louisa Duncan Putnam, May 23, 1890, DANS; on schoolteachers at the American Museum of Natural History, see Clark Wissler, "Survey of the American Museum of Natural History: Made at the Request of the Management Board," unpublished typescript, 1943, p. 5, AMNH.

3. Frank M. Chapman, "The Educational Value of Bird Study," *Educational Review* 17 (1899): 242.

4. George Brown Goode, "The Museums of the Future," in *A Memorial of George Brown Goode: together with a selection of his papers on museums and on the history of science in America,* eds. Samuel Pierpont Langley, George Brown Goode, Randolph Iltyd Geare, and Joint Commission of the Scientific Societies of Washington. (Washington, DC: U.S. National Museum, Government Printing Office, 1901), 243. F. A. Lucas, "The Evolution of Museums," *PAAM* 1 (1907): 85.

5. Barton W. Evermann, "Modern Natural History Museums and Their Relation to Public Education," *Scientific Monthly* 6 (1918): 9.

6. On the Smithsonian's early nationlist collecting efforts, see Pamela Henson, "A National Science and a National Museum," in *Museums and Other Institutions of Natural History: Past, Present and Future*, ed. Alan E. Leviton and Michele L. Aldrich (San Francisco: California Academy of Sciences, 2004), 34–57; Sally Gregory Kohlstedt, "Place and Museum Space: The Smithsonian Institution, National Identity, and the American West, 1846–1896," in *Geographies of Nineteenth-Century Science*, ed. David N. Livingstone and Charles W. J. Withers (Chicago: University of Chicago Press, 2011); Philip J. Pauly, *Biologists and the Promise of American Life: From Meriwether Lewis to Alfred Kinsey* (Princeton, NJ: Princeton University Press, 2000), 45–70.

7. In this way, Smithsonian naturalists encouraged their colleagues to participate in an American iteration of what Lynn Nyhart has dubbed "civic zoology" in nineteenth-century Germany: Lynn K. Nyhart, "Civic and Economic Zoology in Nineteenth-Century Germany: The 'Living Communities' of Karl Mobius," *Isis* 89, no. 4 (1998): 605–30. See also Daniel Goldstein, " 'Yours for Science': The Smithsonian Institution's Correspondents and the Shape of Scientific Community in Nineteenth-Century America," *Isis* 85, no. 4 (1994): 573–99; Juan Francisco Ilerbaig, "Pride in Place: Fieldwork, Geography, and American Field Zoology, 1850–1920" (PhD diss., University of Minnesota, 2002).

8. Robert Ridgway, letter to Witmer Stone, June 29, 1893, as cited in Kohler, *All Creatures*, 108. Regarding Ward's influence on Lucas, see Mary Anne Andrei, "Nature's Mirror: How the Taxidermists of Ward's Natural Science Establishment Reshaped the American Natural History Museum and Founded the Wildlife Conservation Movement" (PhD diss., University of Minnesota, 2006).

9. On George Brown Goode, see *The Origins of Natural Science in America: The Essays of George Brown Goode*, ed. Sally Gregory Kohlstedt (Washington, DC: Smithsonian Institution Press, 1991); Samuel Pierpont Langley, "Memoir of George Brown Goode," in *A Memorial of George Brown Goode, Together with a Selection of His Papers on Museums and on the History of Science in America*, ed. Samuel Pierpont Langley et al. (Washington, DC: U.S. National Museum, 1901).

10. The quote from George Brown Goode can be found in his "Museum-

History and Museums of History," in *The Origins of Natural Science in America*, ed. Kohlstedt, 307. For Goode's earlier, more comprehensive vision, see George Brown Goode, "Plan of Organization and Regulations," in *Circular of the U.S. National Museum*, no. 1 (1881): 1–58. Though Goode often limited his description of the official aims of natural history museums to "record, research, and publication," his belief that museums should instruct the broader public was widely acknowledged: see F. A. Lucas, "Evolution of the Educational Spirit in Museums," *American Museum Journal* 11, no. 7 (1911): 228.

11. Goode, "The Museums of the Future," 433.

12. South Kensington museum director W. H. Flower coined the phrase "New Museum Idea" in 1893 to describe the public-minded philosophies and practices promoted by J. E. Gray, the British Museum's keeper of zoology. On the impact of the New Museum Idea in Europe, see Annie E. Coombes, "Museums and the Formation of National and Cultural Identities," *Oxford Art Journal* 11, no. 2 (1988): 57–68; Lynn K. Nyhart, *Modern Nature: The Rise of the Biological Perspective in Germany* (Chicago: University of Chicago, 2009), esp. 198–250. On U.S. precursors, see Orosz, *Curators and Culture* (see n. 6 above, "Introduction"); Charles Coleman Sellers, "Peale's Museum and 'the New Museum Idea,'" *Proceedings of the American Philosophical Society* 124, no. 1 (1980): 25–34.

13. Langley, "Memoir of George Brown Goode," 152–54; F. A. Lucas, *Fifty Years of Museum Work: Autobiography, Unpublished Papers and Bibliography* (1911; reprint, New York: American Museum of Natural History, 1933), 16.

14. Dividing the collections and putting them into separate spaces reflected and reinforced the reorganization of Americans' relationship to natural science in this era. Just as lay and professional science had separated outside of the museum, dual arrangement, as Lynn Nyhart has observed, abruptly restructured this continuous spectrum of people using collections of natural objects into two categories of visitors—"experts and laypeople—and two kinds of knowledge—esoteric research knowledge and popular knowledge" *Modern Nature*, 223–39).

15. Lucas, *Fifty Years of Museum Work*, 15–16.

16. Henry L. Ward, "The Aims of Museums, with Special Reference to the Public Museum of the City of Milwaukee," in *Second Annual Meeting of the American Association of Museums* (n. 4 above, this chap.), 100.

17. Langley, "Memoir of George Brown Goode," 154; Goode, "The Museums of the Future," 427.

18. On concerns over the status in field- and collections-based research in this era, see Robert E. Kohler, *Landscapes and Labscapes: Exploring the Lab-Field Border in Biology* (Chicago: University of Chicago Press, 2002), 4–5 and 23–59.

19. On museum leadership before 1890, see Susan Sheets-Pyenson, *Cathedrals of Science: The Development of Colonial Natural History Museums during the Late Nineteenth Century* (Kingston, Ontario: McGill-Queen's University

Press, 1988). Around 1900, museum boards agreed more with naturalist Alfred Mayer, who argued that a strong executive "appear[s] to be necessary to insure the success of American institutions of learning." Alfred Goldsborough Mayer, "The Status of Public Museums in the United States," *Science* 17, no. 439 (1903): 844.

20. On museum preparators, and especially taxidermists, in this era, see Andrei, "Nature's Mirror"; Kohler, *All Creatures*, 210–15; Susan Leigh Star, "Craft vs. Commodity, Mess vs. Transcendence: How the Right Tool Became the Wrong One in the Case of Taxidermy and Natural History," in *The Right Tools for the Job: At Work in Twentieth-Century Life Sciences*, ed. Adele E. Clarke and Joan H. Fujimura (Princeton, NJ: Princeton University Press, 1992), 264.

21. On Osborn's politics, see Constance Clark, *God—or Gorilla: Images of Evolution in the Jazz Age* (Baltimore, MD: Johns Hopkins University Press, 2008); Ronald Rainger, *An Agenda for Antiquity: Henry Fairfield Osborn and Vertebrate Paleontology at the American Museum of Natural History, 1890–1935* (Tuscaloosa: University of Alabama Press, 1991).

22. On Allen's commitment to conservation, see Mark V. Barrow Jr., *Nature's Ghosts: Confronting Extinction from the Age of Jefferson to the Age of Ecology* (Chicago: University of Chicago Press, 2009), 78–107; Frank M. Chapman, "Joel Asaph Allen, 1838–1921," *Memoirs of the National Academy of Science* 21 (1927): 1–20. Also see Joel A. Allen, *Autobiographical Notes and a Bibliography of the Scientific Publications of Joel Asaph Allen* (New York: American Museum of Natural History, 1916).

23. See Frank M. Chapman, "The Educational Value of Bird Study," and *Autobiography of a Bird-Lover* (New York: D. Appleton-Century, 1935). On Dickerson, see Maud Slye, "Mary Cynthia Dickerson, 1866–1923, Her Life and Personality," *Natural History* 23, no. 5 (1923): 507–9. Also: Mary Cynthia Dickerson, "The New African Hall Planned by Carl E. Akeley: Principles of Construction Which Strike a Revolution in Methods of Exhibition and Presage the Future Greatness of the Educational Museum," *American Museum Journal* 14, no. 5 (1914): 175–87, and "The 'Toad Group' in the American Museum: A Word as to Its Composite Construction and Interest," *American Museum Journal* 15, no. 4 (1915): 163–66.

24 Steven Conn, *Museums and American Intellectual Life, 1876–1926* (Chicago: University of Chicago Press, 1998), 32–73; Kohlstedt, "Thoughts in Things" (n. 23 above, "Introduction").

25. N. H. Winchell, "Museums and Their Purposes," *Science* 18, no. 442 (1891): 43. Holland quoted in Louis Pope Gratacap, "Natural History Museums," pt. 2, *Science* 7, no. 185 (1898): 67.

26. Louis Pope Gratacap, "The Development of the American Museum of Natural History," *American Museum Journal* 1, no. 1 (1900): 2–4. Gratacap, "Natural History Museums," pt. 2, 68.

27. Henry L. Ward, "The Anthropological Exhibits in the American Museum of Natural History," *Science* 25, no. 645 (1907): 745–46.

28. Henry L. Ward, letter to Charles Dury, January 20, 1906, DF-MPM.

29. Frederic A. Lucas, "Museums and Libraries," in *Fifty Years of Museum Work* (n. 13 above, this chap.).

30. On the conviction held by curators concerning the virtue of visual knowledge in this period, see Tony Bennett, *Pasts beyond Memory: Evolution, Museums, Colonialism* (London: Routledge, 2004), 160–86.

31. A. H. Griffiths, "Discussion—Do Museum Lectures Pay?" in *PAAM* 2 (1908): 100.

32. On curators' interest in the pedagogical potential of commercial visual culture, see Victoria Cain, " 'Attraction, Attention, Desire': Consumer Culture as Pedagogical Paradigm in Museums in the United States, 1900–1930," *Paedagogica Historica* 48, no. 5 (2012): 215–38.

33. Edward S. Morse, "If Public Libraries, Why Not Public Museums?" *Atlantic Monthly* (July 1893), 114.

34. See George E. DeBoer, *A History of Ideas in Science Education: Implications for Practice* (New York: Teachers College Press, 1991), 1–127; Herbert M Kliebard, *The Struggle for the American Curriculum, 1893–1958* (New York: RoutledgeFalmer, 2004), 4–6; W. B. Kolesnik, *Mental Discipline in Modern Education* (Madison: University of Wisconsin Press, 1958). See also Noah Sobe, "Challenging the Gaze: The Subject of Attention and a 1915 Montessori Demonstration Classroom," *Educational Theory* 54, no. 3 (2004): 281–97.

35. On object lessons and observational training in nineteenth-century American schools, see Sarah Anne Carter, "Object Lessons in American Culture" (PhD diss., Harvard University, 2010); Sally Gregory Kohlstedt, "Parlors, Primers, and Public Schooling: Education for Science in Nineteenth-Century America," *Isis* 81, no. 3 (1990): 424–45; and *Teaching Children Science: Hands-on Nature Study in North America, 1890–1930* (Chicago: University of Chicago Press, 2010), 27–34, 42–47. On the uses of natural objects in the German context, see Nyhart, *Modern Nature* (n. 12 above, this chap.), 161–97. On experience with close observation of specimens by curators during their college or graduate training, see, among others, Allen, *Autobiographical Notes* (n. 22 above, this chap.), 8–9. Henry Fairfield Osborn, *Creative Education in School, College, University, and Museum: Personal Observation and Experience of the Half-Century, 1877–1927* (New York: Charles Scribner's Sons, 1927), 23. On observational practices in the field and the studios of museums, see, among others, Star, "Craft vs. Commodity" (n. 20 above, this chap.).

36. Oliver C. Farrington, "Museum Study by Chicago Public Schools," *Science* 15 (January 31, 1902): 183–84.

37. Elizabeth Carss, letter to Frank Chapman, March 23, 1904, Department of Public Instruction, 1904 folder, Children's Lecture Schools C, Administrative Records, AMNH.

38. William C. Maxwell, "Cooperation in Education," *American Museum Journal* 11, no. 7 (1911): 219.

39. The AAM diversified as the years passed, but members continued to complain that natural history museums dominated the discussion—and the ideology—of this professional association. See William J. Holland, *Organization and Minutes of the First Meeting of the American Association of Museums* (New York: American Association of Museums, 1906), 2, 38; "The American Association of Museums," *Science,* n.s., vol. 23, no. 596 (June 1, 1906), 859.

40. See William Orr, "The Opportunity of the Smaller Museums of Natural History," *Popular Science Monthly* 63 (1903): 52.

41. For a typical description of this kind of emulation, see Evermann, "Modern Natural History Museums" (n. 5 above, this chap.).

42. Harlan G. Smith, letter to Frank M. Taylor, May 18, 1910, Frank Taylor Correspondence, CMNH.

43. Goode, "Museum-History and Museums of History" (n. 10 above, this chap.), 308.

44. On Griffin, see Sally Gregory Kohlstedt, "Innovative Niche Scientists: Women's Role in Reframing North American Museums, 1880–1930," *Centaurus* 55, no. 2 (2013): 153–74. The best source on Dickerson's career remains Charles W. Myers, "A History of Herpetology at the American Museum of Natural History," *Bulletin of the American Museum of Natural History* 252 (2000): 205. On women's struggle to achieve recognition in early twentieth-century museums, see Leslie Madsen-Brooks, "To Study, to Control and to Love: Women Scientists in American Natural History Institutions, 1880–1950" (PhD diss., University of California, Davis, 2006), and "Challenging 'Top-Down' Science? Women's Participation in American Natural History Museum Work, 1870–1920," *Journal of Women's History* 21, no. 2 (2009): 11–38.

45. Quote from Lucas, *Fifty Years of Museum Work* (n. 13 above, this chap.), 4. Contemporaries would later occasionally deploy the term with ire, but it was usually used as high praise.

46. Museum man Frederic Lucas, e.g., credited his interest in museum reform to his "liking for mechanical work" and the pleasure he took in "seeing exhibits grow under my hands." Even the most patrician of them had an artistic bent: curator and paleontologist Henry Fairfield Osborn, heir to a railroad fortune and the president of the American Museum for twenty-five years, had organized a sketching club during his Princeton days and frequently commissioned well-known artists to create pieces for the museum. On Lucas's and Osborn's interests and talents, see, respectively, Lucas, *Fifty Years of Museum Work*, 23; and Clark, *God—or Gorilla* (n. 21 above, this chap.), 21.

47. Langley, "Memoir of George Brown Goode" (n. 9 above, this chap.), 9.

48. Journals would later take the place of these informal networks of knowledge, but early in the twentieth century, the American museum community lacked the kind of publications on which other professions relied. After 1907, the American Association of Museums printed the proceed-

ings from annual meetings and published a regular newsletter by 1918, but prior to the 1910s, American museum workers primarily relied on the British publication, *Museums Journal*, as well as occasional articles and notices in general interest science magazines like *Science* and *Popular Science Monthly*.

49 *CMNHAR* (1944), 5. On the relationship between the study of religion and the study of natural history in this era, see, among others, Donald Worster, *The Wealth of Nature: Environmental History and the Ecological Imagination* (New York: Oxford University Press, 1993), 187–202.

50. *Denver Municipal Facts*, 2, no. 30 (July 23, 1910): 5.

51 J. D. Figgins, letter to Robert C. Murphy, March 9, 1921, JDF-CMNH.

52. Much of the personal information about Figgins comes from an oral history of CMNH curator Betty Holmes Huscher Bachman, conducted by Kris Haglund, November 1994, CMNH.

53 See, e.g., the correspondence from his trip to the Caribbean (J. D. Figgins, letter to E. A. Figgins, August 3, 1917, JDF-CMNH).

54. Figgins was an astonishingly versatile museum worker, and colleagues continually remarked on his ease in field and studio. When exhibit backgrounds needed painting or labels needed writing, Figgins spent days in the field or library researching in preparation, then took up brush or pen without hesitation. When a curator's rifle jammed on a 1918 expedition, it was Figgins who removed its barrels and rechambered it (Barnum Brown, letter to Charles H. Hannington, December 30, 1935, CH-CMNH; Jesse D. Figgins, letter to Wm. C. Bradbury, May 10, 1918, JDF-CMNH).

55. Joel Allen, letter to Frank M. Taylor, May 20, 1910, FT-CMNH.

56. Harlan G. Smith, letter to Frank M. Taylor, May 18, 1910, FT-CMNH; see also J. D. Figgins, letter to John Campion, December 18, 1911, and Figgins, letter to Joel A. Allen, January 19, 1911, both in JDF-CMNH.

57. J. D. Figgins, letter to F. W. Hild, March 6, 1916, JDF-CMNH.

58. G. Stanley Hall, "The Contents of Children's Minds on Entering School," *Pedagogical Seminary* 1, no. (1891): 138–73. On Americans' response to these tests, see Scott L. Montgomery, *Minds for the Making: The Role of Science in American Education, 1750–1990* (New York: Guilford Press, 1994), 138; Peter J. Schmitt, *Back to Nature: The Arcadian Myth in Urban America* (New York: Oxford University Press, 1969), 78.

59. *AMNHAR* (1901), 47.

60. Anna D. Slocum, "Possible Connections between the Museum and the School," *PAAM* 6 (1911): 56.

61. On concerns about Americans' distance from nature in this era, see Kevin Connor Armitage, *The Nature Study Movement: The Forgotten Popularizer of America's Conservation Ethic* (Lawrence: University Press of Kansas, 2009); Michael Lewis, ed. *American Wilderness: A New History* (New York: Oxford University Press, 2007); Schmitt, *Back to Nature*. But American naturalists and nature lovers, AMNH curator Joel Allen among them, were almost

evangelical in their eagerness to share the joy, the feeling of luck, of privilege, that an eye for nature could bring. See Allen, *Autobiographical Notes* (n. 22 above, this chap.), 6.

62. Liberty Hyde Bailey, *The Holy Earth* (1915; reprint, New York: C. Scribner's Sons, 1943), 6.

63. As quoted in Pauly, *Biologists and the Promise of American Life* (n. 6 above, this chap.), 3–4. Also see William E. Ritter, "Biology's Contribution to a System of Morals That Would Be Adequate for Modern Civilization," *Bulletin of the Scripps Institution for Biological Research of the University of California* 2 (1917): 1–8.

64. Frank M. Chapman, letter to Hermon Bumpus, February 9, 1910, as cited in Kennedy, "Philanthropy and Science" (n. 7 above, "Introduction"), 170.

65. George A. Dorsey, "The Municipal Support of Museums," *PAAM* 1 (1907): 131.

66. Frank M. Chapman, letter to Hermon Bumpus, February 9, 1910, as cited in Kennedy, "Philanthropy and Science," 170.

67. Even as new science curricula took hold in schools across the nation, science pedagogy was never standardized, and teachers were regularly criticized for their methods. Nature study classes were often an exception to this rule, but, as Sally Gregory Kohlstedt points out, nature study bloomed largely outside the classroom, in settings like summer camps and after-school programs (*Teaching Children Science* [n. 35 above, this chap.], 26).

68. Lucas, "The Evolution of Museums" (n. 4 above, this chap.), 87.

69. Slocum, "Possible Connections between the Museum and the School," 56.

70. On natural history films and their use in museums in this era, see Alison Griffiths, *Wondrous Difference: Cinema, Anthropology and Turn-of-the-Century Visual Culture*, ed. John Belton, Film and Culture (New York: Columbia University Press, 2002); Gregg Mitman, *Reel Nature: America's Romance with Wildlife on Film* (Cambridge, MA: Harvard University Press, 1999), 5–25; Schmitt, *Back to Nature*, 146–53.

71. For more on the consumption of these texts in the United States, see Ralph H. Lutts, *The Nature Fakers: Wildlife, Science and Sentiment* (Charlottesville: University Press of Virginia, 2001), 28–29; Schmitt, *Back to Nature*, 115–40.

72. Bernard Lightman and James Secord have described similar tensions over the presence of natural theology in English popular science. See, e.g., Bernard Lightman, "'The Voices of Nature': Popularizing Victorian Science," in *Victorian Science in Context*, ed. Bernard Lightman (Chicago: University of Chicago Press, 1997); James A. Secord, *Victorian Sensation: The Extraordinary Publication, Reception, and Secret Authorship of Vestiges of the Natural History of Creation* (Chicago: University of Chicago Press, 2000).

73. As quoted in E. R. Whitney, "Nature Study as an Aid to Advanced Work in Science," *Proceedings of the National Educational Association* (Winona, MN: The Association, 1904): 895.

74. Frank M. Chapman, "The Case of William J. Long," *Science* 19, no. 4 (1904): 387–89.

75. Lucas, "The Evolution of Museums," 88.

76. Stephen C. Simms, *The Field Museum and the Child: An Outline of the Work Carried on by the Field Museum of Natural History among School Children of Chicago through the N. W. Harris Public School Extension and the James Nelson and Anna Louise Raymond Public School and Children's Lectures* (Chicago: Field Museum of Natural History, 1928), 6.

77. J. Kennedy Tod, letter to Frank Trumbull, December 22, 1898, JDF-CMNH.

78. H. L. Ward, letter to C. T. Barnes, December 22, 1904, DF-MPM.

79. Ward, "The Aims of Museums" (n. 16 above, this chap.), 100.

80. Victoria Cain, "From Specimens to Stereopticons" (n. 5 above, chap. 1), 23–25.

81. "The Brooklyn Institute Biological Laboratory," *Science* 17, no. 432 (1891): 271; Thomas Montgomery, "Expansion on the Usefulness of Natural History Museums," *Popular Science Monthly* 79 (July 1911): 42.

82. Lurie, *A Special Style* (n. 7 above, this chap.), 47.

83. Museum educators had often attended the same normal schools as their classroom counterparts, but they were free from the curricular obligations that accompanied compulsory education and so at liberty to improvise more than schoolteachers. On the network of professional female museum educators that emerged in this era, see Kohlstedt, "Innovative Niche Scientists" (n. 44 above, this chap.).

84. Speakers frequently had to offer the same lecture two or three times in a row to accommodate the crowds they attracted. See, e.g., C. B. Weil, letter to H. L. Ward, February 4, 1915, CF-MPM.

85. *CMAR* (1899), 13.

86. C. B. Weil, letter to H. L. Ward, February 4, 1915, CF-MPM.

87. Museum workers were highly conscious of the dynamics of professionalization; for a more extensive discussion of literature addressing the early twentieth-century relationships between amateurs and professionals in science, see the bibliographic essay following these notes.

88. Ward, "The Aims of Museums," 102.

89. On Moore's work at the museum, see Tran-Ngoc Loi, "Catalogue of the Types of Polychaete Species Erected by J. Percy Moore," *PANS* 132 (1980): 121–49.

90. Such curators resisted the belief held by their contemporaries that natural history museums and the laboratories of universities and field stations took what zoologist Thomas Montgomery described as "sharply differentiated" approaches to the biological sciences. But they worried that young biologists saw taxonomy and systematics as old-fashioned pursuits—at Harvard, e.g., no doctoral student in zoology worked on the systematics of living animals between 1878 and 1910. (Paleontologists and botanists did theses on systematic topics in those years, but their numbers were in decline.) In

response, curators sought to align their institutions with a "new" or "scientific" natural history, one that married lab and field practices, and saw these new categories of collection as part of this project. Mary P. Winsor, *Reading the Shape of Nature: Comparative Zoology at the Agassiz Museum*, ed. David L. Hull, Science and Its Conceptual Foundations (Chicago: University of Chicago Press, 1991), 186; Kohler, *Landscapes and Labscapes* (n. 18 above, this chap.), 23–40. See also Barrow, *Nature's Ghosts* (n. 22 above, this chap.), chap. 9.

91. Winslow, a bacteriologist fresh from MIT's sanitary biology program, recognized the difficulties of maintaining what he dubbed a "Museum of Living Bacteria," for living cultures of bacteria were inherently unstable and continued to mutate silently once collected. He was undaunted by the challenge, declaring that "such a biological collection when established will furnish also an exceptional opportunity for studies of the systematic relationships of this group in which a better biological classification is greatly needed." Administrators eventually ousted the collection, in order to refocus the museum's work on more traditional—and more easily displayed—natural history subjects. On Winslow's training, see A. J. Viseltear, "The Emergence of Pioneering Public Health Education Programs in the United States," *Yale Journal of Biology and Medicine* 61, no. 6 (1988): 519–48. On his work at the museum, see "A New Field for Museum Work," *American Museum Journal* 10, no. 7 (1910): 199; and Julie K. Brown, "Connecting Health and Natural History: A Failed Initiative at the American Museum of Natural History, 1909–1922," *American Journal of Public Health*, published online ahead of print, November 8, 2013: e1–e12.

92. F. A. Lucas, "Museum Labels and Labeling," *Printing Art* (August–September 1909), 94.

93. George W. Hunter, "The American Museum's Reptile Groups in Relation to High School Biology," *American Museum Journal* 15, no. 8 (1915): 406.

94. H. L. Ward, letter to Albert Keipper, April 13, 1904, DF-MPM.

95. Ellis L. Yochelson, *The National Museum of Natural History: 75 Years in the Natural History Building*, ed. Mary Jarrett (Washington DC: Smithsonian Institution Press, 1985), 61.

96. J. D. Figgins, letter to J. A. Allen, February 14, 1912, JDF-CMNH; *CMNHAR* (1915), 21.

97. Terraria and aquaria were a staple of nineteenth-century natural history displays in both domestic and public settings, and they rarely failed to fascinate. In 1901, after the American Museum installed a series of aquaria in its Hall of Conchology, board president Morris Jesup announced that "the popularity of this feature . . . indicates the desirability of increasing the number of such exhibits." On terraria and aquaria, see David Allen, "Tastes and Crazes," in *Cultures of Natural History*, ed. Jardine, Secord, and Spary (n. 4 above, "Introduction"): 394–407; Bernd Brunner, *The Ocean at Home: An Illustrated History of the Aquarium* (Chicago: University of Chicago Press, 2011); *AMNHAR* (1901), 30–31.

98. H. L. Ward, letter to C. H. Townsend February 21, 1905, DF-MPM. Ward and Townsend, director of the New York Aquarium, kept in close touch, having trained together at Ward's Natural Science Establishment.

99. Botanical gardens made a similar effort; at the New York Botanical Garden, e.g., see Peter Mickulas, *Britton's Botanical Empire: The New York Botanical Garden and American Botany, 1888–1929*, Memoirs of the New York Botanical Garden (New York: New York Botanical Garden Press, 2007), 118–21.

100. Curator Jurgen Paarmann built on the earlier work of local conchologists, who had already compiled a formidable collection. Extensive shell collections were not uncommon in museums that had grown from local academies of science, for collectors found them easy to transport and preserve. *PDANS 10 (1907), 197.*

101. *CMNHAR* (1911), 6.

102. On the history of such expositions, see Julie K. Brown, *Health and Medicine on Display, 1895–1904* (Cambridge, MA: MIT Press, 2009).

103. The show was wildly popular, attracting more than a million people to the museum during its four month run. On "Jewish Day," more than sixty-three thousand people attended the show, and the museum had to admit them in squadrons of a thousand. "10,000 Attend the Tuberculosis Show," *New York Times*, December 1, 1908.

104. *FMNHAR* (1914), 304.

105. As Paul Brinkman has observed, American vertebrate paleontology shifted from private collections to museums in the early twentieth century. At the Smithsonian's newly established National Museum of Natural History, e.g., curators busily accumulated specimens to create a "hall of extinct monsters," the centerpiece of which was a *Basilosaurus* whale skeleton assembled from multiple specimens. Paul D. Brinkman, *The Second Jurassic Dinosaur Rush: Museums and Paleontology in America at the Turn of the Twentieth Century* (Chicago: University of Chicago Press, 2010), especially chap. 12; Rainger, *Agenda for Antiquity* (n. 21 above, this chap.), 67–215.

106. Patrons considered paleontological mounts showy monuments to their civic generosity, and they liked the idea of funding displays that attracted enormous crowds. The acquisition of complete and well-preserved fossil skeletons fueled a very real sense of competition among the nation's wealthiest men, and by the end of the 1890s, men like Andrew Carnegie and Marshall Field were determined to decorate their eponymous museums with gigantic sauropods and sponsored the travel, time, and labor necessary to bring these beasts to Pittsburgh and Chicago, respectively. See, e.g., *CMAR* (1900–1914); Keith M. Parsons, *Drawing out Leviathan: Dinosaurs and the Science Wars*, ed. James O. Farlow, Life of the Past (Bloomington: Indiana University Press, 2001), 4; Tom Rea, *Bone Wars: The Excavation and Celebrity of Andrew Carnegie's Dinosaur* (Pittsburgh: University of Pittsburgh Press, 2001); Jeremy Vetter, "Cowboys, Scientists, and Fossils: The Field Site and Local Collaboration in the American West," *Isis* 99, no. 2 (2008):

164–67; 173–74; Lukas Rieppel, "Bringing Dinosaurs Back to Life: Exhibiting Prehistory at the American Museum of Natural History," *Isis* 103, no. 3, (2012): 460–90.

107. Paleontology was still considered the province of geologists in some quarters, but curators wrested vertebrate paleontology into the realm of biology in these years. Ronald Rainger, "Vertebrate Paleontology as Biology: Henry Fairfield Osborn and the American Museum of Natural History," in *The American Development of Biology*, ed. Rainger, Benson, and Maienschein (n. 5 above, "Introduction"): 219–56.

108. On the history of fossil display (and the appeal of very large fossils) in this period, see Brinkman, *The Second Jurassic Dinosaur Rush*; Clark, *God—or Gorilla* (n. 21 above, this chap.).

109. Well-publicized discoveries such as American Museum curator Barnum Brown's 1902 excavation of the first documented *Tyrannosaurus rex* skeleton fanned the flames of "dinomania" and led to even larger and more enthusiastic audiences. Charles R. Knight, "Autobiography of an Artist," n.d., American Museum of Natural History Archives, New York, IV-18, IV-19.

110. Fossil reconstructions and attendant illustrations often pushed a sociobiological subtext, reminding viewers that species that failed to adapt to their surroundings or otherwise misbehaved went extinct—a message many museum trustees hoped immigrant and laboring visitors would take to heart. On the ideological underpinnings of dinosaur display, see, among others, Clark, *God—or Gorilla*; Rainger, *Agenda for Antiquity* (n. 21 above, this chap.), 54–60.

111. Museum staff and patrons saw the destruction of natural habitats and the devastation wrought by commercial hunting whenever they collected specimens or hunted for sport. So from the 1890s on, museums joined forces with other conservationists to educate the American public about species' decline and collected, as object lessons, eastern elk, monk seals, and other animals on the edge of extinction. Smithsonian staff, charged with chronicling the country's flora and fauna, had incorporated conservation concerns into their work as early as 1876, when mammalogist J. A. Allen published a monograph and five popular articles chronicling the decline of the bison and other vertebrates. Taxidermist William Temple Hornaday also took up this cause, building exhibits and pleading for the bison's preservation in the Smithsonian's 1889 annual report. On participation by museums in the conservation movement, see, among others, Andrei, "Nature's Mirror" (n. 8 above, this chap.); Barrow, *Nature's Ghosts* (n. 22 above, this chap.); Hanna Rose Shell, "Skin Deep: Taxidermy, Embodiment and Extinction in W. T. Hornaday's Buffalo Group," in *Museums and Other Institutions of Natural History: Past, Present, and Future*, ed. Alan E. Leviton and Michele L. Aldrich (San Francisco: California Academy of Sciences, 2004).

112. As Robert Kohler, the source for these numbers, points out, these expeditions didn't include the kind of local jaunts of which museum curators were so fond. Kohler, *All Creatures* (n. 2 above, this chap.), 118.

113. As Philip Pauly has observed, sexual reproduction was central to the biology education offered by museums in this era; see Pauly, "The Development of High School Biology, New York City, 1900-1925," *Isis* 60, no. 4 (1991): 662–88. On the pleasures and complexities of looking at animals in various settings, see Arnold Arluke and Clifford R. Sanders, *Regarding Animals* (Philadelphia: Temple University Press, 1996); Lorraine Daston and Gregg Mitman, eds., *Thinking with Animals: New Perspectives on Anthropomorphism* (New York: Columbia University Press, 2005), 137–223; Liv Emma Thorsen, Karen A. Rader, and Adam Dodd, eds., *Animals on Display: The Creaturely in Museums, Zoos, and Natural History* (University Park, PA: Penn State University Press, 2013).

114. Though museums did not officially track visitors' routes through their halls, they carefully observed and reported which exhibits attracted the most attention from various groups of visitors. See, e.g., F. C. Lincoln, letter to Jesse D. Figgins, December 30, 1916, JDF-CMNH.

115. Museum professionals saw themselves as naturalists, scientists, and educators, not mere entertainers, and they actively resented being grouped with sideshow barkers and swindlers spouting humbug. See, e.g., Harlan G. Smith, letter to Frank M. Taylor, May 18, 1910, FT-CMNH.

116. H. L. Ward, letter to editor, *Evening Wisconsin*, November 23, 1905, DF-MPM.

117. Ibid.

118. Montgomery, "Expansion on the Usefulness of Natural History Museums" (n. 81 above, this chap.).

119. *PDANS* 10 (1907): 198.

120. According to Ward, the books "relating to local birds have been much used by the public." They were so popular, in fact, that people regularly stole them, a transgression he dismissed as an inherent cost of public access. "Director's Monthly Report to the Board of Trustees," February 18, 1908, p. 138, DF-MPM.

121. World War I led to an outpouring of donations and inquiries, as enthusiastic enlistees sent unfamiliar specimens back from training camps. E. Shepherd, letter to Henry L. Ward, October 8, 1917; T. E. B. Pope, letter to E. Shepherd, October 8, 1917—both in CF-MPM.

122. Such donations weren't limited to specimens that represented local species, of course. Responding to the historical role that museums had played as places for the rare and curious, people donated more exotic finds as well. On finding live opossums in bunches of bananas, Milwaukee grocers donated the wriggly beasts to the museum, hoping they would be stuffed and mounted for posterity. *CMAR* (1908), 51-60; Henry L. Ward, letter to George Shrosbee 30 July, 1911, DF-MPM.

123. Regardless of the smell of the gift or the rationale of the supplier, curators and directors of museums welcomed such donations, for innocent questions and donations periodically resulted in important finds or new collecting relationships. See, e.g., the correspondence between Mrs. M. O. Potter and Jesse D. Figgins, February 1921, JDF- CMNH, as well as that between

D. B. Barlow and Frank Daggett, September 1915, Frank Daggett Correspondence, Archives of the Los Angeles County Museum of History, Science and Art, LAM.

124. Though visitors found them less beguiling, scaly specimens were easier to care for than the furred or feathered. H. L. Ward, letter to F. J. Walsh, February 7, 1914, CF-MPM; J. D. Figgins to Steve Elkins, May 6, 1918, JDF-CMNH.

125. *CMNHAR* (1914), n.p.; Chas. D. Griffith, letter to Figgins, December 19, 1910, JDF-CMNH.

126. D. B. Barlow, letter to Frank Daggett, September 11, 1915, LAM.

127. H. L. Ward, letter to Jessie M. Zehl, January 2, 1914, CF-MPM.

128. H. L. Ward, letter to C. B. Weil, June 23, 1915, CF-MPM.

129. The nineteen-year-old son of financier Jay Gould, e.g., proudly donated the body of his Saint Bernard to the American Museum. Geoffrey Hellman, *Bankers, Bones and Beetles: The First Century of the American Museum of Natural History* (Garden City, NY: Natural History Press, 1968), 102.

130. James Noland, letter to J. D. Figgins April 4, 1911, JDF-CMNH.

131. H. L. Ward, letter to Gustav Friedrich July 29, 1908, DF-MPM.

132. Museums typically used specimens lacking accompanying scientific information for educational programs.

133. J. D. Figgins, letter to John Campion January 31, 1914, JDF-CMNH.

134. *CMAR* (1899), 7–8.

135. This is not to say curators' relationships with professional collectors were entirely straightforward, as Paul Brinkman's *Second Jurassic Dinosaur Rush* (n. 105 above, this chap.) and Jeremy Vetter's article ("Cowboys, Scientists, and Fossils" [n. 106 above, this chap.]) have demonstrated.

136. Though museums increasingly left the study of botany to botanical gardens in this era, museum staff agreed it was essential to represent such an important and popular topic of study. But botanical specimens were terribly difficult to display. Live flowers, especially delicate wildflowers, tended to die almost immediately, and dried specimens were, according to one museum botanist, "discolored, dead, ghastly, of no general resemblance to nature and of little interest to the laity." Wax and glass plants, created by skilled artisans like the Blaschka family, seemed preferable. Charles Millspaugh, "Botanical Installation," in *PAAM* 5 (1911): 55. On botany's shift from museums to botanical gardens, see Mickulas, *Britton's Botanical Empire* (n. 99 above, this chap.), 101–54.

137. Early twentieth-century curators concurred with the 1889 observation by the Smithsonian's George Brown Goode that "a gift coupled with conditions is, except in very extraordinary cases, far from a benefaction." As quoted by H. L. Ward in a letter to William Temple Hornaday, July 5, 1902, DF-MPM.

138. Goode, "The Museums of the Future" (n. 4 above, this chap.), 336–37; Lurie, *A Special Style* (n. 7 above, this chap.), 45. Supremely suspicious of gifts, Ward believed the benefits of such a policy well compensated for the

occasional loss of a gift. When the secretary of the Saint Paul Institute of Arts and Sciences, asked Ward's advice on running a natural history museum, Ward warned him to avoid conditional donations, noting that these handicapped museums more than the alternative of empty shelves. H. L. Ward, letter to J. B. Leaman, September 15, 1916, DF-MPM.

139. Monthly Report to the Board of Trustees, October 21, 1902, p. 83, DF-MPM.

140. Lucas, "Museum Labels and Labeling" (n. 92 above, this chap.), 273.

141. *AMNHAR* (1915), 43–44.

142. Though Dana is best known for his writings on art museums, he directed the Newark Public Museum, a general museum with a popular natural history component, and wrote extensively on natural history displays within museums. John Cotton Dana, "A Small American Museum: Its Efforts toward Public Utility," in *The New Museum: Selected Writings of John Cotton Dana*, ed. William A. Peniston (Newark, NJ: Newark Museum Association, 1921), 160.

143. Frank C. Baker, "Some Instructive Methods of Bird Installation," *PAAM* 1 (1907): 89.

144. F. A. Lucas, "The Work of Our Larger Museums as Shown by Their Annual Reports," *Science* 27, no. 679 (1908): 35.

145. J. D. Figgins, letter to Frank Taylor, November 18, 1917, JDF-CMNH.

146. Sharon Kingsland, *The Evolution of American Ecology, 1890–2000* (Baltimore, MD: Johns Hopkins University Press, 2005).

147. J. H. Paarmann, letter to Mary Louisa Duncan Putnam, April 6, 1901, DANS.

148. Davenport Academy of Natural Sciences, *PDANS* 9 (1900–1903): n.p.

149. *CMNHAR* (1914), n.p.

150. Lucas, "Museum Labels and Labeling" (n. 92 above, this chap.), 95.

151. See, e.g., F. A. Lucas, "A Note Regarding Human Interest in Museum Exhibits," *American Museum Journal* 11, no. 6 (1911): 187–88.

152. Henry L. Ward, "The Labeling in Museums," *PAAM* 1 (1907): 42–43.

153. J. D. Figgins, letter to John Campion, October 13, 1913, JDF-CMNH.

154. Yochelson, *National Museum of Natural History* (n. 95 above, this chap.), 56.

155. Sample labels, n.d., folder 2, FAL.

156. S. E. Meek, "A Contribution to Museum Technique," *American Naturalist* 36, no. 421 (1902): 55.

157. Henry L. Ward, letter to Carl E. Akeley, March 8, 1915, DF-MPM.

158. Carl Akeley was the most visible advocate of the movement to improve specimen preservation and taxidermy in the United States, and his sculptural examples inspired a generation of museum preparators to improve the quality of the mounted specimens in their collections. On Akeley, see Stephen Christopher Quinn, *Windows on Nature: The Great Habitat Dioramas of the American Museum of Natural History* (New York: American Museum of Natural History, in association with Harry N. Abrams, 2006), 158–63.

159. Dickerson, "The 'Toad Group' in the American Museum" (n. 23 above, this chap.), 164–65.

160. See Andrew McClellan, "P. T. Barnum, Jumbo the Elephant, and the Barnum Museum of Natural History at Tufts University," *Journal of the History of Collections* 24, no. 1 (2012): 45–57; Mitman, *Reel Nature* (n. 70 above, this chap.), 9–10.

161. Penelope Bodry-Sanders, *Carl Akeley: Africa's Collector, Africa's Savior* (New York: Paragon House, 1991), 104.

162. On contemporary concerns about this kind of manipulation, see Lutts, *Nature Fakers* (n. 71 above, this chap.), chaps. 3–4.

163. H. L. Ward, letter to N. Annandale, December 21, 1912, DF-MPM.

164. Figgins didn't need to explain himself to Bailey. A year earlier, Bailey had purchased the parts of several passenger pigeons and sewed them together to create a few fine, whole birds for his displays. Bailey's mentor, Homer Dill, had once combined twenty odd parts of three damaged passenger pigeons to create a single uninjured bird. J. D. Figgins, letter to A. M. Bailey, November 4, 1918, JDF-CMNH; Ilva Jones, "Alfred M. Bailey," unpublished manuscript, n.d., p. 6, Bailey Biography Files, CMNH; Homer R. Dill, "The Correlation of Art and Science in the Museum," *PAAM* 10 (1916): 126.

165. See, e.g., Dickerson, "The 'Toad Group' in the American Museum" (n. 23 above, this chap.).

166. F. A. Lucas, "The Story of Museum Groups: Part 2," *American Museum Journal* 14, no. 2 (1913): 51–52.

167. H. L. Ward, letter to S. A. Barrett, July 21, 1915, DF-MPM.

168. Oliver C. Farrington, "The Rise of Natural History Museums," *Science* 42, no. 1076 (1915): 207.

169. Though Meyer cautioned American museums about excessive artifice, he endorsed many of the display reforms they advocated. See A. B. Meyer, *Studies of the Museums and Kindred Institutions of New York City, Albany, Buffalo, and Chicago, with Notes on Some European Institutions* (Washington, DC: Government Printing Office, 1905), 339.

170. Lucas, "The Story of Museum Groups," 51–52.

171. German museum workers reached an identical conclusion. See Lynn K. Nyhart, "Science, Art and Authenticity in Natural History Displays," in *Models: The Third Dimension of Science*, ed. Nick Hopwood and Soraya De Chadarevian (Stanford, CA: Stanford University Press, 2004).

172. H. L. Ward, letter to S. A. Barrett, July 21, 1915, DF-MPM.

173. See, e.g., Homer E. Keyes, "Commercial Tendencies and an Esthetic Standard in Education," in *Annual Meeting of the American Association of Museums*, ed. Paul M. Rea (New York: Waverly Press, 1917); George Kunz, "To Increase the Number of Visitors to Our Museums," *PAAM* 5 (1911): 87–89. On Keyes's involvement with the AAM, see his obituary: "Homer E. Keyes Dies; Editor of Antiques: Also Was a Former Professor of Art at Dartmouth," New York Times, October 9, 1938, http://query.nytimes.com/mem/archive/pdf?res=FA0817F93B581A7A93CBA9178BD95F4C8385F9.

174. See Cain, "'Attraction, Attention, Desire'" (n. 32 above, this chap.), 16–24.
175. Ibid., 10–11.
176. J. D. Figgins, letter to F. W. Hild, March 6, 1916, JDF-CMNH.
177. Carla Yanni, *Nature's Museums: Victorian Science and the Architecture of Display* (London: Athlone Press, 1999), 99; Dwight Franklin, "A Recent Development in Museum Groups," *PAAM* 10 (1916): 111.
178. On the increased use of standardized glass cases, in the United States and abroad, and the commercial and domestic design influences this entailed, see Brita Brenna, "The Frames of Specimens: Glass Cases in the Bergen Museum around 1900," in *Animals on Display*, ed. L. E. Thorsen, K. A. Rader, and A. Dodd (State College, PA: Penn State University Press, 2013), 37–57.
179. Arthur C. Parker, "Habitat Groups in Wax and Plaster," *Museum Work* 1, no. 3 (1918): 81.
180. On the pictorial revolution in the paleontology department of the American Museum, see Victoria Cain, "'The Direct Medium of the Vision': Visual Education, Virtual Witnessing and the Prehistoric Past at the American Museum of Natural History, 1890–1923," *Journal of Visual Culture* 10, no. 3 (2010): 284–303.
181. On taxidermy in these decades, see, among others, Andrei, "Nature's Mirror" (n. 8 above, this chap.); Mark V. Barrow Jr., "The Specimen Dealer: Entrepreneurial Natural History in America's Gilded Age," *Journal of the History of Biology* 33, no. 3 (2000), 493–534; Maxine Benson, *Martha Maxwell, Rocky Mountain Naturalist*, Women in the West (Lincoln: University of Nebraska, 1986); Robert Wilson Shufeldt, *Scientific Taxidermy for Museums* (Washington, DC: Government Printing Office, 1893); Karen Wonders, *Habitat Dioramas*, Figura Nova Series 25, Acta Universitatis Uppsaliensis (Uppsala: Almqvist and Wiksell, 1993), 115–16.
182. Wonders, *Habitat Dioramas*, 134.
183. Lucas, *Fifty Years of Museum Work* (n. 13 above, this chap.), 50.
184. Such displays are now uniformly described as habitat dioramas, but they went by a number of terms in this era, including "life groups," "environmental groups," "ecologic groups." Though there is an extensive literature on the development of the habitat diorama, the best overview of their development continues to be Wonders, *Habitat Dioramas*.
185. Swedish and German naturalists were engaged in similar experiments. In Swedish "biological museums" (*biologiska museet*), e.g., taxidermist and designer Gustaf Kolthaff featured displays of one or more mounted animals against large circular painted backgrounds of their natural environment. His scientist-colleague Olaf Gyilling later modified Kolthaff's panoramic approach by breaking such scenes down into regional displays of animal ecology designed to enable viewers to understand the plight of endangered species and landscapes. In Germany, Otto Lehmann experimented with habitat groups arranged in "visually and emotionally charged" configurations but no background paintings, whereas in Berlin,

designers "emphasized the relations of animals to their physical environment, placing animals in scenery but not necessarily posing them in any visually obvious way as engaging with other animals." Nyhart, *Modern Nature* (n. 12 above, this chap.), 272–79; Wonders, *Habitat Dioramas*, 55–58.

186. William Hornaday and William J. Holland, *Taxidermy and Zoological Collecting: A Complete Handbook for the Amateur Taxidermist, Collector, Osteologist, Museum-Builder, Sportsman, and Traveller* (New York: C. Scribner's sons, 1891), ix–x.

187. For more on Chapman's contributions to ornithology in this period, see Mark V. Barrow Jr., *A Passion for Birds: American Ornithology after Audubon* (Princeton, NJ: Princeton University Press, 1998); William King Gregory, "Frank Michler Chapman, 1864–1945," *Biographical Memoirs: National Academy of Sciences of the United States of America* 25, no. 5 (1945): 109–44.

188. Frank M. Chapman, "What Constitutes a Museum Collection of Birds?" in *Proceedings of the Fourth International Ornithological Congress*, ed. Richard Bowdler Sharpe (London: Dulau and Co., 1907), 152; Wonders, *Habitat Dioramas*, 127.

189. Chapman, *Autobiography of a Bird-Lover* (n. 23 above, this chap.), 71–72, 81; Wonders, *Habitat Dioramas*, 164.

190. As Wonders points out, Chapman was right to be concerned. Egg collecting and wanton killing on the island "had already resulted in the extinction of the great auk, still abundant when Audubon visited the island in 1833." The colony of gannets nesting on the summit of the island had fallen from a hundred thousand birds in 1869 to fifty birds some twelve years later as a result of lighthouse keepers who tried to annihilate the noisy flocks. Wonders, *Habitat Dioramas*, 127; Chapman, *Autobiography of a Bird-Lover*, 81.

191. Chapman, *Autobiography of a Bird-Lover*, 164. Depicting birds in their natural habitat was a hallmark of nineteenth-century ornithological illustration, and Fuertes's and Chapman's frequent contemplation of such illustrations may well have influenced their displays. See Ann Shelby Blum, *Picturing Nature: American Nineteenth-Century Zoological Illustration* (Princeton, NJ: Princeton University Press, 1993).

192. Chapman, *Autobiography of a Bird-Lover*, 166.

193. Ibid.

194. Jesse Figgins, then a preparator at the American Museum, worked closely with Chapman on the displays. See Frank Chapman, "The Department of Birds, American Museum: Its History and Aims," *Natural History* 22, no. 4 (1922): 315; Wonders, *Habitat Dioramas*, 126. For more on bird taxidermy, see Karen Wonders, "Bird Taxidermy and the Origin of the Habitat Diorama," in *Non-Verbal Communication in Science Prior to 1900*, ed. Renato Mazzolini (Florence: Instituto e Museo Di Storia Della Scienza, 1993).

195. Milwaukee Public Museum Director's Monthly Report to Board of Trustees, September 22, 1903, MPM.

196. *AMNHAR* (1901), 30–31.

197. Dickerson, "The 'Toad Group' in the American Museum" (n. 23 above, this chap.), 164.

198. Oliver C. Farrington, "Some Relations of Science and Art in Museums," *PAAM* 10 (1916): 26.

199. Wissler, "Survey of the American Museum of Natural History" (n. 2 above, this chap.), 122.

200. Museum workers in other nations had also reached this conclusion, and visitor preference impelled certain European and English museums to abandon typology for group displays. See William Ryan Chapman, "Arranging Ethnology: A.H.F.L. Pitt Rivers and the Typological Tradition," in *Objects and Others: Essays on Museums and Material Culture*, ed. George W. Stocking Jr. (Madison: University of Wisconsin Press, 1985); Nyhart, "Science, Art and Authenticity" (n. 171 above, this chap.).

201. See *PAAM*, vols. 1–11 (1907–17). As the AAM also counted staff from art, history, and commercial museums among its members, the fact that such a significant percentage of the papers were devoted to a phenomenon largely reserved for natural history museums in this era makes this proportion even more remarkable.

202. H. L. Ward, letter to Charles Dury, January 20, 1906, DF-MPM.

203. Lucas, *Fifty Years of Museum Work* (n. 13 above, this chap.), 32.

204. Ibid., 29.

205. Frank C. Baker, "Some Observations on Museum Administration," *Science* 26, no. 672 (1907): 668. On Baker's life and scientific work, see Maggie Jarvis, "Frank C. Baker, 1867 to 1942: Invertebrate Paleontology," Illinois State Geological Survey, updated July 12, 2013, http://www.isgs.illinois.edu/?q=frank-c-baker.

206. George A. Dorsey, "The Aim of a Public Museum," *PAAM* 1 (1907): 97–100. Given his institutional home, Dorsey's assertion came as no surprise. When it first opened in 1902, the Field Museum had found it difficult to attract curators, for accomplished scientists balked when director Frederick Skiff insisted they prioritize the institution's needs over their own research agendas. Brinkman, *The Second Jurassic Dinosaur Rush* (n. 105 above, this chap.), esp. 172–75.

207. Charles Doolittle Walcott, letter to Henry L. Ward, January 26, 1912, DF-MPM.

208. Women, often the wives or daughters of curators, increasingly dominated the rank and file of education departments in museums of this period, although men continued to lead these departments. A 1918 AAM survey captured the resulting disparity in pay: on average, women in museums earned $1,250 a year, while men made nearly twice that. The highest reported salary in the nation for a woman was $3,250; her male counterpart was paid more than 50 percent more. Part-time work paid even worse; two museums reported paying female employees $300 annually. FitzRoy Carrington, Henry L. Ward, and Margaret T. Jackson Rowe, "Instruction in Museum Work," *Museum Work* 1, no. 3 (1918): 90.

209. On the considerable budgets of education departments, see, e.g., the overview data provided in Wissler, "Survey of the American Museum of Natural History" (n. 2 above, this chap.).

210. See Hermon Carey Bumpus Jr., *Hermon Carey Bumpus, Yankee Naturalist* (Minneapolis: University of Minnesota Press, 1947), 117; "Dr. Bumpus Is out of History Museum—Director's Controversies with President Osborn and Others End with His Resignation—Going to Wisconsin University— Had Worked to Popularize Museum and Clashed with Scientists," *New York Times*, January 21, 1911. Also see "Bumpus Coming Back as Museum's Head," *New York Times*, November 15 1910, 15.

211. Montgomery, "Expansion on the Usefulness of Natural History Museums" (n. 81 above, this chap.), 43.

212. Paul M. Rea, "One Hundred and Fifty Years of Museum History," *Science* 57, no. 1485 (1923): 680.

CHAPTER TWO

1. Richard E. Dahlgren, "Nesting Colony of Flamingos Group, from Andros Island, Bahamas, West Indies," 1905, Negative Logbook 18, neg. 3939, AMNH.

2. H. E. Anthony, "The Vernay-Faunthorpe Hall of South Asiatic Mammals," *Scientific Monthly* 33, no. 3 (1931): 281.

3. Lewis Mumford, "The Marriage of Museums," *Scientific Monthly* 7, no. 3 (1918): 252.

4. This enthusiasm was not limited to the United States. See Karen Wonders, "Habitat Dioramas as Ecological Theatre," *European Review* 1, no. 3 (1993): 285–300.

5. On the exorbitant costs of dioramas, Lucas confessed to fellow members of the AAM that he was "afraid first-class groups are not within the reach of ordinary museums." Frederic A. Lucas, commentary, after paper presented by Henry L. Ward, "The Exhibition of Large Groups," *PAAM* 1 (1907): 72.

6. On the emergence of the biological perspective, see Lynn Nyhart, "Teaching Community via Biology in Late-Nineteenth-Century Germany," *Osiris* 17 (2002): 141–70; Nyhart, *Modern Nature* (n. 12 above, chap. 1).

7. For an alternative German case, see Carsten Kretschmann, *Räume Öffnen Sich: Naturhistorische Museen Im Deutschland Des 19. Jahrhunderts*, Wissenskultur Und Gesellschaftlicher Wandel (Berlin: Akademie Verlag, 2006).

8. By the late 1880s, there is some evidence that international networks of curators had begun to discuss the exchange of museum materials and the "standardization" of museum practices: see, for instance, Sally Gregory Kohlstedt, "Otis T. Mason's Tour of Europe: Observation, Exchange, and Standardization in Public Museums, 1889," *Museum History Journal* 2 (2008): 181–208.

9. Adams, a curator at the University of Michigan and Cincinnati museums in these years, was an early member of the AAM. For more on his work, see

Charles C. Adams, "Some of the Advantages of an Ecological Organization of a Natural History Museum," *PAAM* 1 (1907): 170, 173.

10. Farrington, "The Rise of Natural History Museums" (n. 168 above, chap. 1), 207.

11. Best known for his work on ants at Harvard, Wheeler was himself a former museum man: he had begun his career at the Milwaukee Public Museum, where he worked closely with Carl Akeley, then assumed a post as the curator of invertebrate zoology at the AMNH and remained a research associate there long after his move to Cambridge. William Morton Wheeler, "Carl Akeley's Early Work and Environment," *Natural History* 27 (1927): 141, as cited in Wonders, *Habitat Dioramas* (n. 182 above, chap. 1), 223. On Wheeler's work with museums and his own peculiar biological perspective, see Charlotte Sleigh, *Six Legs Better: A Cultural History of Myrmecology* (Baltimore, MD: Johns Hopkins University Press, 2007).

12. On curators' participation in the era's larger conservation movement, see, among others, Andrei, "Nature's Mirror" (n. 8 above, chap. 1); Barrow, *A Passion for Birds* (n. 188 above, chap. 1), and *Nature's Ghosts* (n. 22 above, chap. 1); Wonders, *Habitat Dioramas* (n. 182 above, chap. 1).

13. Early museum conservationists focused their efforts on curbing public behaviors that threatened species' survival; the contents and labels of dioramas reflected this strategy, while also acknowledging its perils, especially that it required curators themselves to obtain and display the desirable specimens. In 1908 and 1909, e.g., Milwaukee Public Museum director Henry Ward built a diorama featuring Wisconsin shorebirds, intended to call attention to the ways that egg collecting and spring shooting endangered local populations. Some museum staff became quite concerned, however, that "a display of this nature excites a desire for [the specimen in question], rather than a sentiment against their use," as Colorado Museum director Jesse Figgins put it in 1919. H. L. Ward, letter to C. H. Van Hise, March 27, 1909, DF-MPM; J. D. Figgins, letter to T. Gilbert Pearson, January 6, 1920, JDF-CMNH.

14. H. F. Osborn, "Preservation of the World's Animal Life," *American Museum Journal* 12, no. 4 (1912): 124.

15. A. R. Crook, "The Museum and the Conservation Movement," *PAAM* 9 (1915): 92–93.

16. Henry L. Ward, "The Desirability of Irrelevant Accessories in Groups," *Museum News: Magazine Section* 9, no. 16 (1932): 253.

17. As cited in Mary Cynthia Dickerson, "Some Methods and Results in Herpetology," *American Museum Journal* 11, no. 6 (1911): 208–9. Also see Myers, "A History of Herpetology" (n. 44 above, chap. 1), 72.

18. Dickerson, "Some Methods and Results in Herpetology," 208–9.

19. Charles C. Adams, "The New Natural History—Ecology," *American Museum Journal* 17 (1917): 491–92.

20. William E. Ritter, "Feeling in the Interpretation of Nature," *Popular Science Monthly* 79 (1911): 130. For more on Ritter's efforts to popularize science

and nature, see Pauly, *Biologists and the Promise of American Life* (n. 6 above, chap. 1), 201–13.

21. All Wilson quotes from Edmund B. Wilson, "Aims and Methods of Study in Natural History," *Science* 13, no. 314 (1901): 22. Throughout the 1910s, both curators and university-based scientists called for the reintroduction of some of the amateur ardor they had so dutifully excised from their own profession. Kohler, *Landscapes and Labscapes* (n. 18 above, chap. 1), 23–59.

22. Ritter, "Feeling in the Interpretation of Nature," 128–29.

23. Psychologists like G. Stanley Hall maintained that playing to the emotions was the most efficient way—sometimes the only way—to awaken curiosity and, therefore, learning. Popular commentators interpreted this as an authoritative mandate for sentiment in education. Edgar James Swift, "The Genesis of the Attention in the Educative Process," *Science* 34, no. 862 (1911): 4–5; "The Value of Sentiment," *Atlantic Monthly* (August 1913), 279–80.

24. Chapman, *Autobiography of a Bird-Lover* (n. 23 above, chap. 1), 166.

25. "Art in a Natural History Museum: With Special Reference to Mural Decorations in the Indian Halls," *American Museum Journal* 13, no. 3 (1913): 99.

26. Lucas, "The Work of Our Larger Museums as Shown by Their Annual Reports" (n. 144 above, chap. 1), 35.

27. *AMNHAR* (1924), 21–22.

28. Francis Lee Jaques, letter to A. M. Bailey, November 1, 1938, box 501, Bailey, 12 folder Exhibits, CMNH.

29. This was especially true in the 1910s and early 1920s, when habitat dioramas were still enclosed in large cases; by the end of the 1920s and the early 1930s, recessed wall spaces housed the displays.

30. According to Sally Metzler, the Field's dioramas opened on the following dates: *White-Tailed Deer* (1902), *American White Pelicans* (1910), *Wild Turkeys* (1910), *Winter Lake Birds* (1912), *Sandhill and Whooping Cranes* (1912), *Beaver* (1912), *Mexican Grizzlies* (1912), *Rocky Mountain Goat* (1912), *American Herons* (1912), *Bering Sea Birds* (1914). Pictures of the beaver and the cranes can be found at "Urban Landscapes from the Field Museum," Illinois Digital Archives, http://www.idaillinois.org/cdm/search/collection /fmnh2/searchterm/1912./field/all/mode/any/conn/and/order/subjec/ad/f /cosuppress/. Sally Metzler, *Theatres of Nature: Dioramas at the Field Museum* (Chicago: Field Museum, 2008).

31. "Art in a Natural History Museum," 99.

32. Myers, "A History of Herpetology" (n. 44 above, chap. 1), 80–86. See also Wonders, *Habitat Dioramas* (n. 182 above, chap. 1), 131–56.

33. Andrei, "Nature's Mirror," (n. 8 above, chap. 1).

34. Museum archives often retained biographical files of their staff members, however, and we have often been able to recover the names and work of otherwise invisible preparators from these files. Useful biographical sketches of exhibit makers can also be found at the end of Quinn, *Windows*

on Nature (n. 158 above, chap. 1), 170–73; Wonders, *Habitat Dioramas*, 231–34.

35. H. F. Osborn, "The Museum of the Future," *American Museum Journal* 11, no. 7 (1911): 225.

36. William James, letter to Jesse D. Figgins, 1912, JDF-CMNH.

37. J. A. McGuire, letter to B. M. Caraway, June 1, 1920, JDF-CMNH.

38. McGuire, the editor of *Outdoor Life*, didn't hesitate to give Caraway the hard sell, noting that the grizzly bears he had donated formed "one of the most conspicuous cases in the museum and it is admired everywhere and commented upon by everyone seeing it. If you could get a group of brown bears it would be even more wonderful and more interesting than this grizzly group of mine" (ibid.).

39. Donors were especially happy to provide funds for expeditions to exotic locales. East Africa was a popular destination with the yachting set, though South America, South Asia, and Alaska were close seconds. Patrons in smaller cities were more likely to sponsor expeditions that remained in the Western Hemisphere, likely because these were less expensive. Imperialist and racist ideologies and actions underscored many of these trips and, cultural theorists have shown, some of the dioramas themselves. For more on the social ideologies of dioramas, please see the "Science, Technology, and Nationalism" section of the bibliographic essay.

40. On dioramas and field research, see Osborn, "The Museum of the Future," 225; Charles C. Adams, letter to Alexander Ruthven, May 22, 1909, as cited in Kohler, *All Creatures*, (n. 2 above, chap. 1), 107–23.

41. Kohler, *All Creatures* (n. 2 above, chap. 1), 107.

42. Ibid., 108–9.

43. Frank Chapman, letter to Witmer Stone, May 9, 1912, cited in ibid., 113–14.

44. Such life histories helped to lay the groundwork for the modern study of ethology in the 1930s and 1940s and helped to link systematic zoology with other biological disciplines. See Barrow, *Nature's Ghosts* (n. 22 above, chap. 1), 260–301; Richard W. Burkhardt Jr., *Patterns of Behavior: Konrad Lorenz, Niko Tinbergen, and the Founding of Ethology* (Chicago: University of Chicago Press, 2005), 17–186; Kohler, *All Creatures*, 227–70; Nyhart, *Modern Nature* (n. 12 above, chap. 1), 293–354. For an excellent discussion of animal biographies and how they relate to institutional display transitions, especially the tension between museum and zoo approaches in this period, see Samuel J. M. M. Alberti, "Introduction: The Dead Ark," in *The Afterlives of Animals*, ed. Samuel J. M. M. Alberti (Charlottesville: University Press of Virginia, 2011).

45. As cited in Myers, "A History of Herpetology" (n. 44 above, chap. 1), 80.

46. See Noble's report of the Department of Herpetology, *AMNHAR* (1924), 75, AMNH. New "life history" dioramas were not the only way to increase visitor interest that Noble sanctioned. To "appeal to a great variety of visitors,"

some new exhibits were aimed at New Yorkers "interested in identifying particular reptiles or amphibians which they have met in their tramps," which meant displaying more local snakes and frogs, often acquired from Boy Scout troops or amateur naturalists (75).

47. Director's Monthly Report to Board of Trustees of the Milwaukee Public Museum, March 19, 1907, MPM.

48. Baker, "Some Observations on Museum Administration" (n. 206 above, chap. 1), 667.

49. One exception to this untidy record keeping was the American Museum, which published exacting attendance numbers in its annual reports, probably because so much municipal funding depended on them.

50. J. D. Figgins, letter to George L. Cannon, October 5, 1916, JDF-CMNH.

51. Homer R. Dill, "Building an Educational Museum as a Function of the University," in *Tenth Annual Meeting of the American Association of Museums*, ed. Paul M. Rea (San Francisco: Waverly Press, 1915), 84.

52. The population of Johnson County (the location of Iowa City) in 1910 was 25,914, according to Richard L. Forstall, ed., *Population of Counties by Decennial Census: 1900 to 1990* ([Washington, DC]: Bureau of the Census, Population Division, 1994). For more on the history of this cyclorama, see Wonders, *Habitat Dioramas* (n. 182 above, chap. 1), 137–39.

53. Warren H. Miller, "Aquarelles of Our Common Woodlands," *American Museum Journal* 15, no. 4 (1915): 167.

54. Ibid., 175.

55. R. W. Shufeldt, "Combining Art and Museum Exhibits," *Bulletin of the Pan-American Union* 48 (1919): 682.

56. On children in natural history museums, see Rebecca Onion, "Picturing Nature and Childhood at the American Museum of Natural History and the Brooklyn Children's Museum, 1899–1930," *Journal of the History of Childhood and Youth* 4, no. 3 (2011), 434–69.

57. On biology curricula in the Progressive Era, see DeBoer, *History of Ideas* (n. 34 above, chap. 1), 92–98; Kohlstedt, *Teaching Children Science* (n. 35 above, chap. 1), 37–58, 111–14; Pauly, *Biologists and the Promise of American Life* (n. 6 above, chap. 1), 171–93. Simultaneously, such lessons had found their way into primary and middle schools in Germany. Nyhart, *Modern Nature* (n. 12 above, chap. 1), 161–97, 296–322. It should be noted that a wide variety of other museum displays presented similar messages; Constance Clark has chronicled how paleontological displays did so (*God—or Gorilla* [n. 21, chap. 1]).

58. J. D. Figgins, letter to P. P. Claxton, n.d., 1914, JDF-CMNH.

59. On the history of this group's construction, see Wonders, *Habitat Dioramas* (n. 182 above, chap. 1), 131–32.

60. On the use of dioramas in New York City high school biology classes, e.g., see Pauly, *Biologists and the Promise of American Life,* 186. Sally Kohlstedt discusses how some nature study teachers believed a trip to a museum or

a botanical garden was a "means for bringing science to children." See
Kohlstedt, *Teaching Children Science* (n. 35 above, chap. 1), 173. On classroom culture of American schools in this period, see, among others, Larry
Cuban, *How Teachers Taught: Constancy and Change in American Classrooms,
1890–1990* (New York: Teachers College Press, 1993); Ian Grosvenor, Martin Lawn, and Kate Rousmaniere, eds., *Silences and Images: The Social History
of the Classroom*, History of Schools and Schooling (New York: Peter Lang
Publishing, 1999).

61. Beginning in the late 1920s, the American Museum's education department produced "school service leaflets," offering New York City teachers
assistance in teaching geography, history, civics, nature study, biology,
chemistry, physics, and hygiene. A film, *The School Service at the American
Museum of Natural History*, describes the program (New York: American
Museum of Natural History, 1927), Film Collection, no 271, AMNH.

62. Cain, "From Specimens to Stereopticons" (n. 5 above, chap. 1), 25–29.

63. *YBANS* (1929), 1–2.

64. H. L. Ward, letter to James L. Clark, May 3, 1917, Directors' Files, 1917,
MPM; Wonders, *Habitat Dioramas* (n. 182 above, chap. 1), 144.

65. On Figgins's work with Rowley, see Denver Public Library Clipping
Files, folder "Denver Museums and Cultural Centers, DMNH, Exhibits,"
DPL; Charles P. Johnson, "1300 Specimens from South America Will Be
Mounted Here," *Rocky Mountain News*, July 25, 1926.

66. Wonders, *Habitat Dioramas*, 145.

67. Shufeldt, "Combining Art and Museum Exhibits" (n. 52 above, this chap.),
690.

68. On visual inspirations for habitat dioramas, see Wonders, *Habitat Dioramas*, 12–45. On the history of the precinematic view in natural history
museums, see Michael Anderson, "Painting Actuality: Diorama Art of James
Perry Wilson," Yale Peabody Museum of Natural History, http://peabody
.yale.edu/james-perry-wilson (2012); Griffiths, *Wondrous Difference* (n. 70
above, chap. 1).

69. On artists' struggle for control over exhibition work in natural history
museums, see Michele Bogart, "Lowbrow/Highbrow: Charles R. Knight,
Art Work and the Spectacle of Prehistoric Life," in *American Victorians and
Virgin Nature*, ed. T. J. Jackson Lears (Boston: Isabella Stewart Gardner Mu-
seum, 2002), 39–63; Victoria Cain, "The Art of Authority: Exhibits, Exhibit
Makers and the Contest for Scientific Status at the American Museum of
Natural History, 1920–1940," *Science in Context* 24, no. 2 (2011), 215–38.

70. Sally Metzler, "The Artists," in her *Theatres of Nature* (n. 30 above, this
chap.).

71. Laurence Vail Coleman, "Some Principles of Group Construction," *Museum
Work* 3, no. 4 (1921): 121–22.

72. As early as 1904, Milwaukee Public Museum director Henry L. Ward wrote
that "we have goods to dispose of (ideas regarding nature) the same as

Gimbel Bros. have." H. L. Ward, letter to A. G. Wright, July 22, 1904, Director's Files, 1904, MPM. On the introduction of commercial techniques into museums, see Cain, "'Attraction, Attention, Desire'" (n. 32 above, chap. 1); Neil Harris, *Cultural Excursions: Marketing Appetites and Cultural Tastes in Modern America* (Chicago: University of Chicago Press, 1990); William Leach, *Land of Desire: Merchants, Power, and the Rise of a New American Culture* (New York: Vintage Books, 1993).

73. Franklin, "A Recent Development in Museum Groups" (n. 178 above, chap. 1), 111.

74. Ibid.

75. On the development of new taxidermic techniques in this period, see John Rowley, *Taxidermy and Museum Exhibition* (New York: D. Appleton and Company, 1925).

76. Henry L. Ward, letter to C. E. Akeley, June 28, 1912, General Correspondence, folder 1912, MPM.

77. This phenomenon occurred decades earlier in European museums and zoos: see Nyhart, *Modern Nature* (n. 12 above, chap. 1), 35–78. Also see John Berger, *About Looking* (New York: Pantheon Books, 1980); Nigel Rothfels, ed., *Representing Animals*, Theories of Contemporary Culture, ed. Daniel J. Sherman (Bloomington: Indiana University Press, 2002); Wonders, *Habitat Dioramas* (n. 182 above, chap. 1).

78. Coleman, "Some Principles of Group Construction," 121–22.

79. Wonders, *Habitat Dioramas*, 201.

80. Quinn, *Windows on Nature* (n. 158 above, chap. 1), 150–57. See also *The Field Museum Handbook* (Chicago: Field Museum, 1931), 27.

81. Keyes, "Commercial Tendencies" (n. 173 above, chap. 1).

82. Advance Theatrical Exchange, letter to Henry L. Ward, August 19, 1914; Wm. Potter and Sons, letter to Henry L. Ward, July 12, 1915; Chas. Polacheck & Bro. Co., letter to Henry L. Ward, February 14, 1913; Robert Brand & Sons Co., letter to Henry L. Ward, April 3, 1914 (all in General Correspondence, folders 1913–15, MPM).

83. A. R. Crook, "A Survey of State Natural History Museums," *PAAM* 3 (1909): 141–44.

84. On nationalism in representations of American landscape, see, among others, Angela Miller, "The Fate of Wilderness in American Landscape Art: The Dilemmas of 'Nature's Nation,'" in *American Wilderness: A New History*, ed. Michael Lewis (New York: Oxford University Press, 2007); and Joel Snyder, "Territorial Photography," in *Landscape and Power*, ed. W. J. T. Mitchell (Chicago: University of Chicago Press, 2002). On the ongoing presence of Romanticism and nationalism in American dioramas, see Wonders, *Habitat Dioramas* (n. 182 above, chap. 1), 182–91.

85. Wonders, *Habitat Dioramas*, 187–91.

86. Frank M. Chapman, "The Bird-Life and the Scenery of a Continent in One Corridor: The Groups in the American Museum of Natural History—a New Method in Museum Exhibition," *World's Work* 17 (March 1909): 11374.

87. While the nation's most famous diorama halls centered on exotic locales, many museum directors limited diorama construction to iconic or
endangered species in North America. In Milwaukee, e.g., Ward explained
that his museum was "not of a size to warrant our attempting to show
the mammals of the world," so would "expect ultimately to have a fairly
strong collection of North American mammals with some few foreign
ones" thrown into the array of groups and dioramas he was building. (He
did place orders with Carl Akeley to acquire rhinoceros, hippopotamus,
and bull elephants for the museum's displays, selecting those animals
"because of their place in literature, making a knowledge of them essential
to ordinary people.") Henry L. Ward, letter to James L. Clark, April 4, 1911,
Director's Files, 1911, MPM.
88. Niedrach trained under taxidermist Al Rogers, who had worked at the
American Museum before his former colleague, Jesse Figgins, persuaded
him to come to Colorado. It's likely that Figgins himself intended to paint
the background, but he left the museum in the mid-1930s, and the accomplished Love, who began as a Works Progress Administration employee
before being hired as a full-time staff member, executed the work instead.
On Love, see Wonders, *Habitat Dioramas*, 233.
89. On ways in which gender roles influenced visualizations of the biological
sciences, see, e.g., Donna Haraway, *Primate Visions: Gender, Race, and Nature
in the World of Modern Science* (New York: Routledge, 1989); Ludmilla Jordanova, *Sexual Visions: Images of Gender in Science and Medicine between the
Eighteenth and Twentieth Centuries* (Madison: University of Wisconsin Press,
1989); Londa Schiebinger, *Nature's Body: Gender in the Making of Modern
Science* (Boston: Beacon Press, 1993); Ann B. Shteir and Bernard Lightman,
eds., *Figuring It Out: Science, Gender, and Visual Culture* (Hanover, NH: University Press of New England, 2006).
90. William Hornaday, "Masterpieces of American Taxidermy," *Scribner's Magazine* (July 1922), 13.
91. "Natural History Museum Attracts Record Crowds," *New York Times*,
August 3 1924, X11.
92. *YBANS* (1930), 4–5; *YBANS* (1931) 5.
93. *YBANS* (1929), 1–2.
94. Wissler, "Survey of the American Museum of Natural History" (n. 2 above,
chap. 1), 9.
95. Jesse D. Figgins, letter to George L. Cannon, October 6, 1916, JDF-CMNH.
96. Panoramas, theatrical spectacles, aquaria, and expositions had long attempted to re-create worlds distant in time and place. On the history of
such virtual environments, see Brunner, *The Ocean at Home* (n. 97 above,
chap. 1); Robert W. Rydell, *All the World's a Fair: Visions of Empire at American International Expositions, 1876–1916* (Chicago: University of Chicago
Press, 1984); Vanessa R. Schwartz, *Spectacular Realities: Early Mass Culture in
Fin-de-Siècle Paris* (Berkeley: University of California Press, 1998); James A.
Secord, "Monsters at the Crystal Palace," in *Models: The Third Dimension*

of Science, ed. Nick Hopwood and Soraya de Chadarevian (Stanford, CA: Stanford University Press, 2004).

97. "Sets New Pace in Bird Groups in Full Flight: American Museum Achieves Unique Sky Mounting," *Museum News* 3, no. 11 (1925): 1, 3.

98. The American Museum's African Hall, designed by Carl Akeley and completed by James Clark, is the most famous example of this, and historians have extensively chronicled its development. See, e.g., Andrei, "Nature's Mirror," (n. 8 above, chap. 1); Bodry-Sanders, *Carl Akeley* (n. 161 above, chap. 1); Donna Haraway, "Teddy Bear Patriarchy," in *Primate Visions: Gender, Race and Nature in the World of Modern Science* (New York: Routledge, 1989); Quinn, *Windows on Nature* (n. 158 above, chap. 1), 25–78. Timothy W. Luke, "Museum Pieces: Politics and Knowledge at the American Museum of Natural History," *Australasian Journal of American Studies* 16 (1997): 15.

99. The decor must have delighted donor Arthur Vernay, who ran an interior design and antique business. Douglas J. Preston, *Dinosaurs in the Attic: An Excursion into the American Museum of Natural History* (New York: St. Martin's Press, 1986), 81.

100. H. E. Anthony, "Southern Asia in the American Museum: The New Vernay-Faunthorpe Hall of South Asiatic Mammals," *Natural History* 30, no. 6 (1930): 590, 592.

101. On the relationship of the cathedral to the diorama hall, see Alison Griffiths, *Shivers Down Your Spine: Cinema, Museums, and the Immersive View* (New York: Columbia University Press, 2008).

102. Osborn, *Creative Education* (n. 35 above, chap. 1), 230. Dioramas depicted a very particular vision of wilderness, one that was as much a product of nineteenth-century Romanticism as any kind of reality. See N. Katherine Hayles, "Simulated Nature and Natural Simulations: Rethinking the Relation between the Beholder and the World," in *Uncommon Ground: Toward Reinventing Nature*, ed. William Cronon (New York: W. W. Norton, 1995); Miller, "Fate of Wilderness" (n. 83 above, this chap.).

103. Dickerson, "The 'Toad Group' in the American Museum" (n. 23 above, chap. 1), 166.

104. Mary Cynthia Dickerson and Bashford Dean, "Recent and Extinct Fishes, Existing Reptiles and Batrachians," in *AMNHAR* (1918), 72–76.

105. Willard G. Van Name, "Zoological Aims and Opportunities," *Science* 50, no. 1282 (1919): 81–82. In *Nature's Ghosts*, Mark Barrow describes how Van Name's colleague, associate curator of ornithology Waldron DeWitt Miller, took a similarly aggressive approach but targeted serious bird-watchers along with his scientific colleagues (n. 22 above, chap. 1), 234–59.

106. Conservation in this period derived from many motives, not all of them benign, and some of the most active conservationists in museums—e.g., American Museum trustee Madison Grant—urged the preservation of wilderness and wildlife for racist and nationalist reasons.

107. On the history of the American Committee for International Wild Life Protection, see Barrow, *Nature's Ghosts* (n. 22 above, chap. 1), 135–67.

108. William Temple Hornaday, *A Wild Animal Round-Up* (New York: C. Scribner's Sons, 1925), 274.

109. Barrow, *Nature's Ghosts*, 201–33; Gregg Mitman, *The State of Nature: Ecology, Community, and American Social Thought, 1900–1950* (Chicago: University of Chicago Press, 1992).

110. The cyclorama's popularity was not enough to save the island's habitat. In fact, by the time it was built, there is evidence that, of the three species that went extinct on the island—the flightless Laysan rail, the honeycreeper, and the Millerbird—at least one had already been wiped out. On these extinctions, see Storrs L. Olson, "History and Ornithological Journals of the Tanager Expedition of 1923 to the Northwestern Hawaiian Islands, Johnston and Wake Islands," *Atoll Research Bulletin* 433 (1996): 1–210. On the production of the cyclorama, see Wonders, *Habitat Dioramas* (n. 18? above, chap. 1), 136–40.

111. This was a bit self-congratulatory on Dill's part. Though the assistant professor of zoology eventually became the director of the University of Iowa's museum, at the time he was himself a full-time taxidermist and the chief of exhibits. Dill, "Correlation of Art and Science" (n. 164 above, chap. 1).

112. *CMNHAR* (1936), n.p.

113. The American Museum was famous for paying its artists well, though their salaries often depended on the department and the curator with whom they were affiliated. (Frank Chapman, e.g., kept a tight rein on artists working for the Department of Ornithology, and an even tighter fist around their paychecks.) High pay for preparators ended in the 1930s, as the makeup of the museum's administration changed, and the Depression severely constrained museum finances. See Anderson, "1934: Joining the American Museum of Natural History," chap. 5 of "Painting Actuality" (n. 67 above, this chap.).

114. Lawrence Vail Coleman, "A Newspaper for Museumists," *Science* 59, no. 118 (1924): 126.

115. James L. Clark, *Good Hunting: Fifty Years of Collecting and Preparing Habitat Groups for the American Museum* (Norman: University of Oklahoma Press, 1966), 11.

116. On the history of taxidermists' scientific knowledge, see among others, Andrei, "Nature's Mirror," (n. 8 above, chap. 1); Cain, "The Art of Authority" (n. 68 above, this chap.); Nyhart, *Modern Nature* (n. 12 above, chap. 1); Star, "Craft vs. Commodity" (n. 20 above, chap. 1). On the broader history of the scientific knowledge of artisans and artists, see Lorraine Daston and Peter Galison, *Objectivity* (New York: Zone Books, 2007); Pamela H. Smith, *The Body of the Artisan: Art and Experience in the Scientific Revolution* (Chicago: University of Chicago Press, 2004).

117. Jaques's statements, however hyperbolic, were not entirely unfounded; the museum's ornithologists admitted he was a formidable observer and admired his knowledge of birds. Even Chapman gave considerable room to Jaques, explaining that his combination of field experience, artistic

skill, and scientific knowledge was "quite unique" and enabled him to "paint both backgrounds and the birds that occupy them in a manner that please equally the scientists and the artist." Francis Lee Jaques, "Francis Lee Jaques," unpublished manuscript (1967), chap. 17, p. 4, Rare Books Collection, AMNH. R. C. Murphy, minutes of the forty-fifth meeting of the Departmental Committee on the Whitney Wing, February 5, 1942, pp. 1–2, DMAP-AMNH.

118. Carl Akeley, "African Hall, a Monument to Primitive Africa," *Mentor* 12, no. 12 (1926): 22, as cited in Wonders, *Habitat Dioramas* (n. 182 above, chap. 1), 172.

119. Scientists eventually made use of dioramas in precisely the way Akeley predicted they might. Researchers at the American Museum's Center for Biodiversity and Conservation, engaged in a long-term study of the Andros Coral Reef in the Bahamas, have consulted the museum's diorama of the reef and the data used to construct it to glimpse the rich biodiversity that characterized it in the 1920s. Quinn, *Windows on Nature* (n. 158 above, chap. 1), 145.

120. Throughout the 1910s and the 1920s, e.g., American Museum president Henry Fairfield Osborn explicitly directed the museum's publicists to dramatize the exploits and personalities of the museum's "naturalist-explorers" and their adventures in the "wild places of the world." American Museum of Natural History press releases, March 22, 1928, and September 17, 1924, as cited in Kennedy, "Philanthropy and Science" (n. 7 above, "Introduction"), 157.

121. Though their presence was only rarely acknowledged, women routinely accompanied their husbands on these expeditions, working alongside other expedition members, locating, shooting, and trapping specimens, recording environs and habits of the animals on film or paper. Blair and William Beebe, Frank and Fanny Chapman, Sally and James Clark, Carl and Delia—and later Mary Jobe—Akeley, Yvette and Roy Chapman Andrews, and Muriel and Alfred M. "Bill" Bailey were just a few of the many couples who traveled together on museum-sponsored collecting and research expeditions. (Sportsmen expressed special admiration for the marksmanship of Sally Clark and both Akeley wives, who brought down some of the largest elephants on record.)

122. These stories were not exaggerations. Carl Akeley, e.g., was attacked by a wounded leopard, which he strangled with his bare hands. Later on, a bull elephant crushed his chest, and he lay in the jungle, listening to the yelps of circling hyenas, until rescued the next day. He sweated through bouts of blackwater fever and, eventually, died in the jungle. A "macho, 'studso'" bunch, as painter Thanos Johnson put it, exhibit makers prided themselves on their physical and mental toughness. As cited in Anderson, "1934," chap. 5 of "Painting Actuality" (n. 67 above, this chap.). On Akeley, see Andrei, "Nature's Mirror," (n. 8 above, chap. 1); Bodry-Sanders, *Carl Akeley* (n. 161 above, chap. 1); Dickerson, "New African Hall" (n. 23 above,

chap. 1); Haraway, "Teddy Bear Patriarchy" (n. 97 above, this chap.). On Pritchard, see Malcolm L. M. Vaughan, "Painting Beauty under the Sea," *Los Angeles Times*, July 8 1928, K12.

123. For more on the relationship between physical risk or suffering and scientific credibility, see R. M. Herzig, *Suffering for Science: Reason and Sacrifice in Modern America* (New Brunswick, NJ: Rutgers University Press, 2005); Naomi Oreskes, "Objectivity or Heroism? On the Invisibility of Women in Science," in "Science in the Field," ed. Henrika Kuklick and Robert Kohler, special issue, *Osiris* 2nd ser., 11, (1996): 87–113.

124. James L. Clark, diary of the British East Africa Expedition, 1909, p. 6, AMNH.

125. James L. Clark, diary of the Morden-Clark Asiatic Expedition, February 22, 1926, n.p., AMNH.

126. Bodry-Sanders, *Carl Akeley* (n. 161 above, chap. 1), 229.

127. Clark, *Good Hunting* (n. 114 above, this chap.), 97.

128. As Francis Lee Jaques later observed, changes to the U.S. tax code likely added to museums' considerable haul in these years—after 1917, any money a patron gave to a museum expedition was considered a tax deduction, a financial incentive that made donors considerably more generous. See Jaques, "Francis Lee Jaques" (n. 116 above, this chap.), chap. 24, p. 1.

129. Press sponsorship of such events was not uncommon. The practice had occurred as early as the 1870s, when the *New York Herald* sponsored Henry Morton Stanley's trip to Zanzibar (now Tanzania). The Sunday issues of local newspapers and popular illustrated periodicals such as *Survey Graphic*, *Century Illustrated*, and *Popular Science Monthly* served as popular outlets for the writing and artwork of many curators and exhibit makers; with the rise of the picture press in the late 1920s and 1930s, natural history museums' expeditions and exhibits were well chronicled by magazines like *Fortune*, *Life*, *Popular Mechanics*, and *Boys' Life*.

130. Memorandum from Field Museum director D. Dwight Davis to zoology curator Wilfred Osgood, July 1, 1926, as cited in Metzler, "The Artists" (n. 69 above, this chap.), n.p.

131. Though eminent paleontologist Walter Granger accompanied Andrews on his many explorations of Central Asia and was responsible for finding most of the fossils there, it was Andrews, a handsome but only mildly accomplished scientist that the American public loved. The swashbuckling, pistol-packing curator won heroic acclaim rivaling Lindbergh's, especially after his location of dinosaur eggs on a 1925 trip to Mongolia's Flaming Cliffs. American moviegoers excitedly watched Andrews discover the eggs—an act that must have been cleverly and promptly restaged, as the khaki-clad Andrews was not the first member of the expedition to find the eggs, nor were the cameras rolling when the discovery was called to his attention. Michael J. Novacek, "Foreword," in *Dragon Hunter: Roy Chapman Andrews and the Central Asiatic Expeditions* (New York: Viking, 2001), xi. Also see Charles Coles and Charles Russell, *Men of Science*, film (New York: American Museum of Natural History, 1938).

132. T. D. A. Cockerell, letter to H. F. Osborn, February 16, 1927, Henry Fairfield Osborn Papers, box 5, folder 12, AMNH.
133. Kennedy, "Philanthropy and Science" (n. 7 above, "Introduction"), 158.
134. Lurie, *A Special Style* (n. 7 above, chap. 1), 63.
135. Clark Wissler, "Survey of the American Museum of Natural History" (n. 2 above, chap. 1), 204
136. See, e.g., the exchange of heated letters between Harold Anthony and James Clark in October 1933, in AMNH Mammalogy Correspondence, Archival Collections, Department of Mammalogy, Division of Vertebrate Zoology, DM-AMNH.
137. Clark Wissler, "The Museum Exhibition Problem," *Museum Work* 7, no. 6 (1925): 173.
138. Yochelson, *National Museum of Natural History* (n. 95 above, chap. 1), 57.
139. The nation's largest museums might send out ten or more expeditions at a time. See Kohler, *All Creatures* (n. 2 above, chap. 1), 122.
140. Mary Cynthia Dickerson, "A Note on the Giant Salamander Group. Some Problems in Panoramic Group Construction," *American Museum Journal* 12, no. 8 (1912): 313.
141. Myers, "A History of Herpetology" (n. 44 above, chap. 1), 65, 216.
142. This was particularly true in smaller or understaffed museums, where everyone needed to be a jack-of-all-trades. On the desired attributes of a curator, see A. R. Crook, "The Training of Museum Curators," *PAAM* 5 (1911), 59–63.
143. Henry L. Ward, "Director's Monthly Report to the Board of Trustees," November 4, 1920, 476, MPM; appointment book of Henry Ward, 1920, DF-MPM. Carlos E. Cummings, *East Is East and West Is West* (Buffalo, NY: Buffalo Museum of Science, 1940), 151.
144. Chapman, *Autobiography of a Bird-Lover* (n. 23 above, chap. 1), 72.
145. Henry L. Ward, letter to A. L. Kroeber, January 6, 1909, Directors' Files 1909, MPM . Also see Henry L. Ward, letter to Paul Radin, March 24, 1916, DF-MPM.
146. Cain, "The Art of Authority" (n. 68 above, this chap.), 224–25.
147. Minutes from a meeting of the Council of Scientific Staff, August 5, 1935, as prepared by H. E. Anthony, Department of Mammalogy Administrative Papers, box 1, folder Council of Scientific Staff: Correspondence I, 1934–35, DM-AMNH.
148. Clark, *Good Hunting* (n. 114 above, this chap.), 96.
149. Minutes from meeting of Division 2, February 17, 1928, prepared by H. E. Anthony, Department of Mammalogy Administrative Papers, box 2 folder AMNH Departments "Division II," 1922–30, DM-AMNH.
150. At the American Museum of Natural History, e.g., three departments out of fourteen—ornithology, mammalogy and paleontology—received over 50 percent of the American Museum's budget between 1920 and 1930. (Though there were usually fourteen departments, the museum periodically merged or separated departments, so there were occasionally thirteen

or fifteen.) Wissler, "Survey of the American Museum of Natural History" (n. 2 above, chap. 1), 64.

151. George Sherwood, letter to Roy W. Miner, April 28, 1926, as cited in Kennedy, "Philanthropy and Science" (n. 7 above, "Introduction"), 204; Wissler, "Survey of the American Museum of Natural History."

152. Herpetology was a section in a unified department of ichthyology and herpetology until 1920, when the department was divided in two. Mary Cynthia Dickerson, letter to Henry Fowler, March 11, 1919, Department of Herpetology Archives, Dickerson Collection, AMNH; Myers, "A History of Herpetology" (n. 44 above, chap. 1), 13, 15.

153. After Dickerson's departure, G. K. Noble became the head of herpetology, and under his leadership, herpetological research at the American Museum improved further. Gladwyn Kingsley Noble, "Her Studies of Reptiles and Amphibians," *American Museum Journal* 23, no. 5 (1923): 516.

154. "Summary of Replies to Request for a Statement as to the Outcome If No Increases Are Made in the Budget of 1935," from Report on Scientific Plant, Equipment and Personnel, December 1935, box 1, folder AMNH Committees—Budget, DMAP-AMNH.

155. As cited in Kohler, *All Creatures* (n. 2 above, chap. 1), 122.

156. *CMAR* (1913), 10.

157. Ibid., 11.

158. Robert E. Kohler, "Finders/Keepers: Collecting Sciences and Collecting Practice," *History of Science* 45 (2007), 1–27. On paleontology collecting practices in a slightly earlier period, see Brinkman, *Second Jurassic Dinosaur Rush* (n 105 above, chap. 1).

159. See, e.g., Rockwell, letter to Clark, July 26, 1936, in DM-AMNH.

160. Frank Chapman, letter to Wilfred Osgood, October 31, 1928, as cited in Kohler, *All Creatures*, 122.

161. Ray De Lucia, oral history interview by Sharon Zane, pp. 13–14, 1997, American Museum of Natural History Oral History Collection, 1975–2002, AMNH.

162. H. E. Anthony, "Outline of Plans for the Department of Mammals for the Next Few Years," 1934, folder H. E. Anthony XI Administrative, 1919–35, DM-AMNH.

163. Though Holland was the director of the Carnegie Museum, he was a passionate entomologist in his own right and frequently took the side of his research-oriented curators when the board suggested cutting research budgets to fund more displays. See, e.g., *CMAR* (1921), 15–17.

164. For more on the response by anthropologists to this split, see Anna Laura Jones, "Exploding Canons: The Anthropology of Museums," *Annual Review of Anthropology* 22 (1993): 202.

165. H. F. Osborn, letter to W. B. Scott, June 11, 1932, as cited in Kennedy, "Philanthropy and Science" (n. 7 above, "Introduction"), 217.

166. Joseph Grinnell, letter to Annie Alexander, May 26, 1921, as cited in Kohler, *All Creatures* (n. 2 above, chap. 1), 122.

167. They wrote letters to one another about the lack of publication venues and the stacks of work—completed and unfinished—piling up on their desks. *"Younger men can wait, I cannot,"* wrote ichthyologist and associate curator E. W. Gudger to curator William K. Gregory in 1935. "Publication is the mouthpiece of the research of the museum," protested Frank Chapman. "What are we working for? We cannot urge our men to go on." E. W. Gudger, letter to William K. Gregory, June 11, 1935, William King Gregory Papers, box 17, folder 1–4, AMNH; minutes from a meeting of the Council of Scientific Staff, August 5, 1935, as prepared by H. E. Anthony, Department of Mammalogy Administrative Papers, box 1 folder AMNH; Council of Scientific Staff: Correspondence I, 1934–35, AMNH.

168. Kennedy, "Philanthropy and Science" (n. 7 above, "Introduction"), 214.

169. Throughout the 1910s, Ward actually tried to wangle more money for research, but the city representatives responsible for funding the museum regularly refused to provide the necessary funds. MPM director's monthly report to the board of trustees, February 10, 1920, 316–317, MPM.

170. As Ward explained in 1916, "This is a moderate sized museum in which we have about fifty employees all told including the power department and the janitor forces. We are of necessity under-manned in the scientific department, and I have not only the directorship of the museum and the secretaryship of the board of trustees but also am acting curator of birds, mammals, mineralogy, geology, and paleontology. My other duties render it impossible to devote very much time to these particular departments, and this is especially so in the department of mineralogy and geology." Henry L. Ward, letter to Ira Edwards, June 7, 1916, DF-MPM.

171. Barton W. Evermann, letter to F. A. Lucas, February 5, 1923, box 1, folder Steinhardt Aquarium, FAL.

172. Henry L. Ward, letter to C. B. Weil, June 23, 1915; Henry L. Ward, letter to Ira Edwards, June 7, 1916, both CF-MPM.

173. Lucas had given more public vent to his feelings in a satirical speech at the 1919 AAM meeting, when he urged museum directors to hire curators based on "their lack of interest in the public." "They should preferably be engaged in some research of personal interest," he instructed his chuckling audience, "if possible on some abstruse subject that cannot be finished during their lifetime and will be promptly rejected by their successors." Herbert Schwartz, letter to F. A. Lucas, August 4, 1922, FAL; F. A. Lucas, "Modern Principles of Museum Administration," *Museum Work*, vol. 2, no. 1 (1919).

174. Henry L. Ward, letter to Otto von Schlichten, July 23, 1912, CF-MPM.

175. Before 1913, Swarth had worked at the Field Museum, where he had also been discouraged from engaging in field research beyond Chicago's most rural suburbs. Frank Daggett, letter to board of governors, August 31, 1916, LAM.

176. A. W. Anthony, letter to J. D. Figgins, February 12, 1923, DF-CMNH.

177. Vernon Bailey, "Obituary—Alfred Webster Anthony," *Auk* 58 (1941): 440.
178. For a detailed exploration of tensions between these groups, see Kohler, *Landscapes and Labscapes* (n. 18 above, chap. 1), 4–5.
179. On the historical emergence of these fields, see Barrow, *Nature's Ghosts* (n. 22 above, chap. 1); Paul Lawrence Farber, *Finding Order in Nature: The Naturalist Tradition from Linnaeus to E. O. Wilson*, Johns Hopkins Introductory Studies in the History of Science, ed. Mott T. Greene and Sharon Kingsland Greene (Baltimore, MD: Johns Hopkins University Press, 2000); Kingsland, *Evolution of American Ecology* (n. 146 above, chap. 1); Mitman, *The State of Nature* (n. 108 above, chap. 2).
180. Adams, "New Natural History" (n. 18 above, this chap.), 491.
181. Chapman, *Autobiography of a Bird-Lover* (n. 23 above, chap. 1), 71. For Chapman's contemporary justification of his commitment to museum education, see his article "Natural History for the Masses," *World's Work* 5 (November 1902): 2761–70.
182. Chapman, *Autobiography of a Bird-Lover*, 205.
183. Ilva Jones, "Alfred M. Bailey," unpublished manuscript, n.d., p. 6, Bailey Biography Files, CMNH.
184. Barnum Brown, letter to Charles H. Hannington, December 30, 1935, CH-CMNH.
185. J. D. Figgins, letter to Frederick C. Lincoln, August 23, 1917, JDF-CMNH.
186. J. D. Figgins, letter to W. C. Henderson, 20 July 1920, JDF-CMNH.
187. J. D. Figgins, letter to Frederick C. Lincoln, August 23, 1917, JDF-CMNH.
188. Ibid.
189. Lincoln's condescension toward exhibit preparation must have wounded his former mentor deeply. Figgins took great pride in the Colorado Museum's exhibits and dioramas and considered their construction a major public and scientific contribution.
190. An accomplished biologist, writer, and administrator, Lincoln ran the bird-banding program for the U.S. Biological Survey from 1920 until 1946. In his time at the survey, he wrote dozens of articles, chapters, and books on bird banding and migration. Lincoln is also an excellent example of the still porous character of museum and field work at this time; despite the increasing professionalization of the institutions in which he worked, he had neither PhD nor college education and yet was still able to advance quickly at both the Colorado Museum and the Biological Survey. Ira N. Gabrielson, "Obituary: Frederick C. Lincoln," *Auk* 79, no. 3 (1962): 495–99; Frederick C. Lincoln, "The History and Purposes of Bird Banding," *Auk* 38, no. 2 (1921): 217–28; R. M. Wilson, *Seeking Refuge: Birds and Landscapes of the Pacific Flyway* (Seattle: University of Washington Press, 2010), 73–75.
191. Allen R. Phillips, "In Memoriam: Alfred M. Bailey," *Auk* 98, no. 1 (1981): 173–75.
192. Baker wound up in Syracuse, at the New York State College of Forestry. Frank C. Baker, letter to Henry L. Ward, June 10, 1915, CF-MPM.

193. Wissler, "The Museum Exhibition Problem" (n. 136 above, this chap.), 173.

194. Lucas, *Fifty Years of Museum Work* (n. 13 above, chap. 1), 4.

195. Council minutes, April 2, 1935, Department of Mammalogy Administrative Papers, box 1 of 5, folder Council of Scientific Staff, Minutes I, 1935–36, AMNH.

CHAPTER THREE

1. The term "life science" emerged in the late 1920s and early 1930s as an umbrella category that encompassed biology, psychology, biochemistry, biophysics, and medicine. On the development of the term, see, e.g., William E. Ritter, "An Untilled Field for a Revised Kind of Research in Zoology," *Condor* 31, no. 4 (1929): 160–66.

2. Minutes from special meeting of Division II, May 21, 1928, Department of Mammalogy Administrative Papers, box 2 folder "Division II," 1922–30, AMNH.

3. George Sherwood, memo to H. E. Anthony et al., February 15, 1929, Department of Mammalogy Administrative Papers, box 3 folder Exhibitions, 1922–59, AMNH.

4. Harold J. Cook, letter to C. H. Hannington, November 27, 1935, CH-CMNH.

5. Laurence Vail Coleman, U.S. Department of the Interior, Office of Education, *Biennial Survey of Education in the United States, 1928–1930: Chapter 12, Recent Progress and Condition of Museums*, Bulletin of the United States Department of the Interior, no. 20 (Washington, DC: Government Printing Office, 1931), 1:768.

6. Edwin E. Slosson, "Science from the Side-Lines," *Century Magazine* 81 (January 1922): 471–74. On the Science Service, see David Rhees, "A New Voice for Science: Science Service under Edwin E. Slosson, 1921–1929" (MA thesis, University of North Carolina, 1979).

7. The literature on interwar efforts to popularize science is vast; for a list of useful sources, see the popular science section of the bibliographic essay.

8. On the place of science in interwar world's fairs, especially as presented by corporate sponsors, see Joseph P. Cusker, "The World of Tomorrow: Science, Culture, and Community at the New York World's Fair," in *Dawn of a New Day: The New York World's Fair, 1939–194*, ed. Helen A. Harrison and Joseph P. Cusker (New York: Queens Museum/New York University Press, 1980), 211–29; Peter Kuznick, "Losing the World of Tomorrow: The Battle over the Presentation of Science at the 1939 World's Fair," *American Quarterly* 46, no. 3 (1994): 341–73; Roland Marchand, *Creating the Corporate Soul: The Rise of Public Relations and Corporate Imagery in American Big Business* (Berkeley: University of California Press, 1998), 249–311; Fred Nadis, *Wonder Shows: Performing Science, Magic, and Religion in America* (New Brunswick, NJ: Rutgers University Press, 2006), 179–210; Robert W. Rydell, *World of Fairs: The Century-of-Progress Expositions* (Chicago: University of Chicago

Press, 1993), 38–60, 92–114; Warren Susman, "The People's Fair: Cultural Contradictions of a Consumer Society," in *Dawn of a New Day*, ed. Harrison and Cusker, 211–29.

9. The Rockefeller Foundation, e.g., sponsored a study and conference in 1938–39 called the "Interpretation of the Natural Sciences," which focused specifically on the dissemination of scientific information to the public. See Science Service Records, Series 9, box 381–384, SIA. On foundation support for interwar biology education, see, among others, William J. Buxton and Manon Niquette, " 'Sugar-Coating the Educational Pill': Rockefeller Support for the Communicative Turn in Science Museums," in *Patronizing the Public: American Philanthropy's Transformation of Culture, Communication and the Humanities*, ed. William J. Buxton (Lanham, MD: Lexington Books, 2009), 153–94; Lily E. Kay, *The Molecular Vision of Life: Caltech, the Rockefeller Foundation, and the Rise of the New Biology* (New York: Oxford University Press, 1996), 22–50.

10. On the way Americans understood the relationship between biology education and citizenship in the first half of the twentieth century, see Eric Engles, "Biology Education in the Public High Schools of the United States from the Progressive Era to the Second World War: A Discursive History" (PhD diss., University of California, Santa Cruz, 1991).

11. Parr, "On the Functions of the Natural History Museum," Department of Mammalogy Administrative Papers, box 2, folder AMNH CSS Correspondence IV, 1940–41, DMAP-AMNH, 2.

12. Perry Osborn, memo to Department of Mammalogy, Department of Mammalogy Administrative Papers, box 2 folder AMNH CSS Correspondence IV, 1939, AMNH.

13. Benjamin C. Gruenberg, *Science and the Public Mind* (New York: McGraw-Hill, 1935), 138. On Gruenberg's critical role in biology education, see Pauly, *Biologists and the Promise of American Life* (n. 6 above, chap. 1), 171–93; Adam R. Shapiro, "Civic Biology and the Origin of the School Antievolution Movement," *Journal of the History of Biology* 41 (2008): 409–33.

14. The Scopes Trial also prompted educators outside the museum to pay closer attention to the educational effectiveness of displays—as Dartmouth president Ernest Hopkins wrote to American Museum president Henry Fairfield Osborn, "when so much ill-feeling and controversy had arisen over the life of past ages," clarity in museum exhibition was more important than ever, for "the American people are largely dependent for fact on those evidences painstakingly gathered or exhibited by the Museum." As cited in Kevin Dann, *Across the Great Border Fault: The Naturalist Myth in America* (New Brunswick, NJ: Rutgers University Press, 2000), 97. For overviews of the evolution controversy of the 1920s, see, among others, Clark, *God—or Gorilla* (n. 21 above, chap. 1); Edward J. Larson, *Summer for the Gods: The Scopes Trial and America's Continuing Debate over Science and Religion* (New York: Basic Books, 1997).

15. On the controversy over early evolution exhibits at the AMNH, see Julie Homchick, "Objects and Objectivity: The Evolution Controversy at the American Museum of Natural History, 1915-1928," *Science and Education* 19 (2010): 485–503. Religious fundamentalists were not the only ones questioning Darwinism and natural selection in these years. As Ernst Mayr points out, many leading geneticists, zoologists, and botanists embraced alternate theories of evolution in the 1920s and 1930s; his own employer, American Museum board president and paleontologist Henry Fairfield Osborn, never fully reconciled himself to Darwinism. Nonscientists had an even more difficult time with these concepts. Ernst Mayr, "Prologue: Some Thoughts on the History of the Evolutionary Synthesis," in *The Evolutionary Synthesis*, ed. Ernst Mayr and William B. Provine (Cambridge, MA: Harvard University Press, 1998), 3.

16. "Quotations, William Jennings Bryan on Evolution," *Science* 55, no. 1418 (1922): 242.

17. On creationists' challenges to the museum's exhibits in the 1920s, see Clark, *God—or Gorilla* (n. 21 above, chap. 1); Charlotte Porter, "The Rise of Parnassus: Henry Fairfield Osborn and the Hall of the Age of Man," *Museum Studies Journal* 1 (1983), 26–34; Rainger, *Agenda for Antiquity* (n. 21 above, chap. 1).

18. William King Gregory, letter to George Sherwood, April 4, 1924, Closed File 209c, AMNH Central Archives, as cited in Sheila Ann Dean, "What Animal We Came From: William King Gregory's Paleontology and the 1920's Debate on Human Origins" (PhD diss., Johns Hopkins University, 1994), 219.

19. Adams, "The New Natural History" (n. 18 above, chap. 2).

20. Henry Fairfield Osborn, letter to Percy Pyne, 21 April, 1910; and Henry Fairfield Osborn, letter to William Berryman Scott, 2 July, 1927, both as cited in Kennedy, "Philanthropy and Science" (n. 7 above, "Introduction"), 212. Though he stoutly defended modern taxonomic practice, botanical geneticist Edgar Anderson admitted that biologists "frequently considered" taxonomists "hopelessly conservative." Edgar Anderson, *Plants, Man and Life* (Cambridge: Cambridge University Press, 1952), 40–41, as cited in Kohler, *All Creatures* (n. 2 above, chap. 1), 240.

21. See, e.g., Kohler, *Landscapes and Labscapes* (n. 18 above, chap. 1), 23–40.

22. Julian Huxley, "Introduction," in *The New Systematics*, ed. Julian Huxley (London: Oxford University Press, 1940), 1–2. For historical accounts of the new systematics, see Jonathan Harwood, *Styles of Scientific Thought: The German Genetics Community, 1900–1933* (Chicago: University of Chicago Press, 1993); Ernst Mayr, "The Role of Systematics in the Evolutionary Synthesis," in *Evolutionary Synthesis*, ed. Mayr and Provine. On the important organizational role played by systematists-naturalists, see V. Betty Smocovitis, "Organizing Evolution: Founding the Society for the Study of Evolution (1939–1950)," *Journal of the History of Biology* 27, no. 2 (1994): 241–309.

23. Such tools were essential for modern taxonomists, who studied ecology and behavior along with specimen collections in order to understand more

about variation in local populations, transitional forms and speciation, and sibling species On the tools of modern taxonomists, see Kohler, *All Creatures*, 277. See also Ernst Mayr, *Principles of Systematic Zoology* (New York: McGraw-Hill, 1969), 105, 183–85.

24. The Smithsonian, the Field, and the American Museum responded to taxonomists' pleas in the 1920s and 1930s, but most museums did not acquire such facilities until well into the 1950s. On the history of museum-affiliated field stations in the interwar years, see Dann, *Across the Great Border Fault* (n. 14 above, this chap.), 89–99.

25. For a quick overview of the relationship between curators at the American Museum of Natural History and their academic colleagues, e.g., see *AMNHAR* (1930), 10–11.

26. Though researchers at the Field Museum, the Smithsonian, and a handful of smaller natural history museums continued to contribute to systematics during the 1910s, 1920s, and 1930s, the curators at the American Museum of Natural History were at the vanguard of museum-based research in these years and made pivotal contributions to the study of ecology and ethology. On the history of research on life histories, animal behavior, and animal psychology in interwar museums in the United States, see, among others, Barrow, *A Passion for Birds* (n. 188 above, chap. 1), 172–75; Burkhardt, *Patterns of Behavior* (n. 43 above, chap. 2), 361–69; Donald A. Dewsbury, "Frank Ambrose Beach," *Biographical Memoirs, National Academy of Sciences* 73 (1998), 64–85; Erika L. Milam, *Looking for a Few Good Males: Female Choice in Evolutionary Biology* (Baltimore, MD: Johns Hopkins University Press, 2010), 54–79; Gregg Mitman and Richard W. Burkhardt Jr., "Struggling for Identity: The Study of Animal Behavior in America, 1930–1945," in *The Expansion of American Biology*, ed. Keith R. Benson (New Brunswick, NJ: Rutgers University Press, 1991), 164–94.

27. Curators at the American Museum were also at the forefront of the evolutionary synthesis: see Edwin H. Colbert, *William King Gregory, 1876–1970* (Washington, DC: National Academy of Sciences, 1975); Leo F. Laporte, *George Gaylord Simpson* (New York: Columbia University Press, 2000); Mary LeCroy, "Ernst Mayr at the American Museum of Natural History," in "Ernst Mayr at 100: Ornithologist and Naturalist," *Ornithological Monographs* 58 (2005): 30–49; Rainger, *An Agenda for Antiquity* (n. 21 above, chap. 1). For more comprehensive accounts, see Joseph Allen Cain, "Common Problems and Cooperative Solutions: Organizational Activity in Evolutionary Studies, 1936–1947," *Isis* 84 (1993): 1–25; Joseph Allen Cain and Michael Ruse, eds., *Descended from Darwin: Insights into the History of Evolutionary Studies, 1900–1970* (Philadelphia: American Philosophical Society, 2009); Mayr and Provine, eds., *The Evolutionary Synthesis*; V. Betty Smocovitis, *Unifying Biology: The Evolutionary Synthesis and Evolutionary Biology* (Princeton, NJ: Princeton University Press, 1996).

28. On the Carnegie Museum's famous diplodocus display, see Brinkman, *The Second Jurassic Dinosaur Rush* (n. 105 above, chap. 1). On the Smithsonian's

display of Martha, the last surviving passenger pigeon, see Barrow, *Nature's Ghosts* (n. 22 above, chap. 1), 127.

29. *AMNHAR* (1946), 11.
30. Frank M. Chapman et al., *The Distribution of Bird-Life in Ecuador: A Contribution to a Study of the Origin of Andean Bird-Life*, ed. Frank M. Chapman, Bulletin of the American Museum of Natural History, vol. 55 (New York: American Museum of Natural History, 1926); Frank M. Chapman et al., *The Distribution of Bird-Life in Colombia : A Contribution to a Biological Survey of South America*, Bulletin of the American Museum of Natural History, vol. 36 (New York: American Museum of Natural History, 1917); Jaques, "Francis Lee Jaques" (n. 116 above, chap. 2), chap. 26, p. 6.
31. William King Gregory, "The Museum of Things versus the Museum of Ideas," *Science* 83, no. 2164 (1936): 585–88.
32. Robert Cushman Murphy, "Natural History Exhibits and Modern Education," *Scientific Monthly* 45 (1937): 80.
33. Frank Lutz, letter to Hermon Bumpus, August 18, 1925, Lutz Papers, AMNH.
34. As in German museums, representations of biology in interwar American museums were "looser and more capacious" than the strict representations of the system that came before it. See Nyhart, *Modern Nature* (n. 12 above, chap. 1), 4. See also Lynn K. Nyhart, "Natural History and the 'New' Biology," in *Cultures of Natural History*, ed. Jardine, Secord, and Spary, 426–46 (n. 4 above, "Introduction").
35. Murphy, "Natural History Exhibits," 77.
36. John Dewey, "The Supreme Intellectual Obligation," *Science* 79, no. 2046 (1934): 242. On the importance of applied science as a tool for teaching and popularization, see Gruenberg, *Science and the Public Mind* (n. 13 above, this chap.), 9–70; Waldemar Kaempffert, *Science Today and Tomorrow* (New York: Viking Press, 1939), 5. Of course, American biologists themselves made ongoing efforts to apply their science to social and economic issues; on this topic, see Pauly, *Biologists and the Promise of American Life* (n. 6 above, chap. 1).
37. Although the journal *Human Biology* was founded in 1929 by biologist Raymond Pearl, the field of human biology really didn't come into its own until the 1960s; on this history, see Michael H. Crawford, "History of Human Biology (1929–2009)," *Human Biology* 82, no. 3 (2010), article 6, http://digitalcommons.wayne.edu/humbiol/vol82/iss3/6/.
38. On exhibitions of human health and biology in the early twentieth century, see Brown, *Health and Medicine on Display* (n. 102 above, chap. 1); Rydell, *World of Fairs* (n. 8 above, this chap.), 38–60.
39. Cummings, *East Is East* (n. 142, chap. 2), 34.
40. Buxton and Niquette, "'Sugar-Coating the Educational Pill'" (n. 9 above, this chap.), 157; Rydell, *World of Fairs*, 38–58.
41. Barrow, *Nature's Ghosts* (n. 22 above, chap. 1), 208–9.
42. On the Dust Bowl's impact on the study of ecology, see Kingsland, *Evolution of American Ecology* (n. 146 above, chap. 1), 148–54; Donald Worster,

Nature's Economy: A History of Ecological Ideas, ed. Donald Worster and Alfred W. Crosby, 2nd ed., Studies in Environment and History (New York: Cambridge University Press, 1994), 219–54.

43. Conservationists, many of them biologists based in museums, embraced ecology in the interwar years. At the American Museum, e.g., Harold Anthony led the American Society of Mammalogists' efforts to curb the Biological Survey's use of poison to control predators. See Barrow, *Nature's Ghosts* (n. 22 above, chap. 1), 201–33; Christian C. Young, *In the Absence of Predators: Conservation and Controversy on the Kaibob Plateau* (Lincoln: University of Nebraska Press, 2002).

44. Parr, "On the Functions of the Natural History Museum" (n. 11 above, this chap.), 2.

45. Gruenberg, *Science and the Public Mind* (n. 13 above, this chap.), 116.

46. Gregory, "The Museum of Things" (n. 31 above, this chap.), 585–88.

47. Though plenty of display critics were involved in science education, this overlap was more a matter of zeitgeist than consultation with educators—it almost never occurred to interwar museum professionals to ask educational reformers or the classroom teachers who used their halls for advice.

48. DeBoer, *History of Ideas* (n. 34 above, chap. 1), 39–84.

49. Lawrence A. Cremin, *American Education: The Metropolitan Experience, 1876–1980* (New York: Harper & Row, 1988), 229; Raymond B. Fosdick, *Adventure in Giving: The Story of the General Education Board* (New York: Harper & Row, 1962), 240–41, 316–17; John Rudolph, *Scientists in the Classroom: The Cold War Reconstruction of American Science Education* (New York: Palgrave Macmillan, 2002), 18.

50. The titles of the era's most popular textbooks clearly illustrate this emphasis on "usable" science: Hunter's *A Civic Biology* (1914), Smallwood, Reveley, and Bailey's *Practical Biology* (1916), Trafton's *Biology of Home and Community* (1923), Peabody and Hunt's *Biology and Human Welfare* (1924), Gruenberg's *Biology and Human Life* (1925), and Curtis, Caldwell, and Sherman's *Biology for Today* (1934). On school biology in the interwar period, see DeBoer, *History of Ideas* (n. 34 above, chap. 1); Engles, "Biology Education" (n. 10 above, this chap.); Ronald P. Ladouceur, "Ella Thea Smith and the Lost History of American Biology Textbooks," *Journal of the History of Biology* 41, no. 3 (2008): 435–71; Shapiro, "Civic Biology" (n. 13 above, this chap.).

51. Ladouceur, "Ella Thea Smith," 466.

52. On the place of the scientific method in science curricula in the first few decades of the twentieth century, see John Rudolph, "Epistemology for the Masses: The Origins of 'the Scientific Method' in American Schools," *History of Education Quarterly* 45, no. 3 (2005): 380–81, and "Turning Science to Account: Chicago and the General Science Movement in Secondary Education, 1905–1920," *Isis* 96, no. 3 (2005): 353–89.

53. Gruenberg, *Science and the Public Mind* (n. 13 above, this chap.), 115.

54. "The American Museum of Natural History," *Fortune*, April 1937, 124.

55. For more on the historical influence of commercial display on interwar museum pedagogy, see Cain, " "Attraction, Attention, Desire'" (n. 32 above, chap. 1); Cummings, *East Is East* (n. 142, chap. 2).

56. Waldemar Kaempffert, *From Cave-Man to Engineer: The Museum of Science and Industry Founded by Julius Rosenwald; an Institution to Reveal the Technical Ascent of Man* (Chicago: Museum of Science and Industry, 1933), 12.

57. Peter Max Ascoli, *Julius Rosenwald: The Man Who Built Sears, Roebuck and Advanced the Cause of Black Education in the American South* (Bloomington: Indiana University Press, 2006), 265.

58. Though the educational establishment agreed on this point, their pedagogical vision rarely translated neatly into classroom practice. On interwar pedagogy, see Cuban, *How Teachers Taught* (n. 59 above, chap. 2); DeBoer, *History of Ideas* (n. 34 above, chap. 1); Kliebard, *Struggle for the American Curriculum* (n. 34 above, chap. 1); Diane Ravitch, *The Great School Wars: A History of the New York City Public Schools* (Baltimore, MD: Johns Hopkins University Press, 1974); David B. Tyack, *The One Best System: A History of American Urban Education* (Cambridge, MA: Harvard University Press, 1974). On the "project method," see Rudolph, "Turning Science to Account," 382–83.

59. "Tactile Education," *Museum News* 4, no. 1 (1926): 2.

60. Cummings, *East Is East* (n. 142, chap. 2), 169.

61. "Answering Back," *Museum News* 2, no. 6 (1924): 2.

62. Gregory, "The Museum of Things" (n. 31 above, this chap.), 585–88.

63. See, e.g., minutes from meeting of Division II, May 6, 1929, Department of Mammalogy Administrative Papers, box 2, folder Departments "Division II," 1922–30, DM-AMNH.

64. Minutes from a special meeting of the Council of Scientific Staff, May 27, 1935, prepared by H. E. Anthony. Department of Mammalogy Administrative Papers, box 1 folder Council of Scientific Staff: Correspondence I 1934–35, DMAP-AMNH.

65. Frank Chapman, "A Plea for the North American Habitat Bird Groups," January 5, 1942, Department of Mammalogy Administrative Papers, box 1 folder Council of Scientific Staff Minutes IV, 1941–42, DMAP-AMNH.

66. William K. Gregory, letter to H. E. Anthony, May 24, 1935, Department of Mammalogy Administrative Papers, box 1, folder Council of Scientific Staff, Correspondence I, 1934–35, DMAP-AMNH.

67. Frank E. Lutz, "Taking Nature Lore to the Public," *Natural History* 26, no. 2 (1926): 115.

68. Minutes of meeting of Advisory Committee on Biological Aspects of Museum Groups, March 5, 1929, prepared by H. E. Anthony, Department of Mammalogy Administrative Papers, box 3, folder Exhibitions, 1922–59, DM-AMNH.

69. Gladwyn Kingsley Noble, "Probing Life's Mysteries: Some Aspects of the Research Work of the American Museum," *Natural History* 30, no. 5 (1930): 469.

70. Gladwyn Kingsley Noble, "Reptiles and Amphibians," *AMNHAR* (1927), 60–61.

71. As cited in Gregg Mitman, "Cinematic Nature: Hollywood Technology, Popular Culture, and the American Museum of Natural History," *Isis* 84, no. 4 (1993): 647.

72. Robert E. Kohler, *Lords of the Fly: "Drosophilia" Genetics and the Experimental Life* (Chicago: University of Chicago Press, 1994), 31.

73. Lutz collaborated with Cornell University physicist F. K. Richtmayr to determine the ultraviolet reflectivity of certain insect-pollinated plants, and the publication describing their work won the New York Academy of Sciences' A. Cressy Morrison Prize in 1923. Frank E. Lutz, "Apparently Non-Selective Characters and Combinations of Characters, Including a Study of Ultraviolet in Relation to the Flower-Visiting Habits of Insects," *Annals of the New York Academy of Science* 29 (1924): 233–83; Frank E. Lutz and F. K. Richtmyer, "The Reaction of *Drosophilia* to Ultraviolet," *Science* 55 (1922): 519. On Lutz's research, see, among other sources, D. R. Barton, "Attorney for the Insects," *Natural History* (1941): 181–85; Dann, *Across the Great Border Fault* (n. 14 above, chap. 2), 88–100; W. J. Gertsch, "Report of the Thirty-Eighth Annual Meeting," *Annals of the Entomological Society of America* 37 (1944): 132–34.

74. At the museum, Lutz became a generous mentor to younger scientists and amateur naturalists and found a side calling as a popularizer of entomology. From the late 1910s till his death, he published books and produced displays intended to interest a broad public in insect life, most famously *The Field Book of Insects* (1917) and *A Lot of Insects: Entomology in a Suburban Garden* (1941), both of which sold well enough to put his four children through college. According to entomologist W. J. Gertsch, Lutz considered it "a personal triumph . . . to find that he had been instrumental in directing new devotees into the field of natural history. Sincerity in their interest in insects was the only qualification for audience and advice from him to school boy, business executive, or scientist." Mabel Smallwood, letter to Charles Zeleny, August 28, 1904, as cited in Gertsch, "Report of the Thirty-Eighth Annual Meeting," 134; Kohler, *Lords of the Fly*, 31.

75. *AMNHAR* (1917), 52.

76. Herbert Schwarz, "Frank E. Lutz," *Entomological News* 55, no. 2 (1944): 30.

77. *AMNHAR* (1930), 62.

78. Schwarz, "Frank E. Lutz," 30.

79. Lutz's stopwatch experiments led him to conclude "that the most expensive exhibit is not always the best. Anyone who doubts this could time visitors at some of our latest and most expensive vertebrate habitat groups." Of course, stopwatches had limited value in assessing visitor reaction. As Carlos Cummings later pointed out, "We have no way of telling whether a subject who spends a carefully counted number of minutes examining a case is making a very thorough study of its contents or is just plain dumb." Frank E. Lutz, letter to George Sherwood, February 29, 1932, Department

of Mammalogy Administrative Papers, box 3, folder Executive Offices, Director and Deputies, DM-AMNH; Cummings, *East Is East* (n. 142, chap. 2), 376. On Hyde's educational work at the museum, see Dann, *Across the Great Border Fault* (n. 14 above, chap. 2), 52–56.

80. As the museum turned its attention to the exhibition of vertebrates in the 1920s, it took Lutz three years of lobbying before the administration agreed to provide the small amount of money necessary to improve the lighting in the Hall of Insect Life's exhibit cases. Happily, the curator was known for his mechanical ingenuity and frequently built his own lab equipment with humble materials. See Frank E. Lutz, "The Use of Live Material in Museum Work," *Museum News: Magazine Section* 8, no. 6 (1930): 8. On Lutz's proclivity for tinkering, see Dann, *Across the Great Border Fault* (n. 14 above, chap. 2), 64.

81. Lutz, "The Use of Live Material in Museum Work," 8.

82. *AMNHAR* (1930), 62.

83. "Museum Crickets Broadcast Chirps," *New York Times*, October 3, 1936.

84. "A large part of the remainder," he wrote, "is adoption of structure by a creature for its needs—a structure that was really forced upon the creature by its internal forces; we can call it orthogenesis if we must—rather than adaptation of a structure to a creature's needs." Frank E. Lutz, letter to Henry Fairfield Osborn, April 27, 1925, as cited in Dann, *Across the Great Border Fault* (n. 14 above, chap. 2), 96.

85. Lutz was particularly keen to create new collaborations between life and physical scientists. "Biology needs the invigorating effect of outbreeding," he maintained. "Physicists, chemists and mathematicians . . . know their subject and it is a bit embarrassing to see how easily they punch holes in the physics and chemistry of apparently first-class work in which biologists have ventured by themselves to apply those sciences" (as cited in ibid., 98–99).

86. The range of habitat on the site Lutz chose resulted in extraordinary biodiversity: in a nine-week period in 1927 within a half-mile radius of the institute's headquarters, volunteers and scientists collected 540 species of flies, thirty of which were previously unknown to science and ninety-seven of which were new to New York State. On the Station for the Study of Insects, see ibid., 62–74. On the entomology department's relative poverty, see Frank Lutz, letter to Henry Fairfield Osborn, October 16, 1924, folder 765, Central Archives, AMNH. On the site's diversity, see, among others, C. H. Curran, "Report on the Diptera Collected at the Station for the Study of Insects, Harriman Interstate Park, N.Y.," *Bulletin of the American Museum of Natural History* 61 (1930): 21–115.

87. Dann, *Across the Great Border Fault* (n. 14 above, chap. 2), 66.

88. Frank E. Lutz, *A Lot of Insects: Entomology in a Suburban Garden* (New York: G. P. Putnam's Sons, 1941), 282–83.

89. Frank Lutz, letter to William H. Welch, June 18, 1925, as cited in Dann, *Across the Great Border Fault*, 68.

90. Frank Lutz, letter to Hermon Bumpus, October 11, 1925, Lutz Papers, AMNH.

91. Kevin Dann vividly describes a number of Lutz's jerry-rigged laboratory instruments, explaining that necessity was the mother of invention for Lutz, whose department was always short of money. Dann, *Across the Great Border Fault*, 64–65, 89–95.

92. Frank Lutz, letter to Hermon Bumpus, August 18, 1925, Lutz Papers, AMNH.

93. As Gregg Mitman has observed, one of the reasons for Noble's success was his friendly relationship with William Douglas Burden, one of the museum's most active trustees. Both men had grown up in wealthy circles of New York society, and Burden championed Noble's work, funding his research, publicizing his findings, and promoting (and even collaborating on) his displays. Mitman, "Cinematic Nature" (n. 71 above, this chap.), 642–43.

94. After receiving offers from Cornell and Columbia in 1928, Noble wangled new laboratories from trustees as a condition of his agreement to remain at the museum. The laboratories, located on the uppermost floor of the wing built to house the museum's African Hall, had a precarious status in the museum and were almost demolished at the end of the 1930s but survived because of the intervention of Burden. See Myers, "A History of Herpetology" (n. 44 above, chap. 1), *AMNHAR* (1945), 35–37. On Noble's work at the museum, see William King Gregory, "Gladwyn Kingsley Noble," *Science* 93, no. 2401 (1941): 10–11; Milam, *Looking for a Few Good Males* (n. 26 above, this chap.), 54–79; Mitman, "Cinematic Nature" (n. .71 above, this chap.).

95. "Herpetology and Experimental Biology," *AMNHAR* (1932), 29.

96. Noble, "Probing Life's Mysteries" (n. 69 above, this chap.), 469.

97. The rattlesnake exhibit is discussed in *AMNHAR* (1932), 46. See also Myers, "A History of Herpetology" (n. 44 above, chap. 1), 86.

98. For a more detailed description of the hall, see Myers, "A History of Herpetology," 87–88.

99. Ibid., 217.

100. See Ruth Crosby Noble and Gladwyn Kingsley Noble, *The Nature of the Beast: A Popular History of Animal Psychology from the Point of View of a Naturalist* (New York: Doubleday, 1945). Also see "Museum Outfits Five Expeditions," *New York Times*, May 28, 1937; "Museum Visitors Can See as Fish Do," *New York Times*, June 9, 1937.

101. On the Komodo dragon display, see Mitman, "Cinematic Nature" (n. 71 above, this chap.), 641–49. Noble was by no means the only curator at the American Museum to experiment with film; as Alison Griffiths has observed, curators in various departments explored the medium's exhibitionary possibilities in the 1920s, though they rarely coordinated these efforts. See "Film Education in the Natural History Museum: Cinema Lights Up the Gallery in the 1920s," in Marsha Orgeron, Devin Orgeron, Dan Streible,

eds., *Learning with the Lights Off* (New York: Oxford University Press, 2012), 124–44.

102. Ibid., 654–55.

103. Dann, *Across the Great Border Fault* (n. 14 above, chap. 2), 99.

104. Myers, "A History of Herpetology" (n. 44 above, chap. 1), 218.

105. Christian D. Jersabek, "The 'Frank J. Myers Rotifera Collection' at the Academy of Natural Sciences of Philadelphia" *Hydrobiologia* 546 (2005): 139.

106. Lurie, *A Special Style* (n. 7 above, chap. 1), 65–66; Milam, *Looking for a Few Good Males* (n. 26 above, this chap.), 59; "Museum Outfits Five Expeditions" (n. 100 above, this chap.).

107. Cummings, *East Is East* (n. 142, chap. 2), 169.

108. At some museums, the funding for diorama halls had been secured well in advance of the 1929 crash, so new or renovated habitat halls opened throughout the early 1930s. Thanks to the contributions of wealthy patrons, in Los Angeles, the natural history museum opened a Hall of African Mammals in 1930 and, just in time for the 1932 Olympics, a Hall of North American Mammals. Chicago's Field Museum, San Francisco's California Academy of Sciences, and the Philadelphia Academy of Natural Science also opened enormous African halls in these years. See *LAMAR*, 1930–32, LAM. Wonders, *Habitat Dioramas* (n. 182 above, chap. 1), 176.

109. Alfred M. Bailey, *CMNHAR* (1938), 19.

110. As cited in Kennedy, "Philanthropy and Science" (n. 7 above, "Introduction"), 204.

111. J. D. Figgins, letter to A. W. Anthony, February 17, 1923, JDF-CMNH.

112. "The American Museum of Natural History" (n. 54 above, this chap.), 124.

113. Often, however, what was touted as modernization was laughable. The Smithsonian, once a pioneer in exhibition methods, introduced its first lighted case in the 1930s and called it dramatic progress. The museum's overwhelming commitment to scientific research and dependence on federal funding conspired to prevent it from modernizing its display until well after World War II. On the reorganization of halls at the American Museum, see William Douglas Burden et al., *Report of Trustees' Committee on Allocation of Space* (New York: American Museum of Natural History, 1938), 2. On the single lighted case, see Yochelson, *National Museum of Natural History* (n. 95 above, chap. 1), 75.

114. Minutes of meeting of Advisory Committee on Biological Aspects of Museum Groups, March 5, 1929, prepared by H. E. Anthony, Department of Mammalogy Administrative Papers, box 3, folder Exhibitions, 1922–59, DM-AMNH.

115. Laurence Vail Coleman, *The Museum in America: A Critical Study*, 3 vols., (Washington, DC: American Association of Museums, 1939), 1:54.

116. Ibid., 1:3.

117. Alan Friedman, "The Evolution of Science and Technology Museums," *Informal Science Review* 17 (1996): 14–17. See also Alan Friedman, "The

Extraordinary Growth of the Science-Technology Museum," *Curator* 50, no. 1 (2007): 63–75.

118. As cited in Ascoli, *Julius Rosenwald* (n. 57 above, this chap.), 267.

119. The club's president, Samuel Insull, squashed discussion of the proposal, probably because he himself was involved, in conjunction with the Smithsonian, in the planning of a national museum of engineering and industry—an effort that was never realized. Ibid., 265–67.

120. "Editor Kaempffert for Chicago Museum," *New York Times*, April 14, 1928, 19. Ascoli, *Julius Rosenwald*, 328; Jay Pridmore, *Inventive Genius: The History of the Museum of Science and Industry, Chicago* (Chicago: Museum of Science and Industry, 1996), 46.

121. Ascoli, *Julius Rosenwald* (n. 57 above, this chap.), 328.

122. "Editor Kaempffert for Chicago Museum," 19. Pridmore, *Inventive Genius*, 46.

123. Waldemar Kaempffert, "Revealing the Technical Ascent of Man in the Rosenwald Industrial Museum," *Scientific Monthly* 28, no. 6 (1929): 487.

124. "Editor Kaempffert for Chicago Museum," 19.

125. Kaempffert, *From Cave-Man to Engineer* (n. 56 above, this chap.), 18.

126. "Editor Kaempffert for Chicago Museum," 19.

127. "Chicago to Exhibit Industry by Acting," *New York Times*, February 21, 1929.

128. For more on the museum's budget, see "Oomph for Science," *Time*, September 2, 1940, 34–35, http://www.time.com/time/magazine/article/0,9171,764562,00.html.

129. At Rosenwald's own urging, the institution had changed its name from the Rosenwald Museum to the Museum of Science and Industry, Chicago, in 1929, but because the museum was so closely associated with Rosenwald himself, the press often continued to refer to it as "Rosenwald's Industrial Museum" or "The Rosenwald Museum of Science and Industry." Ascoli, *Julius Rosenwald*, 329; "Rosenwald Museum Changes Its Name," *New York Times*, July 7, 1929, 20.

130. "Chicagoans Regret Passing of the Fair, but Are Consoled by Fact That Many Exhibits Will Stay in the City," *New York Times*, November 4, 1934, E6. Many of the fair's most popular medical exhibits had been copied or imported directly from the Deutsches Hygiene Museum and encouraged, sometimes subtly but often overtly, racial purity and eugenicist policy. Such exhibits could also be found in a handful of natural history museums, most notably at the American Museum of Natural History and the Field Museum, where prominent or pliant administrators welcomed such donations. On eugenics exhibits in American museums in the 1930s, see Christina Cogdell, *Eugenic Design: Streamlining America in the 1930s* (Philadelphia: University of Pennsylvania Press, 2004), 94; Rydell, *World of Fairs* (n. 8 above, this chap.), 38–58.

131. Buxton and Niquette, "'Sugar-Coating the Educational Pill'" (n. 9 above, this chap.), 163; Jaume Sastre Juan, "Science in Action: The New York

Museum of Science and Industry and the Politics of Interactivity," paper presented at the annual meeting of the History of Science Society, Montreal, Quebec, November 4–7, 2010.

132. This tradition was well established at the Deustches Museum as well. Ascoli, *Julius Rosenwald* (n. 57 above, this chap.), 268.

133. Buxton and Niquette, " 'Sugar-Coating the Educational Pill,' " 163.

134. Kaempffert, *From Cave-Man to Engineer* (n. 56 above, this chap.), 12.

135. "Chicago to View Rosenwald's 'Toys,' " *New York Times*, June 18, 1933, N2.

136. When Kaempffert urged his curators to publish in academic journals, e.g., Rosenwald complained bitterly, grumbling that the institution "was taking on the aura of a university." On Kaempffert's vision for the museum and his ultimate dismissal for spending that the trustees considered profligate, see Ascoli, *Julius Rosenwald*, 332–34.

137. Robert P. Shaw, *Exhibition Techniques: A Summary of Exhibition Practice, Based on Surveys Conducted at the New York and San Francisco World's Fairs of 1939* (New York: New York Museum of Science and Industry, 1940), 4. Also see Robert P. Shaw, "The Progressive Exhibit Method: A New Technic in the Field of Science Presentation," *American Physics Teacher* 7, no. 3 (1939): 165–72.

138. Evan J. Carey, "Health and Medical Science Exhibits," *Phi Chi Quarterly* 38, no. 1 (1941): 41–57, clipping found in "Prenatal Exhibit, 1933–1998" folder, MSIC.

139. Pridmore, *Inventive Genius* (n. 120 above, this chap.), 113.

140. Robert Bernard Carlson, comp., *Official Handbook of Exhibits in the Division of the Basic Sciences—Hall of Science* (Chicago: Century of Progress, 1934).

141. Buxton and Niquette, " 'Sugar-Coating the Educational Pill' " (n. 9 above, this chap.), 153–94. On the history of the transparent man and his relationship to eugenicist ideology, see Cogdell, *Eugenic Design* (n. 130 above, this chap.), 87–90.

142. Carlos E. Cummings, *Your Guide to the Buffalo Museum of Science* (Buffalo, NY: The Museum, ca. 1936), 14.

143. "Woman without Mystery on Display in New York," *Science Newsletter* 30, no. 803 (1936): 135. Also see Waldemar Kaempffert, "Science: The Evolution of the Science Museum," *New York Times*, February 16, 1936, XX6.

144. Robert P. Shaw, "New Developments in Museum Technique and Procedures," *Scientific Monthly* 48, no. 5 (1939): 445.

145. "Visitors Can Turn Levers and Find Length of Lives," *Science Newsletter* 32, no. 866 (1937): 309.

146. "Brain Demonstration," *Science Newsletter* 35, no. 1 (1939): 10.

147. Physician Helen Buttons had assembled the display when she served on the obstetrics staff of the Cook County Hospital, by persuading women who didn't have the money to bury their miscarried fetuses to donate them to science. The collection was first presented in a museum at the Stritch School of Medicine at Loyola University. For acquisition of the exhibit from Buttons, see folder titled "Prenatal Exhibit," MSIC, including Richard Cowdrey, letter to Rosenwald Museum of Science and Industry, Febru-

ary 19, 1949; and J. W. Hostrup, letter to Richard Cowdrey, March 2, 1949. Also see Pridmore, *Inventive Genius* (n. 120 above, this chap.), 111.

148. Barry Aprison, "The Prenatal Exhibit at the Museum of Science and Industry," *Visitor Behavior* 12, nos. 1–2 (1997): 25.

149. An image of Field Museum taxidermists working on the cows at the Museum of Science and Industry in Chicago can be found at "Taxidermist Working on Cow at Museum of Science and Industry," Wisconsin Historical Society, http://www.wisconsinhistory.org/whi/fullRecord.asp?id=41419 &qstring=http%3A%2F%2Fwww.wisconsinhistory.org%2Fwhi%2Fresults .asp%3Fsubject_narrow%3DModels%2Band%2Bmodelmaking.

150. Kaempffert, *From Cave-Man to Engineer* (n. 56 above, this chap.), 12, 25.

151. As cited in Buxton and Niquette, " 'Sugar-Coating the Educational Pill' " (n. 9 above, this chap.).

152. "The New York Museum of Science and Industry," *Science* 82, no. 2130 (1935): 384. Shaw's first major publication cites Chicago as its source for many demonstrations. See Robert P. Shaw, "Scientific Exhibits and Their Planning," *Scientific Monthly* 35, no. 4 (1932): 373. Shaw also published a periodic *Exhibit Newsletter*, which Manon Niquette and William Buxton credit with disseminating the "communicative turn" in science museums to a broader industrial audience. See also Buxton and Niquette, " 'Sugar-Coating the Educational Pill' " (n. 9 above, this chap.).

153. As cited in New York Museum of Science and Industry, *Exhibition Techniques: A Summary of Exhibition Practice* (New York: New York Museum of Science and Industry, 1940); Shaw, "New Developments in Museum Technique," (n. 144 above, this chap.).

154. Century of Progress International Exposition, *Official World's Fair Weekly* (Chicago: n.p., 1933).

155. Cummings, *East Is East* (n. 142, chap. 2), 11–12.

156. Ibid., 31.

157. The Buffalo Museum of Science's Hermon Bumpus and Carlos Cummings, e.g., were both well known for their artistic eyes and mechanical skills, while Bradford Washburn was an accomplished professional photographer.

158. George F. Goodyear, *Society and Museum* (n. 5 above, "Introduction"), 45–46. Also see Carlos E. Cummings, "Mushrooms and Toadstools," *Natural History* 33, no. 1 (1933): 41–53, and "Flowers Reproduced in Wax: Synthetic Nature Now and Then," *Hobbies* 21, no. 4 (1941): 68–73.

159. Bumpus, *Hermon Carey Bumpus* (n. 211 above, chap. 1), 69. On Bumpus's scientific background and work as chair of the biology department at Brown University, see Martha Mitchell, *Encyclopedia Brunoniana* (Providence, RI: Brown University Press, 1933), n.p.

160. Bumpus, *Hermon Carey Bumpus*, 117.

161. After leaving the American Museum, Bumpus turned to academic administration, serving as the business manager of the University of Wisconsin, then president of Tufts University, before returning to museum work in the 1920s. Ibid.

162. Cummings, *East Is East* (n. 142, chap. 2), 120.

163. Goodyear, *Society and Museum*, 37.

164. On the originality of Cummings's story-centered approach, see his *Your Guide to the Buffalo Museum of Science* (n. 142 above, this chap.), 3. Also see Buxton and Niquette, " 'Sugar-Coating the Educational Pill' " (n. 9 above, this chap.), 183–86.

165. Goodyear, *Society and Museum*, 219.

166. Cummings, *Your Guide to the Buffalo Museum of Science*, 14.

167. Buxton and Niquette, " 'Sugar-Coating the Educational Pill,' " 165.

168. Coleman, *Biennial Survey of Education* (n. 5 above, this chap.), 1:768.

169. Meeting of Division II, February 14, 1929, prepared by Robert Cushman Murphy, Department of Mammalogy Administrative Papers, box 2 folder "Division II," 1922–30, DM-AMNH.

170. Buxton and Niquette, " 'Sugar-Coating the Educational Pill,' " 165.

171. Leaders of the Boston Museum observed that the nearby Museum of Comparative Zoology's research and collections far outpaced their own, and so elected to cede its research program to its Harvard-based competitor; all the society's "exotic" collections were transferred to the Museum of Comparative Zoology around 1920. Capt. Percy R. Creed, *Milestones: The Boston Society of Natural History, 1830–1930* (Boston: Boston Society of Natural History, 1930), 54–55. Also see Sally Gregory Kohlstedt, "From Learned Society to Public Museum: The Boston Society of Natural History," in *The Organization of Knowledge in Modern America, 1860–1920*, ed. Alexandra Oleson and John Voss (Baltimore, MD: Johns Hopkins University Press, 1979).

172. Cummings, *East Is East* (n. 142, chap. 2), xv. On Washburn, see Bradford Washburn and Lew Freedman, *Bradford Washburn: An Extraordinary Life: The Autobiography of a Mountaineering Icon* (Portland, OR: WestWinds Press, 2005). Also Ora Dodd, "From Wings to Stars," *Yankee*, March 1961, 96.

173. Director Bradford Washburn argued that a short name signifying "a dynamic and successful program will produce the proper reaction on the part of everyone who comes here" and led the charge to change the name to the Boston Museum of Science. On name change, see Museum of Science member newsletter dated "Christmas 1947" and signed by Bradford Washburn; see also Bradford Washburn, letter to Douglas Adams, July 5, 1960, box "1960," folder "A," both BMoS.

174. On the efforts of children's museums to provide hands-on opportunities in science, see Kohlstedt, "Innovative Niche Scientists" (n. 44 above, chap. 1); Onion, "Picturing Nature and Childhood" (n. 55 above, chap. 2).

175. Cummings, *East Is East* (n. 142, chap. 2), 33.

176. "Wonders of the Human Body: Reproduced in a Museum By Machines and Electricity," *Popular Science* 127, no. 2 (1935): 36–37.

177. The Rockefeller Foundation was particularly supportive of the introduction of this and other "hygienic" models into the museum's Hall of Man, believing they effectively conveyed the aims of its Science of Man program of research in the biological sciences to a broader public. On the Rockefeller

Science of Man program, see Buxton and Niquette," 'Sugar-Coating the Educational Pill'" (n. 9 above, this chap.), 170–72; Goodyear, *Society and Museum* (n. 5 above, "Introduction"), 40.

178. Buxton and Niquette, " 'Sugar-Coating the Educational Pill,' " 157.

179. Cummings, *East Is East* (n. 142, chap. 2), 5–6.

180. Copied from Dresden's Deustche Hygiene Museum, the panels had earlier hung in the biology section of the Chicago World's Fair's Hall of Science. Their inclusion in an array of displays on human biology lent them scientific credence, despite their obviously racist overtones. The panels were destroyed on government order in 1943 and 1950. Cogdell, *Eugenic Design* (n. 130 above, this chap.), 88–90. For an extensive analysis of the relationship between Cummings's changes to the Buffalo Museum of Science, communication theory, and eugenics, see Buxton and Niquette," 'Sugar-Coating the Educational Pill,' " 164–79.

181. Goodyear, *Society and Museum* (n. 5 above, "Introduction"), 34.

182. Cummings, *East Is East* (n. 142, chap. 2), 44.

183. Coleman, *The Museum in America* (n. 115 above, this chap.), 47–48.

184. Foundation officers actively encouraged museums to experiment with more dynamic displays, providing them with generous grants to do so. In 1937, e.g., the Rockefeller Foundation awarded a delighted Cummings and the other staff members at the Buffalo Museum of Science a grant to train both American and foreign museum workers in "the principles and practices of modern museum techniques." Two years later, thanks to Rockefeller money, Robert Shaw and the staff of the New York Museum of Science and Industry published its well-received *Exhibition Techniques*, a handbook on methods of installation and exhibition at world's fairs that was intended to help museums incorporate their most successful techniques. Ibid.; Cummings, *East Is East* (n. 142, chap. 2), xv; L. N. Diamond, "Interpreting Science to the Public," *Scientific Monthly* 40, no. 4 (1935): 372–74. For a detailed history of the Rockefeller grants, see Buxton and Niquette, " 'Sugar-Coating the Educational Pill'" (n. 9 above, this chap.).

185. Cummings, *East Is East* (n. 142, chap. 2), 124.

186. Murphy, "Natural History Exhibits" (n. 32 above, this chap.), 79.

187. "Experimental Biology," *AMNHAR* (1940), 19. "Private Life of Microbes Is Shown by Projector," *Popular Mechanics*, September 1937, 327. For information on the microvivarium's use in earlier world's fairs, see Waldemar Kaempffert, "The Museum and Social Enlightenment," *Science* 92, no. 2386 (1940): 258. See also *Official Guide Book of the World's Fair of 1934* (Chicago: Century of Progress International Exposition, 1934), 39.

188. Halls of geology and earth science were another place that curators and exhibit designers introduced dynamic displays. Inspired by his own visit to Europe's industrial museums, Philadelphia Academy of Natural Sciences curator of exhibits Harold Green introduced an interactive display of fluorescent minerals and a visitor-activated radium detector to the museum's Hall of Earth History, later adding an "electronic quiz," with a light that glowed

when visitors correctly identified the hall's minerals. Barbara Ceiga, "A Brief History of the Education and Exhibits Departments at the Academy of Natural Sciences" (Academy of Natural Sciences, Philadelphia, 2010), 7.

189. "Stories cannot be effectively told in rows and rows of cases and labels, no matter how this combination may be dressed up with color and artistry," William Douglas Burden wrote. "Labels, although adequate for the simple purposes of identifying, will no longer be capable of carrying the much heavier load that is imposed upon them when museums increase the thought content of their halls." Burden, "A New Vision of Learning," *Museum News* (December 15, 1943), 11; Kaempffert, "The Museum and Social Enlightenment." Burden also presented this idea to the American Museum of Natural History's Committee on Plan and Scope as part of an institution-wide survey conducted in 1942. See William Douglas Burden, "Memorandum for Plane and Scope Committee Concerning Exhibition," 1942, 3–4, Central Archives, AMNH.

190. Burden, "A New Vision of Learning," 11.

191. Murphy, "Natural History Exhibits" (n. 32 above, this chap.), 80.

192. Robert Cushman Murphy, "Whitney Wing: The New Home of the American Museum's Department of Birds," *Natural History* 44, no. 2 (1939): 98.

193. Progressive Era museum men did make use of contemporary ideas and scientific research on pedagogical and visual appeal but did so inadvertently rather than in a systematic or scientific way. And they rarely asked visitors for their opinions directly. See Cain, ""Attraction, Attention, Desire'" (n. 32 above, chap. 1).

194. Cummings, *East Is East* (n. 142, chap. 2), 31.

195. Ibid., 3.

196. Wissler, "The Museum Exhibition Problem" (n. 136 above, chap. 2), 177. This was of a piece with larger efforts to measure and quantify what was beginning to be described as the "science of education." Government and foundation support had allowed museum professionals to take halting steps toward statistical assessment of their work in the 1910s and early 1920s. The U.S. Bureau of Education, the Carnegie Corporation, the Rockefeller Foundation, and the American Association of Museums' Committee for Museum Co-operation had all launched surveys of American museums in the 1910s and 1920s, but their interests were primarily institution building and professionalization, rather than any measure of educational success beyond attendance. For more on these early efforts of institutional assessment, see Paul Rea, "Conditions and Needs of American Museums," in *PAAM* 10 (1916): 8–13. See also Frederick P. Keppel, "American Philanthropy and the Advancement of Learning," *Bulletin of the American Association of University Professors* 20, no. 7 (1934): 457, and "General Policies of the Carnegie Corporation," *Science* 60, no. 1562 (1924): 514–17. Paul M. Rea, "Report of the Commission for Museum Cooperation," in *PAAM* 9 (1915): 5–10, and "The Development of Museum Instruction," *Museum Work: Proceedings of the American Association of Museums* 1, no. 4 (1919):

109–13. For a more critical view, see Roy Miner, "Educational Training of Museum Instructors: How Far Is Pedagogical Training or Teaching Experience Necessary?" *Museum Work: Proceedings of the American Association of Museums* 1, no. 4 (1919): 114–17.

197. Edward S. Robinson, *The Behavior of the Museum Visitor* (Washington, DC: American Association of Museums, 1928). Also see Arthur W. Melton, *Problems of Installation in Museums of Art*, ed. Edward S. Robinson, Studies in Museum Education (Washington, DC: American Association of Museums, 1935); Arthur W. Melton, Nita Goldberg Feldman, and Charles W. Mason, *Experimental Studies of the Education of Children in a Museum of Science*, ed. Edward S. Robinson, Studies in Museum Education (Washington, DC: American Association of Museums, 1936), 1; Edward S. Robinson, "Psychological Problems of the Science Museum," *Museum News: Magazine Section* 8, no. 5 (1930): 9–11. For a discussion of the competing approaches of Robinson and Paul Rea, who had conducted studies for the AAM and the Carnegie Corporation on the demographics of museum audiences, see Danielle Rice, "Balancing Act: Education and the Competing Impulses of Museum Work," *Art Institute of Chicago Museum Studies* 29, no. 1 (2003): 6–19, 90.

198. Melton, *Problems of Installation*, 11. Despite the title of Melton's monograph, its contents are not restricted to art museums.

199. Robinson, *Behavior of the Museum Visitor*, 31.

200. Mildred C. B. Porter, *Behavior of the Average Visitor in the Peabody Museum of Natural History, Yale University* (Washington, DC: American Association of Museums, 1938), 15, 16.

201. Ibid.

202. Cummings, *East Is East* (n. 142, chap. 2), 74.

203. Ibid., 134.

204. Ibid., 175.

205. T. R. Adam, *The Museum and Popular Culture* (New York: American Association for Adult Education, 1939), 129–30.

CHAPTER FOUR

1. For a thorough review of the participation of the Smithsonian Institution in war work, see Pamela Henson, "The Smithsonian Goes to War: The Increase and Diffusion of Scientific Knowledge in the Pacific," in *Science and the Pacific War: Science and Survival in the Pacific 1939–1945*, ed. Roy M. MacLeod (Dordrecht: Kluwer Academic Publishers, 1990), 27–50.

2. "Brief Review of Smithsonian War Activities (December 1, 1942)," as cited in ibid., 29.

3. According to the museum's 1941 annual report, staff provided the military with information "on parasites prevalent in war; the importance of insects in war; rodent disease carriers; the distribution and habits of, and safeguards against, poisonous snakes; illustrations of tests used to determine

the fitness of soldiers; the war terrain of various parts of the globe which have been studied and visited in person by our scientists and explorers; descriptions of inhabitants and tribal customs, of flora and fauna; charts of ocean currents in the Caribbean and other waters, and other topics." *AMNHAR* (1942), 8; *AMNHAR* (1941), 2.

4. "Brief Review of Smithsonian War Activities (December 1, 1942)," as cited in Henson, "Smithsonian Goes to War," 29.

5. Waldo Schmitt Papers, RG 7231, Collection Division 2, box 51, folder 6, "Office of War Information," SIA.

6. Alexander Wetmore, "Report of the Secretary," in *SIAR* (1943), 6. See also Alexander Wetmore, "Report of the Secretary," in *SIAR* (1945), 11.

7. See, e.g., *CMNHAR* (1944), 25. Also see *AMNHAR* (1943), 10. To make it easier for servicemen to visit, the Smithsonian instituted longer weekend hours and inaugurated Sunday tours for military personnel, led by female volunteers for the United Service Organizations. In 1945 alone, more than five thousand members of the armed forces spent Sundays listening to docents explain the scientific significance of elephants and geodes. See *SIAR*, 1945, 34 and 40.

8. *AMNHAR* (1946). 2.

9. Pridmore, *Inventive Genius* (n. 120 above, chap. 3), 74–79.

10. See Chicago Commercial Club, *A Museum Goes to War* (Chicago: Museum of Science and Industry, 1943), as cited in Pridmore, *Inventive Genius*, 76.

11. *CMNHAR* (1941), 10.

12. Clifford C. Gregg, "Field Museum of Natural History," *Science* 95, no. 2477 (1942): 630.

13. Albert Eide Parr, "The Year's Work," in *AMNHAR* (1943), 5.

14. *AMNHAR* (1942), 3.

15. George DeBoer provides a thorough description of the anxieties over science education provoked by World War II. See DeBoer, *History of Ideas* (n. 34 above, chap. 1), 128–46.

16. Albert Francis Blakeslee, "Individuality and Science," *Science* 95, no. 2453 (1942): 10.

17. Edward Barry, "Field Museum, at 50th Year, Changes Name," *Chicago Daily Tribune*, September 16, 1943, 24.

18. Albert Eide Parr, "Address," in *Three Addresses Delivered at the Meeting Commemorating the Fiftieth Anniversary of the Field Museum of Natural History*, by Stanley Field, Albert Eide Parr, and Robert Maynard Hutchins (Chicago: Field Museum of Natural History, 1943), 11.

19. Report on proposal to establish exhibits on human biology and public health, Council of Scientific Staff, December 11, 1941, Department of Mammalogy Administrative Papers, box 2, folder CSS Correspondence IV, 1940–41, AMNH.

20. Ibid.

21. Parr embraced this idea, declaring it would be "of considerable intrinsic value to the human imagination, perhaps particularly to the developing

minds of the young." See untitled draft of the committee report Department of Mammalogy Administrative Papers, box 2, folder AMNH CSS Correspondence IV, 1940–41, DM-AMNH; and Parr, "On the Functions of the Natural History Museum" (n. 11 above, chap. 3), 1.

22. Rudolph, *Scientists in the Classroom* (n. 49 above, chap. 3), 17–23. Dorothy Broder's work on life adjustment provides an early account of this movement immediately following the war. See Dorothy Broder, "Life Adjustment Education: A Historical Study of a Program of the United States Office of Education, 1945–54" (diss., Teachers' College, New York, 1977). Also see William G. Wraga, "From Slogan to Anathema: Historical Representations of Life Adjustment Education," *American Journal of Education* 116, no. 2 (2010): 185–209. On the findings of contemporary government educational commissions, see U.S. Office of Education, *Life Adjustment for Every Youth* (Washington, DC: Government Printing Office, 1948). For a broader history of education, and especially science education, in this era, see DeBoer, *History of Ideas* (n. 34 above, chap. 1), 128–46; Andrew Hartman, *Education and the Cold War: The Battle for the American School* (New York: Palgrave-MacMillan, 2008); Diane Ravitch, *The Troubled Crusade: American Education, 1945–1980* (New York: Basic Books, 1983); Kimberley Tolley, *The Science Education of American Girls: A Historical Perspective* (New York: Routledge-Palmer, 2003), 176–208.

23. As cited in Arthur Bestor, *Educational Wastelands: The Retreat from Learning in Our Public Schools* (Urbana: University of Illinois Press, 1953), 76.

24. Sister Mary Janet, "Life Adjustment: Opens New Doors to Youth," *Educational Leadership* 12 no. 3 (1954): 140.

25. American Institute of Biological Sciences, *The A.I.B.S. Story* (Washington, DC: American Institute of Biological Sciences, 1972).

26. Harl R. Douglass, "Education of All Youth for Life Adjustment," in "Critical Issues and Trends in American Education," special issue of *Annals of the American Academy of Political and Social Science* 265 (1949): 111–14.

27. See Rudolph, *Scientists in the Classroom* (n. 49 above, chap. 3), 18; Audra Jayne Wolfe, "Speaking for Nature and Nation: Biologists as Public Intellectuals in Cold War Culture" (PhD thesis, University of Pennsylvania, 2002), 147. Also see A. J. Carlson, "The Science Core in Liberal Education," *Scientific Monthly* 61, no. 5 (1945): 379–81; Lloyd W. Taylor, "Science in the Postwar Liberal Arts Program," *Scientific Monthly* 62, no. 2 (1946): 113–16.

28. Earl J. McGrath, ed., *Science in General Education* (Dubuque, IA: William C. Brown, 1948), 185. This AAAS committee appears to have grown out of an earlier one, founded in the 1920s, called The Place of Science in Education. See Robert H. Carleton, *The N.S.T.A. Story, 1944–1974: A History of Ideas, Commitments and Actions* (Washington, DC: National Science Teachers Association, 1975), 23.

29. Howard A. Meyerhoff, "The Plight of Science Education," *Bulletin of the Atomic Scientists* 12, no. 9 (1956): 333.

30. Edmund W. Sinnott, "Ten Million Scientists," *Science* 111, no. 2876 (1950):

127–28. On Sinnott's advocacy in biology education, see W. Gordon Whaley, "Edmund Ware Sinnott," *Biographical Memoirs, National Academy of Sciences of the United States of America* 54 (1983): 367–71.

31. Dean A. Clark, letter to B. Washburn, November 21, 1958, BMoS.

32. For more on life adjustment in postwar biology curricula, see DeBoer, *History of Ideas* (n. 34 above, chap. 1); Rudolph, *Scientists in the Classroom* (n. 49 above, chap. 3).

33. Barbara A. McEwan, "Let's Put Life into Life Science," *American Biology Teacher* 12, no. 7 (1950): 156–58.

34. Benjamin C. Gruenberg, "Teaching Biology after the Wars," *American Biology Teacher* 9, no. 4 (1947): 101–2.

35. Wraga, "From Slogan to Anathema" (n. 22 above, this chap.), 197.

36. *AMNHAR*, (1947), 5.

37. Robert Maynard Hutchins, "Address," in *Three Addresses Delivered, by Field, Parr, and Hutchins* (n. 18 above, this chap.).

38. Carey, "Health and Medical Science Exhibits" (n. 138 above, chap. 3), 54–55.

39. Ibid., 48–49.

40. Bradford Washburn, memo to Norman Harris, October 10, 1955, File "1956/Exhibits—Medical," BMoS.

41. "Million-Dollar Exhibit Building Program," *Progress* 3, no. 4 (1952): 3, PMSIC.

42. On the *Cardiac Kitchen*, see Marjorie Warvelle Bear, *A Mile Square of Chicago* (Oak Brook, IL: TIPRAC, 2007), 166; "Cardiac Kitchen," *Progress* 11, no. 8 (1951): 3, PMSIC.

43. On the Atomic Energy Commission's traveling museum exhibits program from 1950 onward, see M. D. Maree and A. J. Williams, "A Thirty-Year Look at the Nuclear Science Programs at the American Museum of Science and Energy," *Journal of Radioanalytical and Nuclear Chemistry* 171, no. 1 (1993): 253–58.

44. *Boston Museum of Science Newsletters* for September 1953 and December 1953, all in Newsletter Files; see also Bradford Washburn to Alexander Marble, 28 February 1957, box "1957–58," folder "S," all BMoS.

45. "The Lowly Frog Demonstrates Radiation," *Progress* 2, no. 7 (1951): 1. See also Harvey B. Lemon, "Atomic Energy Exhibit Reopens," *Progress* 2, no. 7 (1951): 4–5. Lemon reports the following biological and medical exhibits: radioactive elements, both natural and artificial; a luminous dial on an alarm clock; desert sand exposed to a one million degree blast of the Alamogordo bomb; and the "hot" frogs.

46. Traditionalism had its appeal, however. Smithsonian visitor surveys taken in the early 1970s suggested the exhibit was one of the most popular in the museum (email from Pamela Henson to Karen Rader, October 20, 2005)

47. For information on diorama building in museums of science and natural history museums in the postwar period, see Lurie, *A Special Style* (n. 7 above, chap. 1), 80–85; Metzler, "The Artists" (n. 69 above, chap. 2), time-

line appendix; Rock, *Museum of Science, Boston* (n. 1 above, chap. 1), chaps. 10 and 11. Smaller museums like the Putnam Museum seemed to have slowed production of dioramas after 1947 but then resumed building them again in the mid-1960s: see "Putnam Museum Exhibit List," (PM).

48. Because of the museum's logistical struggles, the planning for *How Your Life Began* did not start in earnest until 1950 and it wasn't opened to the public until 1964, but its conception in the 1940s demonstrates the strong influence of the "life adjustment" trend. See Carol F. Mallory, "How Your Life Began" (undated), RR (Exhibit Files), BMoS.

49 *Field Museum of Natural History Handbook*, 12th ed. (Chicago: Field Museum of Natural History, 1943), 62. For a broader history of the Field Museum's education programs, see Bruce Hatton Boyer, *The Natural History of the Field Museum: Exploring the Earth and Its People* (Chicago: Field Museum, 1993) 87–89.

50. See Keith Sheppard and Dennis M. Robbins, "High School Biology Today: What the Committee of Ten Actually Said," *CBE-Life Sciences Education* 6, no. 3 (2007): 198–202.

51. Butler Laughlin, "Science for All," *School Science and Mathematics* 3 (1948): 176. Scientists were equally concerned, so much so that the American Association for the Advancement of Science formed the special Cooperative Committee on School Science and Mathematics. After two years of study, the committee published a stern report on the dismal condition of school science in the United States, warning that "science teaching . . . [simply] is not ready for the responsibilities which it must nevertheless assume." McGrath, ed., *Science in General Education* (n. 28 above, this chap.), 185.

52. R. W. Gerard, "Higher General Education in Biology," *AIBS Bulletin* 6, no. 2 (1956): 20.

53. Following the advice of the Harvard Committee on General Education, the U.S. Office of Education, and other proponents of life adjustment, school districts often reserved physics and chemistry for college-bound students in the postwar period. Biology courses, in contrast, became the coda to a general science curriculum. As Hiden Cox, executive director of the American Institute of Biological Sciences noted: "The situations in biology was different. . . . For each student who took chemistry in high school, there were two who took biology. For every one who took physics, there were four who took biology. For more than half of the American population, the only science course ever taken was biology." Laura Engleman, ed., *The BSCS Story: A History of the Biological Sciences Curriculum Study* (Colorado Springs, CO: BSCS, 2001), 1; Tolley, *Science Education of American Girls* (n. 22 above, this chap.), 199–200.

54. Victor H. Noll, "Science Education in American Schools." *The Forty-Sixth Yearbook of the National Society for the Study of Education* 31, no. 5 (1947): 295–99.

55. Rudolph, *Scientists in the Classroom* (n. 49 above, chap. 3), 101.

56. Richard R. Armacost, "Progress Report of the Implementation of Recommendation of the Southeastern Conference on Biology Teaching: A Digest of the Presentation of Panel Members, A.A.A.S. Meeting, 27–30 December 1955 in Atlanta, Georgia," *American Biology Teacher* 18, no. 7 (1956): 221, 223.

57. McEwan, "Let's Put Life into Life Science" (n. 33 above, this chap.), 156–58.

58. Phillip Foss, "Biology Clubs: A Questionnaire," *American Biology Teacher* 4, no. 6 (1942): 165. See also C. A. Federer, "Nation-Wide Junior Science Clubs," *Science* 88, no. 2292 (1938): 526. Federer notes that the American Institute of the City of New York worked to influence and distribute science news for American youth by publishing a youth science research journal, the *Science Observer*, and holding an annual science fair at the American Museum of Natural History.

59. On the Battle Creek Public Schools Museum, see "A Brief History of the Museum," Kingman Museum, accessed May 23, 2011, http://www.king manmuseum.org/history.cfm. The Battle Creek museum's director saw the human embryo exhibit at Chicago's Museum of Science and Industry and sought advice from Chicago's director of medical exhibits about how to mount one: see Edward M. Brigham Jr., letter to Thomas Hull (director of medical exhibits, MSIC), March 14, 1951; Hull, letter to Brigham, March 29, 1951, both MSIC.

60. It is likely that these relationships were reinforced by postwar increases in city funding. A 1954 survey in *Museum News* found that city support for museums doubled between 1944 and 1954; no doubt museums felt the need to repay the taxpayers for their largesse. Cited in Laurie Wertz, "Some Things Never Change," *Museum News* 62, no. 3 (1984): 29.

61. *DMNHAR* (1954), 48.

62. "The Museum of Science and Industry," *Progress* 1, no. 1 (1950): 3.

63. See the folder "School Visits" in BMoS for more information on the museum's arrangements with the Boston Metropolitan District Commission from the 1940s onward. See also Rock, *Museum of Science, Boston* (n. 1 above, chap. 1), 95.

64. See series of correspondence between Bradford Washburn and the Lancaster (PA) AAAS branch, November 1956, about AAAS lecturer Jane Crile (Mrs. George Crile Jr.), who came and talked about underwater life. The lecture is described (by Robert Moore to Bradford Washburn, June 28, 1956) as "very successful." BW Correspondence, "1956," BMoS.

65. Boston Museum of Science Education Department staff spent the winter of 1947–48 promoting science fair involvement and the museum hosted a local science fair awards program that same year; by 1950 the museum was hosting a gathering of state winners for a two-day exhibition. Rock, *Museum of Science, Boston* (n. 1 above, chap. 1), 71; I. E. Wallen, "Scientists Serve High Schools," *Science* 124, no. 3213 (1956): 168.

66. Wallen, "Scientists Serve High Schools," 168. On the broader history of science fairs and clubs in this era, see Sevan G. Terzian, *Science Education and*

Citizenship: Fairs, Clubs, and Talent Searches for American Youth, 1918–1958 (New York: Palgrave Macmillan, 2012).

67. Increasingly, extracurricular groups devoted to science education were affiliated with nationally orchestrated efforts. In 1946, e.g., the AAAS founded the Committee on Junior Scientists' Assembly intending to "bring together young scientists who are still in the midst of their scientific training, so that they may share their experiences and opinions." "News and Notes," *Science* 104, no. 2709 (1946): 506–9.

68. Rock, *Museum of Science, Boston*, 99.

69. Lurie, *A Special Style* (n. 7 above, chap. 1), 89.

70. Rock, *Museum of Science, Boston*, 100.

71. Goodyear, *Society and Museum* (n. 5 above, "Introduction"), 63.

72. Watson Davis, "The Interpretation of Science through Press, Schools and Radio," *American Biology Teacher* 14, no. 3 (1952): 66.

73. Watson Davis, "The Rise of Science Understanding," *Science* 108, no. 2801 (1948): 239–46.

74. Sinnott, "Ten Million Scientists" (n. 30 above, this chap.), 127–28. On Sinnott's advocacy in biology education, see Whaley, "Edmund Ware Sinnott" (n. 30 above, this chap.).

75. Alma Stephanie Wittlin, *The Museum, Its History and Its Tasks in Education* (London: Routledge & Kegan Paul, 1949), v, 9.

76. A. W. Bell, "Reorganization at the Los Angeles Museum," *Science* 94, no. 2437 (1941): 255–56.

77. Rock, *Museum of Science, Boston* (n. 1 above, chap. 1), 71. See also Photo Files, "Natural History of a Drop of Water," BMoS.

78. John J. Desmond, letter to Bradford Washburn, October 17, 1956; Washburn, letter to Norman Harris, October 24, 1956, both in box "1956," folder "Education," BMoS.

79. *AMNHAR* (1947), 5.

80. *Progress Report: Museum of Science*, no. 6 (May 1950), 1, BMoS; see also Washburn's report on his trip to European museums during the summer of 1957, as quoted in "Minutes of MoS Trustee Annual Meeting, 9 October 1957," in box 1957–58, BMoS.

81. Rock, *Museum of Science, Boston*, 100.

82. Bradford Washburn, "Message from the Director," *BMoSAR*, 1955–56, 1–2, BMoS.

83. Though gifts from individuals remained the primary source of museum funding in this decade, postwar corporations often provided the funds for science museums to experiment some with display. See Wertz, "Some Things Never Change" (n. 60 above, this chap.), 28.

84. Bruno Gebhard, "What Good Are Health Museums?" *American Journal of Public Health* 36, no. 9 (1946): 1012–15, and "The Health Museum as Visual Aid," *Bulletin of the Medical Library Association* 35, no. 4 (1947): 329–33.

85. "Oomph for Science," *Time* 36, no. 10 (September 2, 1940): 42.

86. Jay Pridmore, *Museum of Science and Industry, Chicago* (Chicago: Harry N. Abrams, in association with Museum of Science and Industry, 1997), 69.

87. See entry on Department of Public Health in the *Chicago Daily News Almanac and Yearbook for 1921*, digitized July 21, 2005, http://books .google.com/books/about/The_Chicago_daily_news_almanac_and_yearb .html?id=fRR_mxkNlsgC. For a sense of how Hull's medical exhibition experience informed what he installed at Chicago's Museum of Science and Industry, see "A.M.A. Centenary: Exhibits and Papers," *British Medical Journal* 2, no. 4514 (July 12, 1947): 64–67. The reviewer seemed especially impressed by the scale of the exhibits: "The enormous size of the Convention Hall (its floor space would engulf St. Paul's Churchyard), and the equally enormous wealth of many of the firms exhibiting, combined with the native American flair for advertising, produce an impression of color and luxury which would be startling, even shocking, to post-war English eyes. It is of course impossible to know the total amount of money spent to produce these stands, but unofficial estimates varied between three and five million dollars."

88. See, e.g., Hal Kome (assistant director of publicity for the National Society of Medical Research), letter to MSIC director of exhibits, July 21, 1952, proposing a promedical research "zoo" of live research animals; Merck and Company's Dr. Charles Lyght, letter to Thomas Hull, July 17, 1951, proposing an exhibits on hormones; S. Ruston, letter to T. Hull, April 14, 1952—all at MSIC.

89. Hull, letter to George Percy, April 6, 1953. On Hull's solicitation of exhibits, see his letters to Parke Davis and Company (August 21, 1951), Merck and Company (March 16, 1953), Bauer and Black (April 24, 1953), and Curity (May 5, 1953)—all MSIC.

90. Carey, "Health and Medical Science Exhibits" (n. 138 above, chap. 3), 50. On vitamins in scientific research and popular culture in this period, see Rima Apple, *Vitamania: Vitamins and American Culture* (New Brunswick, NJ: Rutgers University Press, 1997); Nicholas Rasmussen, "The Moral Economy of the Drug Company–Medical Scientist Collaboration in Interwar America," *Social Studies in Science* 34, no. 2 (2004): 161–85.

91. "The Conquest of Pain Opened," *Progress* 6, no. 2 (1955): 4–5, PMSIC.

92. Ibid., 4.

93. Harold J. Passeneau, "G. D. Searle and Abbott Laboratories," *Analysts Journal* 10, no. 3 (1954): 158–59.

94. John W. Beal, letter to B. Washburn, November 1, 1955; Washburn, letter to Beal, November 21, 1955; Washburn, letter to W. Furness Thompson, March 15, 1956; Thompson, letter to Washburn, April 3, 1956—all BMoS.

95. Hal Kome, letter to Daniel McMasters, July 21, 1952, MSIC, "Visitors from around the Globe: Easter Crowds Watch Chickens Hatch on Harvester Farm," *Progress* 4, no. 3 (May–June 1953): 8, MSIC.

96. Ibid.

97. "Swift Moves In," *Progress* 4, no. 5 (1953): 2–3, MSIC.

98. Ibid.

99. "Baby Chicks Make Easter Perfect," *Progress* 9, no. 3 (1958): 5, MSIC.

100. See Arthur O. Baker and Lewis H. Mills, *Dynamic Biology* (New York: Rand McNally, 1933); Ladouceur, "Ella Thea Smith" (n. 50 above, chap. 3). On live animal displays in children's museums in the 1950s, see, e.g., "Animal Lending Library," *Life* 33, no. 2 (1952): 89–92; "Want to Borrow a Porcupine?" *Popular Mechanics* 99, no. 4 (1953): 88–90. On the broader history of children's zoos, see Gail Schneider, *Children's Zoos* (Washington, DC: National Recreation and Parks Association, 1970). Zoos in this period seemed to be moving toward a model of ecological immersion, one that recalled museums' attempts to immerse visitors through the unified decor of diorama halls. On this phenomenon, see William Bridges, *Gathering of Animals: An Unconventional History of the New York Zoological Society* (New York: Harper and Row, 1974); David Hancocks, *A Different Nature: The Paradoxical World of Zoos and Their Uncertain Future* (Berkeley: University of California Press, 2001); Jeffrey Hyson, "Jungles of Eden: The Design of American Zoos," in *Environmentalism in Landscape Architecture*, ed. Michel Conan, Dumbarton Oaks Colloquium Series in the History of Landscape Architecture (Cambridge, MA: Harvard University Press, 2000), 23–44; George B. Rabb, "Education in Zoos," *American Biology Teacher* 30, no. 4 (1968): 291–96.

101. The Boston Museum of Science itself was criticized for not presenting microscopic life in more vivid form: in 1956, Bradford Washburn received a letter from one Mrs. Cannon, who informed hims that the museum's existing microscope displays were inadequate, for the public was more interested in seeing the living microscopic world, than it was in viewing dead specimens under slides. "Memo: Mrs. Cannon's Letter," Bradford Washburn to Norm Harris, December 31, 1956, folder "C," Bradford Washburn Papers, Boston Museum of Science Archive, BMoS.

102. On the snake exhibit and other live animal demonstrations, see Boston Museum of Science Member Newsletters, September 1950; also December 1950 (front page photo), both BMoS.

103. *BMoSAR* (1950), introduction. See also Rock, *Museum of Science, Boston* (n. 1 above, chap. 1), 101.

104. Rock, *Museum of Science, Boston*, 104.

105. "Junior Assistants in the Animal Room," cover of the *Boston Museum of Science Newsletter*, March 1956; "science in action" quote from *Boston Museum of Science Newsletter*, March 1957 issue, BMoS.

106. Rock, *Museum of Science, Boston*, 104.

107. Ceiga, "Brief History of the Education" (n. 188 above, chap. 3), 10.

108. *DMNHAR* (1955), 29. On Alice Gray, see "Alice E. Gray, 79: Museum's 'Bug Lady,'" *Hartford Courant*, May 1, 1994.

109. Bradford Washburn, "Message from the Director," *BMoSAR* (1955–56), 1–2.

110. On Chicago's plans at the time for future life science exhibits, including exhibits on the heart and nutrition, see "Schedule of Exhibit Changes and Additions," April 1, 1951; see also J. W. Hostrup, memo to D. W. Mac-Master, April 17, 1951 regarding the need to update the *Modern Operating Room* exhibit that MSIC had inherited from the Century of Progress: "The Modern Operating Room will have to become a static display with no sound from now on unless a new recording (of an exhibit script) is made." All MSIC.

111. On the history of attempts to approach education in scientific terms, see Ellen Condliffe Lagemann, *An Elusive Science: The Troubling History of Education Research* (Chicago: University of Chicago Press, 2000).

112. Pridmore, *Inventive Genius* (n. 120 above, chap. 3), 74–75.

113. Ibid., 78–79.

114. Ibid., 81.

115. Lucy Nielsen Nedzel, "The Motivation and Education of the General Public through Museum Experiences" (PhD diss., University of Chicago, 1953), 80.

116. Ibid., 82.

117. "Black Bear on Mount Washington," *Boston Museum of Science Newsletter*, June 1954 and " 'Shorebird' Superintendent," *Boston Museum of Science Newsletter*, December 1954, both BMoS.

118. On donation and explanation of the transparent woman, see Museum of Science Ready Reference File, "Transparent Woman"; on Fielder and the exhibit's debut, see *Boston Museum of Science Newsletter*, March 1955 and September 1955: all BMoS.

119. Pridmore, *Museum of Science and Industry* (n. 87 above, this chap.), 58–60.

120. Parr's early writings on museums and their role in education were collected and synthesized by his wife in a collection titled *Mostly about Museums: From the Papers of A. E. Parr* (New York: American Museum of Natural History, 1959).

121. Parr had arrived in New York City in 1926 with a young bride and $135 in his pocket—the rest had gone to pay duty on his microscope—and, desperate for money, he had taken a job sweeping floors and emptying spittoons at the New York Aquarium. He quickly made a name for himself as a marine zoologist, began to oversee collecting expeditions for wealthy naturalists, and in 1928, moved two hours north to New Haven to serve as scientific director of Yale's newly established Bingham Oceanographic Laboratory. He spent the next decade sailing around the Caribbean, the Gulf of Mexico, and the Atlantic and directing the Peabody Museum's program in marine research; eventually, he assumed the directorship of Yale's Peabody Museum of Natural History. In 1942, to the great surprise of the museum's curators, he was appointed the director of the American Museum of Natural History. Glenn Fowler, "Albert E. Parr, Museum Director and Oceanographer, Dies at 90," *New York Times*, July 20, 1991. Also see "Albert Eide Parr," *Current Biography* 3, no. 7 (1948): 640–42.

122. Kennedy, "Philanthropy and Science" (n. 7 above, "Introduction"), 240.

123. Parr, "Address" (n. 18 above, this chap.).

124. Ibid.

125. *AMNHAR* (1942), as cited in Myers, "A History of Herpetology" (n. 44 above, chap. 1), 57.

126. On Simpson's work, see Peter Forey, "Systematics and Paleontology," in *Milestones in Systematics*, ed. Peter Forey and David M. Williams (Boca Raton, FL: CRC Press, 2004), 149–80, especially 165; Joel B. Hagen, "Descended from Darwin? George Gaylord Simpson, Morris Goodman, and Primate Systematics," in *Descended from Darwin: Insights into the History of Evolutionary Studies, 1900–1970*, ed. Joseph Allen Cain and Michael Ruse, Transactions of the American Philosophical Society (Philadelphia, PA: American Philosophical Society, 2009); Laporte. On the participation of the museum's curators in the development of the evolutionary synthesis, see Cain, "Common Problems" (n. 27 above, this chap.).

127. "The American Museum of Natural History: Confidential Report and Recommendations, from the Director," April 1944, 29, Manuscript Collections, Albert Parr Collection, AMNH Central Archives, as cited in Myers, "A History of Herpetology" (n. 44 above, chap. 1), 58.

128. Albert Eide Parr, "Toward New Horizons," *AMNHAR* (1947) 20–21.

129. Thanks to new interest in ecology and ethology and reinvigorated research budgets, a number of American museums of natural history built new field stations in the 1950s and early 1960s or expanded existing ones. The Carnegie Museum of Natural History established a 1,160-acre reserve, Powdermill, in western Pennsylvania in 1956 to promote more ecological study, e.g., while Hawaii's Bishop Museum established a montane field station in Papau New Guinea in 1961.

130. *The Second Annual Report of the National Science Foundation* (Washington, DC: Government Printing Office, 1952), 19.

131. On battles between systematists, see David Hall, *Science as a Process* (Chicago: University of Chicago Press, 1988). On the NSF's role in evolutionary biology as a research discipline, see Smocovitis, *Unifying Biology* (n. 27 above, chap. 3), 65–72. Toby Appel historicizes the NSF funding of biology in the post-Sputnik period in *Shaping Biology: The National Science Foundation and American Biological Research, 1945–75* (Baltimore, MD: Johns Hopkins University Press, 2000), 216–22.

132. Appel, *Shaping Biology*, 148.

133. In 1956, e.g., American Museum curators of entomology Mont A. Cazier and Nicholas S. Obraztsov received funding to revise the taxonomical categories of beetles ($2,500, one year) and moths ($10,200, two years), while G. G. Simpson, back at the museum in the newly reconfigured Department of Geology and Paleontology, was awarded $8,000 for an expedition to Alto Rio, Brazil. J. Linsley Gressitt of the Bishop Museum's Department of Entomology, won $23,000 for a three-year study titled "Zoogeography of Pacific Insects." See National Science Foundation, "Research Support

Programs," app. B in *Sixth Annual Report for the Fiscal Year Ended June 30, 1956* (Washington, DC: Government Printing Office, 1957), 119, 141, http://www.nsf.gov/pubs/1956/annualreports/start.htm?org=DGE.

134. Frank Taylor, oral history interview by Pamela Henson, 1974, RU 9512, SIA. According to Toby Appel, the National Museum of Natural History received NSF grants for their curators' systematics research until Congress forbade the NSF from funding other federal agencies. In 1963, e.g., museum curators received twelve awards totaling $234,000. See Appel, *Shaping Biology*, 210.

135. For biographical information on T. Dale Stewart, see the Smithsonian Institution Research Information System database listings on the Thomas Dale Stewart Papers, http://siris-archives.si.edu/ipac20/ipac.jsp?&profile=all &source=~!siarchives&uri=full%3D3100001~!87801~!0.

136. Lurie, *A Special Style* (n. 7 above, chap. 1), 86–90.

137. See J. Merton England, "The National Science Foundation and Curriculum Reform: A Problem of Stewardship," *Public Historian* 11, no. 2 (1989): 23–36.

138. What is now the Florida Museum of Natural History was called the Florida State Museum when Arnold Grobman was its director from 1952 to 1959; see Nathan Crabbe, "Arnold Grobman, Early Director of Florida Musuem [*sic*] of Natural History, Dies at Age 94," *Gainesville Sun*, July 9, 2012, http://www.gainesville.com/article/20120709/ARTICLES/120709639.

139. "40,000 Canal Zone Worker Ants to Go on Exhibition Here Today," *New York Times*, July 18 1952, http://query.nytimes.com/mem/archive/pdf?res =F00711FA3E5E177B93CAA8178CD85F468585F9; Appel, *Shaping Biology* (n. 132 above, this chap.), 53. *AMNHAR* (1952), 44.

140. Ceiga, "Brief History of the Education" (n. 188 above, chap. 3); Quinn, *Windows on Nature* (n. 158 above, chap. 1), 8–24. Also see *DMNHARs*, 1942–59.

141. As cited in Boyer, *Natural History of the Field Museum* (n. 50 above, this chap.), 84; Martha Weinman, "The Marvels in Our Museums," *Collier's*, February 4, 1955, 63.

142. Design trends fascinated Parr, who frequented international design conferences in the hopes that he could duplicate designers' innovations to improve the museum's exhibits. Confidential Report on Albert E. Parr, 1951, Board of Trustees Papers, Albert Eide Parr, Biographical Files, AMNH.

143. *AMNHAR* (1944), 41–42.

144. A. E. Parr, letter to A. P. Osborn, June 7, 1950; A. E. Parr, letter to C. M. Breder, May 11, 1951—both as cited in Kennedy, "Philanthropy and Science" (n. 7 above, "Introduction"), 250.

145. *AMNHAR* (1957–1958), 3 and 7.

146. On the Hall of North American Forests, see *AMNHAR* (1952), 45 and 50; see also Quinn, *Windows on Nature* (n. 158 above, chap. 1), 138. On Pough's involvement, see Stuart Lavietes, "Richard Pough, 99, Founder of the Nature Conservancy, Dies," *New York Times*, June 27 2003.

147. *AMNHAR* (1957), 8–9.

148. Taylor had joined the Smithsonian staff straight out of high school, working as a laboratory apprentice in the museum's Division of Mechanical Technology beginning in 1921. Knowing he would need more formal education to advance at the institution, he left the museum in the mid-1920s to obtain a degree in engineering from MIT and a law degree from Georgetown University, but returned to the Smithsonian as a junior curator in the Arts and Industries building in 1932. Taylor had a calm but commanding temperament: he spent most of World War II leading troops in the Philippines, but once his tour of duty had ended, Taylor returned to the museum, determined to improve its languishing exhibits. On Taylor, see Robert Post, *Who Owns America's Past?: The Smithsonian and the Problem of History*, (Baltimore: Johns Hopkins University Press, 2013), 22-30; Patricia Sullivan, "Frank Taylor; Founding Director of American History Museum," *Washington Post*, June 30, 2007. See also Frank Taylor Oral History (nos. 11–12), Frank Augustus Taylor Interviews, Record Unit 9512, SIA.

149. See "Observation Beehive Noted in a Brief Guide to the Smithsonian Institution, c.1950," History of the Smithsonian Catalog, SIA Information File, box 14—Guidebooks to the National Museum, ca. 1939–50, http://siris -sihistory.si.edu/ipac20/ipac.jsp?&profile=all&source=~!sichronology&uri =full=3100001~!11763~!0#focus.

150. Yochelson, *National Museum of Natural History* (n. 95 above, chap. 1), 88.

151. Frank Taylor Memo, June 22, 1948, RG 190, box 89, "Exhibits Modernization 1950," SIA. Post notes that Taylor began his attempts to push Smithsonian administrators to support the modernization of displays nearly two years earlier. Post, *Who Owns America's Past?* (n. 149, this chap.), 27.

152. His inquiries included curators' opinions about what made a good exhibit and assessments of which exhibits were most popular, as well as a more specific set of questions on display strategies. "How do you feel about the following?" asked the questionnaire. "Popular (genera) exhibits vs. study exhibits (detailed, specialized); combination?; Life-size versus miniature (anthropology, geology, zoology, period rooms); Dioramas vs. cheaper form of art work (diagrams, charts, paintings, photographs); Spill lighting without hall lighting?; Intensive, forceful, educational type of exhibits vs. just showing what the visitor expects to see? (Hit him or let him browse?)" "Checklist for Inquiries," Frank Taylor to Committee on Modernization of Exhibits, attached to memo of July 1948, RG 190, box 89, "Exhibits Modernization 1950," SIA.

153. See Phillip Rittenhouse, "Biology and the Smithsonian Institution," *BioScience* 17, no. 1 (1967): 25–35. On Schmitt's broader career, see Richard E. Blackwelder, *The Zest for Life; or, Waldo Had a Pretty Good Run: The Life of Waldo Lasalle Schmitt* (Lawrence, KS: Allen Press, 1979).

154. Waldo Schmitt, letter to Frank Taylor and Herbert Friedmann (series of correspondence, dating October 2–3, 1956, memo quoted from October 2, 1956), RG 363, box 1, SIA.

155. E. P. Killip, "Comments on a Preliminary Report of Exhibition Subcommittee for Exhibits," October 1, 1948, p. 2 in RU 155—"Director, NMNH, 1948–70," box 10, SIA.

156. Frank Taylor, "Preliminary Report on Exhibits," October 1948, and Kellogg's response (same folder) in RG 190, box 89, "Exhibits Modernization 1950," SIA. Comments on Kellogg's reputation based on Smithsonian historian Pam Henson's, conversation with Karen Rader, October 19, 2005, about Henson's Frank Taylor Oral History, SIA.

157. See, e.g., "Putnam Museum Exhibit List" (PM).

158. The group visited the Carnegie, the Smithsonian, the American Museum of Natural History, the Philadelphia Academy of Natural Sciences, the Peabody Museum, the Museum of Comparative Zoology, the New York State Museum, the Buffalo Museum of Science, the Field Museum, the Chicago Academy of Sciences, the University of Iowa Museum, and the Hastings Museum. Over the course of the next fifteen years, Bailey oversaw the construction of multiple diorama halls. When the museum opened a new wing in 1952, e.g., its first floor was devoted to dioramas of early man, its second floor was filled with habitat dioramas of the wildlife of Australia, Midway, Fiji, and other regions in the Pacific, and its third floor to habitat dioramas of the birds and mammals of South and Central America. The older part of the building continued to boast habitat dioramas of American and Arctic birds and mammals, as well as a new diorama hall depicting ecological scenes of Colorado. CMNHAR (1940), 6; DMNHAR (1951), 28–29; DMNHAR (1952).

159. In 1947, e.g., the local evening paper, the Milwaukee Journal, sponsored a month-long expedition to Alaska to obtain moose, caribou, and other animals and background materials for exhibits. Taxidermists Owen Gromme and Walter Pelzer commanded the party, which also included a guide, another taxidermist Lester Diedrich, artist Keith Gebhart, and Journal reporter Gordon MacQuarrie, who sent stories to the Journal on the progress of the trip. No curators were involved. Lurie, A Special Style (n. 7 above, chap. 1), 79–80.

160. AMNHAR (1953), 54.

161. SIAR (1957); see also picture in RU 95, box 42, folder 2, SIA.

162. Alexander Wetmore, "Report of the Secretary," in SIAR (1949), 1.

163. Albert Eide Parr, "Report of the Director," AMNHAR (1948), .n.p.

164. Bradford had initially explored the possibilities of attracting more adults to the museum through continuing education courses and events aimed at older audiences, but ultimately elected to focus on children: Albert Parr, letter to Bradford Washburn, April 18, 1956, box 1956, folder "Courses for Older People"; Bradford Washburn, letter to Kay Harrison et al., June 20, 1957, box 1957–58, folder "W"; a sample organization chart can be found in box "1957–58," folder "O"; see also Norm Harris to Caroline Harrison et al., "Master Exhibits List Memo," April 3, 1959, box "1963 [sic]," folder "Master Exhibits List," all BMoS. On Bradford Washburn's integrated edu-

cation plan, see "Aquarium Study Tour," March 6, 1958, box 1958, folder "Trustees—Notices"; "Role of Science Museum," *Boston Post*, December 15, 1955, editorial page, in box 1956, folder "Representative and Senator Correspondence" (with Bradford Washburn's note at bottom: "Spontaneous reaction of the press on receipt of our annual report!"), all BMoS.

CHAPTER FIVE

1. Zuoyue Wang, *In Sputnik's Shadow: The President's Science Advisory Committee and the Cold War* (New Brunswick, NJ: Rutgers University Press, 2008), 81.

2. "Boston's 'Living Museum,'" *Christian Science Monitor*, December 23, 1957, 16. Clippings can be found in Bradford Washburn's files (box "1958," folder "M").

3. Ibid. See also "How Strong Is the United States?" *Boston Herald*, December 24, 1957, 18. Emphasis in the original quote, from Bradford Washburn, "Letter to Editor," *Boston Herald*, December 24, 1957, box "1957–58," folder "H," all BMoS.

4. *SIAR* (1959), 4. In the annual report, Carmichael acknowledged that Junior League volunteers were primarily responsible for teaching the school groups that attended the museum, a practice that persisted well into the 1970s.

5. For more on this debate, see, among others, DeBoer, *History of Ideas* (n. 34 above, chap. 1), 134–80; Rudolph, *Scientists in the Classroom* (n. 49 above, chap. 3); George W. Tressel, "Thirty Years of 'Improvement' in Precollege Math and Science," *Journal of Science Education and Technology* 3, no. 2 (1994): 77–88.

6. For a repudiation of this view, see Stephen Hilgartner, "The Dominant View of Popularization: Conceptual Problems, Political Uses," *Social Studies of Science* 20, no. 3 (1990): 519–39.

7. Joseph J. Schwab, "Science and Civil Discourse: The Uses of Diversity," *Journal of General Education* 9 (1956): 142. Also see Audra Wolfe's "Speaking for Nature" (n. 27 above, chap. 4), which provides an excellent analysis of the BSCS.

8. Warren Weaver, "Science and the Citizen," *Science* 126, no. 3285 (1957): 1226.

9. Hilary J. Deason, "Traveling High-School Libraries," *Science* 122, no. 3181 (1955): 1173.

10. The program is described and illustrated in *The Seventh Annual Report of the National Science Foundation* (Washington, DC: Government Printing Office, 1957), 32.

11. Tressel, "Thirty Years of 'Improvement,'" 81. See also Paul E. Marsh and Ross A. Gortner, *Federal Aid to Science Education: Two Programs*, The Economics and Politics of Public Education (Syracuse, NY: Syracuse University Press, 1963).

12. Bestor, *Educational Wastelands* (n. 23 above, chap. 4). On the historical influence of Bestor's book, see Lagemann, *Elusive Science* (n. 112 above, chap. 4), 160–83.

13. Paul Hurd, "Science Literacy: Its Meaning for American Schools," *Educational Leadership* 16, no. 1 (1958): 15.

14. DeBoer, *History of Ideas* (n. 34 above, chap. 1), 175.

15. Ibid., 176.

16. Contemporaries' awareness of the malleability of the term "scientific literacy" is discussed in Marsh and Gortner, *Federal Aid to Science Education*, ix.

17. See, e.g., Henry W. Syer, "Review of Will French's *Behavioral Goals of General Education in High School*," *Science* 127, no. 3307 (1958): 1171–72. For more on the history of discipline-based science education in this era, see DeBoer, *History of Ideas*, 172.

18. On *Man, a Course of Study*, a later NSF-sanctioned curriculum project employing discovery learning methodology, see Erika L. Milam, "Public Science of the Savage Mind: Contesting Cultural Anthropology in the Cold War Classroom," *Journal of the History of the Behavioral Sciences* 49, no. 3 (2013): 306–60. On how the BSCS focus on interactivity was inscribed into classroom technologies, see John Rudolph, "Teaching Materials and the Fate of Dynamic Biology in American Classrooms after Sputnik," *Technology and Culture* 53 (2012): 1–36. Historian Ronald Ladouceur has argued that the BSCS steering committee's inflated sense of its own responsibility for promoting the evolutionary synthesis has obscured historians' recognition of earlier biology textbook writers like Ella Thea Smith who promoted evolutionary principles as a unifying framework as early as 1938. See Ladouceur, "Ella Thea Smith" (n. 50 above, chap. 3); Ella Thea Smith, *Exploring Biology*, 3rd ed. (New York: Harcourt Brace and World, 1949). The "self-serving legend" of the BSCS's role in promoting evolutionary ideas was recognized by biologists. George Gaylord Simpson, "Evolution and Education," *Science* 187, no. 4175 (1975): 389.

19. Walter P. Taylor, "Is Biology Obsolete?" *AIBS Bulletin* 8, no. 3 (1958): 13.

20. Philip Wylie, "A Layman Looks at Biology," *AIBS Bulletin* 9, no. 3 (1959): 13.

21. Sidney Rosen, "The Origins of High School General Biology," *School Science and Mathematics* 59, no. 6 (1959): 487.

22. Oswald Tippo, "Biology and the Educational Ferment," *AIBS Bulletin* 9, no. 2 (1959): 16.

23. J. Stewart Hunter, "Public Health and Medicine Exhibit: Brussels World's Fair," *Public Health Reports* 73, no. 12 (1958): 1122–24.

24. Howard Simons, "Brussels Fair and Science," *Science News-Letter* 73, no. 2 (1958): 27.

25. Such feelings were evident in their commentary about the Seattle World's Fair, which they described as a chance to redeem science exhibitions

among the American public. See, e.g., Lawrence E. Davies, "Scientists to Aid Fair in Seattle," *New York Times*, July 21, 1958; Erik Ellis, "Dixy Lee Ray, Marine Biology, and the Public Understanding of Science in the United States (1930–1970)" (PhD diss., Oregon State University, 2005), 180.

26. The BSCS and Grobman's views are detailed in Rudolph, *Scientists in the Classroom* (n. 49 above, chap. 3), 186. On NSF support for high school curriculum studies, see Rudolph, 113–92; Wolfe, "Speaking for Nature" (n. 27 above, chap. 4), 135–80. On how the worldview of physicists dominated science education policy, see David Kaiser, *How the Hippies Saved Physics* (New York: W. W. Norton, 2011); Marsh and Gortner, *Federal Aid to Science Education* (n. 11 above, this chap.); Jerrold Zacharias, *Education in the Age of Science* (New York: Basic Books, 1959).

27. *The Tenth Annual Report of the National Science Foundation* (Washington, DC: Government Printing Office, 1960), 83.

28. As early as 1956, this approach ignited anxiety that such efforts would siphon students away from other professions. See, e.g., Alan Waterman, "The Role of the Federal Government in Science Education," *Scientific Monthly* 82, no. 6 (1956): 286–93. On disciplinary-based science training, see DeBoer, *History of Ideas* (n. 34 above, chap. 1), 147–72.

29. On the elitism inherent in the NSF-sponsored efforts in science education, see David Gamson, "From Progressivism to Federalism: The Pursuit of Equal Educational Opportunity, 1915–1965," in *To Educate a Nation: Federal and National Strategies of School Reform*, ed. Carl F. Kaestle and Alyssa Lodewick (Lawrence: University Press of Kansas, 2007).

30. See *BMoSAR* (1957–61).

31. Bradford Washburn, letter to George Rothwell, December 7, 1961, box "1961," folder "Correspondence"; Bradford Washburn, "Memorandum—Visit by Bradford Washburn to NSF, 14 November 1962," box "1962," folder "NSF." See also Norm Harris, letter to Bradford Washburn, January 6, 1959 memo re: "Summer Teachers Courses," box "4–1959," folder "Education Department," all BMoS.

32. *The Tenth Annual Report of the National Science Foundation*, 143.

33. Robert Post describes how history museums, specifically the National Museum History and Technology, used a similar strategy in the 1960s and 1970s. Post, *Who Owns America's Past?* (n. 149 above, chap. 4), 8, 114–34.

34. Though scholars have largely ignored exhibition design in science and natural history museums after the 1920s, practitioners and art historians have recently begun to flesh out the broader history of museum display and exhibition. See, among others, David Dernie, *Exhibition Design* (New York: W. W. Norton, 2006); Victoria Newhouse, *Art and the Power of Placement* (New York: Monacelli Press, 2005); Stuart Silver, "Almost Everyone Loves a Winner," *Museum News* 61, no. 2 (1982): 24–35; Bruce Altshuler, *Salon to Biennial : Exhibitions That Made Art History* (New York: Phaidon, 2008); Bruce Altshuler, *Biennials and Beyond: Exhibitions That Made Art*

History, 1962–2002 (New York: Phaidon Press, 2013); Neil Harris, *Capital Culture: J. Carter Brown, the National Gallery of Art, and the Reinvention of the Museum Experience* (Chicago: University of Chicago Press, 2013); Post, *Who Owns America's Past?* 30–36.

35. See, e.g., Marilyn S. Kushner, "Exhibiting Art at the American National Exhibition in Moscow, 1959: Domestic Politics and Cultural Diplomacy," *Journal of Cold War Studies* 4, no. 1 (2002): 6–26; Jane de Hart Mathews, "Art and Politics in Cold War America," *American Historical Review* 81, no. 3 (1976): 762–87.

36. "Leonhard W. Nederkorn Dead; Designed U.S. Fair Displays," *New York Times*, July 12, 1969, 27.

37. Joseph Shannon, "The Icing Is Good but the Cake Is Rotten," *Museum News* 52, no. 5 (1974): 29.

38. As late as 1961, Bradford Washburn complained to colleagues about exhibit makers' clumsy use of rubber cement to attach labels to exhibits. Ibid., 30. See Bradford Washburn, memo to Ed Taft, "Dry Mounting of Exhibits," July 10, 1961, box 1961, folder "Memos—Misc," BMoS.

39. Goodyear, *Society and Museum* (n. 5 above, "Introduction"), 71.

40. See *DMNHAR* (1959).

41. Brad Washburn, memo to Jerry Porter, "Meeting with V. Johnson," May 30, 1962, BW correspondence, BMoS. Johnson currently runs a successful museum design firm of his own.

42. Shannon, "Icing Is Good," 30.

43. According to the Smithsonian Archives finding aid, the Office of Exhibits in the National Museum of Natural History (NMNH) was created in 1973 after a reorganization established independent exhibit offices in several Smithsonian museums. Prior to the establishment of this office, natural history exhibitions were the responsibility of the Office of Exhibits, 1955–69, and the Office of Exhibits Programs, 1969–73. See "Smithsonian Institution Office of Exhibits Central, Records, 1954–1979," Smithsonian Institution Archives, encoded on July 14, 2011, http://siarchives.si.edu /collections/siris_arc_217068. Years later, Frank Taylor reflected on his reasoning: "I've often said that if you want something done concisely and crisply, go to a poet or an artist and he'll get the job done" (Frank Taylor Oral History, January–February 1974, p. 95, interview 4, SIA).

44. Benjamin Lawless and Marilyn Cohen, "The Smithsonian Style," in *The Smithsonian Experience*, ed. Joe Goodwin and Judy Harkinson (New York: Norton, 1977), 52. Frank Taylor spoke of decentralizing exhibits after he realized that, structurally, this set-up "diluted" the capacity of natural history museums and history and technology museums, but this came much later. See Frank Taylor Oral History, interviews nos. 8–10 (April 11–24, 1974), 812, Record Unit 009512, SIA.

45. Supervising this transition was John Anglim, a graduate of the Chicago Academy of Fine Arts, and Benjamin Lawless, whose experience in stage de-

sign proved extremely useful. As Herbert Friedmann noted in his Smithson-ian Oral History, the exhibits modernization project also turned to artists working at Disney and other corporations for help. See Post, *Who Owns America's Past?* (n. 149 above, chap. 4), 31–32; Herbert Friedmann Oral History, (April 22, 1975), 30, RU009506, SIA.

46. *The Smithsonian Institution, 1965* (film directed by Charles Eames)—copy located in Muman Studies Film Archives of Smithsonian Institution (HSFA 2001.1.1), Smithsonian Museum Support Center, Suitland, MD.

47. See Accession 10-002, Smithsonian Institution, Office of Exhibits Programs, Exhibition Records, SIA; for biography of artist, see his website: http://jrclendening.angelfire.com/johns_bio.html (accessed May 25, 2010).

48. James Mahoney Oral History (SIA), as cited in Steven W. Allison-Bunnell, "Transplanting a Rain Forest: Natural History Research and Public Exhibition at the Smithsonian Institution, 1960–1975" (PhD diss., Cornell University, 1995), 251–52.

49. James Mahoney, letter to Frank Taylor and John Anglim, October 1969, RG 363 ("Office of Exhibits, 1955–1990"), box 18, p. 5, SIA.

50. McLuhan was in high demand at museum conferences in the early 1960s. See, e.g., Marshall McLuhan, Harley Parker, and Jacques Barzun, *Exploration of the Ways, Means, and Values of Museum Communication with the Viewing Public* (New York: Museum of the City of New York, 1969), 23–27; Robert S. Weiss and Serge Bouterline Jr., "The Communication Value of Exhibits," *Museum News* 43, no. 3 (1963): 23–27.

51. Shannon, "Icing Is Good" (n. 36 above, this chap.).

52. Kurt Schmidt, "The Nature of the Natural History Museum," *Curator* 1, no. 1 (1958): 27–28.

53. In 1965, Eugene Ferguson, a former curator of the Smithsonian's mechanical and civil engineering collections, declared that narrative and other "errors of thinking in museum operation" were the fault of Carlos Cummings, who had tried to apply world's fair display techniques in museum settings. See Eugene Ferguson, "Technical Museums and International Exhibitions," *Technology and Culture* 6, no. 2 (1965): 43.

54. Schmidt, "Nature of the Natural History Museum," 27–28.

55. Colbert supported the public mission of the museum: after he became emeritus curator in 1970, he wrote numerous popular dinosaur books and continued to consult on exhibits. But like Schmidt and other scientific curators, he found that juggling research and education, display and collecting at the same time was unsustainable. On Colbert's retirement, see the *AMNHAR* (1970–71), 15. Colbert's popular books included *The Dinosaur Book: The Ruling Reptiles and Their Relatives* (1945), *Dinosaur World* (1977), and *Dinosaurs: An Illustrated History* (1983).

56. Edwin H. Colbert, "On Being a Curator," *Curator* 1 (1958): 7.

57. See Frank Taylor on the reoccurrence of conflict between designers and curators. Frank Taylor Oral History, interviews nos. 11–12, pp. 347–48, SIA.

58. Allison-Bunnell, "Transplanting a Rain Forest" (n. 47 above, this chap.), 54. Frank Taylor remarked that the long-time natural scientists were baffled by efforts to improve and expand on exhibits, observing that curators "looked upon [exhibition] as something they had to administer because it was here but they never looked upon it as anything to enlarge or develop or improve." Frank Taylor Oral History, interviews nos. 15–16, pp. 479–80, SIA.

59. See Frank Taylor (and others) in correspondence with Friedmann, October 2–3, 1956, RG 363, box 1, SIA.

60. The Smithsonian's natural history collections and exhibitions was part of the United States National Museum, a museum that encompassed scientific, historical, and industrial collections. In 1957, the National Museum was divided into two administrative subdivisions: the Museum of Natural History and the Museum of Industry and Technology. As a result of the explosion of Smithsonian-affiliated collections and buildings, individual divisions became independent administrative units in 1967, and the old umbrella organization of the United States National Museum was abolished. The Museum of Natural History was renamed the National Museum of Natural History in 1969, the name it currently retains.

61. Susan L. Jewett and Bruce B. Collette, "Ernest A. Lachner, 1915–1996," Copeia 1997, no. 3 (1997): 650–58.

62. Lachner, letter to Friedmann, September 25, 1956, RG 363, box 1, folder "Oceanic Scripts," SIA.

63. Ibid.

64. David Johnson, "Memorandum," April 27, 1959, RG 363, box 1, "Scripts," SIA.

65. John Anglim, letter to Frank Taylor and John Ewers, June 28, 1961, RG 363, box 13, folder "Misc. Correspondence re: permanent halls," SIA.

66. "Smithsonian Sea Hall to Open," Baltimore Sun, February 17, 1963, 22.

67. In a nod to curators' concerns, designers went to great lengths to ensure the whale was accurately modeled. Michael Rossi notes that the Smithsonian designers obtained templates and data from the British Museum; he characterizes the Smithsonian's choice of whale pose as "not unlike a tremendous fishing trophy." Michael Rossi, "Modeling the Unknown: How to Make the Perfect Whale," Endeavour 32, no. 2 (2008): 60.

68. Preopening reviews include T. Wolfe, "Smithsonian Building Biggest Whale Ever," Washington Post, January 7, 1960: A3; Albert Parr, "Concerning Whales and Museums," RG 363, box 1, folder "Whale . . . ," SIA.

69. John Anglim, letter to Frank Taylor, March 18, 1966, RG 363, box 13, SIA.

70. In 1966, the Smithsonian opened the Hall of Medical Sciences in its Museum of History and Technology that utilized the well-worn display strategy of period rooms, in this case, medical offices and hospital rooms featuring objects ranging from tribal fetishes to artificial heart valves. See "Hall of Medical Science Opens at Smithsonian," Public Health Reports 81, no. 11 (1966): 973.

71. On the building of the grasshopper model, see John Anglim, letter to Frank Taylor, April 3, 1967, RG 363, box 13, folder "Insect Hall," SIA.

72. S. D. Ripley, letter to Frank Taylor, June 16, 1966; R. S. Cowan, letter to S. D. Ripley, July 11, 1966; Donald Squires, letter to R. S. Cowan, August 13, 1966—all from RG 363, box 13, folder "Insect Hall," SIA.

73. See series of correspondence (April–May 1967) between Peter Farb, John Anglim, and Frank Taylor, RG 363, box 13, folder "Insect Hall," SIA.

74. Bradford Washburn, letter to Endicott Peabody, November 7, 1962, box "7–1962," folder "P," BMoS. Others at the Boston Museum of Science were concerned with accommodating adults. See F. Rowlands, memo to Whom It May Concern, "Museum Crowds," July 19, 1962, box 1961, folder "Memos and Misc.," BMoS.

75. See folder "1961—New England Aquarium Corporation" (in BMoS), especially Ned Pearce, memo to Brad Washburn, "Aquarium Report," December 14, 1958; résumé of John Siebnaler, dated August 18, 1958, with attached note (undated) from Ned Pearce (quoted); Bradford Washburn, memo to Ned Pearce, "Aquarium Progress," December 10, 1958; and Brad Washburn, letter to Alexander Bright (member of the New England Aquarium Corporation), December 23, 1959. Washburn later consulted with leaders of the New York Aquarium about their success in spinning off their project from the New York Zoological Society's Bronx Zoo, but he was unable to resurrect the project for the Science Park site: see Bradford Washburn, letter to Fairfield Osborn, September 27, 1960, in BMoS.

76. Richard Sheffield, oral history interview by Karen Rader and Sylvie Gassaway, July 2005, pp. 20–21, currently held by Karen Rader.

77. See G. Ray Higgins, letter to B. Washburn, May 11, 1960, in folder "1962: Hall of Medicine," BMoS; also in the folder "1963 /Exhibits—HEART," see "Fact Sheet: How Your Heart Works," February 18, 1963; letter from "Sherry" Washburn to Brad Washburn, August 4, 1962; B. Washburn, memo to Gib Merrile, "Heart Exhibit," August 12, 1962; B. Washburn, memo to Bob Tillotson and/or Kae Callahan, July 5, 1960—all BMoS.

78. On corporate philanthropy in museums in this era, see Victoria D. Alexander, *Museums and Money: The Impact of Funding on Exhibitions, Scholarship, and Management* (Bloomington: Indiana University Press, 1996).

79. The corporate funding of science museum displays worked to the benefit of the new generation of exhibit designers, who were instructed to create exhibits on specific subjects but then given free rein. Such an arrangement was a vast difference from the kind of contentious collaboration that characterized exhibition planning and design at natural history museums and, also, a shift from past corporate funding arrangements at science museums and world's fairs, which were merely "free advertising." In 1961, e.g., after the California Museum of Science and Industry asked IBM to contribute to its new science wing, IBM asked Charles and Ray Eames to

design an interactive exhibit on mathematics. The resulting exhibit, a three thousand–square-foot extravaganza titled *Mathematica: A World of Numbers . . . and Beyond*, was wildly successful; Chicago's Museum of Science and Industry promptly requested a duplicate, and IBM persuaded the Eameses' office to create yet another copy of it for the 1964–65 New York World's Fair, donating the display to the Pacific Science Center once the fair closed. On *Mathematica*, see Wendy Constantine et al., *Summative Evaluation of "Mathematica: A World of Numbers"* (Boston: Boston Museum of Science, 2004); Pat Kirkham, *Charles and Ray Eames: Designers of the Twentieth Century* (Cambridge, MA: MIT Press, 1998).

80. "A Smoking Cow!" *Progress* 8, no. 5 (1957): 8, PMSIC.

81. "Educate Little Folks Early in Telefun Town," *Progress* 8, no. 4 (July–August 1957): 2, PMSIC.

82. See Bradford Washburn, letter to Abraham Sachar, June 21, 1963, box "1963," folder "Correspondence"; Bradford Washburn, memo to Norm Harris, September 20, 1962, box "1962," folder "Memos and Misc."; Bradford Washburn, letter to Endicott Peabody, November 7, 1962, box "7–1962," folder "P"—all BMoS. Others at the Museum of Science were concerned with accommodating adults; see F. Rowlands, memo to Whom It May Concern, "Museum Crowds," July 19, 1962, box 1961, folder "Memos and Misc.," BMoS.

83. Adult scientists would sometimes write letters to Washburn, recounting happy memories of early years spent at Science Park, and Washburn featured these prominently in the museum's annual reports.

84. See box "Museum 1957–58," folder "Educational Stats," with tables indexing growth of U.S. population, ages 14–17, against "Science Enrollments in Public Schools," gleaned from U.S. Office of Education published reports. See also folder "Museum Stats"—"1952," which includes an attendance summary report from 1950–58, showing slight decline in attendance in 1957–58, and "1960," which includes attendance data as well as group visit attendance broken down by county and state, in BMoS.

85. In 1959, the same year it applied for a Lilly Foundation grant, the Boston Museum of Science initiated an internal study to determine what its visitors learned while there. See George Packer Berry, letter to Bradford Washburn, December 5, 1961; and Bradford Washburn, letter to George Packer Berry, December 7, 1961—both in box "1961," folder "B"; Norm Harris, memo to E. Duane, November 19, 1959, "Suggested Questions for the Proposed Survey," box "1960," folder "Visitor Questionnaires"; typescript "Impact of the Museum of Science: Lilly Survey," in box "4–1959," folder [no name]—all BMoS. Bradford Washburn, letter to George Rothwell, December 7, 1961, box "1961," folder "Correspondence"; Bradford Washburn, "Memorandum—Visit by Bradford Washburn to NSF," November 14, 1962," box "1962," folder "NSF"; cf. Norm Harris, memo to Bradford Washburn, January 6, 1959 "Summer Teachers Courses," box "4–1959," folder "Education Department," all BMoS.

86. Orr Reynolds and Science Advisory Committee, letter to Athelstan Spilhaus, March 27, 1962. box 218, folder 5, Warren G. Magnuson Papers (accession no. 3181-4), University of Washington Archives, as cited in Ellis, "Dixy Lee Ray" (n. 25 above, this chap.), 186–87.

87. Dixy Lee Ray, "Educational Programs: Some Preliminary Thoughts and Some Propositions," box 9, folder "Pac Sci Cen Foundation Board of Trustees, Annual Meeting, 1963," Joseph L. McCarthy Papers (accession no. 3331-84-34), University of Washington Archives and Special Collections, Seattle, as cited in Ellis, "Dixy Lee Ray," 188–89.

88. "News," Pacific Science Center Foundation, July 19, 1963, box "Museum 1963," folder "Pacific Science Center Foundation," private archive, BMoS, as cited in Ellis, "Dixy Lee Ray," 191.

89. Bradford Washburn, letter to Julius Stratton, July 15, 1963, box "Museum 1963," folder "Pacific Science Center Foundation," BMoS, as cited in Ellis, "Dixy Lee Ray."

90. Ellis, "Dixy Lee Ray," 203.

91. As cited in Allison-Bunnell, "Transplanting a Rain Forest" (n. 47 above, this chap.), 212.

92. Ibid.

93. Ibid.

94. John F. M. Cannon, "The New Botanical Exhibition Gallery at the British Museum (Natural History)," *Curator* 5, no. 26 (1962): 26–35. That year, Cannon published this same article in *Taxon*, a scientific journal, hoping curators would get the message. See John F. M. Cannon, "The New Botanical Exhibition Gallery at the British Museum (Natural History)," *Taxon* 11 (1962): 248–52. See also John F. M. Cannon, "Some Problems in Botanical Exhibition Work," *Museums Journal* 62 (1962): 167–73.

95. Ripley noted that support for graduate training "in the broad areas of biology and anthropology . . . represents a continuation of the traditional role of the Institution in the educational field, although perhaps historically it received greater emphasis in the early days of the Smithsonian than it has in recent decades." *SIAR* (1964), 2–3.

96. Wolfgang Saxon, "Charles Blitzer," *New York Times*, February 26, 1999. On Ripley, see Daniel P. Moynihan, "Charles Blitzer," *Biographical Memoirs, Proceedings of the American Philosophical Society* 142, no. 2 (2001): 190–91. For broader view of Blitzer's career, see Charles Blitzer Oral History Interview, 1985–86, RU 9604, SIA.

97. By 1968, the idea of bringing the neighborhood into the museum had evolved into bringing the museum into the neighborhood, a throwback to Progressive Era satellite museums. That year, the Smithsonian worked to found the Anacostia Museum, which featured "a live zoo with green monkeys, a parakeet, and a miscellany of animals on loan from the National Zoological Park." On the Anacostia Museum, see *SIAR* (1968), 8–9. On the failed "slum" exhibit, see Bryce Nelson, "The Smithsonian: More Museums in Slums, More Slums in Museums?" *Science* 154, no. 3753 (1966): 1154.

98. "Sherburne to Head New AAAS Program to Improve Public Understanding," *Science*, vol. 132, no. 3442 (1960). Also see E. G. Sherburne Jr., "Public Understanding of Science," *Science* 149, no. 3682 (1965). Sherburne was a central figure in the public understanding of the science movement of the late 1960s, one who bridged past efforts and future trends. On his influence, see Sally Gregory Kohlstedt, Michael Sokal, and Bruce Lewenstein, *The Establishment of Science in America* (New Brunswick, NJ: Rutgers University Press, 1999), 127–28.

99. Charles Blitzer, "Preface," in *Museums and Education: Papers from a Conference, Burlington, Vt, 1966*, ed. Eric Larrabee (Washington, DC: Smithsonian Institution Press, 1968), v–vi.

100. Grove, as quoted in Larrabee, ed., *Museums and Education*, 208. See also National Endowment for the Arts, *Museums USA* (Washington, DC: Government Printing Office, 1974); McLuhan, Parker, and Barzun, *Exploration of the Ways* (n. 49 above, this chap.); Brian O'Doherty, ed., *Museums in Crisis* (New York: George Braziller, 1972).

101. As quoted in Larrabee, ed., *Museums and Education*, 217.

102. Ibid., 208–9.

103. Ibid., 213–16.

104. Ibid., 216.

105. On Frank Oppenheimer's youth and relationship with his brother, see Kai Bird and Martin Sherwin, *American Prometheus: The Triumph and Tragedy of J. Robert Oppenheimer* (New York: Knopf, 2006).

106. On FBI surveillance of Frank Oppenheimer, see Jessica Wang, *American Scientists in the Age of Anxiety: Scientists, Anti-Communism, and the Cold War* (Chapel Hill: University of North Carolina Press, 1999), 60–62.

107. On Frank Oppenheimer's work on the atomic bomb, see Bird and Sherwin, *American Prometheus*, 305. On Oppenheimer's relationship with his family's wealth, see K. C. Cole, *Something Incredibly Wonderful Happens: Frank Oppenheimer and the World He Made Up* (Boston: Houghton Mifflin Harcourt, 2009), 106.

108. Text of Oppenheimer's 1957 "Teaching and Learning" speech to the Pagosa Spring High School PTA can be found on the Exploratorium website: http://www.exploratorium.edu/files/about/our_story/history/frank/pdfs/teaching_and_learning.pdf.

109. In 1961, he obtained the first of a series of National Science Foundation grants for physics research on "Elementary Particle Interactions from Bubble Chamber Photographs." See the series of letters (between March 16, 1961, and September 22, 1966) between Oppenheimer and various NSF officials regarding these grants (BANC MSS. 98/136, 1:19, BL-UCB).

110. Frank Oppenheimer and Malcolm Correll, "A Library of Experiments," *American Journal of Physics* 32, no. 3 (1964): 220–25.

111. Historians of education argue about whether the Physical Science Study Committee's goal included general science literacy or whether members

sought merely to refine and sharpen disciplinary preparation for future physics researchers; scientist members like Zacharias clearly had different goals than members representing education, which included educational psychologists like Jerome Bruner, as well as representatives from the Educational Testing Service and Eastman Kodak's education film unit. Bruner's ideas about interactivity and student-centered learning as paths to participatory democracy ran afoul of those who championed the rapid acquisition of extensive content, but they were popular with museum designers throughout the 1970s and 1980s. As Gustavo Corral notes, Bruner's discovery learning approach was eventually employed for developing the New Exhibition Scheme at the London Natural History Museum. On the work of the Physical Science Study Committee, see David Turner, "Reform of the Physics Curriculum in Britain and the United States," *Comparative Education Review* 28, no. 3 (1984): 444–53. Also see Norman Ramsey, "Jerrold R. Zacharias, 1905–1986," *Biographical Memoirs, National Academy of Sciences of the United States of America* 68 (1995): 435–50. On Bruner's work with the Physical Science Study Committee and his broader contributions to education, see, among others, J. S. Bruner, *The Process of Education*, rev. ed. (Cambridge, MA: Harvard University Press, 2009); Jerome Bruner, *The Relevance of Education* (New York: W. W. Norton, 1971); D. R. Olson, *Jerome Bruner: The Cognitive Revolution in Educational Theory* (London: Continuum, 2007). For an examination of Bruner's influence on museum exhibition, see Gustavo Corral, "Human Biology: Visual Metaphors in the New Exhibition Scheme of the Natural History Museum" (MA thesis, Autonomous University of Barcelona, 2013).

112. Frank Oppenheimer, "The Importance of the Role of Science Pedagogy in the Developing Nations," *Scientific World* 10, no. 4 (1966): 21–23, 28.

113. The United Nations Educational, Scientific, and Cultural Organization (UNESCO) adopted this recommendation in December 1960 and enlisted art, science, and history museum directors to give advice on the issue. See *Museums, Imagination and Education*, 2nd ed., Museums and Monuments, vol. 15 (Paris: UNESCO, 1973). Oppenheimer was an observer for the Federation of American Scientists at the 1965 World Federation of Scientific Workers meeting; before this, UNESCO and the federation clashed throughout the 1940s and 1950s. See Patrick Petitjean, "A Failed Partnership: WFSW and UNESCO in the Late 1940s," in *Sixty Years of Science at Unesco: 1945–2005*, ed. P. Petitjean et al. (Paris: Recherches Epistémologiques et Historiques sur les Sciences Exactes et les Institutions Scientifiques, 2006).

114. Oppenheimer, "History," BANC MSS. 87/148c: folder 5:3 (no date), BL-UCB.

115. Frank Oppenheimer, "Rationale for a Science Museum," *Curator* 11 (1968): 206–7.

116. Ibid.

117. Rodney T. Ogawa, Molly Loomis, and Rhiannon Crain, "Institutional History of an Interactive Science Center: The Founding and Development of the Exploratorium," *Science Education* 93, no. 2 (2009): 269–92. See also George E. Hein, *Learning in the Museum*, Museum Meanings (New York: Routledge, 1998).

118. Jerrold Zacharias and Stephen White, "The Requirements for Major Curriculum Revision," *School and Society*, 92, no. 67 (1964): 66–72; Cole, *Something Incredibly Wonderful Happens* (n. 106 above, this chap.), 174.

119. On the difficulties of implementing hands-on pedagogy in school classrooms, see John Rudolph, "Ward's Dynavue and the Hybridization of Classroom Apparatus in the Post-Sputnik Era," paper presented at Annual Meeting of the National Academy of Education, New York City, October 21, 2005.

120. Oppenheimer, "Importance of the Role of Science Pedagogy," 21.

121. Albert Parr, letter to Frank Oppenheimer, February 6, 1968, BANC MSS. 87/148c, 4:20, BL-UCB.

CHAPTER SIX

1. As cited in Cole, *Something Incredibly Wonderful Happens* (n. 106 above, chap. 5), 157 and 332.

2. Ibid., 151. On Perlman's critical role in the *San Francisco Chronicle*'s coverage of the Exploratorium, see Ogawa, Loomis, and Crain, "Institutional History of an Interactive Science Center" (n. 116 above, chap. 5), 278–79.

3. Some of the Exploratorium's neighbors were more supportive, signing on as museum volunteers. Cole, *Something Incredibly Wonderful Happens* (n. 106 above, chap. 5), 151–52.

4. Ibid., 152.

5. Oppenheimer had had to work mightily against entrenched San Francisco business interests to gain control of the building. See Frank Oppenheimer, "The Palace of Arts and Science: An Exploratorium at San Francisco, Ca, USA," *Leonardo* 5, no. 4 (1972): 343–346. On obtaining a share of city hotel tax revenues for museum funding, see Robert Glaser, letter to Thomas Mellon (SF's Chief Administrative Officer), March 8, 1971; on ongoing problems with developers, including their objections to special road signage directing visitors to the Exploratorium, see California State Senator George Moscone, leter to San Francisco Department of Transportation's T. R. Lammers, February 28, 1975—both letters in BANC MSS. 87/148c, 4:18, BL-UCB.

6. As cited in Cole, *Something Incredibly Wonderful Happens*, 152.

7. Science centers were better established abroad, the most famous example being Canada's Ontario Science Center, which also opened in 1969, but even in the United States, they were not unknown. Dixy Lee Ray's original vision for the Pacific Science Center, e.g., closely resembled Oppenheimer's Exploratorium, though her plans were never fully realized.

8. For one of the first internal incidents documenting Oppenheimer's use of the phrase "no one ever flunked a museum," see the Exploratorium grant proposal, "No One Ever Flunks a Museum—Science Education for Diverse College Student Populations" (n.d.), BANC MSS. 87/148c, 42:11, BL-UCB. See also Frank Oppenheimer, "Schools Are Not for Sightseeing," 1970, Carton 5, folder 17, Exploratorium records, BANC MSS. 87/148, BL-UCB. For background and additional examples, see Frank Oppenheimer, letter to Lou Branscomb, December 16, 1981; BANC MSS. 87/148c, 5:91, BL-UCB, as well as Robert Semper in William R. Carlsen, "Science Museums: Do-It-Yourself Teachers," *New York Times*, April 22, 1979, 17.

9. For a broader description of the Exploratorium's influence on art museums and children's museums, as well as a more general history of the institution's precursors, see Elaine Heumann Gurian "The Molting of the Children's Museum," in *Civilizing the Museum: The Collected Writings of Elaine Heumann Gurian*, ed. Elaine Heumann Gurian (New York: Routledge, 2006), 19–32; Marjorie Schwarzer, *Riches, Rivals and Radicals: 100 Years of Museums in America* (Washington, DC: American Association of Museums, 2010), 145–50; Richard Toon, "Science Centers: A Museum Studies Approach to Their Development and Possible Future Directions," in *Museum Revolutions: How Museums Change and Are Changed*, ed. Simon Knell, Suzanne MacLeod, and Sheila Watson (New York: Routledge, 2007), 105–16; Ibrahim Yahya, "Mindful Play! Or Mindless Learning!—Modes of Exploring Science in Museums," in *Exploring Science in Museums*, ed. Susan Pearce (Atlantic Highlands, NJ: Athlone Press, 1996), 123–47.

10. We thank the participants of the 2013 Barcelona conference "Dark Matters: The Contents and Discontents of Cold War Science," especially discussions with Susan Lindee and Trevor Barnes, for helping us to develop this insight. On Oppenheimer's diligently apolitical stance, see Rebecca Onion, "How Science Became Child's Play: Science and the Culture of American Childhood" (PhD diss., University of Texas, 2012).

11. See Jon Agar, "What Happened in the Sixties?" *British Journal for the History of Science* 41, no. 45 (2008): 567–600.

12. DeBoer, *History of Ideas* (n. 34 above, chap. 1), 173. See also Cremin, *American Education* (n. 49 above, chap. 3); Cuban, *How Teachers Taught* (n. 59 above, chap. 2).

13. Cuban, *How Teachers Taught*, 151–55.

14. M. Harmon, H. Kirschenbaum, and S. Simon, "Teaching Science with a Focus on Values," *Science Teacher* 37, no. 1 (1970): 17.

15. William V. Mayer, "Biology Education in the United States during the Twentieth Century," *Quarterly Review of Biology* 61, no. 4 (1986): 484.

16. "NSTA Position Statement on School Science Education for the '70s," *Science Teacher* 38, no. 1 (1971): 47.

17. René Dubos, quoted in Pacific Science Center Annual Report, 1967, as cited in Ellis, "Dixy Lee Ray" (n. 25 above, chap. 5), 198.

18. Alma S. Wittlin, *Museums: In Search of a Useable Future* (Boston: MIT Press,

1970), 206–11. Wittlin's opinions were not universally shared. See, e.g., Douglas Davis, "The Idea of a Twenty-First Century Museum," *Art Journal* 35, no. 3 (1976): 253–58.

19. Jon Agar offers a historical frame for understanding the "long 1960s" as a period of "interference" of three separate but overlapping waves change. See Agar, "What Happened in the Sixties?" Marc Rothenberg directed us to NSF coverage of this issue in its publicity magazine: see "Doing What Classrooms Can't," *Mosaic* 5 (Spring 1973): 16–21; "Museums to Teach By," *Mosaic* 10 (Summer 1979): 4, 17–25.

20. Bruce Lewenstein, "The Meaning of 'Public Understanding of Science' in the United States after World War Two," *Public Understanding of Science* 1 (1992): 45–69.

21. Though this anger was directed at natural history museums in general, it burned especially hot over traditions of anthropological collecting and display. See Devon Abbott Mihesuah, ed., *Repatriation Reader: Who Owns American Indian Remains?* (Lincoln: University of Nebraska Press, 2000); David Hurst Thomas, *Skull Wars: Kennewick Man, Archaeology, and the Battle for Native American Identity* (New York: Basic Books, 2000).

22. At the 1970 annual meeting of the AAM in the ballroom of New York City's sedate Waldorf-Astoria, e.g., the press reported that some thirty long-haired protesters, calling themselves the New York Artists Strike against Racism, Sexism, Repression, and War, arrived and proceeded to shout down speaker Nancy Hanks, chairman of the National Endowment for the Arts. When the protesters confronted another speaker, New York's governor, Nelson Rockefeller, he had to ask them to define "sexism"—he was as unfamiliar with the term. A week later, *Time* magazine provided some clarification in its article on the AAM protest, dismissively explaining to readers that the term referred to "discrimination against women in employment, politics, art and so forth." See "Art: N.Y.A.S.A.R.S.R.W," *Time*, June 15, 1970. Such accusations continued throughout the 1970s. In 1975, e.g., Bay Area lawyer Alfred Knoll authored a pamphlet titled *Museums: A Gunslinger's Dream*, which suggested that museums could reasonably be prosecuted under the 1965 Civil Rights Act for indifference to various ethnic cross-sections of their communities. Arthur Knoll, *Museums: A Gunslinger's Dream* (San Francisco: Bay Area Lawyers for the Arts, 1975). As cited in Kipi Rawlins, "Educational Metamorphosis of the American Museum," *Studies in Art Education* 20, no. 1 (1978): 4–17.

23. Roy M. MacLeod, "'Strictly for the Birds': Science, the Military and the Smithsonian's Pacific Ocean Biological Survey Program, 1963–1970," *Journal of the History of Biology* 34, no. 2 (2001): 315–52. There is a summary of the project in *SIAR* (1965), 24–30.

24. Ultimately, Stock concluded, the American Museum of Natural History was "the rarest of all in these days of corporate avarice and public corruption— a humane institution." Robert W. Stock, "At 100, the Museum of Natural History Is No Fossil," *New York Times Magazine*, June 29, 1969, SM8.

25. Sherman Lee, "The Art Museum as a Wilderness Area," *Museum News* (1972): 11–12.

26. A group of Museum of Modern Art protesters, e.g., walked out in a body to protest Director John Hightower's failure to establish a black-studies center at the museum; see "Art: N.Y.A.S.A.R.S.R.W." Bruce H. Evans, "By Example and by Exhibition: The Museum as Advocate," *Museum News* 58, no. 4 (1980): 23. Race was also a topic at AAAS meetings, e.g., the 1966 meeting hosted a day-and-a-half long session about the "utility of the concept of race." See "U.S. Science in Session," *Science News* 91, no. 1 (1967): 4.

27. Evans, "By Example and by Exhibition," 23.

28. Milam, "Public Science of the Savage Mind" (n. 18 above, chap. 5).

29. The agency's federal funding for educational projects suffered as a result of the public relations debacle. By 1977, with increased public attention and commentary from creationism advocates, all three versions of the BSCS texts had failed to win renewed adoption in Texas, a result that had broader implications for their commercial production and adoption nationwide.

30. Responding to creationist Duane Gish's tirade against evolution, Ernst Mayr, then at the Museum of Comparative Zoology at Harvard, argued in a 1971 issue of *American Biology Teacher* that Gish's demand to see simple intermediate fossil archetypes was just silly. Gish had naively ignored the scientific findings of museum systematists, who had shown that intermediate forms were "always a mosaic of advanced and ancestral characteristics," he wrote. Entomology curator Richard Alexander of the University of Michigan Museum of Zoology also weighed in, reassuring classroom teachers that teaching evolution through the scientific method gave their students all they needed to know. "As long as biology teachers conduct their courses in the spirit of free inquiry, open debate, and self-correcting searches for predictive theories and repeatable results, no parent need fear that his or her children are being subjected to anything but the best kind of preparation for life in the technologically complex and socially demanding society in which we live," he wrote. Richard D. Alexander, "Evolution, Creation, and Biology Teaching," *American Biology Teacher* 40, no. 2 (1978): 104.

31. Duane T. Gish, "A Challenge to Neo-Darwinism," *American Biology Teacher* 32, no. 8 (1970): 495–97; Ernst Mayr, "Evolution v. Special Creation," *American Biology Teacher* 33, no. 1 (1971): 49–51. For an overview of this controversy from the perspective of the late 1970s, see Stanley L. Weinberg, "Two Views on the Textbook Watchers," *American Biology Teacher* 40, no. 9 (1978): 541–45 and 560. See also Edward J. Larson, *Trial and Error: The American Controversy over Evolution* (New York: Oxford University Press, 2003), 125–84; Dorothy Nelkin, *The Creation Controversy: Science or Scripture in the Schools* (Boston: Beacon, 1984).

32. A ticket for an adult cost a dollar, and children were admitted for fifty cents. Goodyear, *Society and Museum* (n. 5 above, "Introduction"), 77.

33. *AMNHAR* (1970), 2.
34. In 1972, Congress finally funded the 1966 National Museum Act, which provided money for the Smithsonian "to report on all U.S. museums and their activities each year . . . and advise and cooperate with the U.S. government in aiding [them] . . . solving their problems and training programs . . . and helping improve museum techniques." In 1974, the House Committee on Education and Labor introduced and passed the Museum Services Act, which established the Institute for Museum Services, a federal entity that allowed museums to apply for federal funding. Administered by the Smithsonian, the Institute of Museum Services funded everything from flashy multimedia exhibits to office supplies. Museum professionals who rarely agreed on exhibit design and education presented a unanimous front as they lobbied for the passage of these laws. Directors, museum board presidents, trustees, and even student docents from museums as wide-ranging as the Los Angeles Natural History Museum, the California Museum of Science and Industry, the Boston Museum of Fine Arts, and the National Center of Afro-American Art in Roxbury, MA, all testified in hearings for the Museum Services Act, and they were exultant when they passed. These laws, Smithsonian secretary Dillon Ripley proudly declared, were the best form of "legislative recognition to the cultural, educational and scientific significance of the nation's museums." National Endowment for the Arts, *Museums USA* (n. 99 above, chap. 5), 13, 15. "National Museum Act Could Increase Status," *Science Newsletter* 87, no. 14 (1965): 219. See U.S. Congress, Senate, Committee on Rules and Administration, Subcommittee on the Smithsonian Institution, *Smithsonian Institution: National Museum Act of 1965: Hearings Before the United States Senate Committee on Rules and Administration, Subcommittee on the Smithsonian Institution,* 89th Cong. (1965). See transcripts in U.S. Congress, House, Committee on Education and Labor, Select Subcommittee on Education, *Museum Services Act: Hearings Before the Select Subcommittee on Education of the Committee on Education and Labor, House of Representatives, Ninety-third Congress, second session on H.R. 332 . . . Hearings held in Los Angeles, Calif., May 17 and 18, 1974; San Francisco, Calif., May 20, 1974; and Boston, Mass., June 14 and 15, 1974* (Washington, DC: Government Printing Office, 1974), 1–101, esp. iii–vi for lists of museums attending. Barbara Tufty, "A New Life for Museums," *Science News* 88, no. 8 (1965): 122.
35. M. W. Straight, *Nancy Hanks, an Intimate Portrait: The Creation of a National Commitment to the Arts* (Durham, NC: Duke University Press, 1988), 86.
36. See American Association of Museums, *A Statistical Survey of Museums in the United States and Canada* (1965; repr., New York: Arno Press, 1976).
37. American Association of Museums, *America's Museums: The Belmont Report,* ed. Michael Robbins (Washington, DC: American Association of Museums, 1969).
38. Throughout the late 1960s and early 1970s, art museum reformers also questioned the Exploratorium's populist impulse. In the 1972 volume *Mu-*

seums in Crisis, commissioned by the National Endowment for the Arts, art critic Bryan Robertson railed against "the democratic fallacy" that increased attendance at museums like the Exploratorium led to a genuine expansion of knowledge. "Only an exceedingly small number of those who enter a museum are even remotely prepared for what awaits them and only a total revolution in our educational system could remedy this deficiency," Robertson wrote. Bryan Robertson, "The Museum and the Democratic Fallacy," in *Museums in Crisis*, ed. Brian O'Doherty (New York: George Braziller/Art in America, 1972), 77. Other article titles from this same volume—"The Failed Utopia," "The Beleaguered Director," "The Contemporary Museum: Under the Corporate Wing"—reveal an American museum world under siege.

39. Frank Oppenheimer, "Some Special Features of the Exploratorium" (n.d.), BANC MSS. 87/148C, 5;8, BL-UCB

40. The broader public, which had recently developed a taste for drug-induced psychedelic hallucinations, pop art, and other alternative visual experiences, found the topic fascinating, and the Exploratorium was soon blurbed everywhere from *Playboy* to *Artweek*. Though these first exhibits emphasized vision, Oppenheimer made sure to create components that would appeal to the blind. Indeed, creating exhibits for visitors with poor or no sight was a challenge Oppenheimer welcomed. Science "must be developed as a series of experiments and demonstrations which elucidate the topic without benefit of many words," he wrote. "Museums" in *Playboy* (April 1972): 44–48. See also Philip Morrison, "The Palace of Arts and Sciences" (1971), BANC 87/148c, 21 36, BL-UCB. On the media's role in the Exploratorium's success, see Ogawa, Loomis, and Crain, "Institutional History of an Interactive Science Center" (n. 116 above, chap. 5), 278–79. On the museum's attempts to reach the blind, Frank Oppenheimer, "The Content of the Museum" (n.d.), BANC MSS. 87/148C, 5:2, BL-UCB.

41. Andrew Barry discusses how Oppenheimer made arrangements with London's Institute of Contemporary Art to transfer some elements of their successful *Cybernetic Serendipity* exhibit (August–October 1968) to San Francisco for the Exploratorium opening. See Andrew Barry, "On Interactivity: Consumers, Citizens and Culture," in *The Politics of Display: Museums, Science, Culture*, ed. Sharon Macdonald (New York: Routledge, 1998), 98–117.

42. Frank Oppenheimer, letter to "Dear Zack," January 6, 1967, BANC MSS. 98/136, 1:36, BL-UCB. This idea came from an exhibit that Oppenheimer has seen at what he called "The Seattle Museum," which presumably referred to the Pacific Science Center. Frank Oppenheimer, "The Relationship between the Exploratorium . . . and other museums in the Bay Area," BANC MSS. 87/148c, 5:6, BL-UCB.

43. Hilde S. Hein, *The Exploratorium: Museum as Laboratory* (Washington, DC: Smithsonian Institution Press, 1990), 75, 241.

44. To explain interference, e.g., he cited the famous X-ray diffraction pictures of DNA, generated by Rosalind Franklin, and the wave patterns he

observed in the San Francisco Bay during his daily ride on the Sausalito ferry. See Cole, *Something Incredibly Wonderful Happens* (n. 106 above, chap. 5), 227–28.

45. Explainers typically worked four hours per day, twenty hours per week. See "Student Science Project Program," September 30, 1971, BANC MSS. 87/148c, 42:3, BL-UCB. This folder also includes the list of nineteen students who participated; one woman and eighteen men who (judging from their last names) came from varying national and regional origins: Asian, Dutch, British, etc.

46. "Auxilliary Program," ca. 1970, BANC MSS. 87/148c, 5:13, BL-UCB, but see also Oppenheimer, "Palace of Arts and Science," 345 (n. 5 above, this chap.) and "Some Special Features" (n. 39 above, this chap.), 96–97

47. Statement by Joshua Callman (explainer [February–September 1975] and bio lab assistant [July 1975–September 1976]), BANC MSS. 87/148c, 42:45, BL-UCB.

48. Oppenheimer, "Some Special Features." At least one study demonstrating the effectiveness of the explainer program was conducted in the 1980s. See Judy Diamond et al., "The Exploratorium's Explainer Program: The Long-Term Impacts on Teenagers of Teaching Science to the Public," *Science Education* 71, no. 5 (1987): 643–56.

49. David Ucko, "NSF Influence on the Field of Informal Science Education," paper presented at the eighty-third annual International Conference of the National Association for Research in Science Teaching, Philadelphia, March 22, 2010.

50. Quote from Bryce F. Zerner, March 4, 1973 (4:27) but see also Vivian Bernhart to Frank Oppenheimer, June 13, 1977 (4:2); Paul Houts, letter to Frank Oppenheimer May 20, 1974; Lotus Dix, letter to Frank Oppenheimer, September 11, 1963 (4:6); and from a group of school children, Cheryl Woodward, letter to "Exploratorium," (n.d.; 4:26)—all BANC 87/148c, BL-UCB.

51. "Exploratorium Attendance Report, 15 November 1969," BANC MSS. 87/148c, 36:1, BL-UCB.

52. It isn't clear when or how the cow's eye dissection demonstration originated at the Exploratorium but by 1976, teachers were discussing it in their letters to Oppenheimer (see, e.g., Alfonso Iniquez, letter to Lynn Rankin, June 14, 1976, BANC MSS. 87/148c, 4:14, BL-UCB). Rae Carpenter, a founding board member of the Science Museum of Virginia, remembers seeing the materials for it being carried into the museum when he visited in 1976 (phone conversation with Karen Rader, May 2010).

53. Frank Oppenheimer, letter to Mrs. Joseph Caverly, July 3, 1974, BANC MSS. 87/148c, 11:2, BL-UCB.

54. "School at the Exploratorium," May 13, 1978, BANC MSS. 87/148c, 42:13, BL-UCB.

55. Fritz Mulhauser, letter to Frank Oppenheimer, BANC MSS. 87/148c, 4:18, BL-UCB.

56. Ogawa, Loomis, and Crain, "Institutional History of an Interactive Science Center" (n. 116 above, chap. 5).

57. Ron Hipschman, oral history interview by Karen Rader and Sylvie Gassaway, February 2006.

58. Frank Oppenheimer, letter to Steven White, March 24, 1969; see also Frank Oppenheimer, letter to Thomas Ford, February 19, 1970, BANC MSS. 87/148c, 4:1, BL-UCB.

59. Frank Oppenheimer, "Learning with Opportunity for Choice," paper presented at AAAS Annual Meeting, January 7, 1980, session titled "Science Learning Outside the Classroom: Environments for Science Learning," as cited in Marcel LaFollette, "Understanding, Ignorance, and Euphemisms: Reflections on the 146th Annual Meeting of the American Association for the Advancement of Science," *Science, Technology, and Human Values* 5, no. 30 (1980): 23–31.

60. Frank Oppenheimer, "Everyone Is You . . . or Me," *Technology Review* 78, no. 7 (1976): 6–7.

61. See e.g., Robert B. Lewis, letter to Frank Oppenheimer, April 26, 1973 BANC 87/148C, 4:16; Peter Dahl, letter to Exploratorium staff, November 1, 1970; 87/148c, 4:6, all BL-UCB.

62. Arnold Arons, letter to Frank Oppenheimer, February 22, 1972, BANC MSS. 87/148c, 4:1, BL-UCB.

63. Albert Bartlett, letter to Frank Oppenheimer, n.d., BANC MSS. 87/148c, 4:2, BL-UCB.

64. Robert Weinberg, letter to Chamber of Commerce, February 10, 1972, BANC MSS. 87/148c, 4:26, BL-UCB.

65. On the written inquiry from BMoS to Exploratorium, see A. William Konchanczyk, December 13, 1973, BANC MSS. 87/148c, 4:15; Boston Museum of Science educators had visited the Exploratorium earlier: see Valerie Wilcox, letter to Frank Oppenheimer, November 7, 1972, BANC MSS. 87/148c, 4:26, all BL-UCB.

66. Lee Kimche, "Science Centers: A Potential for Learning," *Science* 199, no. 4326 (1978): 270.

67. Ibid., 270–73.

68. R. Binswanger, letter to Frank Oppenheimer, October 6, 1975. BANC MSS. 87/148c, 4:1; Joseph Califano, letter to Frank Oppenheimer, March 9, 1979, BANC MSS. 87/148c, 39:58, all BL-UCB.

69. Katherine Coleman (Exploratorium Education Department), letter to Walter Gillespie (NSF), November 3, 1970, Record Group 190, box 89, SIA.

70. Mary Ellen Munley, *Catalysts for Change: The Kellogg Projects in Museum Education* (Washington, DC: Smithsonian Institution, 1986). Also see Janet Kamien, "An Advocate for Everything: Exploring Exhibit Development Models," *Curator* 44, no. 1 (2002): 114–28.

71. Hein, *The Exploratorium* (n. 43 above, this chap.), 48, 72.

72. Andrew Barry has noted that Oppenheimer's interest in perception was shaped by the work of psychologist Richard Gregory. Barry, "On

Interactivity" (n. 41 above, this chap.), 103. See also Richard Gregory, "Perception in Museums: The Natural History Museum London," *Perception* 7, no. 1 (1978): 1–2. Gustavo Corral argues that the consulting scientists who worked on similar exhibits in the 1970s at the Natural History Museum in London featured cognitive psychology and neuroscience examples because they sought to "validate artificial intelligence and, consequently, the disciplines whose results influenced its origins"; see Corral, "Human Biology" (n. 110 above, chap. 5). See also the draft of an article based on this thesis (submitted for publication to the *European School on the History of Science and Popularization*, precirculated to the authors), p. 8.

73. Hein, *The Exploratorium*, 92.

74. Charles Carlson, oral history interview by Karen Rader and Sylvie Gassaway, 2005, currently held by Karen Rader (nonarchived). This differs from Hilde Hein's assertion that "no funding had been obtained specifically for the biological exhibits . . . the animal behavior section had to be built largely out of the museum's operating budget" (*The Exploratorium*, 240).

75. Hein, *The Exploratorium*, 91–93. See also Exploratorium NSF application, BANC 87/148c, 35:17, BL-UCB. On earlier NSF funding for science education of the physics film program through the section on Scientific Curriculum Improvement, James Strickland, letter to Frank Oppenheimer, January 20, 1971, BANC MSS. 87/148c, 7:44, BL-UCB.

76. Ray Delgado, "Children's Writer, Biologist Evelyn Shaw-Wertheim," *San Francisco Chronicle*, May 23, 2003, http://www.sfgate.com/default/article /Children-s-writer-biologist-Evelyn-Shaw-Wertheim-2615364.php.

77. For examples of her earlier scientific work, see Phyllis Cahn and Evelyn Shaw, "A Method for Studying Lateral Line Cupular Bending in Juvenile Fishes," *Bulletin of Marine Science* 15, no. 4 (1965): 1060–71; Ethel Tobach, Lester Aronson, and Evelyn Shaw, eds., *The Biopsychology of Development* (New York: Academic Press, 1971). On her later work, see Charles Carlson, oral history interview by Karen Rader and Sylvie Gassaway, 2005, currently held by Karen Rader.

78. Cole, *Something Incredibly Wonderful Happens* (n. 106 above, chap. 5), 167.

79. This concept was first introduced in 1943 by Robert J. Havinghurst, a physicist turned general science education advocate. ·

80. Hein, *The Exploratorium* (n. 43 above, this chap.), 95.

81. Ibid., 85. See also Charles Carlson Oral History (Karen Rader and Sylvie Gassaway), 2005

82. Cole, *Something Incredibly Wonderful Happens* (n. 106 above, chap. 5), 158.

83. Delgado, "Children's Writer."

84. Richard Tabor Greene, "60 Models of Creativity" (2004), paper published at http://www.scribd.com/doc/2162318/A-Model-of-60-Models-of-Creativity. See also Hein, *The Exploratorium*.

85. Frank Oppenheimer, letter to Grass Foundation, (n.d.) 1974, BANC MSS. 87/148c, 35:38, BL-UCB.

86. On *The Watchful Grasshopper* exhibit, see Charles Carlson, Bonnie Jones, and Wayne La Rochelle, "Two Simple Electrophysiological Preparations Using Grasshoppers," *Association for Biology Laboratory Education Proceedings* 2 (1980): 181–95.

87. E. Grass, letter to C. Carlson, June 18, 1976, BANC MSS. 87/148c, 4:11—BL-UCB. The "Grass Instrument Company" file includes a trade brochure called "The Watchful Grasshopper," which was likely developed by the Exploratorium (e.g., some of the text matches Carlson's write-up in his own publication). See also Robert Collins, letter to C. Carlson, May 5, 1989 (BANC MSS., BL-UCB), where it appears that the Comprehensive Health Education Foundation frequently consulted with Carlson.

88. Carlson, Jones, and La Rochelle, "Two Simple Electrophysiological Preparations," 182.

89. *The Watchful Grasshopper*'s graphics are described in Ron Hipschman, *Exploratorium Cookbook 2: A Construction Manual for Exploratorium Exhibits*, rev. ed. (San Francisco: The Exploratorium, 1983).

90. Hein, *The Exploratorium* (n. 43 above, this chap.), 107. Some of the earliest visitors who heaped praise on the Exploratorium were the American Museum's Albert Parr, who visited in 1969, and the Smithsonian's John Anglim, who visited in 1970. See Albert Parr, letter to Frank Oppenheimer, 11, June 1969 (4:20) and John Anglim, letter to the Honorable Joseph Alito (Mayor of San Francisco), June 10, 1970, (4:2), both BANC MSS. 87/148c, BL-UCB.

91. See Frances Clark, letter to Frank Oppenheimer, June 29, 1979, BANC MSS. 87/148c, 4:4; also see letter from two teaching assistants in the Department of Applied Behavioral Sciences at the University of California, Davis, to Frank Oppenheimer, April 7, 1977, detailing their students' successful "experiential learning" at the Exploratorium (BANC MSS. 87/148c, 4:4)—all BL-UCB.

92. Harris H. Shettel and Pamela C. Reilly, "Evaluation of Existing Crteria for Judging the Quality of Science Education," *Education Technology Research and Development* 14, no. 4 (1966): 479–88. Compare Harris H. Shettel et al., "Strategies for Determining Exhibit Effectiveness," Final Report on Contract 0E-6–10–213, U.S. Department of Health Education and Welfare, April 1968, in "1968" correspondence file, BMoS.

93. Shettel et al., "Strategies for Determining Exhibit Effectiveness," quotes from 159, 156–58, respectively. D. J. de Solla Price, *Little Science, Big Science* (New York: Columbia University Press, 1963).

94. Carlson et al., Report to Grass Foundation (1977), BANC MSS. 87/148c, 35:38; cf. earlier in-house study, which is extant in archives: "Documentation of Exhibits, the Exploratorium, December 1974 to June 1975," BANC 87/148c, 39:8, BL-UCB.

95. Sol Pelavin, letter to Robert Semper, April 27, 1979; BANC MSS. 87/148c, 43:1/. Compare Lee Kimche, letter to Tom Malloy (DHEW), April 14, 1980, BANC MSS. 87/148c, 35:38, BL-UCB.

96. M. Startt, letter to Frank Oppenheimer, BANC MSS. 87/148c, 4:4, BL-UCB.

97. Leelane E. Hines, letter to Frank Oppenheimer, BANC MSS. 87/148c, 4:13, BL-UCB.

98. See, e.g., M. [illegible name] Clausen, letter to "Director of the Exploratorium," October 1981 (4:4); and Frank Oppenheimer, letter to H. Gennrich, March 1981 (3:18), both BANC MSS. 87/148c, BL-UCB.

99. M. Startt, letter to "Executive Director, Exploratorium," July 1984, BANC MSS. 87/148c, 4:23, BL-UCB.

100. Kimche, "Science Centers" (n. 66 above, this chap.), 270.

101. A 1974 survey reinforced these numbers. Science and natural history museums attracted between 38 percent and 45 percent of all museum visitors that year. Rita Reif, "Both Young and Old Learning by Doing," *New York Times*, September 9, 1975, 83; and Victor Danilov, *Science and Technology Centers* (Cambridge, MA: MIT Press, 1981), 5.

102. Reif, "Both Young and Old," 83.

103. See Alma Wittlin's discussion of the importance of object and collections, as well as her visual presentation of "what not to do" with exhibits; a photograph of a case-based natural history display from an unidentified museum, e.g., is captioned "Symphony or cacophony?" Wittlin, *Museums* (n. 18 above, this chap.), 163–93, 218–19. Also see letter from Frances Rew to Frank Oppenheimer, February 21, 1984, about how the Denver Museum of Natural History "missed your vision of a science museum, settling for the less inspired alternatives of their present Children's Museum and Science Pavilion," BANC 87/148c, 4:21, BL-UCB.

104. Helmuth J. Naumer, "A Marketable Product," *Museum News* 50, no. 2 (1971): 14–15.

105. Ibid.

106. Ibid., 16.

107. "AAM Mission Statement, 1972" in BANC MSS. 87/148c, 6:17, BL-UCB.

108. Ibid.

109. V. Danilov, letter to A. Wall, April 23, 1975; V. Danilov, memo to Association of Science-Technology Centers Members, July 15, 1975—both in BANC MSS. 87/148c, 6:17, BL-UCB. The American Association of Museums became active again in the Reagan years when it developed a legislative service to warn its members of federal policy changes that affected museums. See, e.g., the AAM Legislative Service Memo, April 13, 1981, BANC MSS. 87/148c, 6:5, BL-UCB.

110. Robert Inger, letter to Frank Oppenheimer, November 9, 1971, 4:14, BANC MSS. 87/148c, BL-UCB. The Metropolitan Museum of Art was even later to the game, soliciting Oppenheimer's help in a 1984 review of their education programs. "We would like to learn more about the education programs offered and problems faced by other museums of comparable stature," wrote Philippe de Montebello to Frank Oppenheimer, April 12, 1984, 4:18, BANC MSS. 87/148c, BL-UCB.

111. John Drabik, untitled, 25 May 1978; BANC MSS. 87/148c, 39:2, BL-UCB.

112. On the Exploratorium program, see Cole, *Something Incredibly Wonderful Happens* (n. 106 above, chap. 5), 179–204. On Boston's, see Kyra Montagu, letter to Frank Oppenheimer, October 4, 1977, folder "Museum 1977," box "1977," BMoS.

113. Cynthia Mark and Eve Van Rennes, "Bridging the Visitor Exhibit Gap with Computers," *Museum News* 60, no. 1 (1981): 25.

114. Ibid., 21.

115. B. N. Lewis, "The Museum as an Educational Facility," *Museums Journal* 80, no. 3 (1980): 151–55.

116. Shannon, "Icing Is Good" (n. 36 above, chap. 5), 32.

117. Ibid., 30. Shannon's language suggests that this criticism was leveled directly at the educational theories of Jerome Bruner.

118. Ibid., 29.

119. Harris H. Shettel, "Exhibits: Art Form or Educational Medium?" *Museum News* 52, no. 1 (1973): 40.

120. Albert Eide Parr, "Refuge from the Other Direction," *Discovery* 9, no. 1 (1973): 34.

121. Margaret Mead, "Museums in a Media Saturated World," *Museum News* 49, no. 1 (1970): 25.

122. Albert Eide Parr, "Museums: Enriching the Urban Milieu," *Museum News* 56, no. 4 (1978): 47.

123. Mead, "Museums in a Media Saturated World," 25.

124. Victor Danilov, "Public Education with a Pushbutton," *Biomedical Communications* 2, no 1 (1974), 26–37. A photo of Tam appears on page 29.

125. Rock, *Museum of Science, Boston* (n. 1 above, chap. 1), 158–59.

126. National Endowment for the Arts, *Museums USA* (n. 99 above, chap. 5), 25.

127. Shannon, "Icing Is Good" (n. 36 above, chap. 5), 30; P. Gilmour, "How Can Museums Be More Effective in Society?" *Museums Journal* 79, no. 2 (1979): 120–22.

128. Concerns about the apolitical stance of museums persisted for years after the angry 1970s. In 1983, e.g., museum critic and historian Marcel LaFollette observed that "all too often, exhibits neglect to explain the processes and procedures of the scientific enterprise, electing instead to concentrate on the products and artifacts, or to give visitors an exciting experience," she wrote. "Policy questions are not always presented as open for discussion; rather, exhibits tend to focus on what various experts believe." Marcel LaFollette, Lisa M. Buchholz, and John Zilber, "Science and Technology Museums as Policy Tools—an Overview of the Issues," *Science, Technology, and Human Values* 8, no. 3 (1983): 44.

129. "We have to develop new tools to persuade people to act sensibly," Oppenheimer told a reporter in 1979, and "we don't have to rely on coercion. That is the meaning of a free society." See Theodore Sudia, letter to Frank Oppenheimer, November 19, 1975, which suggested that Oppenheimer

consider some exhibits under development with the "Man-in-the-Bioshpere" program of UNESCO (BANC MSS. 87/148c, 4:23, BL-UCB).

130. "Frank Oppenheimer Interview," October 10, 1981, 4, Carton 21, folder 51, Exploratorium records, BANC MSS. 87/148c, BL-UCB.

131. "Accent on Participation in Proliferating Centers," *New York Times*, September 9, 1975, 41. Rebecca Onion also argues that the Exploratorium's efforts to empty "children" and "science" of political context were doomed to fail, as both categories are deeply politicized. Onion, "How Science Became Child's Play" (n. 10 above, this chap.), 388.

132. See also Alice Carnes, "Showplace, Playground or Forum? Choice Point for Science Museums," *Museum News* 64, no. 4 (1986): 103; Anthea Hancocks, "Museum Exhibition as a Tool for Social Awareness," *Curator* 30, no. 3 (1987): 181–92; Hein, *The Exploratorium* (n. 43 above, this chap.).

133. "Point of View: Boston Museum of Science," *Science* 170, no. 3954 (1970): 145.

134. Ibid.

135. Peter B. Flint, "Thomas D. Nicholson Dies at 68; Led Museum of Natural History," *New York Times*, July 11, 1991, B6; Andrew N. Rowan, *Of Mice, Models, and Men: A Critical Evaluation of Animal Research* (Albany: State University of New York, 1984), 60–61; Peter Singer, *Ethics into Action: Henry Spira and the Animal Rights Movement* (Lanham, MD: Rowman & Littlefield Publishers, Inc., 1998), 52–74; Nicholas Wade, "NIH Cat Sex Study Brings Grief to New York Museum," *Science* n.s. 194, no. 4261 (1976): 52–56; Nathaniel Sheppard, "Cats' Mutilation Laid to Museum: Natural History Is Accused of Inhumane Experiments," *New York Times*, July 19, 1976, 31.

136. They also made sincere, if clumsy, attempts to reach out to newer immigrants and traditionally disenfranchised groups. American Museum, e.g., commissioned a large-scale visitors' poll in 1974–75, becoming one of the first museums in the nation to do so. The survey broke down visitors by race, age, gender, income level, and other demographic factors. The data revealed that three in five visitors lived outside New York City, and one in five lived outside the tristate area. More than half of the children in school groups visiting the museum were Latino and black, but the museum attracted shockingly few adult visitors of color. The museum responded quickly, altering services and facilities to accommodate its current audiences. In 1981, the museum commissioned another survey: adult visitors were still young, highly educated, and largely white. But nearly 30 percent of visitors who attended the museum from six to a dozen times a year were minorities, up from 16 percent in the earlier poll, and the museum attracted far more adults than it had six years earlier. National Endowment for the Arts, *Museums USA* (n. 99 above, chap. 5), 47. The National Research Center on the Arts, an affiliate of New York City–based polling firm Louis Harris & Associates, best known as John F. Kennedy's presidential pollsters, compiled and analyzed the information. *AMNHAR* (1976), 8–9. *AMNHAR*

(1975), 6–7, 8–9. Nan Robertson, "At Natural History, Adults Find More to Do," *New York Times*, June 8, 1983, C19.

137. Museums around the nation relied on this approach. For its Native American audience, e.g., the Denver Museum of Natural History launched the photography show *The Urban Indian Experience: A Denver Portrait* and an accompanying lecture series, "Moccasins on Pavement: Off-Reservation Indians" in 1979. Newer science museums, though they rarely possessed anthropological collections, made similar efforts to host exhibitions on ethnic culture. *DMNHAR* (1979).

138. Throughout the 1970s, California's Museum of Science and Industry, e.g., supported exhibitions of African American, Japanese, and Polynesian life and ran a series of career symposia and summer science workshops for local residents. Attempting to make up for past sins, science museums that had evolved from natural history museums set up topics in their halls and shook out the deerskin robes that had sat folded in their storage rooms for decades: see Goodyear, *Society and Museum* (n. 5 above, "Introduction"), 71; National Endowment for the Arts, *Museums USA* (n. 99 above, chap. 5).

139. *AMNHAR* (1971), 6, 30.

140. William C. Steere, "Preface," in *Museums and the Environment: A Handbook for Education*, ed. American Association of Museums Environmental Committee (New York: Arkville Press, 1971), ii.

141. Goodyear, *Society and Museum*, 79.

142. For a comprehensive treatment of environmental research and displays in museums, see Peter Davis, *Museums and the Natural Environment: The Role of Natural History Museums in Biological Conservation* (London: Leicester University Press, 1996). On environmental efforts by European museums in the later 1970s, see G. Stansfield, "Environmental Education," *Museums Journal*, vol. 72 (1972). Jeffrey Stine reviews the history of the AAM project and offers a prescriptive overview of twenty-first century prospects for environmental displays in museums in "Placing Environmental History on Display," *Environmental History* 7, no. 4 (2002): 566–88.

143. Robert Smithson, "Can Man Survive? (1969)," in *Robert Smithson, the Collected Writings*, ed. Jack D. Flam (Berkeley: University of California Press, 1996), 367–68.

144. Steere, "Preface," ii This project both revealed and exploited the extent to which contemporary museum personnel "crossed over" between the varied institutional sites of biological displays. For example, James Oliver, one of the project leaders, joined the American Museum as an assistant curator of herpetology in 1942. After leaving to teach at the University of Florida from 1948 to 1951, he returned to New York as curator of reptiles at the Bronx Zoo, becoming the zoo's director in 1958. Only one year later, Oliver was named director of the American Museum, where he served until 1969. Myers, "A History of Herpetology" (n. 44 above, chap. 1), 5–6; James A. Oliver, "Museum Education and Human Ecology," *Museum News* 48 (1970):

28–30; Walter Waggoner, "James A. Oliver, 67: Zoologist, Teacher, Administrator," *New York Times*, December 4, 1981. The AAM contract process is detailed in a letter from William C. Steere to Jack Hardesty on February 11, 1970, reproduced in American Association of Museums, Environmental Committee, *Museums and the Environment: A Handbook for Education* (New York: Arkville Press, 1971), in app. B, 215.

145. The *Population* exhibit emerged from a conversation between Smithsonian curators and Woodrow Wilson Fellows about the political relevance of the AMNH's Bicentennial Exhibits: see RG 363, box 13, NMNH, Ecology 200 (ca. 1973), SIA. Likewise, *It All Depends*, a limply titled rainforest exhibit that had started out as the Hall of Living Things, was initiated in 1970 and eventually mounted at the Smithsonian in the 1980s: see RG 263, box 28 ("Rainforest Exhibit"). A few years earlier, the Smithsonian had developed a successful *Endangered Species* exhibit. See RG 503, box 20, SIA. *Endangered Species* (1968–70) also eventually had a traveling component developed. (Thanks to Mary Anne Andrei for directing us to these records.)

146. *AMNHAR* (1970), 29.

147. Ibid., 12–13.

148. R. Moore, *Evolution in the Courtroom: A Reference Guide* (Santa Barbara, CA: ABC-CLIO, 2002), 70.

149. U.S. Court of Appeals, District of Columbia Circuit, ed., *Dale Crowley, Jr., Individually & in His Capacity as Executive Director of the National Foundation for Fairness in Education, Et Al., Appellants, V. Smithsonian Institution Et Al.* (St. Paul, MN: West Publishing Co., 1980), on case 636 F.2d 738. Luther J. Carter, "Briefing," *Science* 204, no. 4396 (1979): 925. For a contemporary reflection that puts attacks on museums in the broader context of creationists' political strategies of this era, see Dorothy Nelkin, "From Dayton to Little Rock: Creationism Evolves," *Science, Technology, and Human Values* 7, no. 40 (1982): 47–53.

150. The suit's dismissal was upheld by a court of appeals in 1980. Nelkin, "From Dayton to Little Rock," 50.

151. Susan West, "Evolution of an Exhibit," *Science News* 116, no. 1 (1979): 10.

152. Similar dynamics tested natural museum professionals outside the United States. After British Museum staff members mounted two new displays, *Man's Place in Evolution* (1978) and *Dinosaurs and Their Living Relatives* (1979), systematist L. Beverly Halstead criticized the new exhibits for privileging cladistics, then a controversial method of biological classification, as their didactic framework for presenting evolution. By emphasizing what lay visitors might interpret as the deficiencies of existing scientific evidence for particular chains of evolutionary descent, Halstead argued, the museum had abdicated its public educational responsibilities. Exhibit designers R.S. Miles and C. G. S. Clarke defended their approach, arguing that the designs employed appropriate "educational principles" that highlighted "clear and overt reasoning" about scientific evidence in the

study of phylogeny. Ultimately *Nature*'s editors weighed in: "The difficulty, with which everybody must sympathize, is that all museums of natural history are used not merely by those who wish to answer questions in their minds about evolution and the like, but also by those who believe Darwinism an abomination." Simplifying exhibits too much was "to ask for trouble" from scientists, and "to fail to make explicit the principles on which . . . exhibitions were designed" was to leave unexamined the as-yet unexamined pedagogical assumptions of new display strategies: "Is the push-button public exhibition the only acceptable modern style?" For the initial complaint about the displays from L. Beverly Halstead, see L. Beverly Halstead, "Letters," *Nature* 275, no. 5682 (1978): 683. Also see L. Beverly Halstead, "Letters," Nature 276, no. 5690 (1978): 759–60, and "Halstead's Defense against Irrelevancy," *Nature* 292 (1981): 403–4. For *Nature*'s own editorial response, see "Cladistics and Evolution on Display," *Nature* 292 (1981): 395.

153. West, "Evolution of an Exhibit," 11.

154. In 1980, American Museum curators and designers began internal conversations about an exhibit that would be mounted in 1984, *Ancestors: Four Million Years of Humanity*, but even then, reported *Science News*, Lucy was "conspicuous in its absence" from the project. In part, this was because the fossil specimen itself was too fragile to travel, but also, Lucy's new owner, paleoanthropologist Richard Leakey, son of Mary and Louis Leakey, did not want natural history museums to get further embroiled in what he saw as an on-going specialized scientific debate. D. Johanson, M. Edey, and M. A. Edey, *Lucy: The Beginnings of Humankind* (New York: Simon & Schuster, 1990). On the Lucy controversy from 1977 to 1983, see John H. Douglas, "The Origins of Culture," *Science News* 115, no. 15 (1979): 252–54; Wray Herbert, "Lucy's Family Problems," *Science News* 124, no. 1 (1983): 8–11. On the Leakeys and their connection to Lucy and debates on human evolution, through Richard Leakey and beyond, see John Noble Wilford, "The Leakeys: A Towering Reputation," *New York Times Magazine*, October 30, 1984. Except in the case of this exhibit, relations between the American Museum and Lucy's scientific stewards were quite cordial: in May of 1982, e.g., Johanson appeared alongside Jane Goodall and Dian Fossey at a symposium on primatology at the American Museum, cosponsored by the Leakey Foundation.

155. The Anthropology Program of the National Science Foundation did agree to sponsor curatorial research based on the specimens, and the Federal Council for the Arts and Humanities indemnified the exhibit—no small contribution when transporting invaluable and inordinately controversial scientific specimens from all over the world. Both agencies did their best to keep a low profile throughout. Ian Tattersall, John A. Van Couvering, and Eric Delson, "Ancestors: An Expurgated History," in *Ancestors: The Hard Evidence*, ed. Eric Delson (New York: Alan R. Liss, Inc., 1985), 2.

156. Daniel Greenberg, "Research Priorities: New Program at N.S.F. Reflects Shift in Values," *Science* 170, no. 3954 (1970): 146.

157. Craig C. Black, "The Case for Research," *Museum News* 58, no. 5 (1980): 52–53.

158. Jason Lillengraven and Tod Stuessy, "Summary of Awards from the Systematic Biology Program of the National Science Foundation, 1973–1977," *Systematic Zoology* 28, no. 2 (1979): 123–31.

159. Black, "The Case for Research," 53.

160. Craig C. Black, "New Strains on Our Resources," *Museum News* 56, no. 3 (1978): 20.

161. Ibid., 21.

162. Ibid., 22.

163. See the series of documents on the Senator Proxmire Golden Fleece award in April 1977, and the public response, located in Record Group 309, box 2, folder "Public Affairs—Radio and TV," SIA.

164. Vincent J. Doyle, letter to Mr. Warner, October 2, 1970, Record Group 190, box 89, SIA.

165. See Victoria D. Alexander, "From Philanthropy to Funding: The Effects of Corporate and Public Support on American Art Museums," in "Museum Research," special issue, *Poetics*, no. 24 (1996): 87–129.

CHAPTER SEVEN

1. Thomas D. Nicholson, "Why Museum Accreditation Doesn't Work," *Museum News* 60, no. 1 (1981): 5.

2. Throughout the 1990s, questions about museums' defining purposes and responsibilities played out in arguments about programming, as a new generation of scholars and artists contemplated the ways in which museums could or should blur once hard distinctions between their educational programs and social activism. On these debates, see, e.g., Richard Sandell, "Museums as Agents of Social Inclusion," *Museum Management and Curatorship* 17, no. 4 (1998): 401–18.

3. See, e.g., Migs Grove, Sybil Walker, and Alexandra Walsh, "The Uses of Adversity," *Museum News* 61, no. 3 (1983): 26–37; Eric Scigliano, "Popular Science: What Next for the Science Center after the 'China' Bonanza?" *Weekly*, October 10–16, 1984; Rudolph Unger, "Chicago Museums Doing Well," *Chicago Tribune*, November 17, 1982, B3.

4. The price was relatively low—$2.50 for a six-month pass, which entitled them to an unlimited number of visits and free admission for accompanying children—but staff members still felt it was a sad departure from the institution's commitment to democratic access.

5. Cole, *Something Incredibly Wonderful Happens* (n. 106 above, chap. 5), 289. See also Tom Freudenheim, letter to Frank Oppenheimer, April 30, 1981, BANC MSS. 87/148c, 4:9, BL-UCB.

6. Eleanor Charles, "Museums Assess Their Future Roles," *New York Times*, September 26, 1982, A6.

7. Ruth Dean, "In Transition: The Federal Scene under Carter and Reagan," *Museum News* 60, no. 3 (1982): 53–57. A similar phenomenon occurred in Great Britain during the 1980s and 1990s. For a brief overview of this history, see Sandell, "Museums as Agents of Social Inclusion."

8. Peter N. Kyros, "Our Own Best Advocates," *Museum News* 59, no. 7 (1981): 18.

9. James V. Toscano, "Development Comes of Age," *Museum News* 60, no. 3 (1982): 61–67.

10. See the article in *Buffalo Courier Express*, June 6, 1968, 16-1-5, as well as Buffalo Museum of Science, *Collections* 48 (1968): 41–43, as cited in Goodyear, *Society and Museum*, 78.

11. On corporate philanthropy in museums, see, among others, Alexander, *Museums and Money* (n. 77 above, chap. 5); Richard Sandell and Robert Janes, eds., *Museum Management and Marketing*, Leicester Readers in Museum Studies (New York: Routledge, 2007).

12. The AMNH Corporate Program was, by 1989, raising more than $900,000 in unrestricted support from more than two hundred companies. *AMNHAR* (1989), 63.

13. Lorraine Glennon, "The Museum and the Corporation: New Realities," *Museum News* (1988): 39.

14. Ibid., 41.

15. *AMNHAR* (1989), 63. On the relationship of oil companies to paleontology, see G. Arthur Cooper, "The Science of Paleontology," *Journal of Paleontology* 32, no. 5 (1958): 1010–18; Roger L. Kaesler, "A Window of Opportunity: Peering into a New Century of Paleontology," *Journal of Paleontology* 67, no. 3 (1993): 329–33.

16. Sharon MacDonald has chronicled how similar dynamics played out in London's Science Museum in *Behind the Scenes at the Science Museum* (London: Bloomsbury Academic Press, 2002), 34–38.

17. Howard Learner, *White Paper on Science Museums* (Washington, DC: Center for Science in the Public Interest, 1979).

18. As cited in ibid.

19. For complaints from educators and corporations, respectively, see Terry Erwin, letter to E. Miller and E. Snyder, November 8, 1976 [folder 10] and Porter Kier, letter to R. Blawitt, March 5, 1976 [folder 8], both RG 309, box 3, SIA. List of sponsors invited to opening of the Insect Zoo, letter from Secretary Dillon Ripley, August 1977, in RG 371, box 2, folder "September 1977," SIA. The Milwaukee Public Museum's display of living bugs directly inspired the Insect Zoo: see Porter Kier, letter to Harry Hart, November 1, 1974, RG 363, box 27, folder "Permanent Insect Zoo," SIA.

20. Glennon, "The Museum and the Corporation," 41.

21. U.S. Congress Senate Committee on Rules Smithsonian Institution, U.S.

Congress House Committee on Interior Administration, and Affairs Insular, *Smithsonian Institution Appropriations Hearings* (Washington, DC: Government Printing Office, 1994), 118–20.

22. Ibid.

23. Victor Danilov, "Promoting Museums through Advertising," *Museum News* 64, no. 6 (1986): 36–37.

24. Ibid.

25. The phrase is Neil Harris's. See Neil Harris, "Polling for Opinions," *Museum News* 69, no. 5 (1990): 49.

26. Jeffrey Birch, "A Museum for All Seasons," *Museum News* 60, no. 4 (1982): 26.

27. Ibid.

28. Bob Flasher, "A Recipe for Survival," *Museum News* 61, no. 5 (1983): 27.

29. Porter Kier, letter to Dillon Ripley, September 7, 1977, RG 309, box 14, folder "Exhibits 1977," SIA.

30. "'Hey Mom, That Exhibit's Alive': A Study of Visitor Perceptions of the Coral Reef Exhibit, National Museum of Natural History," August 1–2, 1981, NMNH archives, RG 564, box 1, folder "Free Floating," SIA.

31. Ibid., introduction.

32. Ibid., 22, 12, 24, and 19.

33. See unlabeled folder in Office of Exhibits records (RG 363, box 1, folder ——) for *Washingtonian* blurb, dated 28 October 1981, SIA.

34. Melvin Kranzberg, "Review of Victor Danilov's *Science and Technology Centers* (1982)," *4S Review* 2, no. 1 (1984): 19.

35. Neil Postman, "Museum as Dialogue," *Museum News* 69, no. 5 (1989): 58.

36. See, e.g., Edward P. Alexander, *Museums in Motion: An Introduction to the History and Functions of Museums* (Nashville: American Association for State and Local History, 1979), 75; George Basalla, "Museums and Technological Utopianism," *Curator* 18, no. 2 (1974): 210–11.

37. Carnes, "Showplace, Playground or Forum?" (n. 132 above, chap. 6), 30.

38. Ibid., 35.

39. See, e.g., Peggy Cole, "Piaget in the Gallery," *Museum News* 63, no. 1 (1984): 9–15; Neil E. Deaver and Robert A. Matthai, "Child-Centered Learning," *Museum News* 54, no. 4 (1976): 15–19; Nina Jensen, "Children, Teenagers and Adults in Museums: A Developmental Perspective," *Museum News* 60, no. 5 (1982): 25–30.

40. Deaver and Matthai, "Child-Centered Learning," 16.

41. In 1986, *Museum News* devoted an entire issue to visitor surveys and exhibit evaluation, and in 1988, after nearly two decades worth of research in the field, University of Wisconsin–Milwaukee psychology professor Chandler Screven and a host of other museum professionals and scholars founded the Visitor Studies Association, a nonprofit organization devoted to research on visitors' experience in the museum. Cole, "Piaget in the Gallery," 9–10.

42. For a brief history of museum surveys, see Harris, "Polling for Opinions."

43. Harris H. Shettel, "Time—Is It Really of the Essence?" *Curator* 40, no. 4 (1997): 246. For a brief history of museum surveys, see Harris, "Polling for Opinions."

44. Museum educators were largely women throughout the twentieth century, and as George Hein has pointed out, their increasing visibility and new demands for recognition within the museum world both coincided with and were likely influenced by the rise of feminism in the 1970s and 1980s. George E. Hein, "Museum Education," in *A Companion to Museum Studies*, ed. Sharon Macdonald (Malden, MA: Blackwell Publishing, 2010); Stephen E. Weil, "From Being about Something to Being for Somebody: The Ongoing Transformation of the American Museum," *Daedalus* 128, no. 3 (1999): 234.

45. American Association of Museums, *Museums for a New Century: A Report of the Commission on Museums for a New Century* (Washington, DC: American Association of Museums, 1984). Also see Weil, "From Being about Something," 234.

46. Janet Raloff, "The Science of Museums," *Science News* 154, no. 12 (1998): 186.

47. Munley, *Catalysts for Change* (n. 70 above, chap. 6), 4, 6.

48. Museums also sought money from other sources. In May 1978, e.g., the Boston Museum of Science's director of exhibits, John Drabik, appealed to the Society of Neuroscientists on behalf of the Association of Science-Technology Centers, requesting support for a collaborative effort in museum exhibit research and planning. According to Drabik, more funding would allow scientists and exhibit designers to create and share state-of-the-art experimental biology displays. John Drabik, untitled, May 25, 1978, BANC MSS. 87/148c, 39:2, BL-UCB.

49. Munley, *Catalysts for Change*. Also see Kamien, Advocate for Everything" (n. 70 above, chap. 6).

50. Mary Ellen Munley, "Educational Excellence for American Museums," *Museum News* 65, no. 2 (1986): 51–57.

51. Minda Borun et al., *Planets and Pulleys: Studies of Class Visits to Science Museums* (Washington, DC: Association of Science-Technology Centers, 1983), 10.

52. "The Profession Honors S. Dillon Ripley," *Museum News* 64, no. 1 (1985): 62.

53. Frank Oppenheimer, "A.A.M. Award for Distinguished Service to Museums Acceptance Speech," *Museum News* 61, no. 2 (1982): 39.

54. See, e.g., Herbert J. Walberg, "Scientific Literacy and Economic Productivity in International Perspective," in "Scientific Literacy," special issue, *Daedalus* 112, no. 2 (1983): 1–28. The United States was not alone. Policymakers and scientists in Britain, e.g., expressed similar concern over the woeful performance of students in science. See citations in SRG, "Preface," in "Scientific Literacy," special issue, *Daedalus* 112, no. 2 (1983): v–vi.

55. William V. Mayer, president emeritus of the Biological Sciences Curriculum Study, observed that conversation about science education in the 1980s

echoed that of the 1890s—in both eras, critics complained that science education was feeble, plagued by unqualified teachers, and poorly integrated with the rest of the curriculum. Mayer, "Biology Education in the United States" (n. 15 above, chap. 6). For reports on science education from the 1980s, see among others, B. M. Vetter et al., *Educating Scientists and Engineers: Grade School to Grad School* (Washington, DC: U.S. Congress, Office of Technology Assessment, 1988); National Commission on Excellence in Education, *A Nation at Risk: The Imperative for Educational Reform: A Report to the Nation and the Secretary of Education, United States Department of Education* (Washington, DC: National Commission on Excellence in Education, 1983); U.S. Department of Education and National Science Foundation, *Science and Engineering Education for the 1980s and Beyond* (Washington, DC: National Science Foundation, 1980).

56. On spending cuts to NSF science education programs, see Tressel, "Thirty Years of 'Improvement'" (n. 5 above, chap. 5).
57. Semper is quoted in Anne Lowrey Bailey, "Ending Science Illiteracy," *Museum News* 66, no. 5 (1988): 52.
58. Ibid.
59. The Pacific Science Center program was a resounding success, especially by the measurements most popular in the 1980s—test scores. At one elementary school where four museum-trained teachers taught, scores rose from the seventy-seventh to the eighty-fifth percentile in two years. Ibid., 50.
60. Ibid.
61. Ibid., 53.
62. Goodyear, *Society and Museum* (n. 5 above, "Introduction"), 98.
63. Bailey, "Ending Science Illiteracy," 53.
64. According to the Oxford English Dictionary, the term first emerged in a 1983 *Fortune* article on video games that "attempt to make learning fun."
65. For an especially astute assessment of this phenomenon, see Martin Hall, "The Reappearance of the Authentic," in *Museum Frictions: Public Cultures/Global Transformations*, ed. Ivan Karp et al. (Durham, NC: Duke University Press, 2006).
66. On this history, see Cain, ""Attraction, Attention, Desire'" (n. 32 above, chap. 1).
67. Neil Harris made a similar observation about blockbusters' impact on art museums. See Neil Harris, "The Divided House of the American Art Museum," *Daedalus* 128, no. 3 (1999): 33–56.
68. Danilov, "Promoting Museums through Advertising," 36–37.
69. Robertson, "At Natural History" (n. 136 above, chap. 6), C19. Hilliard Harper, "Evening Experiments at S.D. Museums," *Los Angeles Times*, August 7, 1985, 1.
70. Museums' aggressive turn toward food, gifts, and services was probably inspired by for-profit "educational" sites like Sea World, which had opened parks in 1964, 1970, and 1973, and Walt Disney's Epcot Center, which

opened in 1982. Naumer, "A Marketable Product" (n. 104 above, chap. 6), 14–15.

71. The evolution of the ground floor of the Buffalo Museum of Science is a good example of this trend. In 1974, the Buffalo Museum of Science moved its sales shop, established with funding generated by the women's committee and dubbed "The Cabinet" as a nod to the museum's history, to the ground floor near the west entrance. The shop was enlarged in 1977, and in 1991, a much grander version of the museum's shop was constructed just inside the museum's recently built atrium, which served as its main entrance. Goodyear, *Society and Museum* (n. 5 above, "Introduction"), 79.

72. Harry R. Albers, " . . . And Visions of Royalties Danced in Their Heads," *Museum News* 52, no. 1 (1973): 23.

73. See E. Henry Lamkin Jr., July 13, 1982, 87/128c, 4:16. On franchising the store, see Frank Oppenheimer, letter to Robert Glaser, April 2, 1981, 87/148c, 13:19. Both BL-UCB.

74. As one museum marketing director pointed out, this money disproportionately benefited a handful of museums. Between ten and twelve museums were responsible for approximately $300 million of that pie, while another sixty-seven hundred museums nationwide divvied up the rest. Rena Zurofsky, "Sharp Shop Talk," *Museum News* 68, no. 4 (1989): 45.

75. On this phenomenon, see, among others, Douglas Davis, "The Museum Impossible," *Museum News* 61, no. 5 (1983): 33–37; Thomas Vonier, "Museum Architecture: The Tension between Function and Form," *Museum News* 66, no. 5 (1988): 24–30.

76. Even then, of course, the stance had been flexible. Just a few months after they turned down the college juniors, the science center board agreed that the Seattle Symphony could use the grounds for its annual fundraising ball but gruffly declared they would only sponsor "one social event of this type each year." Minutes of the Executive Committee of the Pacific Science Center, January 31, 1964 (PSC).

77. See standard meeting planner letter (n.d.) in BANC 87/148c, 36:11 (Brochures); see also G. Rubin et al., "Summary: Estimated Expenses for Leasing Our Space to Others to Host Receptions or Parties," January 9, 1981, BANC 87/148c, 36:10, all BL-UCB.

78. The first international blockbuster, *Treasures of Tutankhamun*, featured fifty-five objects from the boy king's tomb, many of them in solid gold or encrusted with precious gems. Funded by a combination of corporate, foundation, and federal grantmakers, the show attracted more than eight million visitors, many of whom waited for seven or eight hours to see it.

79. Grace Glueck, "Art People," *New York Times*, April 20, 1979, C22.

80. San Diego got its first blockbuster in 1991, e.g., and Milwaukee didn't have the space to bring one in until 2005. See Mike Boehm, "Nature Museums Nearly Relics Themselves," *Los Angeles Times*, June 3, 2007; Eliot Marshall, "Hard Times for San Diego Museum," *Science* 251, no. 4992 (1991): 375.

81. Blockbusters like *Ancestors* were only possible as a result of the passage of the Arts and Artifacts Indemnity Act in 1975. This act authorized the federal government to insure against loss or damage to works of art and artifacts in international exhibitions of humanistic value and curtailed American museums' need to purchase private insurance for objects, which meant that museums in the United States could more easily borrow the biological and cultural treasures of other countries. Maria Papageorge, "Opportunities for International Exhibitions," *Museum News*, vol. 60, no. 2 (1981); Tattersall, Van Couvering, and Delson, "Ancestors" (n. 155 above, chap. 6), 4.

82. See Roger Lewin, "Ancestors Worshiped," *Science* 224, no. 4648 (1984): 477–79.

83. Scigliano, "Popular Science" (n. 3 above, this chap.), 36. Allan R. Gold, "Ramses 2 Visits a Boston Museum," *New York Times*, May 1, 1988, 70.

84. Goodyear, *Society and Museum* (n. 5 above, "Introduction"), 77–78.

85. Ibid., 77.

86. Rental and Attendance Records, 1987–2009, Internal Records, Pacific Science Center (PSC).

87. Ibid.

88. Felicia Maffia, interview by Victoria Cain, 2009; *Whales: Giants of the Deep* Exhibition Information Packet, PSC; Whales Tour Schedule, PSC; Tom Paulu, "Whale Exhibit Surfaces at Center," *Seattle Daily News*, June 11, 1990.

89. Though the 1979 exhibition on Pompeii was partly financed by the Xerox Corporation and the National Endowment for the Humanities, it cost the American Museum a little more than $1.1 million, a shocking sum for most institutions.

90. Julie Tamaki, "Really Big Show Features 5 Life-Size Robot Whales at Natural History Museum," *Los Angeles Times* January 21, 1992, 1.

91. Glueck, "Art People" (n. 78 above, this chap.), C22.

92. Ibid.

93. "An exhibition is an event—akin to a theatrical show—and as such more readily identifiable as entertainment, something to do with one's time," declared *Museum News* in 1989. "In today's event-oriented culture, exhibitions are more familiar, more understandable entities. The artificially imposed time boundaries also work well with a public conditioned to seek the novel and occasional." Danielle Rice, "Examining Exhibits," *Museum News* 68, no. 6 (1989): 47.

94. Zurofsky, "Sharp Shop Talk" (n. 73 above, this chap.), 45.

95. A Museum Store Association had been founded in 1955 and puttered along relatively quietly for two decades. By the late 1970s and early 1980s, however, pressure to generate revenue had become so powerful that the association was compelled to establish its own professional code of ethics in 1982. By the 1980s and 1990s, museum stores had become a major source

of income for most museums. In 1995, e.g., the San Diego Natural History Museum reported that its store sales constituted 13.74 percent of its total budget, an amount roughly the equivalent to what the museum spent on scientific research, collecting and preserving specimens, scientific publications, computerization of records, and its research library. Beverly Barsook, "A Code of Ethics for Museum Stores," *Museum News* 60, no. 3 (1982): 50. See *Annual Report* (San Diego, CA: San Diego Natural History Museum, 1995), 17.

96. Gordon Ambach, "Museums as Places of Learning," *Museum News* 65, no. 2 (1986): 36.

97. Harris, "Divided House" (n. 66 above, this chap.), 42–43.

98. Flasher, "Recipe for Survival" (n. 27 above, this chap.), 5.

99. Silver, "Almost Everyone Loves a Winner" (n. 33 above, chap. 5), 26.

100. Scigliano, "Popular Science" (n. 3 above, this chap.), 40.

101. Flasher, "Recipe for Survival," 5.

102. Oppenheimer, "A.A.M. Award for Distinguished Service" (n. 52 above, this chap.), 43.

103. Ibid., 41.

104. "The museum itself cannot be used as a salesroom," Ward flatly declared to an errant staff member in 1904. J. D. Figgins, letter to·John Campion, October 13, 1913 (JDF- CMNH); H. L. Ward, letter to Maria Herndl, December 19, 1904 (MPM).

105. Charles Alan Watkins, "Fighting for Culture's Turf," *Museum News* 70, no. 2 (1991): 63.

106. See Ann Mintz, "That's Edutainment!" *Museum News* 73, no. 6 (1994): 33–35.

107. James Wines, "Exhibition Design and the Psychology of Situation," *Museum News* 67, no. 2 (1988): 58. In 1990, exhibit designer Larry Klein, perhaps best known for his stint as design director of the 1984 Olympic Games in Los Angeles, heralded "the emergence of a cadre of exhibit professionals who will change the way we see and experience exhibits." See Larry Klein, "Team Players: Now That Designers Are Full Members of the Exhibition Team, They Can Concentrate on Conceptual Design—but It Wasn't Always So . . ." *Museum News* 70, no. 2 (1991): 45.

108. For a retrospective on the "rebalancing" of education and entertainment in museums from the 1960s through the 1990s, see Emlyn Koster, "If We Could Start Again . . . ," *ASTC Dimensions* 7, no. 3 (May–June 2004): 3.

109. One Field Museum administrator blamed this on departmental territoriality and curators' focus on research and collections management. "Context and Commitment," *Museum News* 69, no. 5 (1990): 68.

110. Frank Taylor, letter to S. Dillon Ripley, June 16, 1976; R. S. Cowan, letter to Ripley, July 11, 1976—both RG 309, box 3, folder 8, SIA.

111. See image of National Museum of Natural History intern Nora Besansky feeding a tarantula in the Insect Zoo, featured in *Torch*, the Smithsonian's

Membership Magazine, in September of 1977. RU 371, box 2, folder "September 1977," SIA.

112. Richard Froeschner, letter to Porter Kier (including a twenty-four-page handwritten set of notes that Froeschner wrote, reviewing the Insect Zoo script), January 16, 1976, RG 309, box 3 folder 8–NMNH, SIA

113. Margaret J. King, "Theme Park Thesis," *Museum News* 69, no. 5 (1990): 60, 62.

114. Ibid., 62. On Disney imagineering, see Margaret J. King, "The Theme Park: Aspects of Experience in a Four-Dimensional Landscape," *Material Culture* 34, no. 2 (2002): 1–15; Sharon Zukin, *Landscapes of Power: From Detroit to Disney World* (Berkeley: University of California Press, 1991).

115. For example, Neil Kotler, "Delivering Experience: Marketing the Museum's Full Range of Assets," *Museum News* 78, no. 3 (1999): 30–39, 58–61.

116. On the development and meaning of immersion in display, see Griffiths, *Shivers Down Your Spine* (n. 100 above, chap. 2).

117. James Kelly, "'Gallery of Discovery,'" *Museum News* 70, no. 2 (1991): 52.

118. Jeff Barnard, "'Edu-Tainment' Finds Creative Outlet in New Pacific Northwest Museum Natural History," *Los Angeles Times*, February 20, 1994, 4.

119. Ibid.

120. Ibid.

121. Emlyn Koster, "In Search of Relevance: Science Centers as Innovators in the Evolution of Museums," *Daedalus* 128, no. 3 (1999): 282. Koster also discusses the role of what he calls "themed entertainment centers."

122. Dennis P. Doordan, "Nature on Display," *Design Quarterly* 155 (1992): 35.

123. Ibid., 36.

124. "Context and Commitment" (n. 108 above, this chap.), 69.

125. K. Elaine Hoagland, "Socially Responsible: In the 1990s, Natural History Museums Will Focus Their Efforts on Maintaining the Diverse Life of a Healthy Planet," *Museum News* 68, no. 5 (1989): 50–51.

126. Elizabeth Culotta, "Museums Cut Research in Hard Times," *Science* 256, no. 5061 (1992): 1271.

127. Marshall, "Hard Times for San Diego Museum" (n. 79 above, this chap.), 375.

128. Culotta, "Museums Cut Research," 1271.

129. Ibid., 1269.

130. Ibid., 1271.

131. Ibid.

132. See NSF Annual Reports for this period, 1989 to 1992, which are available on-line at https://www.nsf.gov/about/history/annual-reports.jsp. The United States was not alone in this trend. In May of 1990, curators at London's Natural History Museum went on strike to protest threats of massive layoffs of scientific researchers in the name of market economics. Culotta, "Museums Cut Research," 1268–69; "In the United Kingdom, Museum Controversy Strikes yet Again," *Museum News* 69, no. 5 (1990): 17–19.

133. Lillengraven and Stuessy, "Summary of Awards" (n. 158 above, chap. 6), 123–24.

134. Los Angeles had a recent history of similar cuts. As part of a 1992 county-wide reduction in spending, the County Board of Supervisors had told the museum to cut its budget by $2.175 million, the equivalent of forty-two staff positions. The museum responded by laying off three of the four curators at a satellite site, the city's George C. Page Museum. The Page curators managed one of the most important collections of Ice Age fossils in the world, excavated prehistoric remains of saber-toothed cats, mastodons, and other animals disinterred from the site's forty thousand–year-old tar pits, and supervised more than fifty volunteers as they helped clean the fossils. The layoffs brought this work to a standstill, appalling scientists around the world. When the director of Australia's new National Dinosaur Museum heard the news, he sent a disbelieving fax to Page curators: "What's wrong with the administrators? Surely they must realize the value to science of what is surely the most spectacular Pleistocene deposit of all." Patricia Ward Biederman, "Staffers Irate over Possible Layoffs of 3 Scientists at La Brea Tar Pits," *Los Angeles Times*, December 16, 1992, B1.
135. Carla Rivera, "Bones of Contention " *Los Angeles Times*, August 16, 1994, 1.
136. Ibid.
137. "Natural History in New York," *Science* 263, no. 5154 (1994): 1688.
138. See, e.g., Frederick R. Schram, "Museum Collections: Why Are They There?" *Science* 256, no. 5063 (1992): 1502.
139. Robert J. Baker and Terry L. Yates, "Net-Wielding Anachronisms?" *Science* 282, no. 5391 (1998): 1049.
140. Harvard entomologist Edward O. Wilson warned the president's Council of Advisors on Science and Technology in 1991 that such budget cuts posed a serious threat to systematic biology, especially research involving the evolution of large animal species. Rather than cutting their scientific staffs, natural history museums ought to be planning "a tenfold increase in both capacity and manpower in the next decade," the biologist argued. Marshall, "Hard Times for San Diego Museum" (n. 79 above, this chap.), 375; Culotta, "Museums Cut Research," (n. 125 above, this chap.), 1268.
141. Andrew V. Z. Brower and Darlene D. Judd, "Net-Wielding Anachronisms?" *Science* 282, no. 5391 (1998): 1048.
142. Baker and Yates, "Net-Wielding Anachronisms?" 1049.
143. Debbie Tuma, "Dinosaur Club Stomps Locally," *New York Daily News*, February 4, 1996. On scientists as consultants for the film, and how they managed the uncertain science of paleontology in the face of demands to put animatronic dinosaurs on screem, see David Kirby, *Lab Coats in Hollywood: Science, Scientists, and Cinema* (Cambridge, MA: MIT Press, 2011), esp. 16–34, 127–32, 137–39.
144. "The Dinosaur Society Wants a Bite," *New York Times*, June 21, 1993.
145. William Grimes, "More Dinosaurs! Stalk On," *New York Times*, June 25, 1993, C1.
146. *AMNHAR* (1994), 77.

147. See, e.g., Larry Brown, "Real and Imagined—New Exhibits at Seattle Center Give Kids a Mix of Fantasy, Reality," *Seattle Times*, October 22, 1994; Kevin M. Williams, "Larger Than Life—'Jurassic Park' Dinos Exhibited at Museum," *Chicago Sun-Times*, May 31, 1996.

148. Grimes, "More Dinosaurs!" C1.

149. Ibid.

150. Tuma, "Dinosaur Club Stomps Locally."

151. "Fabulous Sum for a Fearsome Fossil," *Science* 278, no. 5336 (1997): 218. Also see Richard Monastersky, "Paleontology," *Science News* 140, no. 19 (1991): 303, and "For the Sake of Sue: What Will Happen to the World's Best *T. rex*?" *Science News* 148, no. 20 (1995): 316–17. On the collaboration between the Field Museum and its financial partners, see Steve Fiffer, *Tyrannosaurus Sue* (New York: W. H. Freeman, 2000), esp. chap. 12.

152. Robert Siegel, "Analysis: Field Museum of Natural History in Chicago Unveils Largest Complete *Tyrannosaurus rex* Skeleton Ever Found," transcript of radio broadcast, *All Things Considered*, May 17, 2000, http://business .highbeam.com/6346/article-1G1-166108492/analysis-field-museum -natural-history-chicago-unveils.

153. Bill Mullen, "Sue Pays Off for Field Museum: 10 Years on, *T. rex* Proving to Be $8.3 Million Bargain," *Chicago Tribune*, May 16, 2010.

154. Fiffer, *Tyrannosaurus Sue*, chap. 12; Mullen, "Sue Pays Off."

155. Harris, "Divided House" (n. 66 above, this chap.), 53.

156. Ibid.

157. Stephen Weil astutely chronicles this phenomenon from the perspective of a practitioner. Stephen E. Weil, *Making Museums Matter* (Washington DC: Smithsonian Institution Press, 2002), 28–52. Also see Kenneth Hudson, "The Museum Refuses to Stand Still," *Museum International*, vol. 197 (1998).

158. Quoted in Gratacap, "Natural History Museums," pt. 2 (n. 25 above, chap. 1), 67.

159. For a set of strategies for "doing" this value, devised by and for museum practitioners, see the recent article by Wendy Crone, Sharon Dunwoody, Raelyn Rediske, Steven Ackerman, Greta Zenner Petersen, and Ronald Yaros, "Informal Science Education: A Practicum for Graduate Students," *Innovative Higher Education* 36, no. 5 (2011): 291–304.

160. Koster, "In Search of Relevance" (n. 120 above, this chap.).

161. Practitioners often refer to such projects as developing "Museum 2.0," an umbrella concept first named in 2006 by art museum director Nina Simon on her influential and like-named blog. Simon has since elaborated on and synthesized some of these museum efforts in her self-published book. See Nina Simon, *The Participatory Museum* (Santa Cruz, CA: Museum 2.0, 2010).

162. Judy Diamond is quoted in Cornelia Dean, "Challenged by Creationists, Museums Answer Back," *New York Times*, September 20, 2005. For a more thorough description of the NSF-sponsored project, see Judy Diamond and E. Margaret Evans, "Museums Teach Evolution," *Evolution* 61, no. 6 (2007):

1500–1506; Judy Diamond and Judy Scotchmoor, "Exhibiting Evolution," *Museums and Social Issues* 1, no. 1 (2006): 21–48.

163. See, e.g., the Field Museum's recent controversial decision to shrink its scientific staff and merge departments, as reported even on local Chicago newspaper front pages, such as Tammy Webber, "No Bones about It: Museum Struggling," *Hammond Times*, South Lake County ed., July 6, 2013, A1, A6. See also the national reporting on the Exploratorium's cutbacks of exhibit design staff in the summer of 2013: Kenneth Chang, "Exploratorium Forced to Cut Back," *New York Times*, August 26, 2013).

164. On how different institutional contexts shape different visions of public and popular science, see Katherine Pandora and Karen Rader, "Science in the Everyday World: Why Perspectives from the History of Science Matter," *Isis* 99, no. 2 (2008): 350–64.

165. According to the AAM, nearly 850 million people visited American museums in 2011, while only 483 million attended theme parks and sporting events. See "Musuem Facts," American Alliance of Museums, http://www .aam-us.org/about-museums/museum-facts.

166. Museums retained "their more serious purpose for specialists," Chesterton conceded, but they had also been "afflicted with the modern idea of collecting all sorts of totally different things, with totally different types of interest, including a good many of no apparent interest at all, and stuffing them all into one building." For this, and the in-text quotes from Chesterton, see his *All Is Grist* (n. 1 above, "Introduction"), chap. 31. On the ways Chesterton's trip to America shaped his views, see Ker, *G. K. Chesterton* (n. 1 above, "Introduction"), chap. 4.

Bibliographic Essay on Sources

To assist scholars, we have provided a list of the archives we consulted and cited specific sources in the book's endnotes. These sources are also listed in a cumulative bibliography. In this bibliographic essay, we offer more detailed historiographical reflections on some broad domains of scholarship from which we drew important insights. Though it is by no means comprehensive, we hope these discussions offer further context for our account and enrich interested readers' understandings of our key methods and themes.

As historians are book driven, books represent the majority of our references. But we also recognize the intersection of history of science, museum studies, and museum history as a growing area, as evidenced by the development of new publication outlets, such as *Museum History Journal* (founded in 2008), and the increasing numbers of articles on museums in established journals like *Isis*, *Endeavour*, and the *Archives of Natural History*.

Museum Institutional Histories

We deliberately employ a wide-angle lens to view museum history in the twentieth century in order to demonstrate how scientific research, pedagogical reforms, and display trends intersected within and between institutions over the course of time. Historians have not always recognized these intersections, partly because museum histories so often focus on single institutions or even single departments. These soundly researched accounts, usually written by former employees or insiders with special access to archives and personnel, identify obscure museum actors and establish chronologies of institutions' development, offering details of museums' leadership, strategic missions, funding, and illustrations of displays in galleries. Often prompted by anniversary commemorations, these microinstitutional histories are usually celebratory, so they only rarely contain significant historiographical arguments. Nonetheless, we have obtained much important material from them.

Histories of this type that proved especially useful to us include Kris Haglund's book on the Denver Museum of Science (*Denver Museum of Natural History: The First Ninety Years* [Denver: Museum, 1990]); Julie McDonald's history of the Putnam Museum (*The Odyssey of a Museum: A Short History of the Putnam Museum of Science, 1867–1992* [Davenport, IA: Putnam Museum, 1992]); George Goodyear's *Society and Museum: A History of the Buffalo Society of Natural Sciences, 1861–1993, and the Buffalo Museum of Science, 1928–1993* (Buffalo: The Society, 1994); Nancy Lurie's book on the Milwaukee Museum (*A Special Style: The Milwaukee Public Museum, 1882–1982* [Milwaukee: Milwaukee Public Museum, 1983]); Bruce Boyer's book on the Field Museum (*The Natural History of the Field Museum: Exploring the Earth and Its People* [Chicago: Field Museum, 1993]), alongside Sally Metzler's on the Field Museum dioramas (*Theatres of Nature* [Chicago: Field Museum of Natural History, 2008]); Ellis Yochelson's *The National Museum of Natural History: Seventy-Five Years in the Natural History Building* (Washington, DC: Smithsonian Institution Press, 1985); Mary Desmond Rock's account of the Boston Museum of Science (*Museum of Science, Boston: The Founding and Formative Years: The Washburn Era, 1939–1980* [Boston: Museum of Science, 1989]); and Jay Pridmore's two books on the Chicago Museum of Science and Industry (*Inventive Genius: The History of the Chicago Museum of Science and Industry* [Chicago: Museum Books, 1996] and its lavishly illustrated coffee table cousin, *Museum of Science and Industry* [New York: Harry Adams, 1997]). Two books have been written about the history and philosophy of the Exploratorium, both by former staff members: Hilde Hein's *The Exploratorium: Museum as Laboratory* (Washington, DC: Smithsonian Press, 1990) and K. C. Cole's *Something Incredibly Wonderful Happens: Frank Oppenheimer and His Astonishing Exploratorium* (New York: Houghton Mifflin Harcourt Trade, 2009). Several important insider accounts of the American Museum of Natural History include exhibit designer Stephen Quinn's beautiful monograph on the museum's dioramas (*Windows on Nature* [New York: AMNH/Harry Abrahams, 2006]), curator emeritus Charles W. Myers's *A History of Herpetology at the American Museum* (Bulletin of the American Museum of Natural History, no. 252 [New York: American Museum of Natural History, 2000]); and writer Douglas Preston's *Dinosaurs in the Attic* (New York: St. Martin's Griffith, 1993).

A handful of practitioners have also authored more sweeping histories of the development of museums or museum types in the United States. Marjorie Schwarzer's *Riches, Rivals and Radicals: 100 Years of Museums in America*, a readable narrative of American museum history, provides a helpful overview for those unfamiliar with the topic (Washington, DC: American Association of Museums, 2006). The development of science centers and science museums in the United States has been thoroughly described by Emlyn Koster ("In Search of Relevance: Science Centers as Innovators in the Evolution of Museums," in "America's Museums," *Daedalus* 128, no. 3 (1999): 277–96), Alan J. Friedman ("The Evolution of Science and Technology Museums," *Informal Science Review* [no. 17, March–April 1996]: 1, 14–17), and former Chicago Museum of Science and Industry director Victor Danilov (*America's Science Museums* [Westport, CT: Greenwood Press, 1990]).

Few academic historians have addressed museums in great depth, and even fewer have confronted the history of U.S. museums of nature and science in the twentieth century. Neil Harris's work on art museums, including his essays in *Cultural Excursions: Marketing Appetites and Cultural Tastes in Modern America* (Chicago: University of Chicago Press, 1990) and his article "The Divided House of the American Art Museum" (in "America's Museums," special issue, *Daedalus* 128, no. 3 (1999): 33–56) has provided us with valuable points of comparison. For a more recent treatment of similar themes, see also Harris's *Capital Culture: J. Carter Brown, the National Gallery of Art, and the Reinvention of the Museum Experience* (Chicago: University of Chicago Press, 2013). Historian Steven Conn first explored the history of U.S. natural history museums in the late nineteenth and early twentieth centuries in *Museums and American Intellectual Life* (Chicago: University of Chicago Press, 1998) and later, investigated their postwar turn toward children's education in the essay, "Where Have All the Grown-Ups Gone?" in his *Do Museums Still Need Objects?* (Philadelphia: University of Pennsylvania Press, 2010). Conn's work, especially, has inspired our own attempts to further develop and grapple with historical phenomena he identified.

History of Biology in the Twentieth Century

Only four decades ago, the now-flourishing history of biology was a small, undervalued subdiscipline in the history of science. Moving beyond the long and important tradition of scientists and natural historians writing retrospective evaluations and reflections from inside their chosen field, historian Garland Allen kick-started a new era of work in this arena with his classic account, *Life Sciences in the Twentieth Century* (New York, Wiley, 1975; Cambridge: Cambridge University Press, 1978). Allen argued that twentieth-century life science researchers broke from the descriptive and speculative tradition that had dominated much of late nineteenth-century biology. The so-called Allen thesis provoked a range of responses from his contemporaries (see Frederick B. Churchill's commentary, "In Search of the New Biology: An Epilogue," *Journal of the History of Biology* 14, no. 1 [1981]: 171–91) and helped set the scholarly agenda for historians of biology over the next several years. For a broad overview of the initial debate, see Jane Maienschein, "History of Biology," in "Historical Writing on American Science," eds.

Sally Gregory Kohlstedt and Margaret Rossiter, themed issue, *Osiris* 1 (1985): 147–62; for more detailed case-study responses by a variety of Science and Technology Studies scholars, there are a number of important edited collections: *The American Development of Biology*, ed. R. Rainger, K. Benson, and J. Maienschein (Philadelphia: University of Pennsylvania Press, 1988; New Brunswick, NJ: Rutgers University Press, 1991) and *The Expansion of American Biology*, ed. J. Maienschein, Keith Benson, and Ronald Rainger (New Brunswick, NJ: Rutgers University Press, 1991), as well as *The Right Tools for the Job: At Work in Twentieth-Century Life Sciences*, ed. Adele Clarke and Joan H. Fujimura (Princeton, NJ: Princeton University Press, 1992).

More recent scholarship on the changing organizational and cultural forces shaping twentieth-century life science work extends and complicates Allen's core claims; we hope our book contributes to this ongoing conversation. Sally Gregory Kohlstedt's early research, documenting the emergence of a disparate institutional landscape for American biological research, education, and popularization in the nineteenth century, laid important scholarly groundwork for this discussion: see, for instance, her *Isis* essays "Curiosities and Cabinets: Natural History Museums and Education on the Antebellum Campus" (vol. 79, no. 3 [1988]: 405–26) and "Parlors, Primers, and Public Schooling: Education for Science in Nineteenth-Century America" (vol. 81 [1990]: 424–45). Phillip Pauly's ambitious *Biologists and the Promise of American Life* (Princeton, NJ: Princeton University Press, 2000) demonstrated that late nineteenth- and early twentieth-century biologists developed a new academic culture in museums, field stations, and universities, even while they also saw greater possibilities for their work to "culture American life," by helping to improve agricultural production, science education, and public health (a theme he extended in *Fruits and Plains: The Horticultural Transformation of America* [Boston: Harvard University Press, 2007]). Lynn Nyhart's book, *Modern Nature: The Rise of the Biological Perspective in Germany* (Chicago: University of Chicago Press, 2009) masterfully demonstrates the European origins of these cultural dynamics, which, in turn, shaped the views of many American natural history museum founders and builders. Nyhart argues that modern German biological thinking "trickled up" in the late nineteenth century, emerging from the work of a collection of naturalists, teachers, and leaders of local natural history and other museums who, for a variety of reasons, sought to display the lives of animals and people through their connections to shared regional communities. Carsten Kretschmann's argument that Mobius's own Berlin dioramas embodied both new ecological thinking and more traditional "readings" of display further enriches understandings of the relationship between German biological science and museum exhibits of this period—see his *Räume öffnen sich: Naturhistorische Museen im Deutschland des 19. Jahrhunderts, Wissenskultur und gesellschaftlicher Wandel* (Berlin: Akademie Verlag, 2006).

Historians of zoology have also established that life scientists' early focus on whole organisms, a focus shared by the natural history museum curators and exhibit designers we describe, did not fade but remained a force in American science throughout the twentieth century, albeit one with many practical variants. Gregg

Mitman (*State of Nature: Ecology, Community, and American Social Thought, 1900–1950* [Chicago: University of Chicago Press, 1992]) showed how "Chicago school" animal ecologists of the 1930s conceptualized both their field sites and American society as "superorganisms" on which the natural forces of evolution acted; studying social behavior in other animals, they argued, provided a window into human society. Richard Burkhart's comprehensive study of the history of ethology (*Patterns of Behavior: Konrad Lorenz, Niko Tinbergen, and the Founding of Ethology* [Chicago: University of Chicago Press, 2005]) illuminates how this field first institutionalized the study of animal behavior, giving it both theoretical concepts and experimental methods. Erika Milam (*Looking for a Few Good Males: Female Choice in Evolutionary Biology* [Baltimore, MD: Johns Hopkins University Press, 2010]) demonstrated that animal behavior research on sexual selection employed biologists of all methodological stripes—from the experimental to the observational—well into the mid-1950s. Even in the late 1960s, as Tania Munz has shown, animal behavior studies continued to propel arguments about the behavior of and relationships among individual animals versus animal populations ("The Bee Battles: Karl von Frisch, Adrian Wenner and the Honey Bee Dance Language Controversy," *Journal of the History of Biology* 38, no. 3 [2005]: 535–70, and " 'My Goose Child Martina': The Multiple Uses of Geese in the Writings of Konrad Lorentz," *Historical Studies in the Natural Sciences* 41, no. 4 [2011]: 405–46). For a similar take that focuses on the social consequences of such research, see Marga Vicedo, "The Father of Ethology and the Mother of Ducks: Konrad Lorentz as an Expert on Motherhood" (*Isis* 100, no. 2 [2009]: 263–91). Historians of biology, scientists, and museums have collaborated to develop a broader understanding of the relationship among taxonomy, systematics, and evolutionary thinking in museums during the late nineteenth and early twentieth centuries. For a useful prehistory, with a wide chronological and geographical span, see Nicholas Jardine, James Secord, and Emma Spary's *Cultures of Natural History* (Cambridge: Cambridge University Press, 1996). Mary P. Winsor's path-breaking work *Starfish, Jellyfish, and the Order of Life: Issues in Nineteenth Century Science* (New Haven, CT: Yale University Press, 1976) first focused attention on classification as an important mode of doing natural history in museums of the nineteenth century. Winsor's more recent account of Harvard's Agassiz Museum of Comparative Zoology showed how later comparative zoology research failed to thrive in university museums because curators failed to acknowledge and incorporate newer evolutionary thinking (see *Reading the Book of Nature: Comparative Zoology at the Agassiz Museum* [Chicago: University of Chicago Press, 1991]). Winsor's approach exemplifies the value of historical knowledge for contemporary science; she herself later took both historians and scientist-practitioners of natural history to task for not employing a broader cross-disciplinary lens ("The Practitioner of Science: Everyone Her Own Historian," *Journal of the History of Biology* 34, no. 2 [2001]: 229–45).

Ronald Rainger's and Charlotte Porter's books tackled different museums but extended Winsor's intellectual and institutional history of the collection-centered fields of museum natural history: see Porter's *The Eagle's Nest: Natural History and*

American Ideas and Rainger's *An Agenda for Antiquity: Henry Fairfield Osborn and Vertebrate Paleontology at the American Museum of Natural History, 1890–1935* (both published in Tuscaloosa by the University of Alabama Press, in 1986 and 1991, respectively). Porter's focus on the management of collections is important, and her lively descriptions of the relationships between naturalists and collectors in early American museums are especially insightful. Kristin Johnson's more recent *Ordering Life: Karl Jordan and the Naturalist Tradition* (Baltimore, MD: Johns Hopkins University Press, 2012) further explores these same themes, through the lens of one curator's tireless efforts to secure the status of taxonomic study and natural history museums against the background of early twentieth-century Britain's changing scientific and social landscape.

The history and social study of museum-based evolutionary research from the mid- to late twentieth century remains a lively and diverse scholarly field. Betty Smocovitis's microhistorical study of botanical systematics ("Disciplining Botany: A Taxonomic Problem," *Taxon* 41, no. 3 [1992]: 459–70), as well as her masterful overview of what scientists labeled "the evolutionary synthesis" (*Unifying Biology: The Evolutionary Synthesis and Evolutionary Biology* [Princeton, NJ: Princeton University Press, 1996]) offered early and definitive challenges to simplistic views of how systematics and taxonomy accommodated both new researchers and new research methods in the twentieth century. Smocovitis shows how the synthesis involved discipline formation, requiring historians to reexamine not just particular ideas but the entire social and intellectual apparatus of modern biology as well. Joe Cain, in turn, builds on this approach with his attention to structural problems of professionalization in natural history—what he, together with Ernst Mayr, called the "borderlands" of genetics, paleontology, and systematics in early twentieth-century biology. For an overview of this research, much of which was accomplished in or through museums, see Cain's "Towards a 'Greater Degree of Integration': The Society for the Study of Speciation, 1939–41" in *The British Journal for the History of Science* (vol. 33, no. 1 [2000]: 85–108), Cain and Mayr's early coauthored article "Exploring the Borderlands: Documents of the Committee on Common Problems in Genetics, Paleontology, and Systematics," in *Transactions of the American Philosophical Society* (vol. 94, no. 2 [2004]: i, iii–v, vii, ix–xlii, 1–17, 19–39, 41–107, 109–23, 125–33, 135–37, 139–51, 153–60), and, finally, Cain and Michael Ruse's more recent edited collection *Descended from Darwin: Insights into the History of Evolutionary Studies* (Philadelphia: American Philosophical Society, 2009).

Other broad overviews of the science of natural history include Paul Farber's book *Finding Order in Nature* (Baltimore, MD: Johns Hopkins University Press, 2000), which synthesized the history of American research and collecting activities with earlier European efforts in order to trace a persistent "naturalist tradition" in the history of biology. Robert Kohler's *All Creatures: Naturalists, Collectors, and Biodiversity, 1850–1950* (Princeton, NJ: Princeton University Press, 2006) chronicles how systematics and biogeography continued to experience dynamic growth through the expansion of survey collecting, a practice that reached its zenith in

American museums during the 1920s. Kohler argues that museums revived collecting to serve scientists' taxonomical agendas and capitalize on the recreational interests of upper-middle-class museum patrons, whose outdoor pursuits resulted in the enthusiastic collection of diorama specimens. Of course, the fact that a large number of the new specimens flooding American museums came from places like Panama and the Philippines also spoke to America's imperialist ambitions in the Pacific; for more on this history, see the relevant sections in chapters 5 and 6 of this book, as well as Mark Barrow's *Nature's Ghosts: Confronting Extinction from the Age of Jefferson to the Age of Ecology* (Chicago: University of Chicago Press, 2009).

Other historians have highlighted the importance of local workplace dynamics and strong personalities in shaping debates over museum-based systematics research. James Griesemer and Susan Leigh Star's classic article on boundary objects describes the work of Berkeley's Joseph Grinnell and details the circulation of evolutionary ideas within and between all the "social worlds" in which early twentieth-century university natural history museums existed ("Institutional Ecology, 'Translations,' and Boundary Objects: Amateurs and Professionals in Berkeley's Museum of Vertebrate Zoology, 1907–39," *Social Studies of Science* 19, no. 3 [1989]: 387–420). Mark Barrow's book, *A Passion for Birds: American Ornithology after Audubon* (Princeton, NJ: Princeton University Press, 2000) includes a useful discussion of Ernst Mayr's museum work, as does Erika Milam's more recent article, "The Equally Wonderful Field: Ernst Mayr and Organismic Biology" (*Historical Studies in the Natural Sciences* 40, no. 3 [2010]: 279–317). On more recent heated debates in the science of systematics, see Joshua Buhs's study of ant systematics ("Building on the Bedrock," *Journal of the History of Biology* 33, no. 1 [2000]: 27–70) and Mary Sunderland's "Collections-Based Research at Berkeley's Museum of Comparative Zoology" (*Historical Studies in the Natural Sciences* 42, no. 2 [2012]: 83–113). With an eye toward understanding the contemporary scientific resurgence of systematics and taxonomical investigations thanks to DNA sampling and "barcoding" of collections, Andrew Hamilton and Quentin Wheeler have made an impassioned plea for the contemporary methodological significance of better understanding the history of these vigorous debates ("Taxonomy and Why History of Science Matters for Science," *Isis* 99, no. 2 [2008]: 331–40). David Takacs makes a similar case, but cautions, in his *The Idea of Biodiversity* (Baltimore, MD: Johns Hopkins University Press, 1999), that reading contemporary views about biodiversity—gleaned from modern systematics and taxonomy—back into the older conservation biology practices risks flattening important differences between past and present science.

Some new monographs on popular American natural history fields have emphasized, as we do, the cross-communication and collaboration between actors and institutions doing natural history–based life science work throughout the last century. Paul Brinkman's *The Second Jurassic Dinosaur Rush: Museums and Paleontology in America at the Turn of the Century* (Chicago: University of Chicago Press, 2010) skillfully demonstrates how the reputations of natural history museums and the individuals that ran them have both shaped and been shaped by public understandings of paleontology. Brinkman's work highlights power struggles among the field collectors, curators, and administrators in American museums of natural history and captures museums' violent competition for specimens. His discussion of these developments in the context of the assembly of dinosaur bones in the 1890s shows how "life on display" was a messy and contentious business even before the era on which we focus. Likewise, Mary Anne Andrei's study of influential early twentieth-century taxidermists reveals the ways in which those who worked on preparing displayed animals took pride in their aesthetic talents but also saw themselves as contributing to early wildlife preservation science ("Nature's Mirror: How the Taxidermists of Ward's Natural Science Establishment Transformed Wildlife Display in American Natural History Museums and Fought to Save Endangered Species" [PhD thesis, University of Minnesota, 2006]).

The professionalization of the biological sciences, once confidently located by historians in the late Victorian period, has been reexamined in different contexts and historical eras, with a focus on what Adrian Desmond has called "professionalization in the making" ("Redefining the X-Axis: 'Professionals,' 'Amateurs,' and the Making of Mid-Victorian Biology—a Progress Report," *Journal of the History of Biology* 34, no.1 [2001]: 3–50). This scholarship has revealed that distinctions between professionals and amateurs in science were actively created and policed well into the twentieth century, in museums and in society more broadly. Still helpful are David Allen's *The Naturalist in Britain: A Social History* (New York: Pelican Books, 1976) and Susan Leigh Star and James R. Griesemer's article "Institutional Ecology, 'Translations,' and Boundary Objects: Amateurs and Professionals in Berkeley's Museum of Vertebrate Zoology, 1907–1939" (*Social Studies of Science* 19 [1989]: 387–420). We also leaned on Mark Barrow's analysis of relationships between amateurs and professionals, and the way these relations framed the development of American ornithology and natural history more broadly (see his earlier-mentioned *A Passion for Birds*). On the historical collaboration between amateurs and a nascent professional class of botanists, Elizabeth Keeney's *The Botanizers: Amateur Scientists in North America* (Chapel Hill: University of North Carolina Press, 1992) is useful; on entomology, see Conner Sorensen's *Brethren of the Net: American Entomology, 1840–1880* (Tuscaloosa: University of Alabama Press, 1995) and William Leach's *Butterfly People: An American Encounter with the Beauty of the World* (New York: Pantheon, 2013). For examinations of how both cooperative and contentious interactions between lay participants and specialists shaped museum work, see Jeremy Vetter's introduction to the special section "Lay Participation in Scientific Observation" in *Science in Context* (vol. 24, no. 2 [June 2011]: 127–41).

The turn to "practice" in science studies (see Andrew Pickering, *Science as Practice and Culture* [Chicago: University of Chicago Press, 1992]) seemed to have inspired another flurry of work on field biology's history. This work has further illuminated the role that natural history museums and their collections played in creating what Robert Kohler called "border sciences" in his *Landscapes and Lab-*

scapes: Exploring the Lab-Field Border in Biology (Chicago: University of Chicago Press, 2002), such as ecology and conservation biology. In his focused historical treatment of field zoology, Juan Ilerbaig examines shifts in institutionalization resulting from methodological reassessment in the twentieth century, especially in freshwater biological stations and regional natural history museums ("Pride in Place: Fieldwork, Geography, and American Field Zoology, 1850–1920" [PhD diss., University of Minnesota, 2002]). The wide variety of essays in the edited collection *Knowing Global Environments: New Perspectives on the Field Sciences* (ed. J. Vetter [New Brunswick, NJ: Rutgers University Press, 2010]) collectively demonstrate the continued vitality of this topical focus. Likewise, Mark Barrow's engaging prehistory of conservation biology, *Nature's Ghosts*, reveals how nineteenth- and twentieth-century ideas about extinction developed, paradoxically, through the simultaneous practices of collecting endangered specimens and trying to develop value-free knowledge about them, practices that became increasingly difficult to combine as the environmental movement took hold in the 1960s and 1970s. Frederick Davis explores this same tension in his more intimate biographical account of Archie Carr and his work in *The Man Who Saved Sea Turtles: Archie Carr and the Origins of Conservation Biology* (Oxford: Oxford University Press, 2007).

Science, Technology, and Nationalism

Though our book rarely engages directly with nationalist ideologies espoused by museum displays, most of our historical actors believed wholeheartedly that the nation's health depended on public knowledge of science and nature. They believed, too, that museum display and visual culture more broadly were especially effective ways of disseminating this knowledge. Consequently, we found the vast literature on the intersections among science, technology, and political agendas in museums and other venues of display useful for understanding that historical context. Historian of American science Robert Bruce argued in *The Launching of Modern American Science, 1846–1876* (New York: Knopf, 1987) that, since the nineteenth century, American science has been held together by a continuous ideology of cultural nationalism that shaped both its public and research institutions. Around the same time, Robert Rydell showed how American world's fair displays were intended to bolster public faith in industrial and corporate agendas for science and technological development: see his *All the World's a Fair: Visions of Empire at the American International Expositions* and his *World's Fairs: The Century of Progress Exhibitions* (both published by University of Chicago Press, in 1987 and 1993, respectively). For an updated overview, see Rydell's most recent *Buffalo Bill in Bologna: The Americanization of the World, 1869–1922*, coauthored with Rob Kroes (Chicago: University of Chicago Press, 2005). Two important microhistorical case studies exploring similar themes are Sophie Forgan's essay on the 1951

Festival of Britain ("Festivals of Science and the Two Cultures: Science, Design, and Display in the Festival of Britain, 1961," *British Journal of the History of Science* 31, no. 2 [1998]: 271–40) and Peter Kuznik's work on the 1939 New York World's Fair ("Losing the World of Tomorrow: The Battle over the Presentation of Science at the 1939 New York World's Fair," *American Quarterly* 46, no. 3 [1994]: 41–73). More recently, scholars of Cold War science have illuminated postwar scientific and technocratic nationalism in every public arena in the United States, from science education to mass media popular culture (see Hunter Heyck and David Kaiser's introduction to the focus section "Cold War Science," *Isis* 101, no. 2 [2010]: 362–66). See, for instance, Ruth Oldenziel and Karin Zachmann's coedited collection *Cold War Kitchen: Americanization, Technology and European Users* (Cambridge, MA: MIT Press, 2009) and Audra Wolfe's work on the history of exobiology ("Germs in Space: Joshua Lederberg, Exobiology, and the Public Imagination, 1958–1964," *Isis* 93, no. 2 [2002]: 183–205).

The outpouring of cultural history in the late 1980s and early 1990s produced several landmark works on nationalism in natural history museum displays in the United States, Great Britain, and Europe. Donna Haraway's interdisciplinary exegesis of Carl Akeley's dioramas in the American Museum's African Hall argues that the museum's displays represent the natural world in ideological rather than scientific terms, espousing the values of an elite, white patriarchy in the early twentieth century ("Teddy Bear Patriarchy: Taxidermy in the Garden of Eden, New York City, 1908–36," which was first published in *Social Text* 11 [1984–85]: 19–64, but became better known when Haraway published it as a chapter in her *Primate Visions* [New York: Routledge, 1990]). In her superb history of habitat dioramas, Karen Wonders has argued that there was a similar dynamic animating the diorama craze in Swedish, American, and German museums (*Habitat Dioramas: Illusions of Wilderness in Museums of Natural History* [Uppsala: Almqvuiest and Wiksell, 1993]). *Objects and Others: Essays on Museums and Material Culture* (ed. George Stocking [Madison: University of Wisconsin Press, 1985]) helped to generate an enormous literature investigating how museum anthropology perpetuated nationalist and imperialist agendas in the United States. Tony Bennett's *Birth of the Museum* and *Pasts beyond Memories: Evolution, Museums, Colonialism* (both published in London by Routledge, in 1995 and 2004, respectively) offer a broader cross-cultural analysis of national and colonial ideologies in museums. Other important contributions to this area of scholarship include Steven Conn's book on the nineteenth century history of representing Native Americans in museums and the media (*History's Shadow: Native Americans and Historical Consciousness* [Chicago: University of Chicago, 2004]); *Peoples on Parade*, by Sadia Qureshi, which explores displays of diverse humans at nineteenth- and twentieth-century exhibitions and fairs (Chicago: University of Chicago Press, 2011); Alison Griffith's *Wondrous Difference: Cinema, Anthropology, and Visual Culture* (New York: Columbia University Press, 2002), which documents the importance of ethnographic films in museums and culture; Timothy Luke's *Museum Politics* (Minneapolis: University of Minnesota Press, 2002), a Foucaultian exploration of cultural politics across a variety of American museums; and finally, many of the essays in Sharon MacDonald and Gordon Fyfe's edited collection *Theorizing Museums: Representing Identity and Diversity in a Changing World* (Oxford: Blackwell Sociological Review, 1996), as well as those in MacDonald's anthology *The Politics of Display: Museums, Science, Culture* (New York: Routledge, 1998). While Felix Driver's *Geography Militant: Cultures of Exploration and Empire* (London: Wiley-Blackwell, 2000) and *Tropical Visions in an Age of Empire* (coedited with Luciana Martins [Chicago: University of Chicago Press, 2005]) touch only sporadically on museums, their discussion of the relationship among imperialism, visual culture, and field science has helped illuminate our own.

Eugenicist ideologies likewise shaped displays of public health, medicine, and biology throughout the early twentieth century. Julie Brown's *Health and Medicine on Display: International Expositions in the United States, 1876–1904* (Cambridge, MA: MIT Press, 2010) provides historical context for this phenomenon in her descriptions of health displays of nineteenth-century international expositions. Steven Selden and Paul Lombardo offer detailed case studies and a general overview of how educational, political, and legal institutions reflected eugenic representations, especially at the American Museum of Natural History, which hosted two eugenics congresses, and at state fairs. See, for instance, Selden's *Inheriting Shame* (New York: Teachers College Press, 1999) and Lombardo's book on Carrie Buck, *Three Generations, No Imbeciles* (Baltimore, MD: Johns Hopkins University Press, 2010), as well as his more recent *A Century of Eugenics in America* (Bloomington: Indiana University Press, 2011). Eva Åhrén (in *Death, Modernity, and the Body: Sweden 1870–1930*, Rochester Studies in Medical History [Rochester, NY: University of Rochester Press, 2010) examines medicalized eugenic ideals of "normal" bodies in various forms of display in early twentieth-century Sweden. For a film studies perspective on Australian eugenic displays mounted with funding from patrons in the United States, see Eric Stein's "Colonial Theatres of Proof: Representation and Laughter in 1930s Rockefeller Foundation Hygiene Cinema in Java" (*Health and History* 8, no. 2 [2006]: 14–44).

Some of the most important recent work on the history of eugenics displays has come from public historians and art historians. Although much of this history is regionally or institutionally narrow, taken as a whole, this body of scholarship addresses the broader relationship between science and popular entertainment, which broadened our understanding of ongoing conversations between museums and consumer culture, as well as museums' presentation of the human form. See, among others, Carma Gorman's "Educating the Eye: Body Mechanics and Streamlining in the United States, 1925–1950" (*American Quarterly* 58, no. 3 [2006]: 839–68); Laura Lovett's " 'Fitter Families for Future Firesides': Florence Sherbon and Popular Eugenics" (*Public Historian* 29, no. 3 [2007]: 69–85); and Christina Cogdell's *Eugenic Design: Streamlining America in the 1930s* (Philadelphia: University of Pennsylvania Press, 2004).

Public Science, Popular Science

The history of popular science has exploded over the last two decades, encouraged by science studies scholars Roger Cooter and Stephen Pumphrey's provocative call to take such work seriously ("Separate Spheres and Public Places: Reflections on the History of Science Popularization and Science in Popular Culture," *History of Science* 32 [1994]: 237–67; for updated overviews, see the essays of the *Isis* focus section on popular science in the June 2009 issue). More recent work emphasizing the complex social, political, and institutional conditions of popular science's production is relevant to both the history of museums and the changing role of science in American culture. Accounts of how popular, and especially visual, media of twentieth-century America has intersected with both academic and lay concepts of science include Marcel LaFollette's books on popular science media, including *Science on the Air: Popularizers and Personalities on Radio and Early Television* (Chicago: University of Chicago Press, 2008) and *Making Science Our Own: Public Images of Science, 1910–1955* (Chicago: University of Chicago Press, 1990); Katherine Pandora's case study of Luther Burbank and "vernacular science" ("Knowledge Held in Common: Tales of Luther Burbank and Science in the American Vernacular," *Isis* 92 [2001]: 484–516); Gregg Mitman's analysis of the use of film and popular culture in the American Museum of Natural History ("Cinematic Nature: Hollywood Technology, Popular Culture, and the American Museum of Natural History," *Isis* 84, no. 4 [1993]: 637–61); and finally, Bruce Lewenstein's lucid edited collection *When Science Meets the Public* (Washington, DC: American Association for Advancement of Science, 1992).

Though we were surprised to find that museum displays did not engage as frequently and as directly with the topic of evolution as we had anticipated, popular ideas about evolution certainly served as an important backdrop for museum work in the twentieth century. In her article, "The 1959 Darwin Centennial Celebration in America" (in "Commemorative Practices in Science: Historical Perspectives on the Politics of Collective Memory," ed. Pnina G. Abir-Am and Clark A. Elliot, themed issue, *Osiris* 14 [1999]: 274–323), Smocovitis describes the displays of evolution and Darwin featured at the Darwin Centennial of 1959 in the United States. Two vivid accounts of popular displays and press coverage of evolution in museums are Constance Clark's *God—or Gorilla: Images of Evolution in the Jazz Age* (Baltimore, MD: Johns Hopkins University Press, 2008) and Julie Homchick's "Displaying Controversy: Evolution, Creation, and Museums" (PhD diss., University of Washington, 2009). Edward Larson's *Summer for the Gods: The Scopes Trial and America's Continuing Debate over Science and Religion* (New York: Basic Books, 2006), George Webb's *The Evolution Controversy in America* (Lexington: University Press of Kentucky, 2002), Ronald Numbers's *The Creationists: From Scientific Creationism to Intelligent Design* (expanded ed. [Cambridge, MA: Harvard University Press, 2006]), and Jeffrey Moran's *American Genesis: The Evolution Controversies from Scopes to Creation Science* (Oxford: Oxford University Press, 2012) provided helpful overviews of the ongoing popular debate over evolution; Larson's short but comprehensive *Evolution: The Remarkable History of a Scientific Theory* (New York: Modern Library Chronicles, 2004) and Michael Ruse's *From Monad to Man: The Concept of Progress in Evolutionary Biology* (Cambridge, MA: Harvard University Press, 1997) combine the scientific history of evolutionary biology with a history of popular ideas about evolution.

Science Education, Museum Education

To help us explain how, over the course of the century, museum staffs alternately worked to remedy, augment, and collaborate with science educators in American schools, we leaned heavily on the few existing histories of twentieth-century primary and secondary school education in the life sciences. Though historians of science have thoroughly explored U.S. university-based study and training in the twentieth century, they have largely ignored other levels of life science education. One exception to this rule was Philip Pauly, who chronicled the emergence of secondary school biology in an early *Isis* article ("The Development of High School Biology: New York City, 1900–1925," *Isis* 82, no. 314 [1991]: 662–88) that was eventually expanded into a chapter in his *Biologists and the Promise of American Life*. Audra Wolfe built on this work from the perspective of scientists' involvement with curriculum, specifically focusing on how Cold War–era research biologists attempted to insert themselves into discussions of national science policy by rewriting high school biology textbooks (see her "Speaking for Nature and Nation: Biologists as Public Intellectuals in Cold War Culture" [PhD diss., University of Pennsylvania, 2002]). Sally Kohlstedt has long urged historians of science to pay closer attention to this topic, and her excellent *Teaching Children Science: Hands-On Nature Study in North America, 1880–1930* (Chicago: University of Chicago Press, 2010), which describes how scientists and educators used the nature study movement to introduce science into early-twentieth century schools in North America, makes a definitive case for even more sustained historical attention.

Science educators and historians of education have confronted the history of life sciences in schools more directly. George DeBoer's 1991 *A History of Ideas in Science Education: Implications for Practice* (New York: Teacher's College Press), Scott Montgomery's 1994 *Minds for the Making: The Role of Science in American Education, 1750–1990* (New York: Guilford Press), and Kim Tolley's 2002 *The Science Education of American Girls: A Historical Perspective* (New York: Routledge) remain excellent overviews of twentieth-century trends and debates in school science teaching. John Rudolph's articles explain why and how educators attempted to promote social stability and public interest in science by reforming the pedagogy and curricula of school science courses (see, respectively, "Epistemology for the Masses:

The Origin of the 'Scientific Method' in American Schools," *History of Education Quarterly* 45 [2005]: 341–76, and "Turning Science to Account: Chicago and the General Science Movement in Secondary Education: 1905–1920," *Isis* 96 [2005]: 353–89). Rudolph has also paid close attention to the Cold War period, and his *Scientists in the Classroom: The Cold War Reconstruction of American Science Education* (New York: Palgrave Macmillan, 2002) argues that the members of the Biological Sciences Curriculum Study used secondary school science reform to elude political interference and guarantee ongoing public funding for biology. More recently, Erika Milam has written about that organization's social science counterpart, *Man, a Course of Study* ("Public Science of the Savage Mind: Contesting Cultural Anthropology in the Cold War Classroom," *Journal for the History of the Behavioral Sciences* 49, no. 3 [2013]: 306–30).

As most scholarship on museum education is geared to practitioners, we relied largely on primary sources to understand the history of education in museums of science and nature. A few texts provided helpful frameworks, however. Though intended for current and future museum professionals, two of Eilean Hooper-Greenhill's books, *Museums and Education: Purpose, Pedagogy, Performance* (New York: Routledge, 2007) and *The Educational Role of the Museum* (2nd ed. [New York: Routledge, 1999]), provide thorough introductions to museum-based teaching, public communications, and audience research, while George Hein's *Learning in the Museum* (New York: Routledge, 1998) places the study of visitor learning in historical context. On visitor-centered models of museum learning, offered from the perspective of long-time visitor study professionals, see also John Falk and Lynn Dierking's *Learning from Museums: Visitor Experiences and the Making of Meaning* (New York: Alta Mira Press, 2000). Also useful was Stephen Weil's essay, "From Being about Something to Being for Somebody: The Ongoing Transformation of the American Museum," in which he argues that American museums have, since World War II, turned their attention from warehousing collections to empowering the broadest of publics (in "America's Museums," *Daedalus* 128, no. 3 [1999]: 229–58). Paisely Cato and Clyde Jones's edited collection, *Natural History Museums: Directions for Growth* (Lubbock: Texas Tech University Press, 1991) covers the same topic, moving between analyses of the past and practices of the present.

Bibliography

"Accent on Participation in Proliferating Centers." *New York Times*, September 9, 1975, 1.

Adam, T. R. *The Museum and Popular Culture*. New York: American Association for Adult Education, 1939.

Adams, Charles C. "The New Natural History—Ecology." *American Museum Journal* 17 (1917): 491–94.

———. "Some of the Advantages of an Ecological Organization of a Natural History Museum." *Proceedings of the American Association of Museums* 1 (1907): 170–78.

Agar, Jon. "What Happened in the Sixties?" *British Journal for the History of Science* 41, no. 45 (2008): 567–600.

Akeley, Carl. "African Hall, a Monument to Primitive Africa." *Mentor* 12, no. 12 (1926): 10–22.

Albers, Harry R. ". . . And Visions of Royalties Danced in Their Heads." *Museum News* 52, no. 1 (1973): 23–25.

"Albert Eide Parr." *Current Biography* 3, no. 7 (1942): 640–42.

Alberti, Samuel J. M. M. "Introduction: The Dead Ark." In *The Afterlives of Animals*, edited by Samuel J. M. M. Alberti. Charlottesville: University of Virginia Press, 2011.

Alexander, Edward P. *Museums in Motion: An Introduction to the History and Functions of Museums*. Nashville: American Association for State and Local History, 1979.

Alexander, Richard D. "Evolution, Creation, and Biology Teaching." *American Biology Teacher* 40, no. 2 (1978): 91–96, 101–4, 107.

Alexander, Victoria D. "From Philanthropy to Funding: The Effects of Corporate and Public Support on American Art Museums." In "Museum Research," special issue, *Poetics*, no. 24 (1996): 87–129.

——. *Museums and Money: The Impact of Funding on Exhibitions, Scholarship, and Management.* Bloomington: Indiana University Press, 1996.

"Alice E. Gray, 79: Museum's 'Bug Lady.'" *Hartford Courant,* May 1, 1994.

Allen, David. "Tastes and Crazes." In *Cultures of Natural History,* edited by N. Jardine, J. A. Secord, and E. C. Spary. Cambridge: Cambridge University Press, 1996.

Allen, Joel A. *Autobiographical Notes and a Bibliography of the Scientific Publications of Joel Asaph Allen.* New York: American Museum of Natural History, 1916.

Allison-Bunnell, Steven W. "Transplanting a Rain Forest: Natural History Research and Public Exhibition at the Smithsonian Institution, 1960–1975." PhD diss., Cornell University, 1995.

Altshuler, Bruce. *Biennials and Beyond: Exhibitions That Made Art History, 1962–2002.* New York: Phaidon Press, 2013.

——. *Salon to Biennial: Exhibitions That Made Art History.* New York: Phaidon, 2008.

"A.M.A. Centenary: Exhibits and Papers," *British Medical Journal* 2, no. 4514 (July 12, 1947): 64–67.

Ambach, Gordon. "Museums as Places of Learning." *Museum News* 65, no. 2 (1986): 35–41.

American Association of Museums. *America's Museums: The Belmont Report,* edited by Michael Robbins. Washington, DC: American Association of Museums, 1969.

——. *Museums for a New Century: A Report of the Commission on Museums for a New Century.* Washington, DC: American Association of Museums, 1984.

——. *A Statistical Survey of Museums in the United States and Canada.* 1965. Reprint, New York: Arno Press, 1976.

American Institute of Biological Sciences. *The A.I.B.S. Story.* Washington, DC: American Institute of Biological Sciences, 1972.

"The American Museum of Natural History." *Fortune,* April 1937.

Anderson, Edgar. *Plants, Man and Life.* Cambridge: Cambridge University Press, 1952.

Anderson, Michael, "Painting Actuality: Diorama Art of James Perry Wilson." Yale Peabody Museum of Natural History. 2012. http://peabody.yale.edu/james-perry-wilson.

Andrei, Mary Anne. "Nature's Mirror: How the Taxidermists of Ward's Natural Science Establishment Reshaped the American Natural History Museum and Founded the Wildlife Conservation Movement." PhD diss., University of Minnesota, 2006.

"Animal Lending Library." *Life* 33, no. 2 (1952): 89–92.

"Answering Back." *Museum News* 2, no. 6 (1924): 2.

Anthony, H. E. "Southern Asia in the American Museum: The New Vernay-Faunthorpe Hall of South Asiatic Mammals." *Natural History* 30, no. 6 (1930): 577–92.

——. "The Vernay-Faunthorpe Hall of South Asiatic Mammals." *Scientific Monthly* 33, no. 3 (1931), 278–82.

Antin, Mary. *The Promised Land*. 1912. Reprint, New York: Penguin Books, 1997.

Appel, Toby A. "Science, Popular Culture, and Profit: Peale's Philadelphia Museum," *Journal of the Society for the Bibliography of Natural History* 9 (1980) 619–34.

———. *Shaping Biology: The National Science Foundation and American Biological Research, 1945–75*. Baltimore, MD: Johns Hopkins University Press, 2000.

Apple, Rima. *Vitamania: Vitamins and American Culture*. New Brunswick, NJ: Rutgers University Press, 1997.

Aprison, Barry. "The Prenatal Exhibit at the Museum of Science and Industry." *Visitor Behavior* 12, nos. 1–2 (1997): 25–26.

Arluke, Arnold, and Clifford R. Sanders. *Regarding Animals*. Philadelphia: Temple University Press, 1996.

Armacost, Richard R. "Progress Report of the Implementation of Recommendation of the Southeastern Conference on Biology Teaching: A Digest of the Presentation of Panel Members, A.A.A.S. Meeting, 27–30 December 1955 in Atlanta, Georgia." *American Biology Teacher* 18, no. 7 (1956): 219–24.

Armitage, Kevin Connor. *The Nature Study Movement: The Forgotten Popularizer of America's Conservation Ethic*. Lawrence: University Press of Kansas, 2009.

"Art in a Natural History Museum: With Special Reference to Mural Decorations in the Indian Halls." *American Museum Journal* 13, no. 3 (1913): 99–102.

"Art: N.Y.A.S.A.R.S.R.W." *Time*, June 15, 1970.

Ascoli, Peter Max. *Julius Rosenwald: The Man Who Built Sears, Roebuck and Advanced the Cause of Black Education in the American South*. Bloomington: Indiana University Press, 2006.

Bailey, Anne Lowrey. "Ending Science Illiteracy." *Museum News* 66, no. 5 (1988): 50–53.

Bailey, Liberty Hyde. *The Holy Earth*. 1915. Reprint, New York: C. Scribner's Sons, 1943.

Bailey, Vernon. "Obituary—Alfred Webster Anthony." *Auk* 58 (1941): 439–43.

Baker, Arthur O., and Lewis H. Mills. *Dynamic Biology*. New York: Rand McNally & Company, 1933.

Baker, Frank C. "Some Instructive Methods of Bird Installation." *Proceedings of the American Association of Museums* 1 (1907): 52–57.

———. "Some Observations on Museum Administration." *Science* 26, no. 672 (1907): 666–69.

Baker, Robert J., and Terry L. Yates. "Net-Wielding Anachronisms?" *Science* 282, no. 5391 (1998): 1048–49.

Barnard, Jeff. " 'Edu-Tainment' Finds Creative Outlet in New Pacific Northwest Museum Natural History." *Los Angeles Times*, February 20, 1994.

Barrow Jr., Mark V. *Nature's Ghosts: Confronting Extinction from the Age of Jefferson to the Age of Ecology*. Chicago: University of Chicago Press, 2009.

———. *A Passion for Birds: American Ornithology after Audubon*. Princeton, NJ: Princeton University Press, 1998.

———. "The Specimen Dealer: Entrepreneurial Natural History in America's Gilded Age." *Journal of the History of Biology* 33, no. 3 (2000): 493–534.

Barry, Andrew. "On Interactivity: Consumers, Citizens and Culture." In *The Politics of Display: Museums, Science, Culture*, edited by Sharon Macdonald. New York: Routledge, 1998.

Barry, Edward. "Field Museum, at 50th Year, Changes Name." *Chicago Daily Tribune*, September 16, 1943, 24.

Barsook, Beverly. "A Code of Ethics for Museum Stores." *Museum News* 60, no. 3 (1982): 50–52.

Barton, D. R. "Attorney for the Insects." *Natural History* 48, (1941): 181–85.

Basalla, George. "Museums and Technological Utopianism." *Curator* 18, no. 2 (1974): 210–11.

Bear, Marjorie Warvelle. *A Mile Square of Chicago*. Oak Brook, IL: TIPRAC, 2007.

Bell, A. W. "Reorganization at the Los Angeles Museum." *Science* 94, no. 2437 (1941): 255–56.

Bender, Thomas. "Reforming the Disciplines," *Daedalus* 131, no. 3 (2002): 59-62.

Bennett, Tony. *The Birth of the Museum: History, Theory, Politics*. London: Routledge, 1995.

———. *Pasts beyond Memory: Evolution, Museums, Colonialism*. London: Routledge, 2004.

Benson, Maxine. *Martha Maxwell, Rocky Mountain Naturalist*. Women in the West. Lincoln: University of Nebraska, 1986.

Berger, John. *About Looking*. New York: Pantheon Books, 1980.

Bestor, Arthur. *Educational Wastelands: The Retreat from Learning in Our Public Schools*. Urbana: University of Illinois Press, 1953.

Biederman, Patricia Ward. "Staffers Irate over Possible Layoffs of 3 Scientists at La Brea Tar Pits." *Los Angeles Times*, December 16, 1992.

Birch, Jeffrey. "A Museum for All Seasons." *Museum News* 60, no. 4 (1982): 25–28.

Bird, Kai, and Martin Sherwin. *American Prometheus: The Triumph and Tragedy of J. Robert Oppenheimer*. New York: Knopf, 2006.

Black, Craig C. "The Case for Research." *Museum News* 58, no. 5 (1980): 51–53.

———. "New Strains on Our Resources." *Museum News* 56, no. 3 (1978): 18–22.

Blackwelder, Richard E. *The Zest for Life; or, Waldo Had a Pretty Good Run: The Life of Waldo Lasalle Schmitt*. Lawrence, KS: Allen Press, 1979.

Blakeslee, Albert Francis. "Individuality and Science." *Science* 95, no. 2453 (1942): 1–10.

Blitzer, Charles. "Preface." In *Museums and Education: Papers from a Conference, Burlington, Vt, 1966*, edited by Eric Larrabee. Washington, DC: Smithsonian Institution Press, 1968.

Blum, Ann Shelby. *Picturing Nature: American Nineteenth-Century Zoological Illustration*. Princeton, NJ: Princeton University Press, 1993.

Bodry-Sanders, Penelope. *Carl Akeley: Africa's Collector, Africa's Savior*. New York: Paragon House, 1991.

Boehm, Mike. "Nature Museums Nearly Relics Themselves." *Los Angeles Times*, June 3, 2007.

Bogart, Michele. "Lowbrow/Highbrow: Charles R. Knight, Art Work and the Spectacle of Prehistoric Life." In *American Victorians and Virgin Nature*, edited by T. J. Jackson Lears, 39–63. Boston: Isabella Stewart Gardner Museum, 2002.

Borun, Minda, Barbara K. Flexer, Alice F. Casey, and Lynn R. Baum. *Planets and Pulleys: Studies of Class Visits to Science Museums*. Washington, DC: Association of Science-Technology Centers, 1983.

"Boston's 'Living Museum.'" *Christian Science Monitor*, December 23, 1957.

Boyer, Bruce Hatton. *The Natural History of the Field Museum: Exploring the Earth and Its People*. Chicago: Field Museum, 1993.

"Brain Demonstration." *Science Newsletter* 35, no. 1 (1939): 10.

Bridges, William. *Gathering of Animals: An Unconventional History of the New York Zoological Society*. New York: Harper and Row, 1974.

Brinkman, Paul D. *The Second Jurassic Dinosaur Rush: Museums and Paleontology in America at the Turn of the Twentieth Century*. Chicago: University of Chicago Press, 2010.

Brouer, Dorothy. "Life Adjustment Education: A Historical Study of a Program of the United States Office of Education, 1945–54." PhD diss., Teachers' College, 1977.

"The Brooklyn Institute Biological Laboratory." *Science* 17, no. 432 (1891): 270–71.

Brower, Andrew V. Z., and Darlene D. Judd. "Net-Wielding Anachronisms?" *Science* 282, no. 5391 (1998): 1048–49.

Brown, Julie K. "Connecting Health and Natural History: A Failed Initiative at the American Museum of Natural History, 1909–1922." *American Journal of Public Health* (November 8, 2013): e1–e12, http://ajph.aphapublications.org/doi/pdf/10.2105/AJPH.2013.301384.

———. *Health and Medicine on Display, 1895–1904*. Cambridge, MA: MIT Press, 2009.

Brown, Larry. "Real and Imagined—New Exhibits at Seattle Center Give Kids a Mix of Fantasy, Reality." *Seattle Times*, October 22, 1994.

Bruner, Jerome. *The Process of Education*. Rev. ed. Cambridge, MA: Harvard University Press, 2009.

———. *The Relevance of Education*. New York: W. W. Norton, 1971.

Brunner, Bernd. *The Ocean at Home: An Illustrated History of the Aquarium*. Chicago: University of Chicago Press, 2011.

"Bumpus Coming Back as Museum's Head." *New York Times*, November 15, 1910, 20.

Bumpus, Hermon Carey, Jr. *Hermon Carey Bumpus, Yankee Naturalist*. Minneapolis: University of Minnesota Press, 1947.

Burden, William Douglas. "A New Vision of Learning." *Museum News* 21, no. 2 (December 15, 1943): 9–12.

Burden, William Douglas, Livingston Farrand, Clarence Hay, Leonard C. Sanford, and Frederick Osborn. *Report of Trustees' Committee on Allocation of Space*. New York: American Museum of Natural History, 1938.

Burkhardt, Richard W., Jr. *Patterns of Behavior: Konrad Lorenz, Niko Tinbergen, and the Founding of Ethology*. Chicago: University of Chicago Press, 2005.

Buxton, William J., and Manon Niquette. "'Sugar-Coating the Educational Pill': Rockefeller Support for the Communicative Turn in Science Museums." In *Patronizing the Public: American Philanthropy's Transformation of Culture, Communication and the Humanities*, edited by William J. Buxton. Lanham, MD: Lexington Books, 2009: 153–94.

Cahn, Phyllis, and Evelyn Shaw. "A Method for Studying Lateral Line Cupular Bending in Juvenile Fishes." *Bulletin of Marine Science* 15, no. 4 (1965): 1060–71.

Cain, Joseph Allen. "Common Problems and Cooperative Solutions: Organizational Activity in Evolutionary Studies, 1936–1947." *Isis* 84 (1993): 1–25.

Cain, Joseph Allen, and Michael Ruse, eds. *Descended from Darwin: Insights into the History of Evolutionary Studies, 1900–1970*. Philadelphia: American Philosophical Society, 2009.

Cain, Victoria. "The Art of Authority: Exhibits, Exhibit Makers and the Contest for Scientific Status at the American Museum of Natural History, 1920–1940." *Science in Context* 24, no. 2 (2011): 215–38.

———. "'Attraction, Attention, Desire': Consumer Culture as Pedagogical Paradigm in Museums in the United States, 1900–1930." *Paedagogica Historica* 48, no. 5 (2012): 745–69.

———. "'The Direct Medium of the Vision': Visual Education, Virtual Witnessing and the Prehistoric Past at the American Museum of Natural History, 1890–1923." *Journal of Visual Culture* 10, no. 3 (2010): 284–303.

———. "From Specimens to Stereopticons: The Persistence of the Davenport Academy of Natural Sciences and the Emergence of Scientific Education." *Annals of Iowa* 68, no. 1 (2009): 1–36.

Cameron, Duncan F. "The Museum: A Temple or a Forum?" *Curator* 14, no. 1 (1971): 11–24.

Cannon, John F. M. "The New Botanical Exhibition Gallery at the British Museum (Natural History)." *Taxon* 11 (1962): 248–52.

———. "The New Botanical Exhibition Gallery at the British Museum (Natural History)." *Curator* 5, no. 26 (1962): 26–35.

———. "Some Problems in Botanical Exhibition Work." *Museums Journal* 62 (1962): 167–73.

"Cardiac Kitchen." *Progress* 11, no. 8 (1951): 3.

Carey, Evan J. "Health and Medical Science Exhibits." *Phi Chi Quarterly* 38, no. 1 (1941): 41–57.

Carleton, Robert H. *The N.S.T.A. Story, 1944–1974: A History of Ideas, Commitments and Actions*. Washington, DC: National Science Teachers Association, 1975.

Carlsen, William R. "Science Museums: Do-It-Yourself Teachers." *New York Times*, April 22, 1979.

Carlson, A. J. "The Science Core in Liberal Education." *Scientific Monthly* 61, no. 5 (1945): 379–81.

Carlson, Charles, Bonnie Jones, and Wayne La Rochelle. "Two Simple Electrophysiological Preparations Using Grasshoppers." *Association for Biology Laboratory Education Proceedings* 2 (1980): 181–95.

Carnes, Alice. "Showplace, Playground or Forum? Choice Point for Science Museums." *Museum News* 64, no. 4 (1986): 7.

Carrington, FitzRoy, Henry L. Ward, and Margaret T. Jackson Rowe. "Instruction in Museum Work." *Museum Work* 1, no. 3 (1918): 87–91.

Carter, Luther J. "Briefing." *Science* 204, no. 4396 (1979): 924–25.

Carter, Sarah Anne. "Object Lessons in American Culture." PhD diss., Harvard University, 2010.

Ceiga, Barbara. "A Brief History of the Education and Exhibits Departments at the Academy of Natural Sciences." In "Finding Aid: Academy of Natural Sciences, Philadelphia Exhibits Department Records, 1852–2001," ANSP.2010.004, prepared by E. Rosenszweig and K. Vidumsky, 5–16. Academy of Natural Sciences, Philadelphia, 2010, http://www.ansp.org/~/media/Files/ans/library-archives/finding-aids/ANSP_2010-004_Exhibits_Department_Records_1852_2001.ashx.

Century of Progress International Exposition. *Official World's Fair Weekly*. Chicago: n.p., 1933.

Chapman, Frank M. *Autobiography of a Bird-Lover*. New York: D. Appleton-Century, 1935.

———. "The Bird-Life and the Scenery of a Continent in One Corridor: The Groups in the American Museum of Natural History—a New Method in Museum Exhibition." *World's Work*, no. 17 (March 1909), 11367–74.

———. "The Case of William J. Long." *Science* 19, no. 4 (1904): 387–89.

———. "The Educational Value of Bird Study." *Educational Review*, no 17 (1899), 242–49.

———. "Joel Asaph Allen, 1838–1921." *Memoirs of the National Academy of Science* 21 (1927): 1–20.

———. "Natural History for the Masses," *World's Work* 5 (November 1902): 2761–70.

———. "What Constitutes a Museum Collection of Birds?" In *Proceedings of the Fourth International Ornithological Congress*, edited by Richard Bowdler Sharpe, 144–56. London: Dulau and Co., 1907.

Chapman, Frank M., George Kruck Cherrie, William B. Richardson, Geoffrey Gill, Geoffroy M. O'Connell, G. H. H. Tate, Robert Cushman Murphy, and H. E. Anthony. *The Distribution of Bird-Life in Ecuador: A Contribution to a Study of the Origin of Andean Bird-Life*. Bulletin of the American Museum of Natural History, vol. 55, edited by Frank M. Chapman. New York: American Museum of Natural History, 1926.

Chapman, Frank M., William B. Richardson, Louis Agassize Fuertes, Leo E. Miller, Arthur A. Allen, George Kruck Cherrie, Paul Griswold Howes, Geoffroy M. O'Connell, Thomas M. Ring, and Howarth S. Boyle. *The Distribution of Bird-Life in Colombia: A Contribution to a Biological Survey of South America*. Bulletin of the American Museum of Natural History, vol. 36. New York: American Museum of Natural History, 1917.

Chapman, William Ryan. "Arranging Ethnology: A.H.F.L. Pitt Rivers and the Typological Tradition." In *Objects and Others: Essays on Museums and Material*

Culture, edited by George W. Stocking Jr. Madison: University of Wisconsin Press, 1985, 15–48.

Chang, Kenneth. "Exploratorium Forced to Cut Back," *New York Times*, August 26, 2013, http://www.nytimes.com/2013/08/27/science/exploratorium -forced-to-cut-back.html?_r=0.

Charles, Eleanor. "Museums Assess Their Future Roles." *New York Times*, September 26, 1982, 1.

Chesterton, G. K. *All Is Grist: A Book of Essays*. Freeport, NY: Books for Libraries Press, 1967.

"Chicagoans Regret Passing of the Fair, but Are Consoled by Fact That Many Exhibits Will Stay in the City." *New York Times*, November 4, 1934.

"Chicago to Exhibit Industry by Acting." *New York Times*, February 21, 1929.

"Chicago to View Rosenwald's "Toys."" *New York Times*, June 18, 1933.

"Cladistics and Evolution on Display." *Nature* 292 (1981): 395–96.

Clark, Constance. *God—or Gorilla: Images of Evolution in the Jazz Age*. Baltimore, MD: Johns Hopkins University Press, 2008.

Clark, James L. *Good Hunting: Fifty Years of Collecting and Preparing Habitat Groups for the American Museum*. Norman: University of Oklahoma Press, 1966.

Cogdell, Christina. *Eugenic Design: Streamlining America in the 1930s*. Philadelphia: University of Pennsylvania Press, 2004.

Colbert, Edwin H. "On Being a Curator." *Curator* 1 (1958): 7–12.

———. *William King Gregory, 1876–1970*. Washington DC: National Academy of Sciences, 1975.

Cole, K. C. *Something Incredibly Wonderful Happens: Frank Oppenheimer and the World He Made Up*. Boston: Houghton Mifflin Harcourt, 2009.

Cole, Peggy. "Piaget in the Gallery." *Museum News* 63, no. 1 (1984): 9–15.

Coleman, Laurence Vail. *Biennial Survey of Education in the United States, 1928– 1930: Chapter XXII, Recent Progress and Condition of Museums*. Vol 1. Bulletin of the United States Department of the Interior, no. 20. Washington DC: Government Printing Office, 1931.

———. *The Museum in America: A Critical Study*, vol. 1. 3 vols. Washington DC: American Association of Museums, 1939.

———. "A Newspaper for Museumists." *Science* 59, no. 118 (1924): 126.

———. "Some Principles of Group Construction." *Museum Work* 3, no. 4 (1921): 121–25.

Coles, Charles, and Charles Russell. *Men of Science*. Film. New York: American Museum of Natural History, 1938.

Commercial Club of Chicago. *A Museum Goes to War*. Chicago: Museum of Science and Industry, 1943.

Conn, Steven. *Museums and American Intellectual Life, 1876–1926*. Chicago: University of Chicago Press, 1998.

"The Conquest of Pain Opened." *Progress* 6, no. 2 (1955): 4–5.

Constantine, Wendy, Daniel Elias, Gwen Frankfeldt, Abby Haskell, Bronwyn Low, and Lesley Schoenfeld. *Summative Evaluation of "Mathematica: A World of Numbers."* Boston: Boston Museum of Science, 2004.

"Context and Commitment." *Museum News* 69, no. 5 (1990): 66–70.

Cook, James W. *The Arts of Deception: Playing with Fraud in the Age of Barnum.* Cambridge: Harvard University Press, 2001.

Coombes, Annie E. "Museums and the Formation of National and Cultural Identities." *Oxford Art Journal* 11, no. 2 (1988): 57–68.

Cooper, G. Arthur. "The Science of Paleontology." *Journal of Paleontology* 32, no. 5 (1958): 1010–18.

Corral, Gustavo. "Human Biology: Visual Metaphors in the New Exhibition Scheme of the Natural History Museum." M.A. thesis, Autonomous University of Barcelona 2013.

Creed, Capt. Percy R. *Milestones: The Boston Society of Natural History, 1830–1930.* Boston: Boston Society of Natural History, 1930.

Cremin, Lawrence A. *American Education: The Metropolitan Experience, 1876–1980.* New York: Harper & Row, 1988.

Crook, A. R. "The Museum and the Conservation Movement." *Proceedings of the American Association of Museums* 9 (1915): 92–95.

———. "A Survey of State Natural History Museums." *Proceedings of the American Association of Museums* 3 (1909): 141–44.

———. "The Training of Museum Curators." *Proceedings of the American Association of Museums* 5 (1911): 59–63.

Cuban, Larry. *How Teachers Taught: Constancy and Change in American Classrooms, 1890–1990.* New York: Teachers College Press, 1993.

Culotta, Elizabeth. "Museums Cut Research in Hard Times." *Science* 256, no. 5061 (1992): 1268–71.

Cummings, Carlos E. *East Is East and West Is West.* Buffalo, NY: Buffalo Museum of Science, 1940.

———. "Flowers Reproduced in Wax: Synthetic Nature Now and Then." *Hobbies* 21, no. 4 (1941): 68–73.

———. "Mushrooms and Toadstools." *Natural History* 33, no. 1 (1933): 41–53.

———. *Your Guide to the Buffalo Museum of Science.* Buffalo, NY: The Museum, 1929.

———. *Your Guide to the Buffalo Museum of Science.* Buffalo, NY: The Museum, ca. 1936.

Curran, C. H. "Report on the Diptera Collected at the Station for the Study of Insects, Harriman Interstate Park, N.Y." *Bulletin of the American Museum of Natural History* 61 (1930): 21–115.

Cusker, Joseph P. "The World of Tomorrow: Science, Culture, and Community at the New York World's Fair." In *Dawn of a New Day: The New York World's Fair, 1939–1940*, edited by Helen A. Harrison and Joseph P. Cusker. New York: Queens Museum/New York University Press, 1980, 3–15.

Dana, John Cotton. "A Small American Museum: Its Efforts toward Public Utility." In *The New Museum: Selected Writings of John Cotton Dana*, edited by William A. Peniston. Newark NJ: Newark Museum Association, 1921.

Danilov, Victor. "Promoting Museums through Advertising." *Museum News* 64, no. 6 (1986): 32–39.

————. "Public Education with a Pushbutton." *Biomedical Communications* 2, no. 1 (1974): 26–37.

————. *Science and Technology Centers*. Cambridge, MA: MIT Press, 1981.

Dann, Kevin. *Across the Great Border Fault: The Naturalist Myth in America*. New Brunswick, NJ: Rutgers University Press, 2000.

Daston, Lorraine, and Peter Galison. *Objectivity*. New York: Zone Books, 2007.

Daston, Lorraine, and Gregg Mitman, eds. *Thinking with Animals: New Perspectives on Anthropomorphism*. New York: Columbia University Press, 2005.

Davenport Academy of Natural Sciences. *Proceedings of the Davenport Academy of Natural Sciences*. Vol. 9. Davenport, IA: Davenport Academy of Natural Sciences, 1900–1903.

————. *Proceedings of the Davenport Academy of Sciences*. Vol. 10. Davenport, IA: Davenport Academy of Sciences, 1907.

Davies, Lawrence E. "Scientists to Aid Fair in Seattle." *New York Times*, July 21, 1958, 21.

Davis, Douglas. "The Idea of a Twenty-First Century Museum." *Art Journal* 35, no. 3 (1976): 253–58.

————. "The Museum Impossible." *Museum News* 61, no. 5 (1983): 33–37.

Davis, Peter. *Museums and the Natural Environment: The Role of Natural History Museums in Biological Conservation*. London: Leicester University Press, 1996.

Davis, Watson. "The Interpretation of Science through Press, Schools and Radio." *American Biology Teacher* 14, no. 3 (1952): 65–68.

————. "The Rise of Science Understanding." *Science* 108, no. 2801 (1948): 239–46.

de Solla Price, D. J. *Little Science, Big Science:* Columbia University Press, 1963.

Dean, Cornelia. "Challenged by Creationists, Museums Answer Back." *New York Times*, September 20, 2005.

Dean, Ruth. "In Transition: The Federal Scene under Carter and Reagan." *Museum News* 60, no. 3 (1982): 53–57.

Dean, Sheila Ann. "What Animal We Came From: William King Gregory's Paleontology and the 1920's Debate on Human Origins." PhD diss., Johns Hopkins University, 1994.

Deason, Hilary J. "Traveling High-School Libraries." *Science* 122, no. 3181 (1955): 1173–76.

Deaver, Neil E., and Robert A. Matthai. "Child-Centered Learning." *Museum News* 54, no. 4 (1976): 15–19.

DeBoer, George E. *A History of Ideas in Science Education: Implications for Practice*. New York: Teachers College Press, 1991.

Delgado, Ray. "Children's Writer, Biologist Evelyn Shaw-Wertheim." *San Francisco Chronicle*, May 23, 2003.

Dennett, Andrea Stulman. *Weird and Wonderful: The Dime Museum in America*. New York: New York University Press, 1997.

Dernie, David. *Exhibition Design*. New York: W. W. Norton, 2006.

Dewey, John. "The Supreme Intellectual Obligation." *Science* 79, no. 2046 (1934): 240–43.

Dewsbury, Donald A. "Frank Ambrose Beach." *Biographical Memoirs, National Academy of Sciences* 73 (1998): 64–85.

Diamond, Judy, and E. Margaret Evans. "Museums Teach Evolution." *Evolution* 61, no. 6 (2007): 1500–1506.

Diamond, Judy, and Judy Scotchmoor. "Exhibiting Evolution." *Museums and Social Issues* 1, no. 1 (2006): 21–48.

Diamond, Judy, Mark St. John, Beth Cleary, and Darlene Librero. "The Exploratorium's Explainer Program: The Long-Term Impacts on Teenagers of Teaching Science to the Public." *Science Education* 71, no. 5 (1987): 643–56.

Diamond, L. N. "Interpreting Science to the Public." *Scientific Monthly* 40, no. 4 (1935): 370–75.

Dickerson, Mary Cynthia. "The New African Hall Planned by Carl E. Akeley: Principles of Construction Which Strike a Revolution in Methods of Exhibition and Presage the Future Greatness of the Educational Museum." *American Museum Journal* 14, no. 5 (1914): 175–87.

———. "A Note on the Giant Salamander Group. Some Problems in Panoramic Group Construction." *American Museum Journal* 12, no. 8 (1912): 310–14.

———. "Some Methods and Results in Herpetology." *American Museum Journal* 11, no. 6 (1911): 202–12.

———. "The 'Toad Group' in the American Museum: A Word as to Its Composite Construction and Interest." *American Museum Journal* 15, no. 4 (1915): 163–66.

Dill, Homer R. "Building an Educational Museum as a Function of the University." *Proceedings of the American Association of Museums* 9 (1915): 78–86.

———. "The Correlation of Art and Science in the Museum." *Proceedings of the American Association of Museums* 10 (1916): 121–28.

"The Dinosaur Society Wants a Bite." *New York Times*, June 21, 1993.

"Doing What Classrooms Can't." *Mosaic* 5 (Spring 1973): 16–21.

Doordan, Dennis P. "Nature on Display." *Design Quarterly*, no. 155 (1992), 3.

Dorsey, George A. "The Aim of a Public Museum." *Proceedings of the American Association of Museums* 1 (1907): 97–100.

———. "The Municipal Support of Museums." *Proceedings of the American Association of Museums* 1 (1907): 131–33.

Douglas, John H. "The Origins of Culture." *Science News* 115, no. 15 (1979): 252–54.

Douglass, Harl R. "Education of All Youth for Life Adjustment." In "Critical Issues and Trends in American Education," special issue, *Annals of the American Academy of Political and Social Science* 265 (1949): 108–14.

"Dr. Bumpus Is out of History Museum—Director's Controversies with President Osborn and Others End with His Resignation—Going to Wisconsin University—Had Worked to Popularize Museum and Clashed with Scientists." *New York Times*, January 21, 1911, 12.

"Editor Kaempffert for Chicago Museum." *New York Times*, April 14, 1928.

Ellis, Erik. "Dixie Lee Ray, Marine Biology, and the Public Understanding of Science in the United States (1930–1970)." PhD diss., Oregon State University, 2005.

England, J. Merton. "The National Science Foundation and Curriculum Reform: A Problem of Stewardship." *Public Historian* 11, no. 2 (1989): 23–36.

Engleman, Laura, ed. *The BSCS Story: A History of the Biological Sciences Curriculum Study*. Colorado Springs, CO: BSCS, 2001.

Engles, Eric. "Biology Education in the Public High Schools of the United States from the Progressive Era to the Second World War: A Discursive History." PhD diss., University of California, Santa Cruz, 1991.

Evans, Bruce H. "By Example and by Exhibition: The Museum as Advocate." *Museum News* 58, no. 4 (1980): 22–26.

Evermann, Barton W. "Modern Natural History Museums and Their Relation to Public Education." *Scientific Monthly* 6 (1918): 5–36.

"Fabulous Sum for a Fearsome Fossil." *Science* 278, no. 5336 (1997): 218.

Farber, Paul Lawrence. *Finding Order in Nature: The Naturalist Tradition from Linnaeus to E. O. Wilson*. Johns Hopkins Introductory Studies in the History of Science, edited by Mott T. Greene and Sharon Kingsland Greene. Baltimore, MD: Johns Hopkins University Press, 2000.

Farrington, Oliver C. "The Rise of Natural History Museums." *Science* 62, no. 1076 (1915): 197–208.

———. "Some Relations of Science and Art in Museums." *Proceedings of the American Association of Museums* 10 (1916): 21–29.

Federer, C. A. "Nation-Wide Junior Science Clubs." *Science* 88, no. 2292 (1938): 526.

Ferguson, Eugene. "Technical Museums and International Exhibitions." *Technology and Culture* 6, no. 2 (1965): 30–46.

Field Museum of Natural History Handbook. 12th ed. Chicago: Field Museum of Natural History, 1943.

Fiffer, Steve. *Tyrannosaurus Sue*. New York: W. H. Freeman, 2000.

Flasher, Bob. "A Recipe for Survival." *Museum News* 61, no. 5 (1983): 5–17.

Flint, Peter B. "Thomas D. Nicholson Dies at 68; Led Museum of Natural History." *New York Times*, July 11, 1991.

Forey, Peter. "Systematics and Paleontology." In *Milestones in Systematics*, edited by Peter Forey and David M. Williams, 149–80. Boca Raton, FL: CRC Press, 2004.

"40,000 Canal Zone Worker Ants to Go on Exhibition Here Today." *New York Times*, July 18, 1952.

Fosdick, Raymond B. *Adventure in Giving: The Story of the General Education Board*. New York: Harper & Row, 1962.

Foss, Phillip. "Biology Clubs: A Questionnaire." *American Biology Teacher* 4, no. 6 (1942): 165–70.

Fowler, Glenn. "Albert E. Parr, Museum Director and Oceanographer, Dies at 90." *New York Times*, July 20, 1991.

Franklin, Dwight. "A Recent Development in Museum Groups." *Proceedings of the American Association of Museums* 10 (1916): 110–12.

Friedman, Alan. "The Evolution of Science and Technology Museums." *Informal Science Review*, no. 17 (1996): 14–17.

————. "The Extraordinary Growth of the Science-Technology Museum." *Curator* 50, no. 1 (2007): 63–75.

Gabrielson, Ira N. "Obituary: Frederick C. Lincoln." *Auk* 79, no. 3 (1962): 495–99.

Gamson, David. "From Progressivism to Federalism: The Pursuit of Equal Educational Opportunity, 1915–1965." In *To Educate a Nation: Federal and National Strategies of School Reform*, edited by Carl F. Kaestle and Alyssa Lodewick. Lawrence: University Press of Kansas, 2007, 177–201.

Gebhard, Bruno. "The Health Museum as Visual Aid." *Bulletin of the Medical Library Association* 35, no. 4 (1947): 329–33.

————. "What Good Are Health Museums?" *American Journal of Public Health* 36, no. 9 (1946): 1012–15.

Gerard, R. W. "Higher General Education in Biology." *AIBS Bulletin* 6, no. 2 (1956): 20–22.

Gertsch, W. J. "Report of the Thirty-Eighth Annual Meeting." *Annals of the Entomological Society of America* 37 (1944): 133–35.

Gilmour, P. "How Can Museums Be More Effective in Society?" *Museums Journal* 79, no. 2 (1979): 120–22.

Gish, Duane T. "A Challenge to Neo-Darwinism." *American Biology Teacher* 32, no. 8 (1970): 495–97.

Glennon, Lorraine. "The Museum and the Corporation: New Realities." *Museum News* 66 (1988): 36–43.

Glueck, Grace. "Art People." *New York Times*, April 20, 1979.

Gold, Allan R. "Ramses 2 Visits a Boston Museum." *New York Times*, May 1, 1988.

Goldstein, Daniel. " 'Yours for Science': The Smithsonian Institution's Correspondents and the Shape of Scientific Community in Nineteenth-Century America." *Isis* 85, no. 4 (1994): 573–99.

Goode, George Brown. "Museum-History and Museums of History." In *The Origins of Natural Science in America: The Essays of George Brown Goode*, edited by Sally Gregory Kohlstedt. Washington, DC: Smithsonian Institution Press, 1991.

————. "The Museums of the Future." In *A Memorial of George Brown Goode: Together with a Selection of His Papers on Museums and on the History of Science in America*, by Samuel Pierpont Langley, George Brown Goode, Randolph Iltyd Geare, Joint Commission of the Scientific Societies of Washington, and U.S. National Museum. Washington, DC: Government Printing Office, 1901.

————. "Plan of Organization and Regulations." *Circular of the U.S. National Museum*, no. 1 (1881): 1–58.

Goodyear, George F. *Society and Museum: A History of the Buffalo Society of Natural History 1861–1993 and the Buffalo Museum of Science 1928–1993. Bulletin of the Buffalo Society of Natural History*, 34. Buffalo, NY: Bulletin of the Buffalo Society of Natural History, 1994.

Gratacap, Louis Pope. "The Development of the American Museum of Natural History." *American Museum Journal* 1, no. 1 (1900): 2–4.

————. "Natural History Museums." Pt. 2. *Science* 7, no. 185 (1898): 61–68.

Greenberg, Daniel. "Research Priorities: New Program at N.S.F. Reflects Shift in Values." *Science* 170, no. 3954 (1970): 144–46.

Gregg, Clifford C. "Field Museum of Natural History." *Science* 95, no. 2477 (1942): 629–31.

Gregory, Richard. "Perception in Museums: The Natural History Museum London." *Perception* 7, no. 1 (1978): 1–2.

Gregory, William King. "Frank Michler Chapman, 1864–1945." *Biographical Memoirs: National Academy of Sciences of the United States of America* 25, no. 5 (1945): 109–44.

———. "Gladwyn Kingsley Noble." *Science* 93, no. 2401 (1941): 10–11.

———. "The Museum of Things versus the Museum of Ideas." *Science* 83, no. 2164 (1936): 585–88.

Griffiths, A. H. "Discussion—Do Museum Lectures Pay?" *Proceedings of the American Association of Museums* 2 (1908): 99–108.

Griffiths, Alison. "Film Education in the Natural History Museum: Cinema Lights Up the Gallery in the 1920s." In *Learning with the Lights Off,* edited by Marsha Orgeron, Devin Orgeron, and Dan Streible (New York: Oxford University Press, 2012), 124–44.

———. *Shivers Down Your Spine: Cinema, Museums, and the Immersive View.* New York: Columbia University Press, 2008.

———. *Wondrous Difference: Cinema, Anthropology and Turn-of-the-Century Visual Culture.* Film and Culture, edited by John Belton. New York: Columbia University Press, 2002.

Grimes, William. "More Dinosaurs! Stalk On." *New York Times,* June 25, 1993, 1.

Grosvenor, Ian, Martin Lawn, and Kate Rousmaniere, eds. *Silences and Images: The Social History of the Classroom.* History of Schools and Schooling, vol. 7. New York: Peter Lang Publishing, 1999.

Grove, Migs, Sybil Walker, and Alexandra Walsh. "The Uses of Adversity." *Museum News* 61, no. 3 (1983): 26–37.

Gruenberg, Benjamin C. *Science and the Public Mind.* New York: McGraw-Hill, 1935.

———. "Teaching Biology after the Wars." *American Biology Teacher* 9, no. 4 (1947): 101–4.

Gurian, Elaine Heumann. "The Molting of the Children's Museum." In *Civilizing the Museum: The Collected Writings of Elaine Heumann Gurian,* edited by Elaine Heumann Gurian, 19–32. New York: Routledge, 2006.

Hagen, Joel B. "Descended from Darwin? George Gaylord Simpson, Morris Goodman, and Primate Systematics." In *Descended from Darwin: Insights into the History of Evolutionary Studies, 1900–1970,* edited by Joseph Allen Cain and Michael Ruse, 93–109. Transactions of the American Philosophical Society Held at Philadelphia for Promoting Useful Knowledge, 99, pt. 1. Philadelphia: American Philosophical Society, 2009.

Hall, David. *Science as a Process.* Chicago: University of Chicago Press, 1988.

Hall, G. Stanley. "The Contents of Children's Minds on Entering School." *Pedagogical Seminary* 1 (1891): 138–73.

Hall, Martin. "The Reappearance of the Authentic." In *Museum Frictions: Public Cultures/Global Transformations*, edited by Ivan Karp, Corinne A. Kratz, Lynn Szwaja, and Tomas Ybarra-Frausto. Durham, NC: Duke University Press, 2006.

"Hall of Medical Science Opens at Smithsonian." *Public Health Reports* 81, no. 11 (1966): 972–76.

Halstead, L. Beverly. "Halstead's Defense against Irrelevancy." *Nature* 292 (1981): 403–4.

———. "Letters." *Nature* 275, no. 5682 (1978): 683.

Hancocks, Anthea. "Museum Exhibition as a Tool for Social Awareness." *Curator* 30, no. 3 (1987): 181–92.

Hancocks, David. *A Different Nature: The Paradoxical World of Zoos and Their Uncertain Future*. Berkeley: University of California Press, 2001.

Haraway, Donna. *Primate Visions: Gender, Race, and Nature in the World of Modern Science*. New York: Routledge, 1989.

———. "Teddy Bear Patriarchy." In *Primate Visions: Gender, Race and Nature in the World of Modern Science*. New York: Routledge, 1989.

Harmon, M., H. Kirschenbaum, and S. Simon. "Teaching Science with a Focus on Values." *Science Teacher* 37, no. 1 (1970): 16–20.

Harper, Hilliard. "Evening Experiments at S.D. Museums." *Los Angeles Times*, August 7, 1985, 1.

Harris, Neil. *Capital Culture: J. Carter Brown, the National Gallery of Art, and the Reinvention of the Museum Experience*. Chicago: University of Chicago Press, 2013.

———. *Cultural Excursions: Marketing Appetites and Cultural Tastes in Modern America*. Chicago: University of Chicago Press, 1990.

———. "The Divided House of the American Art Museum." In "America's Museums," special issue, *Daedalus* 128, no. 3 (1999): 33–56.

———. *Humbug: The Art of P. T. Barnum*. Phoenix ed. Chicago: University of Chicago Press, 1981.

———. "Polling for Opinions." *Museum News* 69, no. 5 (1990): 46–53.

Harrison, Les. *The Temple and the Forum: The American Museum and Cultural Authority in Hawthorne, Melville, Stowe, and Whitman*. Tuscaloosa: University of Alabama Press, 2007.

Hartman, Andrew. *Education and the Cold War: The Battle for the American School*. New York: Palgrave-MacMillan, 2008.

Harwood, Jonathan. *Styles of Scientific Thought: The German Genetics Community, 1900–1933*. Chicago: University of Chicago Press, 1993.

Hayles, N. Katherine. "Simulated Nature and Natural Simulations: Rethinking the Relation between the Beholder and the World." In *Uncommon Ground: Toward Reinventing Nature*, edited by William Cronon. New York: W. W. Norton, 1995, 409–25.

Hazen, Barbara Shook, and Mel Crawford. *A Visit to the Children's Zoo*. New York: Golden Books, 1963.

Hein, George E. *Learning in the Museum*. Museum Meanings. New York: Routledge, 1998.

————. "Museum Education." In *A Companion to Museum Studies*, edited by Sharon Macdonald, 340–52. Malden, MA: Blackwell Publishing, 2010.

Hein, Hilde S. *The Exploratorium: Museum as Laboratory*. Washington, DC: Smithsonian Institution, 1990.

Hellman, Geoffrey. *Bankers, Bones and Beetles: The First Century of the American Museum of Natural History*. Garden City, NY: Natural History Press, 1968.

Henson, Pamela M. "A National Science and a National Museum." In *Museums and Other Institutions of Natural History: Past, Present and Future*, edited by Alan E. Leviton and Michele L. Aldrich, 34–57. San Francisco: California Academy of Sciences, 2004.

————. "The Smithsonian Goes to War: The Increase and Diffusion of Scientific Knowledge in the Pacific." In *Science and the Pacific War: Science and Survival in the Pacific 1939–1945*, edited by Roy M. MacLeod, 27–50. Dordrecht: Kluwer Academic Publishers, 1990.

Herbert, Wray. "Lucy's Family Problems." *Science News* 124, no. 1 (1983): 8–11.

Herzig, R. M. *Suffering for Science: Reason and Sacrifice in Modern America*. New Brunswick, NJ: Rutgers University Press, 2005.

Hilgartner, Stephen. "The Dominant View of Popularization: Conceptual Problems, Political Uses." *Social Studies of Science* 20, no. 3 (1990): 519–39.

Hipschman, Ron. *Exploratorium Cookbook 2: A Construction Manual for Exploratorium Exhibits*. Rev. ed. San Francisco: The Exploratorium, 1983.

Hoagland, K. Elaine. "Socially Responsible: In the 1990s, Natural History Museums Will Focus Their Efforts on Maintaining the Diverse Life of a Healthy Planet." *Museum News* 68, no. 5 (1989): 50–52.

Holland, William J. *Organization and Minutes of the First Meeting of the American Association of Museums*. New York City: American Association of Museums, 1906.

Homchick, Julie. "Displaying Controversy: Evolution, Creation, and Museums." PhD diss., University of Washington, 2009.

————. "Objects and Objectivity: The Evolution Controversy at the American Museum of Natural History, 1915–1928," *Science and Education* 19 (2010): 485–503.

Hornaday, William. "Masterpieces of American Taxidermy." *Scribner's Magazine*, July 1922, 3–17.

Hornaday, William, and William J. Holland. *Taxidermy and Zoological Collecting: A Complete Handbook for the Amateur Taxidermist, Collector, Osteologist, Museum-Builder, Sportsman, and Traveller*. New York: C. Scribner's Sons, 1891.

Hornaday, William Temple. *A Wild Animal Round-Up*. New York: C. Scribner's Sons, 1925.

Horowitz, Helen Lefkowitz. *Culture and the City: Cultural Philanthropy in Chicago*. Chicago: University of Chicago Press, 1976.

"How Strong Is the United States?" *Boston Herald*, December 24, 1957.

Hudson, Kenneth. "The Museum Refuses to Stand Still." *Museum International* 197 (1998): 43.

Hughes, Everett. *The Sociological Eye: Selected Papers*. New Brunswick, NJ: Transaction Books, 1984.

Hunter, George W. "The American Museum's Reptile Groups in Relation to High School Biology." *American Museum Journal* 15, no. 8 (1915): 405–9.

Hunter, J. Stewart. "Public Health and Medicine Exhibit: Brussels World's Fair." *Public Health Reports* 73, no. 12 (1958): 1121–25.

Hurd, Paul. "Science Literacy: Its Meaning for American Schools." *Educational Leadership* 16, no. 1 (1958): 13–16.

Hutchins, Robert Maynard. "Address." In *Three Addresses Delivered at the Meeting Commemorating the Fiftieth Anniversary of the Field Museum of Natural History*. By Stanley Field, Albert Eide Parr, and Robert Maynard Hutchins. Chicago: Field Museum of Natural History, 1943.

Huxley, Julian "Towards the New Systematics." In *The New Systematics*, edited by Julian Huxley, 1–46. London: Oxford University Press, 1940.

Hyson, Jeffrey. "Jungles of Eden: The Design of American Zoos." In *Environmentalism in Landscape Architecture*, edited by Michel Conan, 23–44. Cambridge, MA: Harvard University Press, 2000.

Ilerbaig, Juan Francisco. "Pride in Place: Fieldwork, Geography, and American Field Zoology, 1850–1920." PhD diss., University of Minnesota, 2002.

"In the United Kingdom, Museum Controversy Strikes yet Again." *Museum News* 69, no. 5 (1990): 17–19.

Janet, Sister Mary. "Life Adjustment: Opens New Doors to Youth." *Educational Leadership* 12, no. 3 (1954): 137–41.

Jensen, Nina. "Children, Teenagers and Adults in Museums: A Developmental Perspective." *Museum News* 60, no. 5 (1982): 25–30.

Jersabek, Christian D. "The 'Frank J. Myers Rotifera Collection' at the Academy of Natural Sciences of Philadelphia." *Hydrobiologia* 546 (2005): 137–40.

Jewett, Susan L., and Bruce B. Collette. "Ernest A. Lachner, 1915–1996." *Copeia* 1997, no. 3 (1997): 650–58.

Johanson, D., M. Edey, and M. A. Edey. *Lucy: The Beginnings of Humankind*. New York: Simon & Schuster, 1990.

John, Richard R. "Why Institutions Matter," *Common-place*, vol. 9, no. 1 (2008).

Johnson, Charles P. "1300 Specimens from South America Will Be Mounted Here." *Rocky Mountain News*, July 25, 1926.

Jones, Anna Laura. "Exploding Canons: The Anthropology of Museums." *Annual Review of Anthropology* 22 (1993): 201–20.

Jordanova, Ludmilla. *Sexual Visions: Images of Gender in Science and Medicine between the Eighteenth and Twentieth Centuries*. Madison: University of Wisconsin Press, 1989.

Juan, Jaume Sastre. "Science in Action: The New York Museum of Science and Industry and the Politics of Interactivity." Paper presented at the annual meeting of the History of Science Society, Montreal, Quebec, November 4–7, 2010.

Kaempffert, Waldemar. *From Cave-Man to Engineer: The Museum of Science and Industry Founded by Julius Rosenwald; an Institution to Reveal the Technical Ascent of Man*. Chicago: Museum of Science and Industry, 1933.

———. "The Museum and Social Enlightenment." *Science* 92, no. 2386 (1940): 258–59.

———. "Revealing the Technical Ascent of Man in the Rosenwald Industrial Museum." *Scientific Monthly* 28, no. 6 (1929): 481–98.

———. "Science: The Evolution of the Science Museum." *New York Times*, February 16, 1936.

———. *Science Today and Tomorrow*. New York: Viking Press, 1939.

Kaesler, Roger L. "A Window of Opportunity: Peering into a New Century of Paleontology." *Journal of Paleontology* 67, no. 3 (1993): 329–33.

Kaiser, David. *How the Hippies Saved Physics*. New York: W. W. Norton, 2011.

Kamien, Janet. "An Advocate for Everything: Exploring Exhibit Development Models." *Curator* 44, no. 1 (2002): 114–28.

Kay, Lily E. *The Molecular Vision of Life: Caltech, the Rockefeller Foundation, and the Rise of the New Biology*. New York: Oxford University Press, 1996.

Kelly, James. "'Gallery of Discovery.'" *Museum News* 70, no. 2 (1991): 49–52.

Kennedy, John Michael. "Philanthropy and Science in New York City: The American Museum of Natural History, 1868–1968." PhD diss., Yale University, 1968.

Keppel, Frederick P. "American Philanthropy and the Advancement of Learning." *Bulletin of the American Association of University Professors*, 20, no. 7 (1934): 456–58.

———. "General Policies of the Carnegie Corporation." *Science* 60, no. 1562 (1924): 514–17.

Keyes, Homer E. "Commercial Tendencies and an Esthetic Standard in Education." *Proceedings of the American Association of Museums* 11 (1917): 113–19.

Kimche, Lee. "Science Centers: A Potential for Learning." *Science* 199, no. 4326 (1978): 270–73.

King, Margaret J. "The Theme Park: Aspects of Experience in a Four-Dimensional Landscape." *Material Culture* 34, no. 2 (2002): 1–15.

———. "Theme Park Thesis." *Museum News* 69, no. 5 (1990): 60–62.

"Kingman Museum: A Brief History of the Museum." The New Kingman Museum http://www.kingmanmuseum.org/history.cfm.

Kingsland, Sharon. *The Evolution of American Ecology, 1890–2000*. Baltimore, MD: Johns Hopkins University Press, 2005.

Kirby, David. *Lab Coats in Hollywood: Science, Scientists, and Cinema*. Cambridge, MA: MIT Press, 2011.

Kirkham, Pat. *Charles and Ray Eames: Designers of the Twentieth Century*. Cambridge, MA: MIT Press, 1998.

Klein, Larry. "Team Players: Now That Designers Are Full Members of the Exhibition Team, They Can Concentrate on Conceptual Design—but It Wasn't Always So . . ." *Museum News* 70, no. 2 (1991): 44–45.

Kliebard, Herbert M. *The Struggle for the American Curriculum, 1893–1958.* New York: RoutledgeFalmer, 2004.

Knight, Charles R. "Autobiography of an Artist." American Museum of Natural History Archives, New York, n.d.

Knoll, Arthur. *Museums: A Gunslinger's Dream.* San Francisco: Bay Area Lawyers for the Arts, 1975.

Kohler, Robert E. *All Creatures: Naturalists, Collectors and Biodiversity, 1850–1950.* Princeton, NJ: Princeton University Press, 2006.

———. "Finders/Keepers: Collecting Sciences and Collecting Practice." *History of Science* 45 (2007): 1–27.

———. *Landscapes and Labscapes: Exploring the Lab-Field Border in Biology.* Chicago: University of Chicago Press, 2002.

———. *Lords of the Fly: "Drosophilia" Genetics and the Experimental Life.* Chicago: University of Chicago Press, 1994.

Kohlstedt, Sally Gregory. "Curiosities and Cabinets: Natural History Museums and Education on the Antebellum Campus," *Isis* 79, no. 3 (1988): 405–26

———. "From Learned Society to Public Museum: The Boston Society of Natural History." In *The Organization of Knowledge in Modern America, 1860–1920,* edited by Alexandra Oleson and John Voss. Baltimore, MD: Johns Hopkins University Press, 1979, 386–406.

———. "Innovative Niche Scientists: Women's Role in Reframing North American Museums, 1880–1930." *Centaurus* 55, no. 2 (2013): 153–74.

———. "Institutional History." In "Historical Writing on American Science," edited by Sally Gregory Kohlstedt and Margaret W. Rossiter. Themed issue, *Osiris,* 2nd ser., vol. 1, no. 1 (1985): 17–39.

———. "Otis T. Mason's Tour of Europe: Observation, Exchange, and Standardization in Public Museums, 1889," *Museum History Journal* 2 (2008): 181–208.

———, ed. *The Origins of Natural Science in America: The Essays of George Brown Goode.* Washington, DC: Smithsonian Institution Press, 1991.

———. "Parlors, Primers, and Public Schooling: Education for Science in Nineteenth-Century America." *Isis* 81, no. 3 (1990): 424–45.

———. "Place and Museum Space: The Smithsonian Institution, National Identity, and the American West, 1846–1896." In *Geographies of Nineteenth-Century Science,* edited by David N. Livingstone and Charles W. J. Withers, 399–437. Chicago: University of Chicago Press, 2011.

———. *Teaching Children Science: Hands-on Nature Study in North America, 1890–1930.* Chicago: University of Chicago Press, 2010.

———. "Thoughts in Things: Modernity, History, and North American Museums." *Isis* 96, no. 4 (2005): 586–601.

Kohlstedt, Sally Gregory, Michael Sokal, and Bruce Lewenstein. *The Establishment of Science in America*. New Brunswick, NJ: Rutgers University Press, 1999.

Kolesnik, W. B. *Mental Discipline in Modern Education*. Madison: University of Wisconsin Press, 1958.

Koster, Emlyn. "If We Could Start Again. . . ." *ASTC Dimensions* 7, no. 3 (May/June 2004): 3.

———. "In Search of Relevance: Science Centers as Innovators in the Evolution of Museums." In "America's Museums," special issue, *Daedalus* 128, no. 3 (1999): 277–97.

Kotler, Neil. "Delivering Experience: Marketing the Museum's Full Range of Assets." *Museum News* 78, no. 3 (1999): 30–39, 58–61.

Kranzberg, Melvin. "Review of Victor Danilov's *Science and Technology Centers* (1982)." *4S Review* 2, no. 1 (1984): 18–20.

Kretschmann, Carsten *Räume Öffnen Sich: Naturhistorische Museen Im Deutschland Des 19. Jahrhunderts*. Wissenskultur Und Gesellschaftlicher Wandel. Berlin: Akademie Verlag, 2006.

Kunz, George. "To Increase the Number of Visitors to Our Museums." *Proceedings of the American Association of Museums* 5 (1911): 87–89.

Kushner, Marilyn S. "Exhibiting Art at the American National Exhibition in Moscow, 1959: Domestic Politics and Cultural Diplomacy." *Journal of Cold War Studies* 4, no. 1 (2002): 6–26.

Kuznick, Peter. "Losing the World of Tomorrow: The Battle over the Presentation of Science at the 1939 World's Fair." *American Quarterly* 46, no. 3 (1994): 341–73.

Kyros, Peter N. "Our Own Best Advocates." *Museum News* 59, no. 7 (1981): 16–21.

Ladouceur, Ronald P. "Ella Thea Smith and the Lost History of American Biology Textbooks." *Journal of the History of Biology* 41, no. 3 (2008): 435–71.

LaFollette, Marcel. "Understanding, Ignorance, and Euphemisms: Reflections on the 146th Annual Meeting of the American Association for the Advancement of Science." *Science, Technology, and Human Values* 5, no. 30 (1980): 23–31.

LaFollette, Marcel, Lisa M. Buchholz, and John Zilber. "Science and Technology Museums as Policy Tools—an Overview of the Issues." *Science, Technology, and Human Values* 8, no. 3 (1983): 41–46.

Lagemann, Ellen Condliffe. *An Elusive Science: The Troubling History of Education Research*. Chicago: University of Chicago Press, 2000.

Langley, Samuel Pierpont. "Memoir of George Brown Goode." In *A Memorial of George Brown Goode: Together with a Selection of His Papers on Museums and on the History of Science in America*, by Samuel Pierpont Langley, George Brown Goode, Randolph Iltyd Geare, Joint Commission of the Scientific Societies of Washington, and U.S. National Museum. Washington, DC: Government Printing Office, 1901.

Langlois, Richard N. "*The Institutional Revolution*: A Review Essay," *Review of Austrian Economics* 26 (2013): 383–95.

Laporte, Leo F. *George Gaylord Simpson*. New York: Columbia University Press, 2000.

Larrabee, Eric, ed. *Museums and Education: Papers from a Conference, Burlington, Vt, 1966*. Washington, DC: Smithsonian Institution Press, 1968.

Larson, Edward J. *Summer for the Gods: The Scopes Trial and America's Continuing Debate over Science and Religion*. New York: Basic Books, 1997.

———. *Trial and Error: The American Controversy over Evolution*. New York: Oxford University Press, 2003.

Laughlin, Butler. "Science for All." *School Science and Mathematics* 3 (1948): 169–76.

Lavietes, Stuart. "Richard Pough, 99, Founder of the Nature Conservancy, Dies." *New York Times*, June 27, 2003.

Lawless, Benjamin, and Marilyn Cohen. "The Smithsonian Style." In *The Smithsonian Experience*, edited by Joe Goodwin and Judy Harkinson, 52–59. New York: Norton, 1977.

Leach, William. *Land of Desire: Merchants, Power, and the Rise of a New American Culture*. New York: Vintage Books, 1993.

Learner, Howard. *White Paper on Science Museums*. Washington, DC: Center for Science in the Public Interest, 1979.

LeCroy, Mary. "Ernst Mayr at the American Museum of Natural History." In "Ernst Mayr at 100: Ornithologist and Naturalist," *Ornithological Monographs* 58 (2005): 30–49.

Lee, Sherman. "The Art Museum as a Wilderness Area." *Museum News* 51, no. 2 (1972): 11–12.

Lemon, Harvey B. "Atomic Energy Exhibit Reopens." *Progress* 2, no. 7 (1951): 4–5.

"Leonhard W. Nederkorn Dead; Designed U.S. Fair Displays." *New York Times*, July 12, 1969.

Lewenstein, Bruce. "The Meaning of 'Public Understanding of Science' in the United States after World War II." *Public Understanding of Science* 1 (1992): 45–69.

Lewin, Roger. "Ancestors Worshiped." *Science* 224, no. 4648 (1984): 477–79.

Lewis, B. N. "The Museum as an Educational Facility." *Museums Journal* 80, no. 3 (1980): 151–55.

Lewis, Michael, ed. *American Wilderness: A New History*. New York: Oxford University Press, 2007.

Lightman, Bernard. "'The Voices of Nature': Popularizing Victorian Science." In *Victorian Science in Context*, edited by Bernard Lightman, 187–211. Chicago: University of Chicago Press, 1997.

Lillengraven, Jason, and Tod Stuessy. "Summary of Awards from the Systematic Biology Program of the National Science Foundation, 1973–1977." *Systematic Zoology* 28, no. 2 (1979): 123–31.

Lincoln, Frederick C. "The History and Purposes of Bird Banding." *Auk* 38, no. 2 (1921): 217–28.

Loi, Tran-Ngoc. "Catalogue of the Types of Polychaete Species Erected by J. Percy Moore." *Proceedings of the Academy of Natural Sciences of Philadelphia* 132 (1980): 121–49.

Lucas, F. A. "The Evolution of Museums." *Proceedings of the American Association of Museums* 1 (1907): 82–90.

———. "Evolution of the Educational Spirit in Museums." *American Museum Journal* 11, no. 7 (1911): 227–28.

———. *Fifty Years of Museum Work: Autobiography, Unpublished Papers and Bibliography.* 1911. Reprint, New York: American Museum of Natural History, 1933.

———. "Modern Principles of Museum Administration." *Museum Work* 2, no. 1 (1919): 17–19.

———. "Museums and Libraries." In *Fifty Years of Museum Work: Autobiography, Unpublished Papers, and Bibliography,* 37–39. 1911. Reprint, New York: American Museum of Natural History, 1933.

———. "Museum Labels and Labeling." Pts. 1 and 2. *Printing Art* (August 1909), 345–48; (September 1909), 25–31.

———. "A Note Regarding Human Interest in Museum Exhibits." *American Museum Journal* 11, no. 6 (1911): 187–94.

———. "The Story of Museum Groups: Part 2." *American Museum Journal* 14, no. 2 (1913): 51–65.

———. "The Work of Our Larger Museums as Shown by Their Annual Reports." *Science* 27, no. 679 (1908): 33–36.

Luke, Timothy W. "Museum Pieces: Politics and Knowledge at the American Museum of Natural History." *Australasian Journal of American Studies* 16 (1997): 1–28.

Lurie, Nancy Oestreich. *A Special Style: The Milwaukee Public Museum 1882–1982.* Milwaukee: Milwaukee Public Museum, 1983.

Lutts, Ralph H. *The Nature Fakers: Wildlife, Science and Sentiment.* Charlottesville: University Press of Virginia, 2001.

Lutz, Frank E. *A Lot of Insects: Entomology in a Suburban Garden.* New York: G. P. Putnam's Sons, 1941.

———. "Apparently Non-Selective Characters and Combinations of Characters, Including a Study of Ultraviolet in Relation to the Flower-Visiting Habits of Insects." *Annals of the New York Academy of Science* 29 (1924): 233–83.

———. "Taking Nature Lore to the Public." *Natural History* 26, no. 2 (1926): 111–23.

———. "The Use of Live Material in Museum Work." *Museum News: Magazine Section* 8, no. 6 (1930): 7–9.

Lutz, Frank E., and F. K. Richtmyer. "The Reaction of *Drosophilia* to Ultraviolet." *Science* 55 (1922): 519.

Macdonald, Sharon. *Behind the Scenes at the Science Museum.* (London: Bloomsbury Academic Press, 2002).

————. "Exhibitions of Power and Powers of Exhibition." In *The Politics of Display: Museums, Science, Culture*, edited by Sharon MacDonald, 1–24. New York: Routledge, 1998.

MacLeod, Roy M. "'Strictly for the Birds': Science, the Military and the Smithsonian's Pacific Ocean Biological Survey Program, 1963–1970." *Journal of the History of Biology* 34, no. 2 (2001): 315–52.

Madsen-Brooks, Leslie. "Challenging 'Top-Down' Science? Women's Participation in American Natural History Museum Work, 1870–1920." *Journal of Women's History* 21, no. 2 (2009): 11–38.

————. "To Study, to Control and to Love: Women Scientists in American Natural History Institutions, 1880–1950." PhD diss., University of California, Davis, 2006.

Marchand, Roland. *Creating the Corporate Soul: The Rise of Public Relations and Corporate Imagery in American Big Business.* Berkeley: University of California Press, 1998.

Maree, M. D., and A. J. Williams. "A Thirty-Year Look at the Nuclear Science Programs at the American Museum of Science and Energy." *Journal of Radioanalytical and Nuclear Chemistry* 171, no. 1 (1993): 253–58.

Mark, Cynthia, and Eve Van Rennes. "Bridging the Visitor Exhibit Gap with Computers." *Museum News* 60, no. 1 (1981): 21–30.

Marsh, Paul E., and Ross A. Gortner. *Federal Aid to Science Education: Two Programs.* The Economics and Politics of Public Education. Syracuse, NY: Syracuse University Press, 1963.

Marshall, Eliot. "Hard Times for San Diego Museum." *Science* 251, no. 4992 (1991): 1.

Mathews, Jane de Hart. "Art and Politics in Cold War America." *American Historical Review* 81, no. 3 (1976): 762–87.

Maxwell, William C. "Cooperation in Education." *American Museum Journal* 11, no. 7 (1911): 219.

Mayer, Alfred Goldsborough. "The Status of Public Museums in the United States." *Science* 17, no. 439 (1903): 843–51.

Mayer, William V. "Biology Education in the United States during the Twentieth Century." *Quarterly Review of Biology* 61, no. 4 (1986): 481–507.

Mayr, Ernst. "Evolution v. Special Creation." *American Biology Teacher* 33, no. 1 (1971): 49–51.

————. *Principles of Systematic Zoology.* New York: McGraw-Hill, 1969.

————. "Prologue: Some Thoughts on the History of the Evolutionary Synthesis." In *The Evolutionary Synthesis*, edited by Ernst Mayr and William B. Provine, 1–48. Cambridge, MA: Harvard University Press, 1998.

————. "The Role of Systematics in the Evolutionary Synthesis." In *The Evolutionary Synthesis*, edited by Ernst Mayr and William B. Provine, 123–36. Cambridge, MA: Harvard University Press, 1980.

Mayr, Ernst, and William B. Provine, eds. *The Evolutionary Synthesis: Perspectives on the Unification of Biology.* Cambridge, MA: Harvard University Press, 1980.

McClellan, Andrew. "P. T. Barnum, Jumbo the Elephant, and the Barnum Museum of Natural History at Tufts University." *Journal of the History of Collections* 24, no. 1 (2012): 45–57.

McEwan, Barbara A. "Let's Put Life into Life Science." *American Biology Teacher* 12, no. 7 (1950): 156–58.

McGrath, Earl J., ed. *Science in General Education*. Dubuque, IA: William C. Brown, 1948.

McLuhan, Marshall, Harley Parker, and Jacques Barzun. *Exploration of the Ways, Means, and Values of Museum Communication with the Viewing Public*. New York: Museum of the City of New York, 1969.

Mead, Margaret. "Museums in a Media Saturated World." *Museum News* 49, no. 1 (1970): 23–25.

Meek, S. E. "A Contribution to Museum Technique." *American Naturalist* 36, no. 421 (1902): 53–71.

Melton, Arthur W. *Problems of Installation in Museums of Art*. Studies in Museum Education, edited by Edward S. Robinson. Washington, DC: American Association of Museums, 1935.

Melton, Arthur W., Nita Goldberg Feldman, and Charles W. Mason. *Experimental Studies of the Education of Children in a Museum of Science*. Studies in Museum Education, edited by Edward S. Robinson. Washington, DC: American Association of Museums, 1936.

Metzler, Sally. *Theatres of Nature: Dioramas at the Field Museum*. Chicago: Field Museum, 2008.

Meyer, A. B. *Studies of the Museums and Kindred Institutions of New York City, Albany, Buffalo, and Chicago, with Notes on Some European Institutions*. Washington, DC: Government Printing Office, 1905.

Meyerhoff, Howard A. "The Plight of Science Education." *Bulletin of the Atomic Scientists* 12, no. 9 (1956): 333–37.

Mickulas, Peter. *Britton's Botanical Empire: The New York Botanical Garden and American Botany, 1888–1929*. Memoirs of the New York Botanical Garden. New York: New York Botanical Garden Press, 2007.

Mihesuah, Devon Abbott, ed. *Repatriation Reader: Who Owns American Indian Remains?* Lincoln: University of Nebraska Press, 2000.

Milam, Erika L. *Looking for a Few Good Males: Female Choice in Evolutionary Biology*. Baltimore, MD: Johns Hopkins University Press, 2010.

———. "Public Science of the Savage Mind: Contesting Cultural Anthropology in the Cold War Classroom." *Journal of the History of the Behavioral Sciences* 49, no. 3 (2013): 306–60.

Miller, Angela. "The Fate of Wilderness in American Landscape Art: The Dilemmas of 'Nature's Nation.'" In *American Wilderness: A New History*, edited by Michael Lewis, 91–112. New York City: Oxford University Press, 2007.

Miller, Warren H. "Aquarelles of Our Common Woodlands." *American Museum Journal* 15, no. 4 (1915): 167–75.

"Million-Dollar Exhibit Building Program." *Progress* 3, no. 4 (1952): 2–3.

Millspaugh, Charles. "Botanical Installation." *Proceedings of the American Association of Museums* 5 (1911): 52–56.

Miner, Roy. "Educational Training of Museum Instructors: How Far Is Pedagogical Training or Teaching Experience Necessary?" *Museum Work: Including the Proceedings of the American Association of Museums* 1, no. 4 (1919): 114–17.

Mintz, Ann. "That's Edutainment!" *Museum News* 73, no. 6 (1994): 33–35.

Mitchell, Martha. *Encyclopedia Brunoniana*. Providence, RI: Brown University Press, 1933.

Mitman, Gregg. "Cinematic Nature: Hollywood Technology, Popular Culture, and the American Museum of Natural History." *Isis* 84, no. 4 (1993): 637–61.

———. *Reel Nature: America's Romance with Wildlife on Film*. Cambridge, MA: Harvard University Press, 1999.

———. *The State of Nature: Ecology, Community, and American Social Thought, 1900–1950*. Chicago: University of Chicago Press, 1992.

Mitman, Gregg, and Richard W. Burkhardt Jr. "Struggling for Identity: The Study of Animal Behavior in America, 1930–1945." In *The Expansion of American Biology*, edited by Keith R. Benson, 164–94, New Brunswick, NJ: Rutgers University Press, 1991.

Monastersky, Richard. "For the Sake of Sue: What Will Happen to the World's Best *T. rex*?" *Science News* 148, no. 20 (1995): 316–17.

———. "Paleontology." *Science News* 140, no. 19 (1991): 303.

Montgomery, Scott L. *Minds for the Making: The Role of Science in American Education, 1750–1990*. New York: Guilford Press, 1994.

Montgomery, Thomas. "Expansion on the Usefulness of Natural History Museums." *Popular Science Monthly* 79, no. July (1911): 36–44.

Moore, R. *Evolution in the Courtroom: A Reference Guide*. Santa Barbara, CA: ABC-CLIO, 2002.

Morse, Edward S. "If Public Libraries, Why Not Public Museums?" *Atlantic Monthly* (July 1893), 112–19.

Moynihan, Daniel P. "Charles Blitzer." *Biographical Memoirs, Proceedings of the American Philosophical Society* 142, no. 2 (2001): 190–91.

Mullen, Bill. "Sue Pays Off for Field Museum: 10 Years on, *T. rex* Proving to Be $8.3 Million Bargain." *Chicago Tribune*, May 16, 2010.

Mumford, Lewis. "The Marriage of Museums." *Scientific Monthly* 7, no. 3 (1918): 8.

Munley, Mary Ellen. *Catalysts for Change: The Kellogg Projects in Museum Education*. Washington, DC: Smithsonian Institution, 1986.

———. "Educational Excellence for American Museums." *Museum News* 65, no. 2 (1986): 51–57.

Murphy, Robert Cushman. "Natural History Exhibits and Modern Education." *Scientific Monthly* 45 (1937): 76–81.

———. "Whitney Wing: The New Home of the American Museum's Department of Birds." *Natural History* 44, no. 2 (1939): 98–106.

"Museum Crickets Broadcast Chirps." *New York Times*, October 3, 1936.

"The Museum of Science and Industry." *Progress* 1, no. 1 (1950): 3.

"Museum Outfits Five Expeditions." *New York Times*, May 28, 1937.

Museums, Imagination and Education. 2 ed. Museums and Monuments, vol. 15. Paris: UNESCO, 1973.

"Museums to Teach By." *Mosaic* 10, no. Summer (1979): 4, 17–25.

Museums USA: Art, History, Science, and Others. Washington, DC: National Endowment for the Arts, 1974.

"Museum Visitors Can See as Fish Do." *New York Times*, June 9, 1937.

Myers, Charles W. "A History of Herpetology at the American Museum of Natural History." *Bulletin of the American Museum of Natural History* 252 (2000): 232.

Nadis, Fred. *Wonder Shows: Performing Science, Magic, and Religion in America*. New Brunswick, NJ: Rutgers University Press, 2006.

National Commission on Excellence in Education. *A Nation at Risk: The Imperative for Educational Reform: A Report to the Nation and the Secretary of Education, United States Department of Education*. Washington, DC: National Commission on Excellence in Education, 1983.

National Endowment for the Arts. *Museums USA*. Washington, DC: Government Printing Office, 1974.

"National Museum Act Could Increase Status." *Science Newsletter* 87, no. 14 (1965): 219.

"Natural History in New York." *Science* 263, no. 5154 (1994): 1.

"Natural History Museum Attracts Record Crowds." *New York Times*, August 3, 1924, X11.

Naumer, Helmuth J. "A Marketable Product." *Museum News* 50, no. 2 (1971): 14–16.

Nedzel, Lucy Nielsen. "The Motivation and Education of the General Public through Museum Experiences." PhD diss., University of Chicago, 1953.

Nelkin, Dorothy. *The Creation Controversy: Science or Scripture in the Schools* Boston: Beacon, 1984.

———. "From Dayton to Little Rock: Creationism Evolves." *Science, Technology, and Human Values* 7, no. 40 (1982): 47–53.

Nelson, Bryce. "The Smithsonian: More Museums in Slums, More Slums in Museums?" *Science* 154, no. 3753 (1966): 1152–54.

"A New Field for Museum Work." *American Museum Journal* 10, no. 7 (1910).

Newhouse, Victoria. *Art and the Power of Placement*. New York: Monacelli Press, 2005.

"News and Notes." *Science* 104, no. 2709 (1946): 506–9.

"The New York Museum of Science and Industry." *Science* 82, no. 2130 (1935): 384.

New York Museum of Science and Industry. *Exhibition Techniques: A Summary of Exhibition Practice*. New York: New York Museum of Science and Industry, 1940.

Nicholson, Thomas D. "Why Museum Accreditation Doesn't Work." *Museum News* 60, no. 1 (1981): 5–10.

Noble, Gladwyn Kingsley. "Her Studies of Reptiles and Amphibians." *American Museum Journal* 23, no. 5 (1923): 514–16.

———. "Probing Life's Mysteries: Some Aspects of the Research Work of the American Museum." *Natural History* 30, no. 5 (1930): 469–82.

Noble, Ruth Crosby, and Gladwyn Kingsley Noble. *The Nature of the Beast: A Popular History of Animal Psychology from the Point of View of a Naturalist.* New York: Doubleday, 1945.

Noll, Victor H. "Science Education in American Schools." *The Forty-Sixth Yearbook of the National Society for the Study of Education* 31, no. 5 (1947): 295–99.

Novacek, Michael J. "Foreword." In *Dragon Hunter: Roy Chapman Andrews and the Central Asiatic Expeditions*, by Charles Gallenkamp, ix–xii. New York: Viking, 2001.

"NSTA Position Statement on School Science Education for the '70s." *Science Teacher* 38, no. 1 (1971): 46–51.

Nyhart, Lynn K. "Civic and Economic Zoology in Nineteenth-Century Germany: The 'Living Communities' of Karl Mobius." *Isis* 89, no. 4 (1998): 605–30.

———. *Modern Nature: The Rise of the Biological Perspective in Germany*. Chicago: University of Chicago, 2009.

———. "Natural History and the 'New' Biology." In *Cultures of Natural History*, edited by N. Jardine, J. A. Secord and E. C. Spary, 426–46. New York: Cambridge University Press, 1996.

———. "Science, Art and Authenticity in Natural History Displays." In *Models: The Third Dimension of Science*, edited by Nick Hopwood and Soraya De Chadarevian, 307–35. Stanford, CA: Stanford University Press, 2004.

———. "Teaching Community via Biology in Late-Nineteenth-Century Germany." In "Science and Civil Society," eds. Lynn K. Nyhart and Thomas Broman. Themed issue, *Osiris* 17 (2002): 141–70.

O'Doherty, Brian, ed. *Museums in Crisis*. New York: George Braziller, 1972.

Oehser, P. H. *The Smithsonian Institution*. 2nd ed. Boulder, CO: Westview Press, 1983.

Official Guide Book of the World's Fair of 1934. Chicago: Century of Progress International Exposition, 1934.

Ogawa, Rodney, Molly Loomis, and Rhiannon Crain. "Institutional History of an Interactive Science Center: The Founding and Development of the Exploratorium." *Science Education* 93, no. 2 (2009): 269–92.

Oliver, James A. "Museum Education and Human Ecology." *Museum News* 48 (1970): 28–30.

Olson, D. R. *Jerome Bruner: The Cognitive Revolution in Educational Theory*. London: Continuum, 2007.

Olson, Storrs L. "History and Ornithological Journals of the Tanager Expedition of 1923 to the Northwestern Hawaiian Islands, Johnston and Wake Islands." *Atoll Research Bulletin* 433 (1996): 1–210.

Onion, Rebecca. "How Science Became Child's Play: Science and the Culture of American Childhood." PhD diss., University of Texas, 2012.

———. "Picturing Nature and Childhood at the American Museum of Natural History and the Brooklyn Children's Museum, 1899–1930." *Journal of the History of Childhood and Youth* 4, no. 3 (2011): 434–69.

"Oomph for Science." *Time*, September 2, 1940.

Oppenheimer, Frank. "A.A.M. Award for Distinguished Service to Museums Acceptance Speech." *Museum News* 61, no. 2 (1982): 39–45.

———. "Everyone Is You . . . or Me." *Technology Review* 78, no. 7 (1976): 6–10.

———. "The Importance of the Role of Science Pedagogy in the Developing Nations." *Scientific World* 10, no. 4 (1966): 21–23, 28.

———. "The Palace of Arts and Science: An Exploratorium at San Francisco, CA, USA." *Leonardo* 5, no. 4 (1972): 343–46.

———. "Rationale for a Science Museum." *Curator* 11 (1968): 206–9.

Oppenheimer, Frank, and Malcolm Correll. "A Library of Experiments." *American Journal of Physics* 32, no. 3 (1964): 220–25.

Oreskes, Naomi. "Objectivity or Heroism? On the Invisibility of Women in Science." In "Science in the Field," edited by Henrika Kuklick and Robert Kohler. Themed issue, *Osiris*, 2nd ser., 11 (1996): 87–113.

Orosz, Joel J. *Curators and Culture: The Museum Movement in America, 1740–1870.* Tuscaloosa: University of Alabama Press, 1990.

Orr, William. "The Opportunity of the Smaller Museums of Natural History." *Popular Science Monthly* 63 (1903): 40–52.

Osborn, H. F. *Creative Education in School, College, University, and Museum: Personal Observation and Experience of the Half-Century, 1877–1927.* New York: Charles Scribner's Sons, 1927.

———. "The Museum of the Future." *American Museum Journal* 11, no. 7 (1911): 223–25.

———. "Preservation of the World's Animal Life." *American Museum Journal* 12, no. 4 (1912): 123–24.

Pandora, Katherine, and Karen Rader. "Science in the Everyday World: Why Perspectives from the History of Science Matter." *Isis* 99, no. 2 (2008): 350–64.

Papageorge, Maria. "Opportunities for International Exhibitions." *Museum News* 60, no. 2 (1981): 9–15.

Parker, Arthur C. "Habitat Groups in Wax and Plaster." *Museum Work* 1, no. 3 (1918): 78–85.

Parr, Albert Eide. "Address." In *Three Addresses Delivered at the Meeting Commemorating the Fiftieth Anniversary of the Field Museum of Natural History.* By Stanley Field, Albert Eide Parr, and Robert Maynard Hutchins. Chicago: Field Museum of Natural History, 1943.

———. *Mostly about Museums: From the Papers of A. E. Parr.* New York: American Museum of Natural History, 1959.

———. "Museums: Enriching the Urban Milieu." *Museum News* 56, no. 4 (1978): 46–51.

———. "Refuge from the Other Direction." *Discovery* 9, no. 1 (1973): 34–36.

Parsons, Keith M. *Drawing out Leviathan: Dinosaurs and the Science Wars*. Life of the Past, edited by James O. Farlow. Bloomington: Indiana University Press, 2001.

Passeneau, Harold J. "G. D. Searle and Abbott Laboratories." *Analysts Journal* 10, no. 3 (1954): 158–59.

Paulu, Tom. "Whale Exhibit Surfaces at Center." *Daily News*, June 11, 1990.

Pauly, Philip J. *Biologists and the Promise of American Life: From Meriwether Lewis to Alfred Kinsey*. Princeton, NJ: Princeton University Press, 2000.

———. "The Development of High School Biology, New York City, 1900-1925." *Isis* 60, no. 4 (1991): 662–88.

Petitjean, Patrick. "A Failed Partnership: WFSW and UNESCO in the Late 1940s." In *Sixty Years of Science at Unesco: 1945–2005*, edited by P. Petitjean, V. Zharov, G. Glaser, J. Richardson, B. de Padirac, and G. Archibald, 78, 80. Paris: Recherches Epistémologiques et Historiques sur les Sciences Exactes et les Institutions Scientifiques, 2006.

Phillips, Allen R. "In Memoriam: Alfred M. Bailey." *Auk* 98, no. 1 (1981): 173–75.

"Point of View: Boston Museum of Science." *Science* 170, no. 3954 (1970): 145.

Porter, Charlotte. "The Rise of Parnassus: Henry Fairfield Osborn and the Hall of the Age of Man." *Museum Studies Journal* 1 (1983): 26–34.

Porter, Mildred C. B. *Behavior of the Average Visitor in the Peabody Museum of Natural History, Yale University*. Washington DC: American Association of Museums, 1938.

Post, Robert. *Who Owns America's Past? The Smithsonian and the Problem of History*. Baltimore: Johns Hopkins University Press, 2013.

Postman, Neil. "Museum as Dialogue." *Museum News* 69, no. 5 (1989): 55–58.

Preston, Douglas J. *Dinosaurs in the Attic: An Excursion into the American Museum of Natural History*. New York: St. Martin's Press, 1986.

Pridmore, Jay. *Inventive Genius: The History of the Museum of Science and Industry, Chicago*. Chicago: Museum of Science and Industry, 1996.

———. *Museum of Science and Industry, Chicago*. Chicago: Harry N. Abrams/Museum of Science and Industry, 1997.

"Private Life of Microbes Is Shown by Projector." *Popular Mechanics* (September 1937), 327.

"The Profession Honors S. Dillon Ripley." *Museum News* 64, no. 1 (1985): 4.

Quinn, Stephen Christopher. *Windows on Nature: The Great Habitat Dioramas of the American Museum of Natural History*. New York: American Museum of Natural History/Henry Adams, 2006.

"Quotations, William Jennings Bryan on Evolution." *Science* 55, no. 1418 (1922): 242–43.

Rabb, George B. "Education in Zoos." *American Biology Teacher* 30, no. 4 (1968): 291–96.

Rainger, Ronald. *An Agenda for Antiquity: Henry Fairfield Osborn and Vertebrate Paleontology at the American Museum of Natural History, 1890–1935*. Tuscaloosa: University of Alabama Press, 1991.

————. "Vertebrate Paleontology as Biology: Henry Fairfield Osborn and the American Museum of Natural History." In *The American Development of Biology*, edited by Ronald Rainger, Keith R. Benson, and Jane Maienschein, 219–56. Philadelphia: University of Pennsylvania Press, 1988.

Raloff, Janet. "The Science of Museums." *Science News* 154, no. 12 (1998): 184–86.

Ramsey, Norman. "Jerrold R. Zacharias, 1905–1986." *Biographical Memoirs, National Academy of Sciences of the United States of America* 68 (1995): 435–50.

Rasmussen, Nicholas. "The Moral Economy of the Drug Company–Medical Scientist Collaboration in Interwar America." *Social Studies in Science* 34, no. 2 (2004): 161–85.

Ravitch, Diane. *The Great School Wars: A History of the New York City Public Schools*. Baltimore, MD: Johns Hopkins University Press, 1974.

————. *The Troubled Crusade: American Education, 1945–1980*. New York: Basic Books, 1983.

Rawlins, Kipi. "Educational Metamorphosis of the American Museum." *Studies in Art Education* 20, no. 1 (1978): 4–17.

Rea, Paul M. "The Development of Museum Instruction: The Outlook." *Museum Work: Including the Proceedings of the American Association of Museums* 1, no. 4 (1919): 109–13.

————. "One Hundred and Fifty Years of Museum History." *Science* 57, no. 1485 (1923): 677–81.

————. "Report of the Commission for Museum Cooperation." *Proceedings of the American Association of Museums* 9 (1915): 5–10.

Rea, Tom. *Bone Wars: The Excavation and Celebrity of Andrew Carnegie's Dinosaur*. Pittsburgh: University of Pittsburgh Press, 2001.

Reif, Rita. "Both Young and Old Learning by Doing." *New York Times*, September 9, 1975, 1.

Rhees, David. "A New Voice for Science: Science Service under Edwin E. Slosson, 1921–1929." M.A. thesis, University of North Carolina, 1979.

Rice, Danielle. "Examining Exhibits." *Museum News* 68, no. 6 (1989): 47–50.

Rieppel, Lukas. "Bringing Dinosaurs Back to Life: Exhibiting Prehistory at the American Museum of Natural History." *Isis* 103, no. 3, (2012): 460–90.

Rittenhouse, Phillip. "Biology and the Smithsonian Institution." *BioScience* 17, no. 1 (1967): 25–35.

Ritter, William E. "Biology's Contribution to a System of Morals That Would Be Adequate for Modern Civilization." *Bulletin of the Scripps Institution for Biological Research of the University of California* 2 (1917): 1–8.

————. "Feeling in the Interpretation of Nature." *Popular Science Monthly* 79 (1911): 126–36.

————. "An Untilled Field for a Revised Kind of Research in Zoology." *Condor* 31, no. 4 (1929): 160–66.

Rivera, Carla. "Bones of Contention" *Los Angeles Times*, August 16, 1994.

Robertson, Bryan. "The Museum and the Democratic Fallacy." In *Museums in Crisis*, edited by Brian O'Doherty, 75–90. New York: George Braziller/Art in America, 1972.

Robertson, Nan. "At Natural History, Adults Find More to Do." *New York Times*, June 8, 1983, 1.

Robinson, Edward S. *The Behavior of the Museum Visitor*. Washington, DC: American Association of Museums, 1928.

———. "Psychological Problems of the Science Museum." *Museum News: Magazine Section* 8, no. 5 (1930): 9–11.

Rock, Mary Desmond. *Museum of Science, Boston: The Founding and Formative Years: The Washburn Era, 1939–1980*. Boston: The Museum, 1989.

Rosen, Sidney. "The Origins of High School General Biology." *School Science and Mathematics* 59, no. 6 (1959): 473–89.

"Rosenwald Museum Changes Its Name." *New York Times*, July 7, 1939.

Rossi, Michael. "Modeling the Unknown: How to Make the Perfect Whale." *Endeavour* 32, no. 2 (2008): 58–63.

Rothfels, Nigel, ed. *Representing Animals*. Theories of Contemporary Culture, edited by Daniel J. Sherman. Bloomington: Indiana University Press, 2002.

Rowan, Andrew N. *Of Mice, Models, and Men: A Critical Evaluation of Animal Research*. Albany: State University of New York, 1984.

Rowley, John. *Taxidermy and Museum Exhibition*. New York: D. Appleton, 1925.

Rudolph, John. "Epistemology for the Masses: The Origins of 'the Scientific Method' in American Schools." *History of Education Quarterly* 45, no. 3 (2005): 341–76.

———. *Scientists in the Classroom: The Cold War Reconstruction of American Science Education*. New York: Palgrave Macmillan, 2002.

———. "Teaching Materials and the Fate of Dynamic Biology in American Classrooms after Sputnik." *Technology and Culture* 53 (2012): 1–36.

———. "Turning Science to Account: Chicago and the General Science Movement in Secondary Education, 1905–1920." *Isis* 96, no. 3 (2005): 353–89.

———. "Ward's Dynavue and the Hybridization of Classroom Apparatus in the Post-Sputnik Era." Paper presented at the Annual Meeting of the National Academy of Education, New York City, October 21, 2005.

Rydell, Robert W. *All the World's a Fair: Visions of Empire at American International Expositions, 1876–1916*. Chicago: University of Chicago Press, 1984.

———. *World of Fairs: The Century-of-Progress Expositions*. Chicago: University of Chicago Press, 1993.

SRG. "Preface." In "Scientific Literacy," special issue, *Daedalus* 112, no. 2 (1983): v–vii.

Sandell, Richard. "Museums as Agents of Social Inclusion." *Museum Management and Curatorship* 17, no. 4 (1998): 401–18.

Sandell, Richard, and Robert Janes, eds. *Museum Management and Marketing*. Leicester Readers in Museum Studies. New York: Routledge, 2007.

San Diego Natural History Museum. *Annual Report*. San Diego, CA: San Diego Natural History Museum, 1995.

Saxon, Wolfgang. "Charles Blitzer." *New York Times*, February 26, 1999.

Schiebinger, Londa. *Nature's Body: Gender in the Making of Modern Science*. Boston: Beacon Press, 1993.

Schmidt, Kurt. "The Nature of the Natural History Museum." *Curator* 1, no. 1 (1958): 20–28.

Schmitt, Peter J. *Back to Nature: The Arcadian Myth in Urban America*. New York: Oxford University Press, 1969.

Schneider, Gail. *Children's Zoos*. Washington, DC: National Recreation and Parks Association, 1970.

Schram, Frederick R. "Museum Collections: Why Are They There?" *Science* 256, no. 5063 (1992): 1502.

Schwab, Joseph J. "Science and Civil Discourse: The Uses of Diversity." *Journal of General Education* 9 (1956): 133–48.

Schwartz, Vanessa R. *Spectacular Realities: Early Mass Culture in Fin-de-Siècle Paris*. Berkeley: University of California Press, 1998.

Schwarz, Herbert. "Frank E. Lutz." *Entomological News* 55, no. 2 (1944): 29–32.

Schwarzer, Marjorie. *Riches, Rivals and Radicals: 100 Years of Museums in America*. Washington, DC: American Association of Museums, 2010.

Scigliano, Eric. "Popular Science: What Next for the Science Center after the 'China' Bonanza?" *Weekly*, October 10–16, 1984, 36–40.

The Second Annual Report of the National Science Foundation. Washington, DC: Government Printing Office, 1952.

Secord, James A. "Monsters at the Crystal Palace." In *Models: The Third Dimension of Science*, edited by Nick Hopwood and Soraya De Chadarevian, 138–69. Stanford, CA: Stanford University Press, 2004.

———. *Victorian Sensation: The Extraordinary Publication, Reception, and Secret Authorship of Vestiges of the Natural History of Creation*. Chicago: University of Chicago Press, 2000.

Sellers, Charles Coleman. "Peale's Museum and 'the New Museum Idea.'" *Proceedings of the American Philosophical Society* 124, no. 1 (1980): 25–34.

"Sets New Pace in Bird Groups in Full Flight: American Museum Achieves Unique Sky Mounting." *Museum News* 3, no. 11 (1925): 1, 3.

The Seventh Annual Report of the National Science Foundation. Washington, DC: Government Printing Office, 1957.

Shannon, Joseph. "The Icing Is Good but the Cake Is Rotten." *Museum News* 52, no. 5 (1974): 29–34.

Shapiro, Adam R. "Civic Biology and the Origin of the School Antievolution Movement." *Journal of the History of Biology* 41 (2008): 409–33.

Shaw, Robert P. *Exhibition Techniques: A Summary of Exhibition Practice, Based on Surveys Conducted at the New York and San Francisco World's Fairs of 1939*. New York: New York Museum of Science and Industry, 1940.

———. "New Developments in Museum Technique and Procedures." *Scientific Monthly* 48, no. 5 (1939): 443–49.

———. "The Progressive Exhibit Method: A New Technic in the Field of Science Presentation." *American Physics Teacher* 7, no. 3 (1939): 165–72.

———. "Scientific Exhibits and Their Planning." *Scientific Monthly* 35, no. 4 (1932): 370–73.

Sheets-Pyenson, Susan. *Cathedrals of Science: The Development of Colonial Natural History Museums during the Late Nineteenth Century.* Kingston, Ontario: McGill-Queen's University Press, 1988.

Shell, Hanna Rose. "Skin Deep: Taxidermy, Embodiment and Extinction in W. T. Hornaday's Buffalo Group." In *Museums and Other Institutions of Natural History: Past, Present, and Future,* edited by Alan E. Leviton and Michele L. Aldrich, 88–113. San Francisco: California Academy of Sciences, 2004.

Sheppard, Keith, and Dennis M. Robbins. "High School Biology Today: What the Committee of Ten Actually Said." *CBE–Life Sciences Education* 6, no. 3 (2007): 198–202.

Sherburne, E. G., Jr. "Public Understanding of Science." *Science* 149, no. 3682 (1965): 381.

"Sherburne to Head New AAAS Program to Improve Public Understanding." *Science* 132, no. 3442 (1960): 1823–24.

Shettel, Harris H. "Exhibits: Art Form or Educational Medium?" *Museum News* 52, no. 1 (1973): 32–41.

———. "Time—Is It Really of the Essence?" *Curator* 40, no. 4 (1997): 246–49.

Shettel, Harris H., and Pamela C. Reilly. "Evaluation of Existing Criteria for Judging the Quality of Science Education." *Education Technology Research and Development* 14, no. 4 (1966): 479–88.

Shteir, Ann B., and Bernard Lightman, eds. *Figuring It Out: Science, Gender, and Visual Culture.* Hanover, NH: University Press of New England, 2006.

Shufeldt, R. W. "Combining Art and Museum Exhibits." *Bulletin of the Pan-American Union* 48 (1919): 682–93.

Shufeldt, Robert Wilson. *Scientific Taxidermy for Museums.* Washington, DC: Government Printing Office, 1893.

Siegel, Robert. "Analysis: Field Museum of Natural History in Chicago Unveils Largest Complete *Tyrannosaurus rex* Skeleton Ever Found." Transcript of radio broadcast, *All Things Considered,* May 17, 2000. http://business.high beam.com/6346/article-1G1-166108492/analysis-field-museum-natural -history-chicago-unveils.

Silver, Stuart. "Almost Everyone Loves a Winner." *Museum News* 61, no. 2 (1982): 24–35.

Simms, Stephen C. *The Field Museum and the Child: An Outline of the Work Carried on by the Field Museum of Natural History among School Children of Chicago through the N. W. Harris Public School Extension and the James Nelson and Anna Louise Raymond Public School and Children's Lectures.* Chicago: Field Museum of Natural History, 1928.

Simon, Nina. *The Participatory Museum.* Santa Cruz, CA: Museum 2.0, 2010.

Simons, Howard. "Brussels Fair and Science." *Science News-Letter* 73, no. 2 (1958): 26–27.

Simpson, George Gaylord. "Evolution and Education." *Science* 187, no. 4175 (1975): 389–90.

Singer, Peter. *Ethics into Action: Henry Spira and the Animal Rights Movement.* Lanham, MD: Rowman & Littlefield, 1998.

Sinnott, Edmund W. "Ten Million Scientists." *Science* 111, no. 2876 (1950): 123–29.

Skocpol, Theda. "Bringing the State Back In: Strategies of Analysis in Current Research." In *Bringing the State Back In,* edited by Theda Skocpol, Peter B. Evans, and Dietrich Rueschemeyer, 3–37. New York: Cambridge University Press, 1985.

Sleigh, Charlotte. *Six Legs Better: A Cultural History of Myrmecology.* Baltimore, MD: Johns Hopkins University Press, 2007.

Slocum, Anna D. "Possible Connections between the Museum and the School." *Proceedings of the American Association of Museums* 9 (1911): 55–67.

Slosson, Edwin E. "Science from the Side-Lines." *Century Magazine* 81 (January 1922): 471–76.

Slye, Maud. "Mary Cynthia Dickerson 1866–1923, Her Life and Personality." *Natural History* 23, no. 5 (1923): 507, 509.

Smith, Ella Thea. *Exploring Biology.* 3rd ed. New York: Harcourt Brace and World, 1949.

Smith, Pamela H. *The Body of the Artisan: Art and Experience in the Scientific Revolution.* Chicago: University of Chicago Press, 2004.

Smithson, Robert. "Can Man Survive? (1969)." In *Robert Smithson, the Collected Writings,* edited by Jack D. Flam, 367–68. Berkeley: University of California Press, 1996.

"Smithsonian Sea Hall to Open." *Baltimore Sun,* February 17, 1963.

Smocovitis, V. Betty. "Organizing Evolution: Founding the Society for the Study of Evolution (1939–1950)." *Journal of the History of Biology* 27, no. 2 (1994): 241–309.

———. *Unifying Biology: The Evolutionary Synthesis and Evolutionary Biology.* Princeton, NJ: Princeton University Press, 1996.

Snyder, Joel. "Territorial Photography." In *Landscape and Power,* edited by W. J. T. Mitchell. Chicago: University of Chicago Press, 2002.

Sobe, Noah. "Challenging the Gaze: The Subject of Attention and a 1915 Montessori Demonstration Classroom." *Educational Theory* 54, no. 3 (2004): 281–97.

Stansfield, G. "Environmental Education." *Museums Journal* 72 (1972): 99–100.

Star, Susan Leigh. "Craft vs. Commodity, Mess vs. Transcendence: How the Right Tool Became the Wrong One in the Case of Taxidermy and Natural History." In *The Right Tools for the Job: At Work in Twentieth-Century Life Sciences,* edited by Adele E. Clarke and Joan H. Fujimura, 257–86. Princeton, NJ: Princeton University Press, 1992.

Steere, William C. "Preface." In *Museums and the Environment: A Handbook for Education,* edited by American Association of Museums Environmental Committee, 5–10. New York: Arkville Press, 1971.

Stine, Jeffrey. "Placing Environmental History on Display." *Environmental History* 7, no. 4 (2002): 566–88.

Stock, Robert W. "At 100, the Museum of Natural History Is No Fossil." *New York Times Magazine*, June 29, 1969.

Straight, M. W. *Nancy Hanks, an Intimate Portrait: The Creation of a National Commitment to the Arts.* Durham, NC: Duke University Press, 1988.

Sullivan, Patricia. "Frank Taylor; Founding Director of American History Museum." *Washington Post*, June 30, 2007.

Susman, Warren. "The People's Fair: Cultural Contradictions of a Consumer Society." In *Dawn of a New Day: The New York World's Fair, 1939–1940*, edited by Helen A. Harrison and Joseph P. Cusker, 211–29. New York: Queens Museum/New York University Press, 1980.

Swift, Edgar James. "The Genesis of the Attention in the Educative Process." *Science* 34, no. 868 (1911): 1 ff.

Syer, Henry W. "Review of Will French's Behavioral Goals of General Education in High School." *Science* 127, no. 3307 (1958): 1171–72.

"Tactile Education." *Museum News* 4, no. 1 (1926): 2.

Tamaki, Julie. "Really Big Show Features 5 Life-Size Robot Whales at Natural History Museum." *Los Angeles Times* January 21, 1992.

Tattersall, Ian, John A. Van Couvering, and Eric Delson. "Ancestors: An Expurgated History." In *Ancestors: The Hard Evidence*, 1–5. New York: Alan R. Liss, Inc., 1985.

Taylor, Lloyd W. "Science in the Postwar Liberal Arts Program." *Scientific Monthly* 62, no. 2 (1946): 113–16.

Taylor, Walter P. "Is Biology Obsolete?" *AIBS Bulletin* 8, no. 3 (1958): 12–14.

The Tenth Annual Report of the National Science Foundation. Washington, DC: Government Printing Office, 1960.

"10,000 Attend the Tuberculosis Show." *New York Times*, December 1, 1908.

Terzian, Sevan G. *Science Education and Citizenship: Fairs, Clubs, and Talent Searches for American Youth, 1918–1958.* New York: Palgrave Macmillan, 2012.

Thomas, David Hurst. *Skull Wars: Kennewick Man, Archaeology, and the Battle for Native American Identity.* New York: Basic Books, 2000.

Thorsen, Liv Emma, Karen A. Rader, and Adam Dodd, eds. *Animals on Display: The Creaturely in Museums, Zoos, and Natural History.* University Park: Penn State University Press, 2013.

Tippo, Oswald. "Biology and the Educational Ferment." *AIBS Bulletin* 9, no. 2 (1959): 16–17.

Tobach, Ethel, Lester Aronson, and Evelyn Shaw, eds. *The Biopsychology of Development.* New York: Academic Press, 1971.

Tolley, Kimberley. *The Science Education of American Girls: A Historical Perspective.* New York: RoutledgePalmer, 2003.

Toon, Richard. "Science Centers: A Museum Studies Approach to Their Development and Possible Future Directions." In *Museum Revolutions: How Museums Change and Are Changed*, edited by Simon Knell, Suzanne MacLeod, and Sheila Watson, 105–16. New York: Routledge, 2007.

Toscano, James V. "Development Comes of Age." *Museum News* 60, no. 3 (1982): 61–67.

Tressel, George W. "Thirty Years of 'Improvement' in Precollege Math and Science." *Journal of Science Education and Technology* 3, no. 2 (1994): 77–88.

Tufty, Barbara. "A New Life for Museums." *Science News* 88, no. 8 (1965): 122.

Tuma, Debbie. "Dinosaur Club Stomps Locally." *New York Daily News*, February 4, 1996.

Turner, David. "Reform of the Physics Curriculum in Britain and the United States." *Comparative Education Review* 28, no. 3 (1984): 444–53.

Tyack, David B. *The One Best System: A History of American Urban Education.* Cambridge, MA: Harvard University Press, 1974.

Unger, Rudolph. "Chicago Museums Doing Well." *Chicago Tribune*, November 17, 1982.

U.S. Congress Senate Committee on Rules Smithsonian Institution, U.S. Congress House Committee on Interior Administration, and Affairs Insular. *Smithsonian Institution Appropriations Hearings.* Washington, DC: Government Printing Office, 1994.

U.S. Court of Appeals, District of Columbia Circuit, ed., *Dale Crowley, Jr., Individually & in His Capacity as Executive Director of the National Foundation for Fairness in Education, Et Al., Appellants, V. Smithsonian Institution Et Al.* St. Paul, MN: West Publishing Co., 1980.

U.S. Department of Education and National Science Foundation. *Science and Engineering Education for the 1980s and Beyond.* Washington, DC: National Science Foundation, 1980.

U.S. Office of Education. *Life Adjustment for Every Youth.* Washington, DC: Government Printing Office, 1948.

"U.S. Science in Session." *Science News* 91, no. 1 (1967): 4.

"The Value of Sentiment." *Atlantic Monthly* (August 1913), 279–81.

Van Name, Willard G. "Zoological Aims and Opportunities." *Science* 50, no. 1282 (1919): 81–84.

Vaughan, Malcolm L. M. "Painting Beauty under the Sea." *Los Angeles Times*, July 8, 1928.

Vetter, B. M., et al. *Educating Scientists and Engineers: Grade School to Grad School.* Washington, DC: U.S. Congress, Office of Technology Assessment, 1988.

Vetter, Jeremy. "Cowboys, Scientists, and Fossils: The Field Site and Local Collaboration in the American West." *Isis* 99, no. 2 (2008): 273–303.

Viseltear, A. J. "The Emergence of Pioneering Public Health Education Programs in the United States." *Yale Journal of Biology and Medicine* 61, no. 6 (1988): 519–48.

"Visitors Can Turn Levers and Find Length of Lives." *Science Newsletter* 32, no. 866 (1937): 308–9.

"Visitors from around the Globe: Easter Crowds Watch Chickens Hatch on Harvester Farm." *Progress* 4, no. 3 (1953): 8.

Vonier, Thomas. "Museum Architecture: The Tension between Function and Form." *Museum News* 66, no. 5 (1988): 24–30.

Wade, Nicholas. "NIH Cat Sex Study Brings Grief to New York Museum." *Science* 194, no. 4261 (1976): 162–67.

Waggoner, Walter. "James A. Oliver, 67: Zoologist, Teacher, Administrator." *New York Times*, December 4, 1981.

Walberg, Herbert J. "Scientific Literacy and Economic Productivity in International Perspective." In "Scientific Literacy," special issue, *Daedalus* 112, no. 2 (1983): 1–28.

Wallen, I. E. "Scientists Serve High Schools." *Science* 124, no. 3213 (1956): 168.

Wang, Jessica. *American Scientists in the Age of Anxiety: Scientists, Anti-Communism, and the Cold War*. Chapel Hill: University of North Carolina Press, 1999.

Wang, Zuoyue. *In Sputnik's Shadow: The President's Science Advisory Committee and the Cold War*. New Brunswick, NJ: Rutgers University Press, 2008.

"Want to Borrow a Porcupine?" *Popular Mechanics* 99, no. 4 (1953): 88–90.

Ward, Henry L. "The Aims of Museums, with Special Reference to the Public Museum of the City of Milwaukee." *Proceedings of the American Association of Museums* 1 (1907): 100–103.

———. "The Anthropological Exhibits in the American Museum of Natural History." *Science* 25, no. 645 (1907): 745–46.

———. "The Desirability of Irrelevant Accessories in Groups." *Museum News: Magazine Section* 9, no. 16 (1932): 1.

———. "The Exhibition of Large Groups." *Proceedings of the American Association of Museums* 1 (1907): 68–71.

———. "The Labeling in Museums." *Proceedings of the American Association of Museums* 1 (1907): 42–44.

Washburn, Bradford, and Lew Freedman. *Bradford Washburn: An Extraordinary Life: The Autobiography of a Mountaineering Icon*. Portland, OR: WestWinds Press, 2005.

Waterman, Alan. "The Role of the Federal Government in Science Education." *Scientific Monthly* 82, no. 6 (1956): 286–93.

Watkins, Charles Alan. "Fighting for Culture's Turf." *Museum News* 70, no. 2 (1991): 61–65.

Weaver, Warren. "Science and the Citizen." *Science* 126, no. 3285 (1957): 1225–29.

Webber, Tammy. "No Bones about It: Museum Struggling," *Hammond Times*, South Lake County ed., July 6, 2013, A1, A6.

Weil, Stephen E. "From Being about Something to Being for Somebody: The Ongoing Transformation of the American Museum." In "America's Museums," special issue, *Daedalus* 128, no. 3 (1999): 229–58.

———. *Making Museums Matter*. Washington DC: Smithsonian Institution Press, 2002.

Weinberg, Stanley L. "Two Views on the Textbook Watchers." *American Biology Teacher* 40, no. 9 (1978): 541–60.

Weinman, Martha. "The Marvels in Our Museums." *Collier's*, February 4, 1955.

Weiss, Robert S., and Serge Bouterline Jr. "The Communication Value of Exhibits." *Museum News* 43, no. 3 (1963): 23–27.

Wertz, Laurie. "Some Things Never Change." *Museum News* 62, no. 3 (1984): 26–29.

West, Susan. "Evolution of an Exhibit." *Science News* 116, no. 1 (1979): 10–11.

Whaley, W. Gordon. "Edmund Ware Sinnott." *Biographical Memoirs, National Academy of Sciences of the United States of America* 54 (1983): 350–73.

Wheeler, William Morton. "Carl Akeley's Early Work and Environment." *Natural History* 27 (1927): 133–41.

Whitney, E. R. "Nature Study as an Aid to Advanced Work in Science" *Proceedings of the National Educational Association* 43 (1904): 889–96.

Wilford, John Noble. "The Leakeys: A Towering Reputation." *New York Times Magazine*, October 30, 1984.

Williams, Kevin M. "Larger Than Life—'Jurassic Park' Dinos Exhibited at Museum." *Chicago Sun-Times*, May 31, 1996.

Wilson, Edmund B. "Aims and Methods of Study in Natural History." *Science* 13, no. 314 (1901): 14–23.

Wilson, R. M. *Seeking Refuge: Birds and Landscapes of the Pacific Flyway:* University of Washington Press, 2010.

Winchell, N. H. "Museums and Their Purposes." *Science* 18, no. 442 (1891): 43–46.

Wines, James. "Exhibition Design and the Psychology of Situation." *Museum News* 67, no. 2 (1988): 58–61.

Winsor, Mary P. *Reading the Shape of Nature: Comparative Zoology at the Agassiz Museum.* Science and Its Conceptual Foundations, edited by David L. Hull. Chicago: University of Chicago Press, 1991.

Wissler, Clark. "The Museum Exhibition Problem." *Museum Work* 7, no. 6 (1925): 173–80.

———. "Survey of the American Museum of Natural History: Made at the Request of the Management Board." Typescript. American Museum of Natural History, New York, 1943.

Witcomb, Andrea. *Reimagining the Museum: Beyond the Mausoleum.* New York: Routledge, 2003.

Wittlin, Alma Stephanie. *The Museum, Its History and Its Tasks in Education.* London: Routledge & K. Paul, 1949.

———. *Museums: In Search of a Useable Future.* Boston: MIT Press, 1970.

Wolfe, Audra Jayne. "Speaking for Nature and Nation: Biologists as Public Intellectuals in Cold War Culture." PhD thesis, University of Pennsylvania, 2002.

"Woman without Mystery on Display in New York." *Science Newsletter* 30, no. 803 (1936): 135.

Wonders, Karen. "Bird Taxidermy and the Origin of the Habitat Diorama." In *Non-Verbal Communication in Science Prior to 1900,* edited by Renato Mazzolini. Florence: Instituto e Museo Di Storia Della Scienza, 1993.

———. *Habitat Dioramas.* Figura Nova Series 25, Acta Universitatis Uppsaliensis. Uppsala: Almqvist and Wiksell, 1993.

———. "Habitat Dioramas as Ecological Theatre." *European Review* 1, no. 3 (1993): 285–300.

Worster, Donald. *Nature's Economy: A History of Ecological Ideas*. 2nd ed. Studies in Environment and History, edited by Donald Worster and Alfred W. Crosby. New York: Cambridge University Press, 1994.

———. *The Wealth of Nature: Environmental History and the Ecological Imagination*. New York: Oxford University Press, 1993.

Wraga, William G. "From Slogan to Anathema: Historical Representations of Life Adjustment Education." *American Journal of Education* 116, no. 2 (2010): 185–209.

Wylie, Philip. "A Layman Looks at Biology." *AIBS Bulletin* 9, no. 3 (1959): 12–15.

Yahya, Ibrahim. "Mindful Play! Or Mindless Learning! Modes of Exploring Science in Museums." In *Exploring Science in Museums*, edited by Susan Pearce, 123–47. Atlantic Highlands, NJ: Athlone Press, 1996.

Yanni, Carla. *Nature's Museums: Victorian Science and the Architecture of Display*. London: Athlone Press, 1999.

Yochelson, Ellis L. *The National Museum of Natural History: 75 Years in the Natural History Building*, edited by Mary Jarrett. Washington DC: Smithsonian Institution Press, 1985.

Young, Christian C. *In the Absence of Predators: Conservation and Controversy on the Kaibob Plateau*. Lincoln: University of Nebraska Press, 2002.

Zacharias, Jerrold. *Education in the Age of Science*. New York: Basic Books, 1959.

Zacharias, Jerrold, and Stephen White. "Requirements for Major Curriculum Revision." *School and Society* 92, no. 67 (1964): 66–72.

Zukin, Sharon. *Landscapes of Power: From Detroit to Disney World*. Berkeley: University of California Press, 1991.

Zurofsky, Rena. "Sharp Shop Talk." *Museum News* 68, no. 4 (1989): 45–47.

Index

Page numbers in italics refer to illustrations.